Lecture Notes in Electrical Engineering

Volume 1039

The book series *Lecture Notes in Electrical Engineering* (LNEE) publishes the latest developments in Electrical Engineering—quickly, informally and in high quality. While original research reported in proceedings and monographs has traditionally formed the core of LNEE, we also encourage authors to submit books devoted to supporting student education and professional training in the various fields and applications areas of electrical engineering. The series cover classical and emerging topics concerning:

- Communication Engineering, Information Theory and Networks
- Electronics Engineering and Microelectronics
- Signal, Image and Speech Processing
- Wireless and Mobile Communication
- Circuits and Systems
- Energy Systems, Power Electronics and Electrical Machines
- Electro-optical Engineering
- Instrumentation Engineering
- Avionics Engineering
- Control Systems
- Internet-of-Things and Cybersecurity
- Biomedical Devices, MEMS and NEMS

For general information about this book series, comments or suggestions, please contact leontina.dicecco@springer.com.

To submit a proposal or request further information, please contact the Publishing Editor in your country:

China

Jasmine Dou, Editor (jasmine.dou@springer.com)

India, Japan, Rest of Asia

Swati Meherishi, Editorial Director (Swati.Meherishi@springer.com)

Southeast Asia, Australia, New Zealand

Ramesh Nath Premnath, Editor (ramesh.premnath@springernature.com)

USA, Canada

Michael Luby, Senior Editor (michael.luby@springer.com)

All other Countries

Leontina Di Cecco, Senior Editor (leontina.dicecco@springer.com)

**** This series is indexed by EI Compendex and Scopus databases. ****

Surender Reddy Salkuti · Papia Ray ·
Arvind R. Singh
Editors

Power Quality in Microgrids: Issues, Challenges and Mitigation Techniques

Springer

Editors
Surender Reddy Salkuti
Department of Railroad and Electrical Engineering
Woosong University
Daejeon, Korea (Republic of)

Papia Ray
Department of Electrical Engineering
Veer Surendra Sai University of Technology (VSSUT)
Burla, Odisha, India

Arvind R. Singh
Department of EECE
University of South Africa
Pretoria, Gauteng, South Africa

ISSN 1876-1100 ISSN 1876-1119 (electronic)
Lecture Notes in Electrical Engineering
ISBN 978-981-99-2065-5 ISBN 978-981-99-2066-2 (eBook)
https://doi.org/10.1007/978-981-99-2066-2

This Springer imprint is published by the registered company Springer Nature Singapore Pte Ltd.
The registered company address is: 152 Beach Road, #21-01/04 Gateway East, Singapore 189721, Singapore

Preface

Power Quality (PQ) is defined as the capability of the electrical devices connected to the power network to consume the supplied energy. Power quality has become a significant matter for electric utilities and is nowadays acquiring a lot of interest. A microgrid (MG) is a single powerful entity with many loads and distributed generators embedded in it. For high power output in MG, a specific standard has to be met, which can be possible if we address the PQ issues properly. Before the power engineers, the main challenge is to eliminate the PQ disturbances like sag, swell, harmonics, spikes, etc., in MGs to get an uninterrupted power supply which is nowadays a research subject. The PQ disturbances in MGs lead to malfunction of protective devices, overloading, etc., leading to high-frequency emission, resulting from which supra harmonics (SH) emission occurs, which is a serious issue. These challenges exist in both standalone and grid-connected MGs. It has been observed from early research that PQ disturbances in MGs are generally mitigated by using external devices like static synchronous compensator (STATCOM), static var compensator (SVC), dynamic voltage restorer (DVR), etc.

The central concept of this book revolves around the PQ issues in MG. This book provides insight into the different challenges faced by MG due to PQ issues and how to mitigate them to provide an uninterrupted power supply. The main objective of this book is to make aware the power and control engineers with different innovative techniques to mitigate the challenges due to PQ issues in MG. This book will also educate postgraduate students and researchers in the field of PQ in MG. The topics covered in this book are PQ disturbances in MG and different recent and innovative schemes to mitigate them. The book consists of 25 chapters. The introduction of this book deals with the basic concept of PQ and the different challenging issues which the Indian power sector is facing in the MG and their solutions.

Unlike many other books, this book does not direct the reader to the manufacturer's documentation; instead, it tends to gather detailed information for better understanding and comparison. Also in this book, the diverse fields of research interest have been covered comprehensively and new concepts in research fields have been discussed. The individual chapters are different from most technical publications. They are journal-type chapters but are not textbooks in nature. Further, they are intended to be overviews providing ready access to needed information while at the same time providing sufficient references to more in-depth coverage of the topic. This is a book intended for researchers, utility engineers, and academicians in power engineering.

Enjoy Reading!

Daejeon, Korea (Republic of) — Surender Reddy Salkuti
Sambalpur, India — Papia Ray
Pretoria, South Africa — Arvind R. Singh

About This Book

This book is a collection of research articles and critical review articles, describing the power quality (PQ) issues and challenges in microgrids and proposing proper mitigation techniques to overcome them. The book emphasizes the technical issues, theoretical background, and practical applications that drive postgraduates, researchers, and practicing engineers with the right advanced skills, vision, and knowledge who will further be capable of leading teams involved in addressing the challenges of power quality issues in a microgrid. This feature strengthens the benefits of the book for the readers who will gain an insightful understanding of power quality issues in microgrids including (i) the formulation of decision-making models, (ii) the familiarization with efficient solution algorithms for such models, and (iii) insights into these problems through the detailed analysis of numerous illustrative examples. This book aims to provide insight into the different challenges of microgrids due to power quality issues and how to mitigate them to provide an uninterrupted power supply.

The book's primary objective is to aware the readers with various power quality challenges in microgrids and to propose different advanced schemes to mitigate them. Further, this book leads toward the application of advanced optimization techniques and IoT and 5G implementation in the microgrid to address various PQ issues. The future scope of this emerging topic includes online detection and classification of PQ events in the microgrid, exploring advanced hardware that is fast enough and compatible with real-time PQ event classification. The results of this book will significantly advance the state of the art in improving PQ in microgrids by proposing several recent and innovative mitigation schemes. This book is an effort to educate the next generation of academicians and researchers proficient in advanced analytics and improve national energy sustainability. This book comprises 25 chapters. Each chapter gives an innovative idea and enormous information about PQ challenges in microgrids and their mitigation techniques. Further, the chapters in this book provide comprehensive coverage of PQ issues in various microgrid applications and their mitigation schemes which are quite different from most technical publications.

Contents

Editors and Contributors

About the Editors

Dr. Surender Reddy Salkuti is working with Woosong University, South Korea as an Associate Professor in the Department of Railroad and Electrical Engineering since April 2014. He received the Ph.D. degree in Electrical Engineering from the Indian Institute of Technology Delhi (IITD), India, in 2013. He was a Postdoctoral Researcher at Howard University, Washington, DC, USA, from 2013 to 2014. His research interests include power system restructuring issues, smart grid development with the integration of wind and solar photovoltaic energy sources, battery storage and electric vehicles, demand response, power system analysis and optimization, soft computing techniques application in power systems, and renewable energy. He has published one edited volume with Springer (LNEE) and more than 200 research articles in peer-reviewed international journals and conference proceedings. He has served as or is serving as Guest Editor for various international journals. He is also an editorial board member for many journals. He is the recipient of the 2016 Distinguished Researcher Award from Woosong University Educational Foundation, South Korea, and the POSOCO Power System Award (PPSA) 2013, India. He is a Member of IEEE and IEEE Power and Energy Society.

Dr. Papia Ray is an Associate Professor and Head at the Department of Electrical Engineering, Veer Surendra Sai University of Technology, Burla, Odisha, India. She completed her Ph.D. in the area of Power System Engineering at the Indian Institute of Technology, Delhi in 2013. Her research area includes Power system protection, Power Quality, Smart grid protection, Renewable Energy, Wide area Measurement systems and application of soft computing techniques in power system protection. She has more than 20 years of experience in teaching and research. She is a senior member of IEEE, Fellow of Institution of Engineer's India, Life Member of Indian Society for Technical Education. She is the recipient of Young Scientist Award from DST, SERB, New Delhi in 2015. She has supervised many Ph.D. scholars and has published numerous papers in various reputed international Journals and Conferences. Till now she has edited 02 Books of Springer Nature.

Dr. Arvind R. Singh is currently working as Postdoctoral Research Fellow at faculty of EBIT, University of Pretoria (UP), South Africa. Prior to his appointment at UP, he has also served as a postdoctoral research fellow for 2 years from 2018–2020 at Shandong University, Jinan, P. R China. He has acquired special First Category Merit-based Funding from Shandong Provincial Government under Provincial Postdoctoral Advance Program in 2019 for his work on Reliable protection schemes for microgrid as PI. He was awarded Excellence Technical Education Quality Improvement Programme (TEQIP) Master's Scholarship by Ministry of Human Resource and Development (MHRD), Government of India for a pursue his Master's Degree Program in Power System of the Department of Electrical Engineering at College of Engineering, Pune (Public Deemed University, Under MHRD, Government of India), India in 2009. He was also awarded with Indian Government MHRD TEQIP Scholarship for Doctoral Degree study at Electrical Engineering, Visvesvaraya National Institute of Technology Nagpur (VNIT) in 2014 respectively. He has a diversified and multidisciplinary research which includes Microgrid Operation, Control and Protection. Optimal operation of distributed energy resources in Microgrids. Smart grids and Integration of Renewable energy sources. Optimal Generation Scheduling and Energy Management of Microgrids. He has made several contributions in form of journal articles, conference papers and book chapters majorly in the area of power system protection, power system compensation, smart grid, microgrid and renewable energy generation, optimization & integration, and soft computing applications in power system relaying, optimization, and analysis. He has also contributed as a role of Outstanding Reviewer for many prestigious journals such as Applied Energy, Renewable Energy, Energy, IEEE/IET journals and also as a Associate Editor of Frontiers in Sustainability Journal and Academic Editor in International Transactions on Electrical Energy System. He is currently a Senior Member of IEEE Industrial Electronics Society, Industry Applications Society, Power & Energy Society USA.

Contributors

Almoataz Y. Abdelaziz Faculty of Engineering and Technology, Future University in Egypt, Cairo, Egypt

Machidi Akshith Department of Electrical and Electronics Engineering, Sreenidhi Insitute of Science and Technology, Hyderabad, India

Sayed Javed Alam Department of Electrical Engineering, Sardar Vallabhai National Institute of Technology, Surat, India

Anil Annamraju Department of Electrical and Electronics Engineering, CBIT, Hyderabad, Telangana, India

Salauddin Ansari Department of Electrical Engineering, NIT Jamshedpur, Jamshedpur, Jharkhand, India

Sabha Raj Arya Department of Electrical Engineering, Sardar Vallabhai National Institute of Technology, Surat, India

Amuda Mahesh Babu Department of Electrical and Electronics Engineering, SR University, Warangal, India

Surakasi Balamurali Department of Electrical and Electronics Engineering, ANITS, Visakhapatnam, Andhra, India

B. V. V. L. Kala Bharathi Department of EEE, Aditya Engineering College, Surampalem, Andhra Pradesh, India

Laxman Bhukya Department of Electrical and Electronics Engineering, Methodist College of Engineering and Technology, Hyderabad, India

Phaneendra Babu Bobba Department of Electrical and Electronics Engineering, GRIET, Hyderabad, Telangana, India

Fossy Mary Chacko Department of Electrical Engineering, Rajiv Gandhi Institute of Technology, (Affiliated to APJ Abdul Kalam Technological University), Kottayam, Kerala, India

Rajasekhara Reddy Chilipi Department of Electrical Engineering, Sardar Vallabhai National Institute of Technology, Surat, India

Solomon Feleke Department of Electrical and Computer Engineering, Debre Berhan University, Debre Berhan, Ethiopia

Sravanthy Gaddameedhi Department of Electrical Engineering, Sreenidhi Insitute of Science and Technology, Hyderabad, India

Om Hari Gupta Department of Electrical Engineering, NIT Jamshedpur, Jamshedpur, Jharkhand, India

Ramjee Prasad Gupta Department of Electrical Engineering, MIT Muzaffarpur, Muzaffarpur, India

Battapothula Gurappa Department of Electrical and Electronics Engineering, BVRIT, Narsapur, Telangana, India

M. V. Jayan Department of Electrical Engineering, Government Engineering College, (Affiliated to APJ Abdul Kalam Technological University), Thrissur, Kerala, India

Smrutisikha Jena Department of Electrical and Electronics Engineering, IIIT Bhubaneswar, Bhubaneswar, Odisha, India

Narender Reddy Kedika Department of Electrical and Electronics Engineering, Institute of Aeronautical Engineering, Hyderabad, India

Seong-Cheol Kim Department of Railroad and Electrical Engineering, Woosong University, Daejeon, Republic of Korea

Prabhu Kiran Department of Electrical and Electronics Engineering, SR University, Warangal, India

Abhishek Kumar Department of Electronics and Communication Engineering, SOET, CMR University, Bengaluru, India

Nishant Kumar Department of Electrical Engineering, BK Birla Institute of Engineering and Technology, Pilani, India

Thattiparthi Pavan Kumar Department of Electrical and Electronics Engineering, Sreenidhi Insitute of Science and Technology, Hyderabad, India

Debani Prasad Mishra Department of Electrical and Electronics Engineering, IIIT Bhubaneswar, Bhubaneswar, Odisha, India

Rambabu Motamarri Department of Electrical and Electronics Engineering, Vignan's Institute of Information Technology, Vishakhapatnam, India

Dhananjaya Mudadla Department of Electrical and Electronics Engineering, Anil Neerukonda Institute of Technologyand Sciences (A), Visakhapatnam, India

V. Nagamalleswari Department of Electrical and Electronics Engineering, Balaji Institute of Technology and Science, Warangal, Telangana, India

Kumari Namrata Department of Electrical Engineering, NIT Jamshedpur, Jamshedpur, India

Bibhu Prasad Nanda Department of Electrical and Electronics Engineering, IIIT Bhubaneswar, Bhubaneswar, Odisha, India

Uday Kumar Neerati Department of Electrical and Electronics Engineering, Vasavi College of Engineering, Hyderabad, India

Sravanthi Pagidipala Department of Electrical Engineering, National Institute of Technology Andhra Pradesh, Tadepalligudem, Andhra Pradesh, India

Partha Sarathi Panuya Department of Railroad and Electrical Engineering, Woosong University, Daejeon, Republic of Korea

Krishna Mohan Reddy Pothireddy Department of Electrical Engineering, National Institute of Technology Andhra Pradesh, Tadepalligudem, Andhra Pradesh, India

Raseswari Pradhan Department of Electrical Engineering, Veer Surendra Sai University of Technology, Burla, Odisha, India

A. Prince Department of Electrical Engineering, Rajiv Gandhi Institute of Technology, (Affiliated to APJ Abdul Kalam Technological University), Kottayam, Kerala, India

Srinivas Punna Division of Motor Controller Design, Mahindra and Mahindra, Bengaluru, India

Balamurali Pydi Department of Electrical and Electronics Engineering, Aditya Institute of Technology & Management, Tekkali, Srikakulam, Andhra, India

Kunchala Raju Department of Electrical and Electronics Engineering, SR University, Warangal, India

Rathlavath Rakesh Department of Electrical and Electronics Engineering, SR University, Warangal, India

Ponamanenni Anoop Rao Department of Electrical and Electronics Engineering, Sreenidhi Insitute of Science and Technology, Hyderabad, India

C. Srinivasa Rathnam Department of Electrical and Electronics Engineering, Vasavi College of Engineering, Hyderabad, Telangana, India

Papia Ray Department of Electrical Engineering, Veer Surendra Sai University of Technology, Burla, Sambalpur, Odisha, India

W. Razia Sultana School of Electrical Engineering, VIT University, Vellore, Tamil Nadu, India

Arun Kumar Sahoo Department of Electrical and Electronics Engineering, IIIT Bhubaneswar, Bhubaneswar, Odisha, India

Sameep Sahu Department of Electrical Engineering, NIT Jamshedpur, Jamshedpur, Jharkhand, India

Surender Reddy Salkuti Department of Railroad and Electrical Engineering, Woosong University, Daejeon, Republic of Korea

V. Sandeep Department of Electrical and Electronics Engineering, National Institute of Technology, Tadepalligudem, Andhra Pradesh, India

Raavi Satish Department of Electrical and Electronics Engineering, Anil Neerukonda Institute of Technology and Sciences (A), Visakhapatnam, Andhra, India

Y. Satyavani Department of Electrical and Electronics Engineering, National Institute of Technology, Tadepalligudem, Andhra Pradesh, India

Ch Sekhar Department of Electrical Engineering, NIT Jamshedpur, Jamshedpur, India

R. Seshu Kumar Department of Electrical and Electronics Engineering, KPR Institute of Engineering and Technology, Coimbatore, Tamil Nadu, India

Kunal Shankar Department of Electrical Engineering, Birla Institute of Technology Mesra, Patna, Bihar, India

Arvind R. Singh Department of Electrical, Electronics and Computer Engineering, University of Pretoria, Pretoria, South Africa

Koganti Srilakshmi Department of Electrical Engineering, Sreenidhi Insitute of Science and Technology, Hyderabad, India

Ishan Srivastava GreyB Services, Sahibzada Ajit Singh Nagar, Punjab, India

Munjampally Sushma Department of Electrical and Electronics Engineering, SR University, Warangal, India

Venkataramana Veeramsetty Center for AI and Deep Learning, SR University, Warangal, India

Rajagopal Veramalla Department of Electrical and Electronics Engineering, Kakatiya Institute of Technology and Science, Telangana, India

A. Viswa Teja School of Electrical Engineering, VIT University, Vellore, Tamil Nadu, India

Vidhya Viswambaran Department of Electrical and Electronic Engineering, University of Bolton, RAK Campus, Ras Al-Khaima, United Arab Emirates

Raveena Voddamalla Department of Electrical and Electronics Engineering, Kakatiya Institute of Technology and Science, Telangana, India

Sandeep Vuddanti Department of Electrical Engineering, National Institute of Technology Andhra Pradesh, Tadepalligudem, Andhra Pradesh, India

Chandrasekhar Yammani Department of Electrical Engineering, NIT Warangal, Warangal, Telangana, India

Introduction to Power Quality in Microgrids

Arvind R. Singh, Papia Ray, R. Seshu Kumar, and Surender Reddy Salkuti

Abstract This chapter presents the conceptual application of power quality (PQ) in the microgrid environment. The distortion in the current and voltage waveform is increased by a spike in the penetration of renewable energy producers containing sophisticated power electronics converter modules. Also, renewable energy-based generation is more attractive than conventional non-renewable-based generators due to enhanced environmental, reduced carbon emission, economical and rural energy electrification benefits. Integrating the multiple microgrids into the distribution system forced utilities to look to preserve the quality of power supplied within standard limits. Nowadays, PQ is more critical due to problems arising in equipment malfunctioning, insulation deterioration, and degraded equipment output performances. This chapter will focus on the standards and methods reported in the scientific literature required to assess the PQ in a microgrid environment operating in isolated and grid-connected modes. Further, the chapter will discuss the essentials of various grid codes and standards available for assessment, monitoring, and improvement.

Keywords Power quality · Microgrid · Renewable energy generators · Mitigation · Harmonics · Sag · Swell

---

A. R. Singh
Department of Electrical, Electronics and Computer Engineering, University of Pretoria, Pretoria, South Africa

P. Ray (✉)
Department of Electrical Engineering, VSSUT, Burla, India
e-mail: papiaray_ee@vssut.ac.in

R. Seshu Kumar
Department of Electrical and Electronics Engineering, KPR Institute of Engineering and Technology, Coimbatore, Tamil Nadu, India

S. R. Salkuti
Department of Railroad and Electrical Engineering, Woosong University, Daejeon, Republic of Korea
e-mail: surender@wsu.ac.kr

S. R. Salkuti et al. (eds.), *Power Quality in Microgrids: Issues, Challenges and Mitigation Techniques*, Lecture Notes in Electrical Engineering 1039,
https://doi.org/10.1007/978-981-99-2066-2_1

Nomenclature

APF	Active Power Filter
APC	Active Power Conditioner
ADALINE	Adaptive Linear Neuron
BES	Battery Energy Storage
DES	Distributed Energy Sources
DER	Distributed Energy Resources
DOE	Department of Energy
DG	Distributed Generation
DVR	Dynamic Voltage Regulator
DB	Database
FACTS	Flexible Alternating Current Transmission System
GA	Genetic Algorithm
HVRT	High Voltage Ride Through
ICT	Information and Communication Techniques
IoT	Internet of Things
LVRT	Low Voltage Ride Through
MG	Microgrid
MPC	Model Predictive Control
NN	Neural Network
PQ	Power Quality
PV	Photo Voltaic
PCC	Point of common coupling
PR	Proportional Resonant
PI	Proportional Integral
PF	Power Factor
P-F	Power-Frequency
P-V	Power-Voltage
RES	Renewable Energy Sources
STATCOM	Static synchronous compensator
SVC	Static Var Compensator
S-Transform	Stockwell Transform
THD	Total Harmonic Distortion
UPQC	Unified Power Quality Conditioner
VUF	Voltage Unbalance Factor
VSI	Voltage Source Inverter

1 Introduction

The swift development of comprehensive energy utilization is the fundamental cause of high conventional power usage of fossil fuels and escalating CO_2 emissions. Many researchers have been extensively focused on the above concern to replace fossil fuel energy usage and reduce environmental issues. To overcome the above problem, integrating renewable energy sources (RES) into the existing traditional power grid is the viable solution, and the conventional grid becomes zero emission.

1.1 Tidal Energy Source

The overall installed capacity of the RES globally is 3146 GW, with the participation of more than 136 countries. The top ten countries to produce a high share of solar power production are tabulated in Table 1. According to [1], how the percentage of RES is dramatically increased by 8% and the share of fossil fuel is depreciated by 6% is illustrated in Fig. 1.

One of the significant problems with RES, like solar and wind, is that they exhibit an uncertain nature and depend on weather conditions. Due to the weather dependency of these RES sources combined with distributed sources and energy storage devices to become distributed networks. Blackouts can be disastrous and expensive even though the traditional power system provides a constant and dependable power supply while the grid is not threatened by natural disasters or security breaches. The centralized structure of the traditional grid has shifted to a distributed network to build flexible and resilient power systems called microgrids. According to the Department of Energy (DOE), the microgrid (MG) is "A cluster of electrically unique, connected loads and distributed energy resources (DER) that can operate both on the grid and off it and that can be controlled as a single unit." MGs are designed to provide an

Table 1 Production of global solar power

S. No.	Country	Power (MW)
1	China	306,973
2	US	95,209
3	Japan	74,191
4	Germany	58,461
5	**India**	**49,684**
6	Italy	22,698
7	Australia	19,076
8	South Korea	18,161
9	Vietnam	16,660
10	Spain	15,952

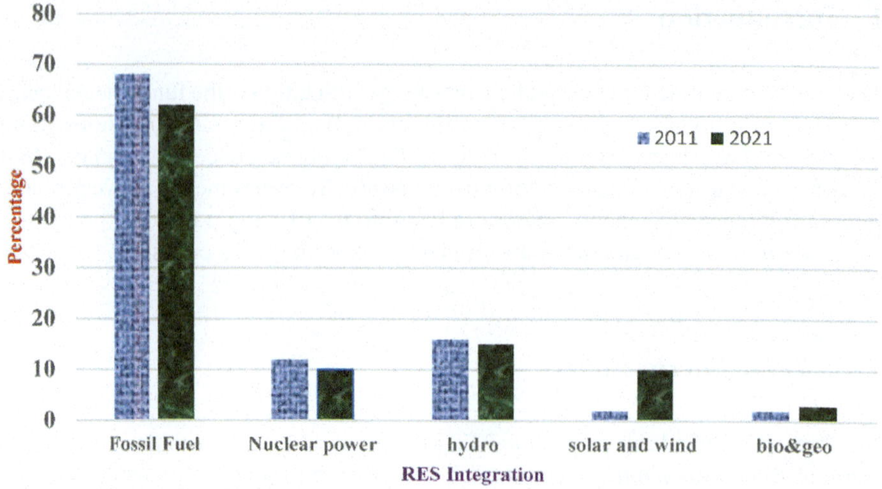

Fig. 1 Scenario of renewable energy and fossil fuel

uninterruptible power supply 24/7 and provide required load demands for industrial, commercial, and domestic loads. Microgrids generate power locally instead of transmitting power from a central utility source, which makes MG safe and secure from cyber threats. The wide range of applications with the RESs and integration with the traditional grid reduces the grid burden and enhances the microgrid operations [2].

The large-scale deployment of RESs and Distributed generation (DGs) associated with the MG yields low carbon emissions and helps to mitigate climate change. The high participation of DGs, RESs, and battery energy storage (BES) in the MG is challenging to control and operate, leading to the primary technical challenge, i.e., PQ. The capacity to maintain a pure sinusoidal voltage waveform with stated frequencies and magnitudes within prescribed limits, without changes in shape or magnitude, is referred to as PQ.

1.2 PQ Events

PQ changes to improve in a system when the magnitude and frequency of the power waveform deviate from the intended range, causing issues for the consumer. PQ disturbances are classified as follows: (1) voltage imbalance, (2) transients, (3) voltage sags and swells, (4) Over and under voltages, (5) outages, (6) harmonic distortion, (7) voltage notching, (8) flicker, (9) noise.

MG can run in a flexible, reliable, and optimal manner, which indicates that the DGs' PQ standards are significant and can provide substantial power to the connected loads in the MG. If the PQ of DG sources is low, then the poor PQ impacts the premature failure of devices, capacitors, motors, transformers, and cables. The MG's efficiency decreases when the PQ is poor and the heating losses explored by the apparatus are more. The low PQ of the MG can trigger the tripping of electronic devices and relays and lead to charges from your interconnected utility. The challenges related to MG PQ are classified into four categories [3]: The first type of PQ issue in the MG is under operating conditions with the integration of RESs like solar and wind. Because of the volatile nature of the RESs, like solar wind integrated with the MG, they are unpredictable that do not meet the required load demand, and the PQ of the MG becomes low. The second category is related to voltage and current harmonics due to the switching operations of the converters in the distributed network. Even though non-linear loads connected in the distributed network can cause voltage and current harmonics, it comes under the third category. The final type comes under voltage unbalance. Voltage unbalance problems occurred in the MG due to unbalanced 1-phase and 3-phase loads, respectively. PQ issues in the MG and PQ mitigation, methods, devices, and control strategies are discussed in this chapter.

In recent years the PQ term has been associated with electromagnetic events in the distributed MG. The PQ of the distributed network is improved with the integration of DGs, RESs, and energy storage units to compensate for transmission and distribution losses. The PQ issue has gained more importance in different technical aspects like operation, planning, protection, reliability, and efficiency. The potential causes of the PQ issue in the MG operation are tabulated in Table 2. Isolated systems face significant difficulty with PQ since the usage of non-linear loads results in voltage issues, fluctuations, and voltage swell/sag [4]. When comparing grid-connected and islanded modes of operation, it is clear that an imbalance in the load causes voltage variations and disturbances in the latter. However, voltage sag and variable utility voltages are the most frequently occurring issues in grid-connected mode. Solar panels, wind turbines, and fuel cells are all examples of intermittent energy sources, and because of their inherent unpredictability, they cannot be linked directly to the microgrid. Different DG causes of PQ issues are enumerated in Table 3.

The advancement of PQ measuring standards is taken into account while analyzing the PQ incidence. According to this view, one of the most influential factors in evaluating PQ is setting standards for measuring it. The two global standards are the IEC and the IEEE, shown in Fig. 2, and several countries with a large deployment of RES, like Germany and United States, have improved their GCs with specifications [5–11]. PQ standards are employed in many research because they describe the acceptable ranges of distortion and variance for numerous electrical variables. Table 4 includes some of the parameters for defining PQ traits.

1.2.1 Issues Related to Swag/Swell in MG

One of the most significant problems with PQ is sag (dip) occurrence, which is brought on by faults and can cause the power grid to become unstable. Voltage sag commonly prevents delicate electrical components from operating, which would be

Table 2 PQ challenges in the MG

Event	PQ issue	Definition	Varying parameters	Reason	Impact
Under voltage	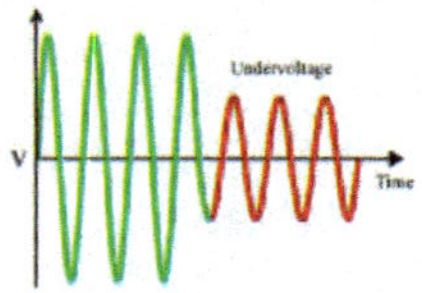	When the RMS voltage falls below 90% of the nominal RMS value and remains there for more than a minute, an **under voltage** occurs	Voltage magnitude, time interval	An oversized transformer, uneven circuit loading, isolation failures	Flashovers, insulation breakdown, failure of the transformer
Over voltage	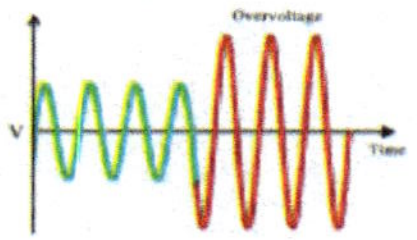	When the RMS voltage exceeds 110% of the nominal RMS voltage & remains there for more than a minute, it is referred to as **over voltage**	Voltage magnitude, time interval	Overloading, loose connections	Malfunctions, components melting
Voltage sag	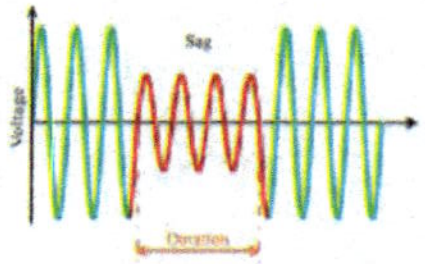	"A reduction in RMS voltage between 0.1 and 0.9 p.u. and a time length between 0.5 cycles and 1 min are the definitions by IEEE 1159"	Voltage magnitude, time interval	Connectivity of high-rating loads, motors starting, faults	Damages MG power electronic devices, High power losses, damage-sensitive equipment

(continued)

Table 2 (continued)

Event	PQ issue	Definition	Varying parameters	Reason	Impact
Voltage swell	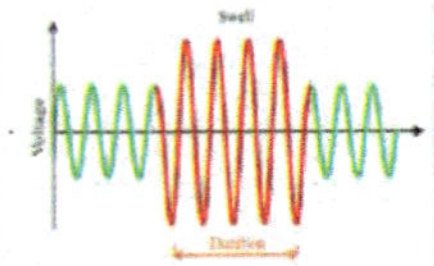	"Definition as per IEEE 1159 voltage swell is short duration rise in magnitude more than 0.1 p.u"	Voltage magnitude, time interval	Sudden load reduction, removal of heavy loads	Breakdown of components and power supplies, hardware failure
Unbalance	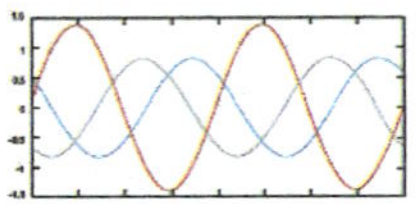	A variation of 3-phase voltages in which phase angles and magnitude are unequal	Phase shift, magnitude	Unequal load distribution, heavy 1-phase loads, fluctuations in the load	Degrading equipment life, high cable loss, poor efficiency of the inverter
Fluctuation	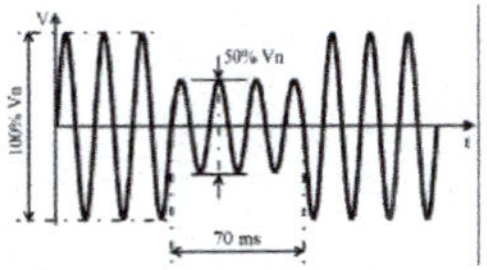	Distortion of the pure sinusoidal waveform is due to the presence of harmonics in the system	Peak magnitude, frequency	Non-linear loads, dynamic motors, inverters	Light flickering, reduces performance, fluctuations in the output wave

Table 3 PQ issues associated with DG units in the MG

PQ issue	Solar PV	Wind	Hydro	Diesel	EV
Voltage (swell/sag)	x	✔	✔	✔	x
Over/under voltage	x	✔	x	✔	✔
Voltage unbalance	✔	x	x	x	x
Voltage transients	x	✔	x	x	x
Voltage harmonics	✔	✔	✔	x	✔
Flicker	✔	✔	x	✔	✔
Current harmonics	✔	✔	✔	x	✔
Interruption	✔	✔	x	x	✔

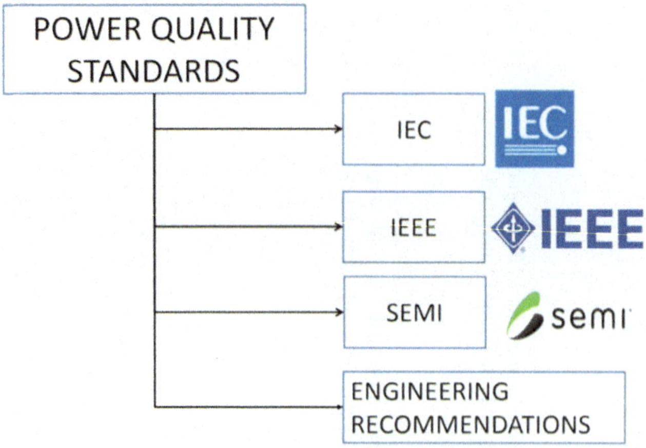

Fig. 2 Classification of specification standards

prevalent in DES systems that include MG [5]. Another significant PQ problem is swell, which exhibits behavior opposite to sag [12]. As the inclusion of DES and MG increases, many specifications and grid codes (GC) are implementing new regulations, such as low-voltage ride-through (LVRT) and high-voltage ride-through (HVRT) in the event of sag and swell, respectively. According to these standards, MG sources must cut off the grid if the sag or swell persists for a predetermined time [13]. Table 5 illustrates the maximum voltage and time duration permitted following various national grid codes and regulatory frameworks in the event of sag and swell incidents [14].

1.2.2 Harmonics-Related Issues in MG

Applications for MGs include devices that produce harmonics, including non-linear loads, electronic converters, computerized systems, and variable-speed motors. Most

Table 4 Specification standards for PQ parameters

Standards	Report
IEEE 519-92	IEEE harmonic control recommendations and requirements for electric power systems
IEEE 1159-95	Recommended practice for electric PQ monitoring
IEEE 1100-99	Recommended practice for powering and grounding electronic equipment
IEEE 1250-95	Recommended for the operation of sensitive equipment from momentary voltage distribution
IEEE 1366-2012	Guide for electric power distribution reliability indices
IEC 61000-2-2	It recommended that covering low-frequency disturbances in industrial and non-public networks
IEC 61000-2-4	Levels of compatibility for non-public and industrial power distribution systems at nominal frequencies of 50 Hz/60 Hz and nominal voltages up to 35 kV
IEC 61000-3-2	Harmonic current emission limits
IEC 61000-4-15	Functional and design requirements for flicker measuring devices that are intended to show the appropriate flicker perception level for all applicable voltage fluctuation waveforms
IEC 50160	Characteristics and specifications of distributed voltage systems

Table 5 Voltage and time duration of PQ events

Country	Voltage swell		Voltage sag	
	Rise	Time (s)	Drop (%)	Time (s)
India	110%	0.02	20	0.5
Germany	120%	0.1	0	0.15
China	–	–	20	0.625
Spain	130%	0.25	0	0.15
Italy	125%	0.1	0	0.2
UK	–	–	15	0.14
USA	140%	1	15	0.6
Japan	–	–	15	1.0
Australia	130%	0.06	0	0.45
South Africa	120%	0.5	0	0.15

electrical systems can withstand harmonics to a certain extent, but if that limit is reached, harmonics cause communication problems, high line losses, overheating, and circuit breaker trips [15]. Several investigations have investigated PQ in low-voltage systems through the harmonics issue. MGs are a low-voltage network, making PQ an especially crucial concern that needs to be studied and addressed [16]. Most MG sources are renewable energy systems that use a power electronics device to add harmonics to the grid. Because of this, new GC regulations necessitate that

MG systems dramatically cut on harmonics emissions to meet these standards [17]. Harmonic distortion requirements are incorporated to make the voltage and current waveform more grid compatible as grid integration standards for MGs develop. As a result, constraints on single and total harmonics distortion (THD) have been set for DES and MG that are associated with the primary power grid [18–20]. According to the standards of IEEE, IEC, and EREC, the harmonic distortion limit of any system should not exceed the range of 0.5%–3%, respectively.

1.2.3 Voltage Unbalance and Fluctuations

The most common form of PQ phenomena is voltage imbalance. The degree of unbalance can be measured by calculating the voltage unbalance factor (VUF) [21], which is the ratio of the positive to the negative sequence of voltage components. The MG power electronics and other power system devices are vulnerable to damage from voltage unbalance. For this reason, it is crucial to have a well-balanced system, especially given various MG sources, as unbalanced systems incur more significant losses and are less stable [22]. As a result, specific GCs and standard criteria have been devised to guarantee the steady and well-balanced incorporation of MGs and other DESs to control the VUF. According to most international norms, a VUF level of 1–2% is acceptable.

1.3 PQ Mitigation Methods and Control Strategies

For the past decade, there has been a lot of focus on finding solutions to the PQ issues plaguing MG, mainly because of the growing importance of renewable energy sources in low-voltage grids. There are a variety of technologies and control strategies that could reduce the impact of PQ on the electricity grid. Specific techniques have been devised to facilitate the connection of MGs to the electrical grid.

1.3.1 Dynamic Voltage Restorer

Overcoming voltage dips and spikes in electrical power distribution is possible through the dynamic voltage restoration (DVR) technique. These are critical because power is consumed during surges and the efficiency of some devices drops during sags. DVR reduces power consumption by injecting voltage into the electrical grid to change the phase and waveform of the incoming current. The typical structure of DVR is represented in Fig. 3. An MG-connected grid's sag and swell can be mitigated with the help of a fuzzy logic-based dynamic voltage regulator (DVR). Using MPC [23] improved the DVR's functionality in an MG with a PV, a supercapacitor, and a battery to withstand the sag and swell. Despite sag and swell, the MG is maintained with the help of the suggested DVR-based MPC system. According to [24], the DVR

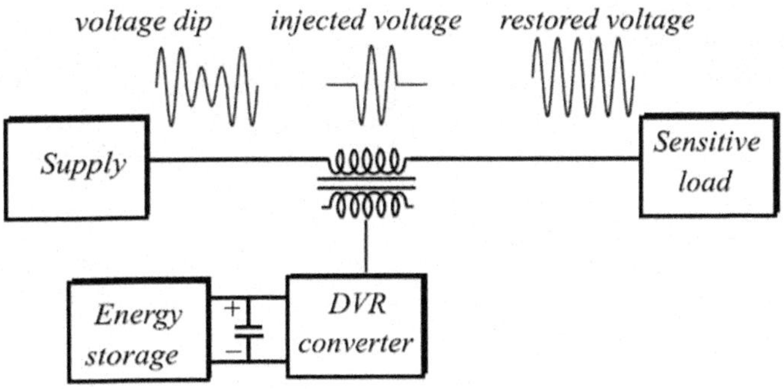

Fig. 3 Typical structure of DVR

detects and compensates for sag in islanded MG. The findings demonstrate that DVR can reduce the sag event in response to various grid faults. DVR is used to minimize voltage fluctuation in MG, which includes RESs. These findings demonstrate the DVR's effectiveness in mitigating the inherent source-dependent variability of an MG that provides for PV and wind systems. With the efficient use of DVR, recommended in [25], voltage imbalance and harmonics at AC bus caused by wind/PV/fuel cells and the primary grid can be tolerated. To improve the PQ [21] employing DVR by the minimum specifications. Current and voltage THD must be less than 5%, and VUF must be less than 1%, according to recent integration standards. DVR is widely regarded as one of the most effective tools for dealing with PQ issues in traditional power grids and MGs.

1.3.2 FACTS Devices

Other tools used to handle PQ issues are STATCOM and SVC. Both devices depicted in Figs. 4 and 5 are examples of flexible AC transmission system devices that have gained widespread utilization in recent years to address PQ problems primarily caused by the incorporation of renewable energy sources (RES), such as LVRT, for mitigating sag events in photovoltaic (PV) and wind power systems. In analyzing how successful SVC and STATCOM are in dealing with sag events, authors of [26] concluded that STATCOM adds more to the temporary margin. High penetration of DREs as MG in [27] necessitated the usage of STATCOM to smooth out voltage swings. Also, STATCOM was implemented in MG to reduce voltage fluctuations and compensate for reactive power [28]. Another study demonstrated that STATCOM could improve voltage management and system power factor while decreasing power fluctuations in MG [29].

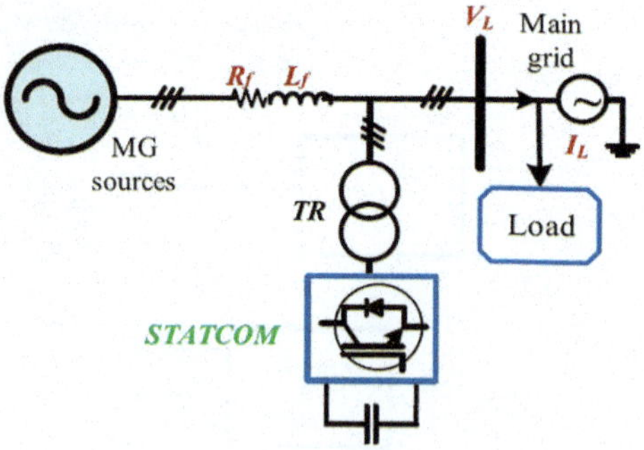

Fig. 4 Typical configuration of STATCOM used for PQ mitigation in MG systems

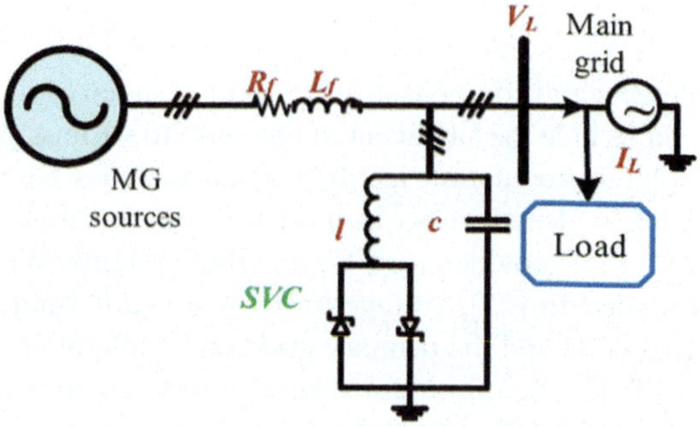

Fig. 5 Typical configuration of SVC used for PQ mitigation in MG systems

1.3.3 Unified Power Quality Conditioner (UPQC)

As shown in Fig. 6, the UPQC is composed of a series controller and a shunt controller linked together via a DC bus. The shunt controller's point of connection allows it to generate or consume reactive power. The MG line's attributes are managed by a series controller connected in series with the line [30]. To reduce voltage and current harmonics on UPQC, the authors of [30] used a fuzzy logic controller. It can be seen from the findings that overall harmonic distortion went down from 8.93 to 3.34%. For reduced harmonic distortion, consider the design of the MG proposed in reference [30], which incorporates an appropriate UPQC. A substantial THD (2.69%) is observed between 0.2 and 0.3 s, corresponding to a voltage sag in the measurements. When analyzing the load current's harmonic spectrum without using UPQC, we see that the total harmonic distortion (THD) is 33.26%. To conform to IEEE 519-1992 requirements for THD to be less than 5% [30], UPQC is used to achieve the present THD of 3.11%.

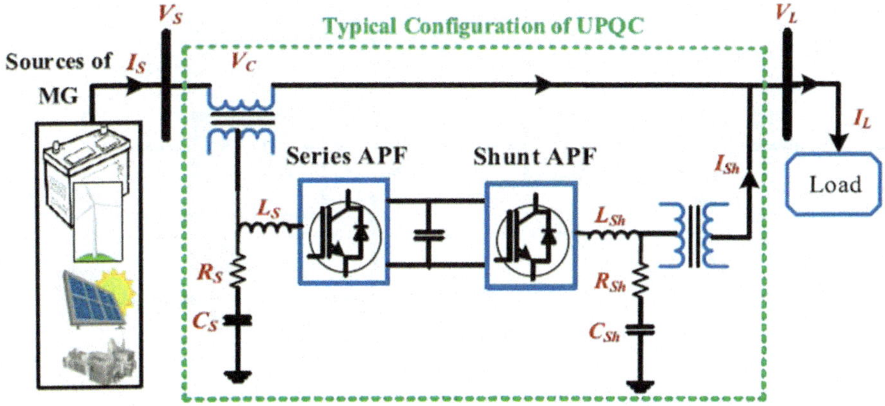

Fig. 6 Typical configuration UPQC used for PQ mitigation in MG systems

2 Control Strategies for Enhancing PQ in MG

Power, voltage, and current control loops make up an interface inverter's controller features. Real and reactive power exhibit droop characteristics in the external power control loop to maintain a constant output voltage of the inverter in terms of amplitude and frequency. A filter is used in conjunction with voltage and current controllers to avoid high-frequency interruptions and, consequently, damp-off fluctuations.

2.1 Droop Control Method

Droop control is a powerful technique primarily used to regulate actual and reactive power better. In either mode of operation, a microgrid may benefit from using a droop controller. In microgrid systems, load balancing and equalization are facilitated using control techniques such as active power-frequency (P-F) and reactive power-voltage (Q–V) droop controllers. Droop controllers play a crucial role in the MG by balancing the actual and reactive power. Figures 7 and 8 represent the active and reactive power compensation block diagrams.

Techniques relying on droop management are presented for an isolated system to reduce reactive power-sharing problems and voltage THD at PCC [31]. The voltage fluctuations at the PCC are dampened by using a virtual impedance loop using capacitance. While injected active and reactive power are adjusted in grid-connected mode [31] using an internal current control loop in response to changes in grid voltage and frequency, the power converter is run in three distinct ways when in autonomous mode: conventional droop mode PQ mode and synchronization mode. Adjusting DC link voltage to achieve desired voltage at inverter output droop control is utilized to achieve power balance in the microgrid. If the voltage goes over the constant power

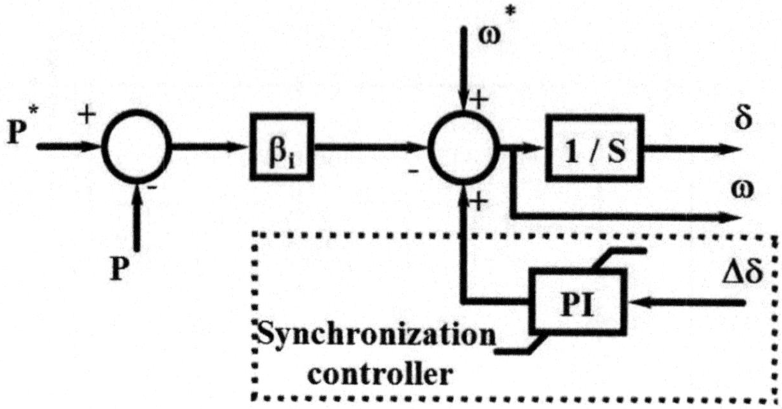

Fig. 7 Compensation of real power

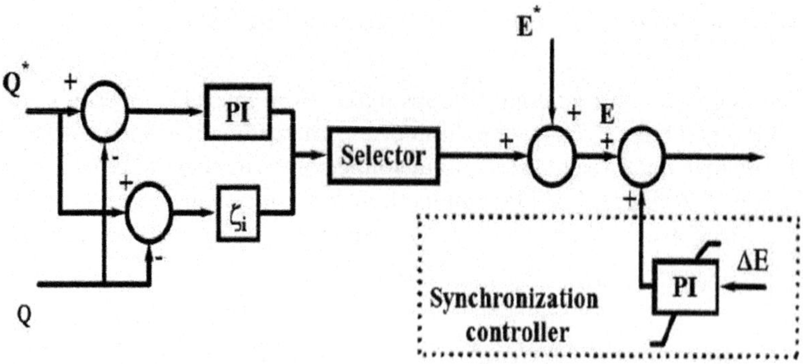

Fig. 8 Compensation of reactive power

band, the P-Vg droop control prevents that. For a VSI-based MG system, where real and reactive powers modify the phase angle and voltage reference, a multilevel control is given [31] that incorporates droop control and a virtual impedance loop based on a synchronous reference frame.

2.2 PQ Improvement with Controllers

Depending on the Distributed generation systems, the VSI of a microgrid can be managed in distinct ways. Employing the controllers improves the system dynamics (stability) and PQ produced, leading to adequate performance. In inverter-based DG, voltage regulation is achieved using both current control loop design and voltage

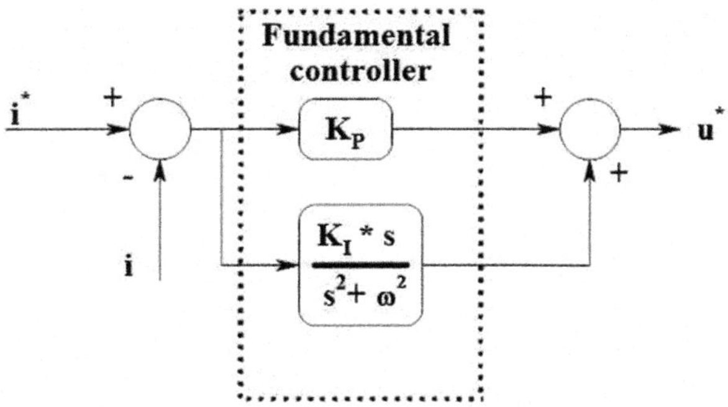

Fig. 9 PR controller

control loop design. This book chapter introduces and discusses the many electrical voltages and current controllers established in the literature.

2.2.1 Proportional–Resonant (PR) Controller

Proportional resonant (PR) controllers are frequently utilized to improve reference tracking performance when the control variables are sinusoidal in a stationary reference frame. Like how PR controllers fix PI controllers' problems, PI controllers aren't perfect. With the help of PR controllers, we can lower the system's overall harmonic content while minimizing the steady-state error. Figure 9 displays the block diagram of a PR controller.

Single-phase and three-phase grid converters use proportional-resonant controllers (PR), which are shown to be suitable for current control [32]. For basic current regulation, power factor (PF) controllers can be tuned either at the grid frequency or the harmonic frequency (for harmonic compensation). Reference [33] combines a proportional-integral (PI) controller and a general harmonic resonant controller into a single system that runs at the harmonic frequency in a rotating frame. Without removing the harmonic or fundamental component, an improved virtual impedance control strategy may accurately share power and reduce voltage harmonics at the point of common coupling (PCC) [32, 33].

2.2.2 H∞ Controller

Using H∞ control as inspiration, designers have developed a current controller incorporating an internal model and a stabilizing compensator [34]. Even when there are disruptions in the grid, H∞ s repeating current controller can inject a steady flow of current. The H repeating controller outperforms the more commonplace PI, PR, and DB controllers [34]. Even with distortions in the grid voltage and despite the non-linear and unbalanced nature of the loads, the output current's THD value is substantially below the acceptable range [34]. Considering both inverter voltage and grid current, cascaded current-voltage control H repeated control technique-based [34] enhances PQ. The nominal and perturbed systems are stabilized by the voltage and current loop for a micro source inverter based on the H loop shaping controller [34].

2.2.3 Fuzzy and Neural-Based Controllers

The capability of a non-linear DG interface to regulate active and reactive power while reducing harmonics, unbalance and voltage fluctuations can be enhanced by employing a control method based on a Fuzzy logic controller and neural networks. Voltage control is handled by a Fuzzy Logic Controller (FLC), while harmonic elimination and imbalance compensation is handled by an Adaptive Linear Neuron (ADALINE) [35]. Sluggish natural frame, three elements make up the suggested interfacing solution for grid-connected DG inverters in [36]: the current controller, an adaptive neural network (NN) based disturbance estimator, and a resilient sensor-less synchronization loop. Because of NN's inherent ability to learn, an adaptive controller design is possible, resulting in a solid performance despite grid-side disturbances.

2.2.4 Filters

Frequency/sequence selective filters [37] are also employed in voltage-source inverter (VSI) based microgrids on enhancing power conditioning performance. Like those in DG units, shunt active power filters can control active power entering the grid and account for non-linear load current harmonics [38]. In this case, there is little to no interference with the grid current; therefore, it continues to look identical to a sinusoidal wave. In [38], a new control method for the grid-connected inverter is developed, allowing it to serve as a (a) power converter and (b) shunt Active power filter (APF). It can add renewable energy source (RES)-generated power to the grid by acting as a converter. Two more functions are filtering out harmonics from the load current and correcting for current imbalance.

2.2.5 PQ Compensators

The administration of electricity is managed using a decentralized algorithm for power sharing [39]. Micro-sources' active and reactive power needs are taken care of by the reference current used to improve PQ; thus, DGs can be used to mitigate PQ problems like imbalance and harmonics in the load. As shown in [40], a 3-phase Active Power Conditioner (APC), which functions as an interface between RES and microgrid, can raise PQ in a microgrid system. An enhanced control approach is required to compensate for harmonics to achieve a balanced and sinusoidal line current despite an unbalanced load. The PQ compensator suggested in [41] has raised the system's flexibility as its primary objective. Instead of the inverter output current, the grid current is balanced, sinusoidal, and brought into phase with the grid voltage by this process. By integrating negative, a series inverter keeps balanced line currents. When the voltage level drops below the limit, zero sequence components tolerate voltage unbalance and stop significant fault currents from flowing (sag). In addition, power is sent to the grid while the shunt inverter is managed to keep a set of balanced, delicate load voltages. Two examples of control algorithms are the power control algorithm (P-f and Q–V droop characteristics) and the voltage (outer controller) and current (inner controller) control (PI) algorithm. The authors also examine the flux-charge control method that regulates the series inverter when the utility voltage drops.

3 Application of Optimization Techniques to Analyze PQ Issues

Earlier researchers have used several techniques to mitigate PQ issues. Some of them are Genetic Algorithm [42], Wavelet transform [42], Fourier Transform [43], S-Transform [43], Hilbert Transform [44], Gabor-Winger Transform [45], etc. These techniques are used to preprocess the raw PQ signal for further analysis. After that, features are extracted from the refined signal, and optimal ones are chosen with the help of various feature selection techniques like feed-forward, GA based, etc. The choice of proper optimization technique for PQ event analysis depends on the accuracy and the simulation time it takes.

4 Applications of IoT and 5G Technology in Microgrid

PQ and dependability are the main problems customers and utilities encounter in conventional power systems. As new technologies such as the internet of things (IoT), 5G, and information and communication techniques (ICT) advance, they offer a dependable remedy for these problems, converting the traditional grid into a microgrid [46, 47]. IoT technology is becoming an essential microgrid component for households and businesses [48]. The microgrid aims to transfer power efficiently at a cheap cost per unit of time while maintaining security. Worldwide adoption of 5G technology is happening quickly, changing every industry. The high reliability, support for machine connections, and incredibly low communication delays. These characteristics of 5G make it an excellent choice for microgrid IoT applications. Another issue related to the conventional grid is the security threat leading to the whole grid's collapse. This issue can be sorted out with the adoption of IoT and 5G. Further, the application of IoT and 5G in microgrids make them more reliable and efficient in handling and mitigating PQ issues.

5 Conclusions

The PQ concerns and their microgrid-related mitigating strategies are covered in detail in this chapter. The compensators and controllers for PQ enhancement in the microgrid are covered in detail in this chapter. Also, a brief overview of various FACTS devices and filters to mitigate PQ issues in a microgrid is presented here. The application of advanced optimization techniques and IoT and 5G implementation in the microgrid to address PQ issues is also discussed here. Further, this chapter facilitates the researchers to select the appropriate method to carry forward the research work in the concerned area. The future scope of this emerging topic includes online detection and classification of PQ events in the microgrid, exploring advanced hardware that is fast enough and compatible with real-time PQ event classification.

References

1. Global Status Report (2022) Renewables 2022 Global Status Report-REN21 [Online]. https://www.unenvironment.org/resources/report/renewables-2022-global-status-report. Accessed 24 Feb 2020
2. Khodayar ME, Barati M, Shahidehpour M (2012) Integration of high reliability distribution system in microgrid operation. IEEE Trans Smart Grid 3(4):1997–2006. https://doi.org/10.1109/TSG.2012.2213348
3. Kamel RM (2016) New inverter control for balancing standalone micro-grid phase voltages: a review on MG power quality improvement. Renew Sustain Energy Rev 63:520–532. https://doi.org/10.1016/j.rser.2016.05.074
4. Li YW, He J (2014) Distribution system harmonic compensation methods: an overview of DG-interfacing inverters. IEEE Ind Electron Mag 8(4):18–31. https://doi.org/10.1109/MIE.2013.2295421
5. Gayatri MTL, Parimi AM, Kumar AVP (2016) Utilization of unified power quality conditioner for voltage sag/swell mitigation in microgrid. In: Proceedings of 2012 international conference on power and energy systems: towards sustainable energy (PESTSE), Bengaluru, India, pp 1–6. https://doi.org/10.1109/PESTSE.2016.7516475
6. Hossain E, Kabalci E, Bayindir R, Perez R (2014) Microgrid testbeds around the world: state of art. Energy Convers Manag 86:132–153. https://doi.org/10.1016/j.enconman.2014.05.012
7. Ghosh A, Ledwich G (2012) Power quality enhancement using custom power devices, 2nd edn. Springer, Cham, pp 136–198. ISBN: 978-1-4615-1153-3
8. Whaite S, Grainger B, Kwasinski A (2015) Power quality in DC power distribution systems and microgrids. Energies 8(5):4378–4399. https://doi.org/10.3390/en8054378
9. Bhavani R, Prabha NR, Kanmani C (2015) Fuzzy controlled UPQC for power quality enhancement in a DFIG based grid connected wind power system. In: Proceedings of 2015 IEEE international conference on circuits, power and computing technologies, Nagercoil, India, pp 1–7. https://doi.org/10.1109/ICCPCT.2015.7159410
10. Yuvaraj V, Deepa SN, Rozario APR, Kumar M (2011) Improving grid power quality with FACTS device on integration of wind energy system. In: Proceedings of 2011 5th Asia modelling symposium, Manila, Philippines, pp 157–162. https://doi.org/10.1109/AMS.2011.38
11. Chavhan ST, Bhattar CL, Koli PV, Rathod VS (2015) Application of STATCOM for power quality improvement of grid integrated windmill. In: Proceedings of IEEE 9th international conference on intelligent system and control (ISCO), Coimbatore, India, pp 1–7. https://doi.org/10.1109/ISCO.2015.7282295

12. Hu WX, Xiao XY, Zheng ZX (2020) Voltage sag/swell waveform analysis method based on multi-dimension characterisation. IET Gener Trans Distrib 14(3):486–493. https://doi.org/10.1049/iet-gtd.2019.1038
13. Zheng F, Chen Y, Zhang Y, Guo M (2019) Low voltage ride through capability improvement of microgrid using a hybrid coordination control strategy. J Renew Sustain Energy 11(3):034102. https://doi.org/10.1063/1.5061690
14. Al-ShetwiAQ, Hannan MA, Jern KP, Mansur M, Mahlia TMI (2020) Grid-connected renewable energy sources: review of the recent integration requirements and control methods. J Clean Prod 253:119831. https://doi.org/10.1016/j.jclepro.2019.119831
15. Alwaz N, Raza S, Ali S, Bhatti MKL, Zahra S (2019) Harmonic power sharing and power quality improvement of droop controller based low voltage islanded microgrid. In: International symposium on recent advances in electrical engineering (RAEE), Islamabad, Pakistan, pp 1–6. https://doi.org/10.1109/RAEE.2019.8887029
16. Astorga OAM, Silveira JL, Damato JC (2006) The influence of harmonics from non-linear loads in the measuring transformers of electrical substations. In: Laboratory of high voltage and electric power quality, optimization group of energy systems, São Paulo, Brazil, vol 1, no 4, pp 275–280. https://doi.org/10.24084/repqj04.411
17. Blooming TM, Carnovale DJ (2006) Application of IEEE STD 519-1992 harmonic limits. In Proceedings of the IEEE conference record of 2006 annual pulp and paper industry technical conference, Appleton, WI, USA, pp 1–9. https://doi.org/10.1109/PAPCON.2006.1673767
18. Cho N, Lee H, Bhat R, Heo K (2019) Analysis of harmonic hosting capacity of IEEE Std. 519 with IEC 61000-3-6 in distribution systems. In: Proceedings of IEEE PES GTD grand international conference and exposition Asia (GTD Asia), Bangkok, Thailand, pp 730–734. https://doi.org/10.1109/GTDAsia.2019.8715918
19. Cleveland FM (2008) IEC 61850-7-420 communications standard for distributed energy resources (DER). In: Proceedings of IEEE power and energy society general meeting—conversion and delivery of electrical energy in the 21st century, July 2008, Pittsburgh, PA, USA, pp 1–4. https://doi.org/10.1109/PES.2008.4596553
20. IEEE Standard for interconnecting distributed resources with electric power systems amendment 1, IEEE Standard 1547. Institute of Electrical and Electronics Engineers, New York, NY (2014)
21. Kim YJ (2018) Development and analysis of a sensitivity matrix of a three-phase voltage unbalance factor. IEEE Trans Power Syst 33(3):3192–3195. https://doi.org/10.1109/TPWRS.2018.2807980
22. Savaghebi M, Jalilian A, Vasquez JC, Guerrero JM (2012) Secondary control scheme for voltage unbalance compensation in an islanded droop-controlled microgrid. IEEE Trans Smart Grid 3(2):797–807. https://doi.org/10.1109/TSG.2011.2181432
23. Li Z, Li W, Pan T (2016) An optimized compensation strategy of DVR for micro-grid voltage sag. Prot Control Mod Power Syst 1(4):1–8. https://doi.org/10.1186/s41601-016-0018-9
24. Avhad MVD, Hadpe SS (2019) Performance analysis of a DC microgrid integrated dynamic voltage restorer with model predictive control strategy. Int Res J Eng Technol (IRJET) 6(8):1579–1584. e-ISSN: 2395-0056
25. Srivatchan NS, Rangarajan P (2018) Half cycle discrete transformation for voltage sag improvement in an islanded microgrid using dynamic voltage restorer. Int J Power Electron Drive Syst 9(3):25–32. https://doi.org/10.11591/ijpeds.v9.i1.pp25-32
26. Molinas M, Aree SJ, Undeland T (2008) Low voltage ride through of wind farms with cage generators: STATCOM versus SVC. IEEE Trans Power Electron 23(3):1104–1117. https://doi.org/10.1109/TPEL.2008.921169
27. Lee TL, Hu SH, Chan YH (2013) D-STATCOM with positive sequence admittance and negative-sequence conductance to mitigate voltage fluctuations in high-level penetration of distributed-generation systems. IEEE Trans Ind Electron 60(4):1417–1428. https://doi.org/10.1109/TIE.2011.2166233
28. Chaudhari P, Rane P, Bawankar A, Shete P, Kalange K, Moghe A, Panda J, Kadrolkar A, Gaikwad K, Bhor N, Nikam V (2015) Design and implementation of STATCOM for reactive

power compensation and voltage fluctuation mitigation in microgrid. In: Proceedings of IEEE international conference on signal processing, informatics, communication and energy systems (SPICES), Feb 2015, Kozhikode, India, pp 1–5. https://doi.org/10.1109/SPICES.2015.7091541
29. Christian LE, Putranto LM, Hadi SP (2019) Design of microgrid with distribution static synchronous compensator (D-STATCOM) for regulating the voltage fluctuation. In: Proceedings of IEEE 7th international conference on smart energy grid engineering (SEGE), Oshawa, ON, Canada, pp 48–52. https://doi.org/10.1109/SEGE.2019.8859860
30. Cauveri S, Nayanatara C (2014) Optimization of PQ problems using FACTS devices. Int J Adv Res Electr Electron Instrum Eng 3(1):154–159. ISSN (Online): 2278-8875
31. Meng X, Liu J, Liu Z (2019) A generalized droop control for grid-supporting inverter based on comparison between traditional droop control and virtual synchronous generator control. IEEE Trans Power Electron 34(6):5416–5438. https://doi.org/10.1109/TPEL.2018.2868722
32. Kuperman A (2015) Proportional-resonant current controllers design based on desired transient performance 30(10):5341–5345. https://doi.org/10.1109/TPEL.2015.2408053
33. Dash AR, Chitti Babu B, Mohanty KB, Dubey R (2011) Analysis of PI and PR controllers for distributed power generation system under unbalanced grid faults. In: Proceedings of IEEE 2011 international conference on power and energy systems, Chennai, India, pp 1–6. https://doi.org/10.1109/ICPES.2011.6156688
34. Harshini J, Manoharan PS, Deepamangai P (2019) H∞ controller for buck boost converter. In: Proceedings of 2019 IEEE international conference on intelligent techniques in control, optimization and signal processing (INCOS), Tamil Nadu, India, pp 1–6. https://doi.org/10.1109/INCOS45849.2019.8951358
35. Garanayak P, Panda G, Mishra S (2019) Harmonic elimination using SW based HSAPF system and evaluation of compensation effect employing ADALINE-DFFRLS algorithm. J Eur Power Electron Drives 29(2):64–81. https://doi.org/10.1080/09398368.2018.1483118
36. Zhang M, Li Y, Liu F, Li W, Peng Y, Wu W, Cao Y (2019) Cooperative operation of DG inverters and a RIHAF for power quality improvement in an integrated transformer-structured grid-connected microgrid. IEEE Trans Ind Appl 55(2):1157–1170. https://doi.org/10.1109/TIA.2018.2882504
37. Illindala M, Venkataramanan G (2012) Frequency/Sequence selective filters for power quality improvement in a microgrid. IEEE Trans Smart Grid 3(4):2039–2047. https://doi.org/10.1109/TSG.2012.2197834
38. Rahmani S, Hamadi A, Al-Haddad K, Dessaint LA (2014) A combination of shunt hybrid power filter and thyristor-controlled reactor for power quality. IEEE Trans Industr Electron 61(5):2152–2164. https://doi.org/10.1109/TIE.2013.2272271
39. Fan X, Crisostomi E, Thomopulos D, Zhang B, Shorten R, Yang S (2021) An optimized decentralized power sharing strategy for wind farm de-loading 36(1):136–146. https://doi.org/10.1109/TPWRS.2020.3008258
40. Das SR, Mishra DP, Ray PK, Salkuti SR, Sahoo AK (2021) Power quality improvement using fuzzy logic-based compensation in a hybrid power system. Int J Power Electron Drive Syst 12(1):576–584. https://doi.org/10.11591/ijpedsv12.i1.pp576-584
41. Xu Y, Chen X, Liu M, Tian F, Yan S (2018) A switch-controlled hybrid power quality compensator for all-day power quality compensation in AC-DC networks at different times of day. In: Proceedings of IEEE 2018 international conference on power system technology (POWERCON), Guangzhou, China, pp 1–6. https://doi.org/10.1109/POWERCON.2018.8602142
42. Ray P, Panigrahi BK, Senroy N (2013) Hybrid methodology for fault distance estimation in series compensated transmission line. IET Gener Transm Distrib 7(5):431–439. https://doi.org/10.1049/iet-gtd.2012.0243
43. Ray P (2017) Power system low frequency oscillation mode estimation using wide area measurement systems. Eng Sci Technol Int J 20(2):598–615. https://doi.org/10.1016/j.jestch.2016.11.019
44. Ray P, Lenka RK, Biswal M (2019) Frequency mode identification using modified masking signal-based empirical mode decomposition. IET Gener Transm Distrib 13(8):1266–1276. https://doi.org/10.1049/iet-gtd.2018.5527

45. Salkuti SR (2022) Emerging and advanced green energy technologies for sustainable and resilient future grid. Energies 15(18):6667. https://doi.org/10.3390/en15186667
46. Beniwal RK, Saini MK, Nayyar A, Qureshi B, Aggarwal A (2021) A critical analysis of methodologies for detection and classification of power quality events in smart grid. IEEE Access 9:83507–83534. https://doi.org/10.1109/ACCESS.2021.3087016
47. Salkuti SR (2020) A survey of big data and machine learning. Int J Electr Comput Eng 10(1):575–580. https://doi.org/10.11591/ijece.v10i1.pp575-580
48. Al-Turjman F, Abujubbeh M (2019) IoT-enabled smart grid via SM: an overview. Future Gener Comput Syst 96:579–590. https://doi.org/10.1016/j.future.2019.02.012

Application of Computational Intelligence Methods for Power Quality Disturbance Detection, Classification and Mitigation in Microgrids

Abhishek Kumar, Ishan Srivastava, and Arvind R. Singh

Abstract Power quality (PQ) is defined as the ability to maintain a pure sinusoidal voltage waveform with the specified amplitude and frequency within the required limit with no changes in shape or magnitude. Both the supply and demand sides are noticing an increase in the importance of power quality. Switching devices are increasingly being used, which inevitably leads to a decline in power quality. At the same time, these devices are also more likely to break down as a result of bad power quality. The primary research focus corresponding to a microgrid is on techniques for detecting and mitigating power quality disturbances. The classification of disturbances can be done in a variety of ways, including by using Artificial Neural Networks (ANN), fuzzy logic, machine learning, and deep learning. Numerous compensation-based methods, controller-based methods, and optimization-based methods have all been reported for the mitigation of power quality issues. The significant portions of the main developments are techniques based on computational intelligence. In this work, an effort has been made to thoroughly review computational methods for power quality disturbance detection and power quality enhancement using computational intelligence methods.

Keywords Power quality · Microgrid · Artificial Neural Networks · Genetic algorithm · Controller · Fuzzy-logic · Particle swarm optimization · Classifier

---

A. Kumar
Department of Electronics and Communication Engineering, SOET, CMR University, Bengaluru, India

I. Srivastava
GreyB Services, Sahibzada Ajit Singh Nagar, Punjab 160062, India

A. R. Singh (✉)
Department of Electrical, Electronic and Computer Engineering, University of Pretoria, Pretoria, South Africa
e-mail: arvindsinghwce@gmail.com

S. R. Salkuti et al. (eds.), *Power Quality in Microgrids: Issues, Challenges and Mitigation Techniques*, Lecture Notes in Electrical Engineering 1039,
https://doi.org/10.1007/978-981-99-2066-2_2

Nomenclature

ANN	Artificial Neural Networks
MG	Microgrid
DER	Distributed Energy Resources
RER	Renewable Energy Resources
PSO	Particle Swarm Optimization
ACO	Ant Colony Optimization
SLDV	Short and Long Duration Variations
VUF	Voltage Unbalance Factor
PQD	Power Quality Disturbances
FFT	Fast Fourier Transform
PQDC	Power Quality Detection and Classification
DFT	Discrete Fourier Transform
STET	Short-Time Fourier Transform
DTFT	Discrete Time Fourier Transform
WT	Wavelet Transform
SPT	Signal Processing Techniques
ST	Stockwell Transform
HHT	Hilbert Huang Transform
HT	Hilbert Transform
EMD	Empirical Mode Decomposition
IMF	Intrinsic Mode Functions
TFR	Time Frequency Representation
GA	Genetic Algorithm
SVM	Support Vector Machine
FL	Fuzzy Logic

1 Introduction

A Microgrid (MG) framework consists of Distributed Energy Resources (DER), load, power electronic devices (PED), and energy storage systems (like supercapacitors, batteries, and flywheels). DERs used by MGs are generally Renewable Energy Resources such as wind turbines, photovoltaic (PV) systems, biogas plants, and many more. The main objective of using MGs are to optimize efficiency, availability, reliability, economic performance, and security in power system [1]. There are two modes in which MGs can work: (i) Islanded and (ii) Grid connected. In general, grid-connected mode MGs are used as it has the advantages of both main network and DERs.

MGs are having very complex architecture due to the connection of various DERs and the use of PEDs. These connections in MGs bring challenges related to stability, power quality, security, and reliability in the grid structure. To overcome these challenges, grid features using proper control architecture can be realized [2]. The rising penetration of DERs indirectly increased diverse power electronic devices in the distribution network, and the existence of now linear loads in the grid is also responsible for deteriorating the power supply quality [3]. As a result, a key decision is required to deploy advanced control technologies to mitigate the negative consequences of the DG-connected grid. Low on-grid electricity prices will come from poor power quality (PQ), particularly in a future power quality-sensitive market [4].

Researchers have proposed various methods to address the poor power quality issue in MGs; these can be categorized as follows [1]:

- Controller-based power quality improvement techniques
- FACTS-based power quality improvement techniques
- Computational methods for power quality enhancement
- Harmonic mitigation technique

A lot of work and improvement has been achieved in microgrid operation since its inception. Various kinds of MG controllers are reviewed in [2] where authors divided MG controllers into four categories: Hieratical, distributed, decentralized, and centralized. These controllers can be used to mitigate above mentioned challenges. Advanced machine learning algorithms like reinforcement learning, where the controller learns by its experiences [5], has also been used by authors to improve energy storage system [6] and energy management system [7] in MGs. Since microgrids use PEDs, there is always a possibility to have poor power quality in MGs. So, maintaining power quality in MGs is a very essential task. Researchers have proposed various schemes to improve the power quality in MGs. Authors [8] have reviewed power quality issues along with other challenges in microgrids and their compensation methods. Recent optimization-based techniques like Particle Swarm Optimization (PSO) [8], Ant Colony Optimization (ACO) [9], Artificial Neural Network (ANN), and Fuzzy Logic based control scheme has been proposed by researchers to mitigate power quality issues in MGs.

In this chapter, our main concern is to look at various computationally intelligent techniques to detect, classify and mitigate power quality issues in micro-grids. We organized this chapter into 6 sections. Section 1 focused on the introduction followed by the basics of Power Quality (PQ) Issues and its standard in Sect. 2. Sections 3 and 4 concentrate on the detection and automatic classification of power quality disturbances respectively. In Sect. 5, details about various intelligent methods to enhance power quality is presented followed by the conclusion in Sect. 6.

2 Power Quality Issues

In general, power quality (AC/DC) can be seen as a combination of current and voltage quality and thus concerned about any deviation in current/voltage from its desired waveform [10]. However, some study says DC power quality focused on voltage only [11]. It has been observed that DC micro-grid is more efficient compared to AC micro-grid and hence standards for power quality issues have been studied by authors [12]. IEC61000 and IEEE std1159 are used extensively in power quality issues. Some of the power quality issues are Transients, Short and Long Duration Variations (SLDV), Imbalance, and Waveform distortion. Voltage fluctuations may also be considered power quality issues but it is a debatable topic [12].

2.1 Transients

Transients are essentially short-time-varying phenomena between successive steady states. There are two kinds of transients namely impulsive transients and oscillatory transients. Both the transients are sudden and there may be a change in current and/ or voltage. Impulsive transients are applicable for unidirectional current/voltage and are characterized by time domain quantities like rise and fall time. On the other hand, oscillatory transients are applicable to the bidirectional (positive and negative) polarity of current/voltage and are described by frequency spectrum. Lighting and abrupt switching may be the cause of transients.

2.2 Short and Long Duration Variations

SLDV are categorized by cause of the disturbances. Faults, abrupt switching, and huge power variation are the cause of Short duration variations, lasting for up to 1 min (according to IEEE Std1159), and up to 3 min (according to EN50160). Power flow variation in steady state is cause for Long term variations and lasts for more than 1 min and 3 min according to IEEE Std1159 and EN50160 respectively. SLDV can be further classified as swell, sag, and interruptions depending on "magnitude" and "change direction" of rms voltages.

Swell can be defined as the increment of rms voltage above 1.1 per unit (p.u.) for a period of 0.5 cycles to a maximum of 1 min. Sag is a drop in rms voltage in the range from 0.1 p.u to 0.9 p.u. over a period of 0.5 cycles to 1 min. Interruption can be seen as a drop in rms voltage below 0.1 p.u. for a period of 0.5 cycles to 1 min. Thus swell, sag and interruption can be seen as short-duration variations. The counterpart of swell, sag, and interruption for long-term variation is named as overvoltage, undervoltage, and sustained interruption. However, the level of magnitude distortion is the same for both short-term and long-term duration variation. IEC61000 uses the terms voltage dip and voltage swell for sag and swell, respectively. Also, there is no different word for an interruption in the IEC61000 standard it is the same as a voltage dip but with a lower threshold for the interruption.

2.3 Imbalance

Voltage imbalance and current imbalance have been noticed in PQ issues. Voltage imbalance is basically a fraction of negative or zero sequence voltage to positive voltage sequence for 3 phase system. IEEE Std159 and EN50160 has used limit of 3% and 2% respectively on Voltage Unbalance Factor (VUF) that characterizes voltage imbalance and is given as,

$$VUF = \frac{V_{UB}}{V_B} = \frac{V_+ - V_-}{V_+ + V_-} \quad (1)$$

Here V_{UB} and V_B are unbalanced and balance voltage respectively and are given as:

$$V_{UB} = \frac{V_+ - V_-}{2} \quad (2)$$

$$V_B = \frac{V_+ + V_-}{2} \quad (3)$$

V_+ and V_- are positive terminals to neutral voltage and neutral to negative terminal voltages respectively.

Current imbalance is characterized by unbalanced current I_{UB} and causes current to flow through a neutral conductor.

2.4 Waveform Distortion

AC offset, Notching, Inter-harmonics, and Noise are some of the major causes of waveform distortion. AC offset may result in very high voltage variation due to a change in reactive current. Notching or voltage notching is an episodic disturbance caused due to switching of current from one phase to another. Notching can result in parallel resonance and hence overvoltage. Period and depth of notch characterize voltage notching. Switch-mode converters generate current at the switching frequency and harmonics of the switching frequency resulting in inter-harmonics. These inter-harmonics add deprivation in filtering capacitors and sometimes hamper the communication signal. Ripples in current and voltage waveform directly affect inter-harmonics. Table 1 summarizes important power quality issues [12–14].

Table1 Power Quality Issues

Category	Visualization	Indicator and typical time	Sources	Outcome
Transients				
Impulsive	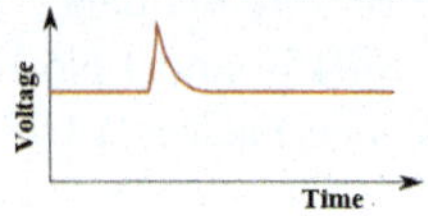	Rise and Fall time Peak magnitude. nSec – mSec	Lightning strikes Inductance switching	Resonance in voltage and Insulation breakdown
Oscillatory	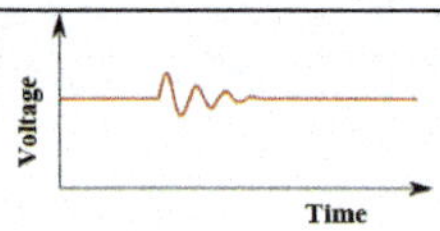	Peak magnitude, Duration, Spectral content. microSec - mSec	Power Variation, Line, and Capacitor switching	Resonance in voltage and Insulation breakdown
Short-duration variations				
Swell	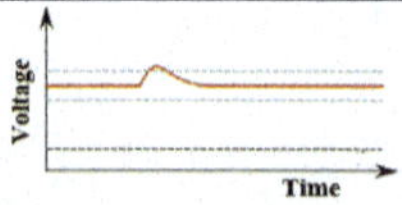	Magnitude and Duration 0.5 cycle – 1 min	Capacitor switching, Faults, and Variation in Power	Protection malfunction, Burdening Apparatus and Insulation breakdown
Sag	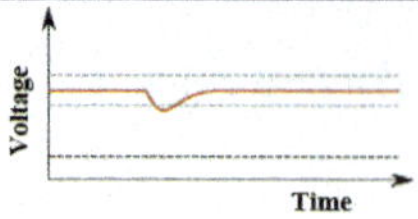	Magnitude and Duration 0.5 cycle – 1 min	Capacitor switching and Variation in Power	Shutdown of Apparatus
Interruption	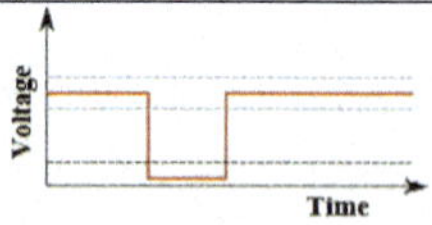	Magnitude and Duration 0.5 cycle – 1 min	Faults	Shutdown of Apparatus
Long-duration variations				
Overvoltage	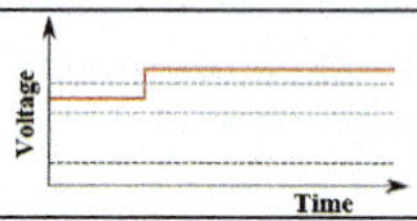	Magnitude 1 min and more	Irregular voltage regulation	Burdening Apparatus
Undervoltage	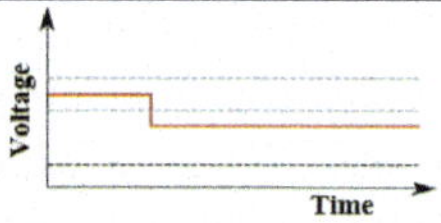	Magnitude 1 min and more	Irregular voltage regulation	Unstable Voltage and losses
Interruption	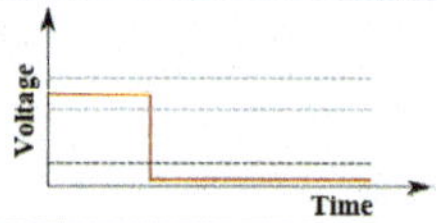	Magnitude 1 min and more	Faults	Burdening Apparatus

(continued)

Table1 (continued)

Imbalance				
Voltage	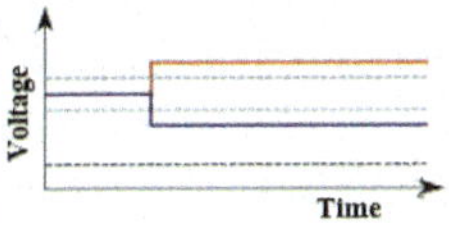	Magnitude steady state	Faults at Pole-neutral and use of Single pole Equipment	Under and Over-voltage at the pole-neutral junction.
Current	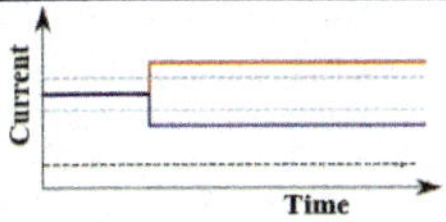	Magnitude steady state	Use of Single pole Equipment	Imbalance in Voltage, Loss increment, and overloading neutral conductor.
Waveform distortion				
AC offset	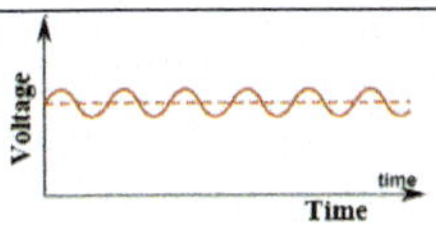	Magnitude steady state	Electromagnetic coupling and Faults	Increment in loss
Notching	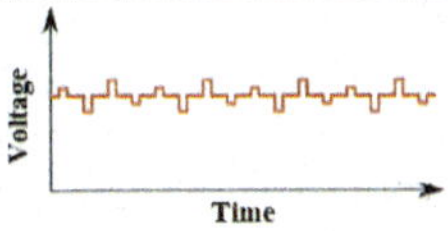	Magnitude steady state	Switch mode power converters	Losses, Irregularity in communication and measurement devices, Increased filter capacitor current
inter harmonics	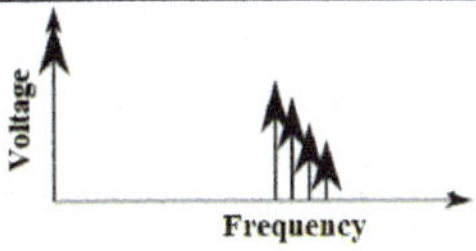	Spectral content steady state	Switch-mode power converters	Losses, Irregularity in communication and measurement devices, Increased filter capacitor current
Noise	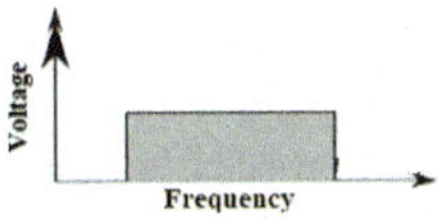	Spectral content steady state	Switch-mode power converters	Losses, Irregularity in communication and measurement devices, Increased filter capacitor current
Voltage fluctuations	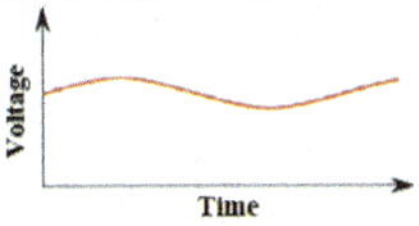	Duration and Spectral content irregular	Low-frequency power oscillations	Fluctuations with low intensity

3 Power Quality Disturbance Detection and Classification

With the advancement in MG, distributed generation (that depends mainly on renewable sources) is widely used thus reducing losses due to distribution resulting in the overall economic power sector. However, the use of renewable resources and power electronics introduces power quality issues/disturbances in MGs as renewable resources may generate unpredictable power due to unpredictable environmental conditions such as varying solar irradiance and wind speed [15]. Hence while using MGs, these Power Quality Disturbances (PQD) must be detected and classified (i.e. sag or swell or another category), as each PQD has its own effect (listed in Table 1). MGs can be operated in islanding mode along with grid-connected mode hence it becomes crucial to detect islanding and PQD positively to minimize faults.

Authors in [15] used Fast Fourier Transform (FFT) to detect islanding and PQD (mainly harmonics) in MG and reduced the PQD using an approach based on the Kalman filter. The authors simulated a hybrid system considering solar energy (photovoltaic array) and wind energy (doubly fed induction generator based wind turbine) as the source and other required peripherals like dc capacitor and inverter. FFT has the quality to localize data in a time-frequency plane and hence is appropriate for analyzing the stationary signal. Stationary signals are generally obtained while analyzing signals under PQ and islanding conditions thus FFT fits to analyze transient current and voltage under islanding and PQ conditions. However, the drawback of FFT is that can't be used to analyze the non-stationary signal. Islanding Detection and PQD Detection with energy sources connected to Micro-grid is observed in [15]. It is obtained that Islanded mode outperforms the grid-connected mode in terms of total harmonic distortion and PQD gives comparatively good results when the load is rejected (i.e. sudden load removal) as compared to the sudden introduction of load.

A lot of research and survey has been done on Power Quality Detection and Classification (PQDC) considering various power system. R. K. Beniwal et al. [16] surveyed the PQDC in a smart grid. The authors mainly focused on signal processing and artificial intelligence (AI) techniques for the same. N. Singh et al. [17] surveyed the same for distributed generation using AI tools like genetic algorithms, fuzzy logic, and neural networks. A detailed study about PQDC, dedicated to digital signal processing and AI, has been carried out in [18] mostly for micro-grid. Authors have also used an intelligent classifier using a probabilistic approach for PQDC in the microgrid.

In general, PQDC consists of two stages. The two stages are pre-processing followed by classification. Pre-processing is basically a feature extraction technique where signal processing tools are used to extract features from disturbances in voltage and/or current waveform [18]. The extracted features are then utilized, in the classification phase, for training and testing of the classifier to categorize the disturbances.

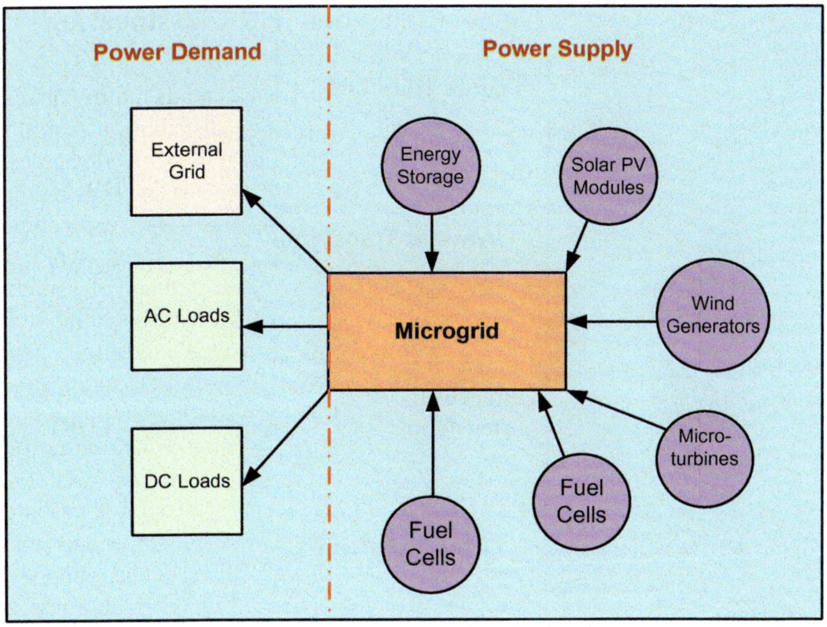

Fig. 1 Basic architecture of a microgrid

3.1 Feature Extraction Techniques

For feature extraction from disturbances various Signal Processing Techniques (SPT) have been used by authors such as Fourier and Wavelet transform along with their variants. SPT was used by researchers Fourier Transform, Wavelet Transform, Stockwell Transform, Gabor Transform, Hilbert-Huang Transform, Kalman Filter, Mathematical Morphology, and some miscellaneous approach. A detailed classification of the PQ detection technique is shown in Fig. 2.

3.2 Fourier Transform and Its Variants

Fourier Transform (FT) is basically frequency domain analysis, where a signal is decomposed into sinusoids with all the frequencies. Moreover using the concept of Inverse Fourier Transform, we can get an original signal in the time domain [16]. FT and Discrete Fourier Transform (DFT) have some shortcomings e.g. these techniques don't give satisfactory feature extraction for non-stationary signals [19]. Fast Fourier Transform (FFT) is a variant of FT used to find DFT and Inverse DFT by breaking or factorizing a large DFT into smaller DFT thus reducing the computation time and complexity. FT is generally used for harmonic analysis of PQ signal and no

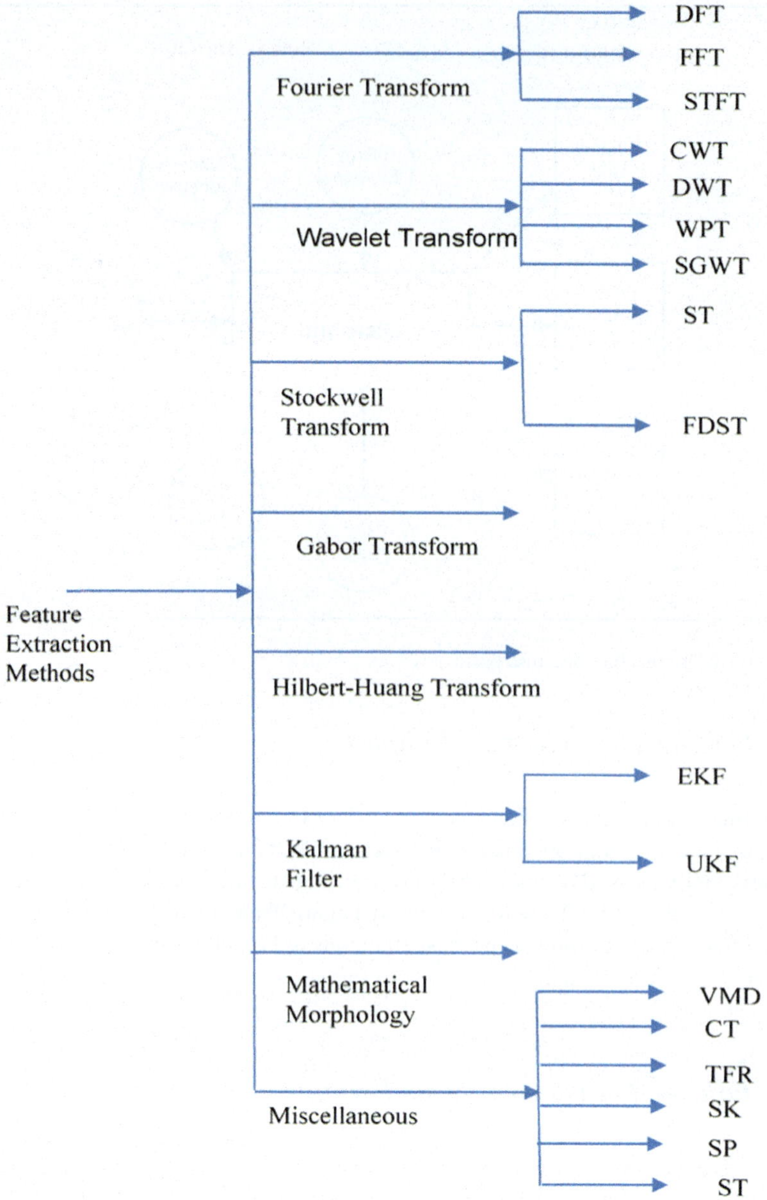

Fig. 2 Feature extraction methods

other PQ issues as other PQ issues require time information [19]. To overcome these issues associated with FT/DFT, Short-time Fourier Transform (STFT) [20] has been utilized that can be used even for non-stationary signals giving better efficacy as compared to FT. STFT uses a window function of size much smaller as compared to the fluctuation rate of the disturbances. It is actually the window size that determines the performance and resolution of STFT. Authors [21] have used STFT for spectral information extraction. Though STFT can be used with non-stationary signal analysis it is very difficult to use STFT for such signal analysis [22]. Windowed FFT [23], which is the fusion of windowing (window with adjustable width) with Discrete Time FT (DTFT) can also be used to extract features from non-stationary signals.

3.3 Wavelet Transform

A relatively new Wavelet Transform (WT) proved to be better compared to STFT [24] experimentally. WT is a popular feature extraction technique and is widely used to analyze PQ issues in the time-frequency plane. WT works on multi-resolution analysis where a signal is decomposed in low and high-frequency components using low and high-pass filters respectively. The decomposition takes place in multiple levels until the desired level is reached for analyzing the signal. The high-frequency component of the decomposed signal is used to detect impulsive changes such as switching operation, transients, and sudden change in voltages. Slow changes like harmonics are detected by a low-frequency component of the decomposed signal. Figure 3 shows the decomposition level (showing two levels only) used in WT. To ease disturbance recognition, the square of the WT coefficient can also be used.

Some of the proposed variants of WT are Discrete WT (DWT), Continuous WT (CWT), Wavelet Packet Transform (WPT), and second-generation wavelet transform (SGWT) [18].

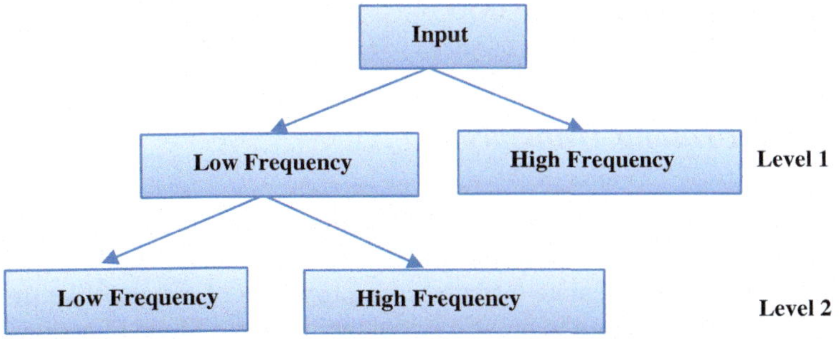

Fig. 3 Wavelet transform

3.4 Stockwell Transform (S-Transform)

Stockwell Transform (ST) utilizes attributes of WT and STFT for processing a signal, thus improving signal representation in both the time and frequency domains. ST provides frequency-dependent resolution as the window used in ST decreases its width with the frequency. ST has the advantage that it can retain the phase of all frequencies due to its multi-resolution property. However, instantaneous indication for disturbance is not available with ST, hence; it is not suitable for harmonics. Moreover, ST suffers from high sensitivity to the Gaussian window and requires complex computation. Some of the variants of ST are used to provide fast ST such as discrete ST (FDST).

3.5 Gabor Transform

Gabor Transform (GT) is used to estimate phasor perfectly and results in a better signal study in terms of time and frequency as compared to FT. Gaussian window function has been used with GT to analyze harmonics in power systems using time series signals only [25].

3.6 Kalman Filter

Kalman filter (KF) is one of the most useful tools for signal processing and is widely used to estimate frequency, phase angle, and magnitude of harmonics in presence of noise. Some of the variants are extended KF (EKF) and unscented KF (UKF). Authors [26] used the fusion of DWT with KF for PQDC using a fuzzy expert system. DWT and KF were used for the detection of noise in voltage signal and a fuzzy system was used to automatically identify the PQD.

3.7 Hilbert-Huang Transform

Hilbert Huang Transform (HHT) is a combination of the Hilbert Transform (HT) and Empirical Mode Decomposition (EMD) technique and hence has advantages of both. HT is used to calculate instantaneous properties of a signal like frequency and amplitude. EMD is used to break a signal into Intrinsic Mode Functions (IMF). Thus in HHT, EMD provides an intrinsic mode function from a signal, and then the function is used to get instant frequency data. HHT can be used to analyze non-linear, non-stationary data and noisy data with limited bandwidth.

3.8 Mathematical Morphology

Mathematical Morphology (MM) utilizes the concept of geometry and set theory to study a signal. It is a non-linear technique and can modify the signal shape, however, the disadvantage of MM is that it can perform processing operations in the time domain only. MM technique can be used with both stationary and non-stationary PQDC [27, 28]. Lu et al. [29] proposed an MM-based technique for feature extraction of PQ disturbance along with the duration and location of the disturbances.

3.9 Miscellaneous Techniques

Apart from the aforementioned standard signal processing techniques, various signal processing tools have been used to extract PQ features. Time-Frequency Representation (TFR) technique proposed in [30] used the kernel technique for PQDC. Chirptel transform (CTs), generalization of WT, STFT, and FT, is proposed by Hu et al. [31] for PQDC. Tables 2 and 3 brief mathematical expression and some of the aspects of various features extraction techniques (FETs) respectively.

Table 4 shows some of the extracted features and techniques used (we considered only four techniques here) to extract that features [22].

So far, we have focused on feature extraction techniques used in PQ issues. However, selecting the most appropriate features is also important and this selection of significant features is generally known as feature selection. In other words, feature selection provides a subset of all features to provide a minimum error in classification [32]. A proper feature selection improves the efficiency of PQ detection and reduces the computational time required in detection and classification by ignoring extraneous features. Authors have proposed various feature selection methods including the simulated annealing approach, norm entropy approach, and many more [33].

4 Classifier for Automatic Classification

AI is now extensively used in a variety of classification problems resulting in intelligent classifiers. AI has been used by researchers to make machines capable of taking intelligent decisions, solving complex problems, and classification have used AI. In [34], Support Vector Machine (SVM) classifier is used to solve the fault classification problem in the distribution network. Various AI techniques like ANN, SVM, Fuzzy Logic, and optimization techniques have proved to be better classifiers for the automatic classification of PQ issues [16]. Figure 4 show some example of intelligent classifier.

Table 2 Mathematical analysis of FETs

Transform used	Underlying expression	Description
FT	$F(W) = \int_{-\infty}^{\infty} u(t)e^{-j\upsilon t}dt$	*u(t)* is the input signal and υ is the angular frequency in rad/s
STFT	$STFT(t, \delta) = \frac{1}{\sqrt{2\pi}} \int_{-\infty}^{\infty} u(t)\delta(t-\sigma)e^{-j\upsilon\sigma}dt$	$\delta(t-\sigma)$ is window function
WT	$WT(x, y) = \frac{1}{\sqrt{\lvert x\rvert}} \int_{-\infty}^{\infty} u(t)\phi(\frac{t-y}{x})dt$	*x*, *y* represents scaling and translational function $\phi(\frac{t-y}{x})$ is mother wavelet
HT	$HT[u(t)] = \frac{1}{\pi} P \int_{-\infty}^{\infty} \frac{u(\sigma)}{t-\sigma}d\sigma$	P represents Cauchy Principal Value
HHT	$HHT = HT + EMD$ $EMD = c_n(t) + \sum_{i=1}^{n} r_i(t)$	$c_n(t)$ is concluding residue, $r_i(t)$ i^{th} component of IMF
ST	$ST(\sigma, f) = \frac{\lvert f\rvert}{\sqrt{2\pi}} \int_{-\infty}^{\infty} u(t)e^{\frac{[(t-\sigma)f]^2}{2}} e^{-j2\pi ft}dt$	f is frequency, $e^{-j2\pi ft}$ is the kernel used
GT	$GT(\sigma, f) = \int_{-\infty}^{\infty} u(t)e^{-\pi(t-\sigma)^2}e^{-j2\pi ft}dt$	$e^{-\pi(t-\sigma)^2}$ is the window used having properties of Gaussian function

4.1 *Artificial Neural Network*

Artificial Neural Network (ANN) mimics the neural network of the human brain to solve many real-world problems including classification. An ANN has sets of interconnected neurons in input, output, and hidden layers. Various types of activation function and NN structure leads to different types of ANN. Authors [35] used a probabilistic neural network (PNN) to classify transients based on their duration and energy. A quantitative comparison among multilayer perceptron NN, SVM, and Naïve Bayes classifier has been done in [18] where Naïve Bayes outperforms NN and SVM. The PQ issues used for the study include sag, swell, interruption and harmonics. Quantum NN [36], Dynamic structural NN [37] and self-adaptive ANN [38] has been proposed by researchers for the successful classification of various PQ events.

Table 3 Features of FETs

Feature extraction methods	Salient feature	Drawback
FT/ DFT	Good for harmonic analysis of PQ disturbances	Not applicable for non-stationary signals and hence not suitable for measuring instantaneous changes in the PQ disturbances
FFT	Same as FT/DFT but is computationally fast and requires less memory compared to FT/DFT	Not applicable to non-stationary signals. The aliasing and leakage effect makes it inefficient for procuring phases, frequencies, and amplitude
STFT	The phase and frequency of the waveform can be determined Better efficacy with the non-stationary signal as compared to FT/FFT using the concept of the window function	Although it can be used for analyzing the non-stationary signal, the analysis process is very hard Time-frequency resolution is limited
WT	Frequency and time domain analysis can be performed simultaneously (which is not possible in normal FT) with better resolution	Computationally more complex, highly sensitive to noise Picket fence effect and spectral leakage deteriorate performance
HHT	Provides good results with the nonlinear and nonstationary signals Able to analyze the distorted signal	Due to hybridization with the EMD technique, this SPT is limited to signals with narrow band and sometimes create problems in selecting the proper IMF function
ST	Improved representation in both time and frequency domains Retain phase of all frequencies PQ issues can be recognized with higher efficacy due to the use of the Gaussian window	Not suitable for harmonics, computationally complex, and sensitive to Gaussian window
GT	Good for spectral and harmonics analysis. Better signal study in terms of time and frequency as compared to FT	Sampling frequency affects computational complexity directly. Cannot be used frequently at high frequency
KF	Extensively used to estimate frequency, phase angle, and magnitude of harmonics in presence of noise	Highly dependent on the initial condition and may result in divergence with an improper selection of the initial condition and this instability may give errors while estimating fundamental and harmonic components. Both frequency and time domain disintegration are not obtainable

(continued)

Table 3 (continued)

Feature extraction methods	Salient feature	Drawback
MM	Applicable for both stationary and non-stationary signal. Computationally not very complex as it requires mainly set theory and simple arithmetic operations	Signal processing is possible in the time domain only

Table 4 Some features and their extraction methods

Extracted feature	Feature extraction methods
Energy	STFT, WT, ST, HHT
Standard deviation	STFT, WT, ST, HHT
Skewness	STFT, WT, ST, HHT
Entropy	STFT, WT, HHT
Kurtosis	STFT, WT, ST
Total harmonic distortion	WT, ST

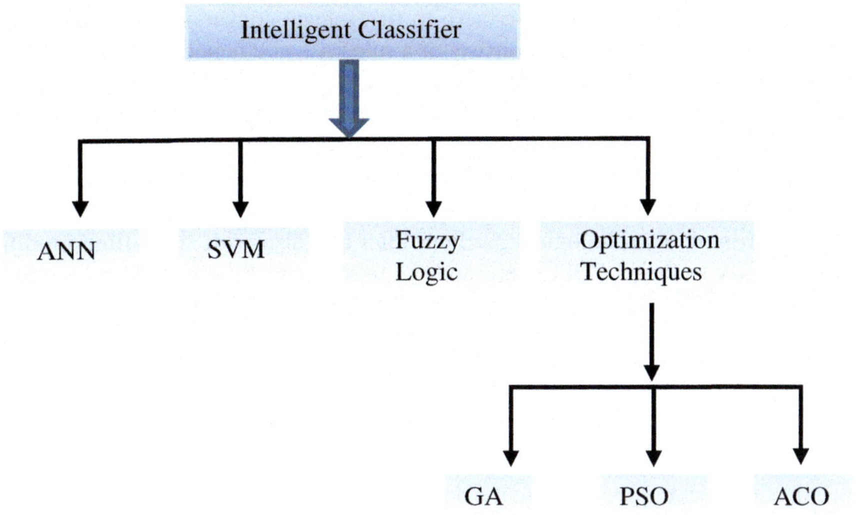

Fig. 4 Intelligent classifier

4.2 Support Vector Machine (SVM)

SVM is widely used for classification and pattern recognition. It is a supervised machine learning and has shown very impressive results in PQ classification. SVM can handle large inputs and outperform conventional classifiers. Authors [39] proposed a robust N-1 SVM technique to identify N PQ issues. To show the efficacy of SVM, SVM is compared with RBF for classification in presence of noise [40]. Random subspace ensemble classifier based on SVM has been proposed in [41] to classify various PQ events (harmonic and voltage-related signals along with different transients) in a microgrid consisting of photovoltaic in an on-grid as well as an off-grid mode of MG system. An independent component analysis SVM is presented in [42] and gives better results as compared to WT-SVM for the classification of PQDs including transients, harmonics, interruptions, flickering, sag and swell. WT-SVM is based on wavelet transform and is, hence, highly sensitive to noise. Authors [42] found that WT-SVM fails with noise levels of more than 25 dB.

4.3 Fuzzy Logic (FL)

Fuzzy logic can be considered an expert system that depends on a simple IF–THEN rule base. FL is a multivalued logic that works with a fuzzy set lying in the range of 0 to 1 [43]. Any FL system consists of three sections fuzzyfier, inference engine, and de-fuzzyfier. Fuzzification i.e. crisp to fuzzy level conversion is performed in fuzzyfier. The inference engine consists of a rule base and hence logical/reasoning-related tasks are done here. Finally, de-fuzzyfier is used to present the final output in the form of the crisp level using de-fuzzification operation. The core of fuzzy logic is its rule base. Owing to its expert system nature, FL can be implemented as a classifier. Zhu et al. [44] have used WT-based fuzzy logic for the classification of distinct PQ issues. Chakravorti et al. [45] incorporated a fuzzy judgment tree with a multi-scale morphological gradient filter and also with short-time modified HT to classify multiple PQD. Here authors used simulation as well as hardware environment to showcase the effectiveness of their proposed technique.

4.4 Optimization Techniques

Optimization Techniques improve the detection and classification efficiency as these techniques try to find an optimal solution for a given problem. GA, PSO, and ACO are some of the optimization techniques used in PQD classification.

The genetic algorithm is an evolutionary algorithm that finds an optimal solution by selecting the best parents (solutions) at each stage. A GA technique involves selection and mutation at each stage. Krishna and Baskaran [46] used Multi-objective GA and the concept of a decision tree for classifying PQD. Authors [47] used GA to classify PQD in a system consisting of power generation from wind and PV. PSO is a stochastic search technique based on population. PSO has been used in conjunction with fuzzy logic to increase the efficiency of FL [48] in PQD classification. Huang et al. [49] proposed Multi-resolution generalized ST using PSO to improve the classification of PQD. ACO is nature inspired algorithm to get an optimal solution for a complex problem. ACO has been implemented in PQD classification in [50].

5 Role of Computational Intelligence Methods in Power Quality Enhancement in Microgrid

In order to solve challenging engineering challenges, computational intelligence-based methods are extremely helpful. These methods employ approaches such as fuzzy logic, neural networks, evolutionary theory, learning theory, and probabilistic theory, making them an effective choice for solving complex problems. Researchers applied a variety of computational intelligence approaches inherent in the primary solution to improve power quality in an MG. The following section examines the role of computational intelligence-based methods for power quality improvement in MGs.

5.1 *Particle Swarm Optimization (PSO)*

As mentioned earlier, the researchers have proposed a controller-based solution for power quality improvement in MG. Based on a real-time self-tuning mechanism, the authors in [8], have proposed an optimum power control strategy for an autonomous MG operation. Both the voltage-frequency (Vf) control technique and the active-reactive power (PQ) control approach is used by this power controller. Real-time self-tuning of the power control settings is accomplished using the intelligent searching technique known as Particle Swarm Optimization (PSO). In the further work reported by the same authors [51], in this case, the PSO algorithm has been implemented into the PQ control mode to implement a real-time self-tuning mechanism in order to spread the load equitably during the load change situation.

In [16], the authors have proposed a novel technique for addressing voltage harmonic elimination. The authors used a system that combines a sine PWM inverter and a PSO-based PWM inverter, eliminating higher-order harmonics first and subsequently lower order harmonics using the selective harmonic elimination methodology.

5.2 *Fuzzy Logic (FL)*

In order to decrease the number of converters in a single ac or dc grid, the authors in [19] suggest a single-stage converter-based microgrid. The presented MG idea may operate in grid-interfaced mode as well as stand-alone mode. By applying control techniques designed for shunt active hybrid filters, such as series and shunt converters, distortions in the power system caused by changes in load or the use of non-linear loads may be prevented. To improve power quality, output power distortions are reduced using fuzzy logic controllers and a basic proportional-integral (PI). Benachaiba et al. [52] used the FL controller to adjust Unified Power Quality

Conditioner (UPQC) settings to enhance the voltage profile of an AC MG system. In order to reduce voltage sag and unbalancing, the performance of the suggested controller was compared with that of the traditional PI controller. When compared to the conventional PI controller, which recorded the same value of 2.23%, the FL-based UPQC lowered the total harmonic distortion (THD) in source current down to 2.11%.

A fuzzy-based PI controller was developed by the authors in [53] for inverter control in a grid-tied MG. To attain the necessary operating point for the MG system, the FL controller fine-tuned the gains of the PI controller. In simulation and experimental results, the THD for the proposed controller was reported as 3.85% and 4.4%, respectively.

5.3 Ant Colony Optimization (ACO)

Real-time optimization for an autonomous MG is presented in [9] using a power controller. Based on a synchronous reference frame, this controller is connected to the active control loop. In this design, a conventional PI regulator is employed. The Vf control mode based on the ACO algorithm (used for real-time self-tuning of control parameters) is used to govern the voltage and frequency regulation of MG when it transitions to an islanding mode or a load change circumstance.

5.4 Artificial Neural Network (ANN)

Researchers have also employed the ANN for AC MGs' reactive power management and power quality improvement in addition to MPPT control. In an autonomous hybrid system, Bansal et al. [54] employed the ANN-based tuning strategy for the PI controller-based static VAR compensator (SVC). The ANN-tuned SVC controller's responsibility was to control the reactive power of the AC MG system [55, 56]. Isolated DVR is one of the most advanced control hardware options for MGs [57]. The DVR works best to alleviate power quality issues if any uncertainty is produced in the MG.

5.5 Other AI-Based Approaches

The use of AI and traditional techniques to reduce power quality issues created by PV grid-tied systems is investigated in [58]. Over the past 50 years, power system monitoring has been employed to keep tabs on the power distribution system. It allows for the quick identification and correction of errors. By incorporating AI into power system monitoring applications, system analyzer workload, and stress are reportedly reduced. Furthermore, it is thought that improved PV system configuration can prevent some disturbances on its own. Additionally, compensator technology demonstrates the ability to resolve a number of power quality problems.

Authors in [59] focused on shunt hybrid filters for PQ enhancement in the MG system (SHF). In order to achieve an effective smart grid functioning under various scenarios of loads and supply voltages, the performance of SHF is explored utilizing an enhanced and sophisticated regulating approach called Adaptive Fuzzy-Neural-Network (AFNN) Control. The suggested controller is contrasted with alternative controlling strategies including adaptive fuzzy sliding (AFS) control and adaptive fuzzy back stepping (AFBS).

6 Conclusion

MG networks have developed into reliable power sources for supplying electricity to remote areas in a secure, reliable, and low-emission manner. The performance of the MG network is frequently impacted by power quality disturbances, which restrict its utilization on a limited scale. The network's numerous power components' reliability, efficiency, and lifespan are often reduced by power quality disturbances. In this chapter, various techniques relating to the mitigation of power quality disturbances are discussed. The aforementioned mitigation strategies, such as controller-based approaches, FACTS-based methods, or harmonic mitigation techniques, use various types of computational intelligence methods, such as fuzzy logic, neural networks, and evolutionary computation, in one way or another.

References

1. Kaur S, Dwivedi B (2016) Power quality issues and their mitigation techniques in microgrid system-a review. In: 2016 7th India International Conference on Power Electronics (IICPE), pp 1–4. https://doi.org/10.1109/IICPE.2016.8079543
2. Sen S, Kumar V (2018) Microgrid control: a comprehensive survey. Annu Rev Control 45:118–151. https://doi.org/10.1016/j.arcontrol.2018.04.012
3. Srivastava I, Bhat S, Reddy AA (2019) Multi-objective service restoration of radial distribution system in the presence of non-linear loads. Electronics 8:369. https://doi.org/10.3390/electronics8030369
4. Li P, Yu X, Zhang J, Yin Z (2015) The H∞ control method of grid-tied photovoltaic generation. IEEE Trans Smart Grid 6:1670–1677. https://doi.org/10.1109/TSG.2015.2409371
5. Kumar A, Sharma R (2015) A stable Lyapunov constrained reinforcement learning based neural controller for non linear systems. In: International Conference on Computing, Communication & Automation. IEEE, Greater Noida, India, pp 185–189. https://doi.org/10.1109/CCAA.2015.7148402
6. Ali KH, Sigalo M, Das S et al (2021) Reinforcement learning for energy-storage systems in grid-connected microgrids: an investigation of online vs. offline implementation. Energies 14:5688. https://doi.org/10.3390/en14185688

7. Sigalo MB, Pillai AC, Das S, Abusara M (2021) An energy management system for the control of battery storage in a grid-connected microgrid using mixed integer linear programming. Energies 14:6212. https://doi.org/10.3390/en14196212
8. Al-Saedi W, Lachowicz SW, Habibi D, Bass O (2012) Power quality enhancement in autonomous microgrid operation using Particle Swarm Optimization. Int J Electr Power Energy Syst 42:139–149. https://doi.org/10.1016/j.ijepes.2012.04.007
9. Chitra N, Prabaakaran K, Senthil Kumar A, Munda J (2013) Ant colony optimization adopting control strategies for power quality enhancement in autonomous microgrid. IJCA 63:34–38. https://doi.org/10.5120/10529-5513
10. Deckmann SM, Ferrira AA (2002) About voltage sags and swells analysis. In: 10th International Conference on Harmonics and Quality of Power. Proceedings (Cat. No.02EX630). IEEE, Rio de Janeiro, Brazil, pp 144–148. https://doi.org/10.1109/ICHQP.2002.1221423
11. IEEE Recommended Practice for 1 kV to 35 kV Medium-Voltage DC Power Systems on Ships. IEEE. https://doi.org/10.1109/IEEESTD.2018.8569023
12. Van den Broeck G, Stuyts J, Driesen J (2018) A critical review of power quality standards and definitions applied to DC microgrids. Appl Energy 229:281–288. https://doi.org/10.1016/j.apenergy.2018.07.058
13. IEEE Recommended Practice for Monitoring Electric Power Quality. IEEE. https://doi.org/10.1109/IEEESTD.2019.8796486
14. Singh B, Chandra A, Al-Haddad K (2015) Power quality problems and mitigation techniques. Wiley, Chichester, West Sussex, United Kingdom. https://doi.org/10.1002/9781118922064
15. Priyadharshini KM, Srinivasan S, Srinivasan C (2014) Power quality disturbance detection and islanding in micro grid connected distributed generation. In: 2014 IEEE International Conference on Computational Intelligence and Computing Research. pp 1–6. https://doi.org/10.1109/ICCIC.2014.7238390
16. Beniwal RK, Saini MK, Nayyar A et al (2021) A critical analysis of methodologies for detection and classification of power quality events in smart grid. IEEE Access 9:83507–83534. https://doi.org/10.1109/ACCESS.2021.3087016
17. Singh N, Ansari MA, Tripathy M, Singh VP (2021) Feature extraction and classification techniques for power quality disturbances in distributed generation: a review. IETE J Res 1–16. https://doi.org/10.1080/03772063.2021.1920849
18. Mishra M (2019) Power quality disturbance detection and classification using signal processing and soft computing techniques: A comprehensive review. Int Trans Electr Energ Syst 29. https://doi.org/10.1002/2050-7038.12008
19. Suganthi ST, Vinayagam A, Veerasamy V (2021) Detection and classification of multiple power quality disturbances in Microgrid network using probabilistic based intelligent classifier. Sustainable Energy Technologies and Assessments 47:101470. https://doi.org/10.1016/j.seta.2021.101470
20. Wright PS (1999) Short-time Fourier transforms and Wigner-Ville distributions applied to the calibration of power frequency harmonic analyzers. IEEE Trans Instrum Meas 48:475–478. https://doi.org/10.1109/19.769633
21. Choong F, Reaz MBI, Mohd-Yasin F (2005) Advances in signal processing and artificial intelligence technologies in the classification of power quality events: a survey. Elect Power Compon Syst 33:1333–1349. https://doi.org/10.1080/15325000590964155
22. Azam MS, Tu F, Pattipati KR, Karanam R (2004) A dependency model-based approach for identifying and evaluating power quality problems. IEEE Trans Power Del 19:1154–1166. https://doi.org/10.1109/TPWRD.2003.822537
23. Jurado F, Acero N, Ogayar B (2002) Application of signal processing tools for power quality analysis. In: IEEE CCECE2002. Canadian Conference on Electrical and Computer Engineering. Conference Proceedings (Cat. No.02CH37373). pp 82–87 vol. 1. https://doi.org/10.1109/CCECE.2002.1015179
24. Heydt GT, Fjeld PS, Liu CC et al (1999) Applications of the windowed FFT to electric power quality assessment. IEEE Trans Power Delivery 14:1411–1416. https://doi.org/10.1109/61.796235

25. Gu YH, Bollen MHJ (2000) Time-frequency and time-scale domain analysis of voltage disturbances. IEEE Trans Power Del 15:1279–1284. https://doi.org/10.1109/61.891515
26. Huang SJ, Huang CL, Hsieh CT (1996) Application of Gabor transform technique to supervise power system transient harmonics. IEE Proc, Gener Transm Distrib 143:461. https://doi.org/10.1049/ip-gtd:19960534
27. Abdelsalam AA, Eldesouky AA, Sallam AA (2012) Classification of power system disturbances using linear Kalman filter and fuzzy-expert system. Int J Electr Power Energy Syst 43:688–695. https://doi.org/10.1016/j.ijepes.2012.05.052
28. Xue C, Hui-jin L, Quan-ming Z, et al. (2008) Power quality disturbances detection and location using mathematical morphology and complex wavelet transformation. In: 2008 3rd IEEE Conference on Industrial Electronics and Applications, pp 2263–2268. https://doi.org/10.1109/ICIEA.2008.4582920
29. Huang Y, Liu Y, Hong Z (2009) Detection and location of power quality disturbances based on mathematical morphology and hilbert-huang transform. In: 2009 9th International Conference on Electronic Measurement & Instruments, pp 2–319–2–324. https://doi.org/10.1109/ICEMI.2009.5274596
30. Lu Z, Turner DR, Wu QH, et al. (2004) Morphological transform for detection of power quality disturbances. In: 2004 International Conference on Power System Technology, 2004. PowerCon 2004, pp 1644–1649, Vol. 2
31. Wang M, Mamishev AV (2004) Classification of power quality events using optimal time-frequency representations—part 1: theory. IEEE Trans Power Delivery 19:1488–1495. https://doi.org/10.1109/TPWRD.2004.829940
32. Hu GS, Zhu FF, Tu YJ (2006) Power quality disturbance detection and classification using Chirplet transforms. In: Wang T-D, Li X, Chen S-H, et al. (eds) Simulated Evolution and Learning. Springer, Berlin, Heidelberg, pp 34–41. https://doi.org/10.1007/11903697_5
33. Jain AK, Duin RPW, Mao J (2000) Statistical pattern recognition: a review. IEEE Trans Pattern Anal Mach Intell 22:4–37. https://doi.org/10.1109/34.824819
34. Ribeiro MV, Pereira JLR (2007) Classification of single and multiple disturbances in electric signals. EURASIP J Adv Signal Process 2007:1–18. https://doi.org/10.1155/2007/56918
35. Srivastava I, Bhat S, Singh AR (2020) Fault diagnosis, service restoration, and data loss mitigation through multi-agent system in a smart power distribution grid. Energy Sources, Part A: Recovery, Utilization, and Environmental Effects 1–26. https://doi.org/10.1080/15567036.2020.1817190
36. Manimala K, Selvi K (2007) Power quality disturbances classification using probabilistic neural network. In: International Conference on Computational Intelligence and Multimedia Applications (ICCIMA 2007), pp 207–211. https://doi.org/10.1109/ICCIMA.2007.204
37. He Z, Zhang H, Zhao J, Qian Q (2012) Classification of power quality disturbances using quantum neural network and DS evidence fusion. Euro Trans Electr Power 22:533–547. https://doi.org/10.1002/etep.584
38. Hsiao YT, Chuang CL, Jiang JA (2005) Recognition of power quality events using wavelet-based dynamic structural neural networks. In: 2005 IEEE International Symposium on Circuits and Systems (ISCAS). 4: 3885–3888. https://doi.org/10.1109/ISCAS.2005.1465479
39. Wijayakulasooriya JV, Putrus GA, Minns PD (2002) Electric power quality disturbance classification using self-adapting artificial neural networks. IEE Proc, Gener Transm Distrib 149:98. https://doi.org/10.1049/ip-gtd:20020014
40 Lv G, Wang X, Zhang H, Zhang C (2005) PQ Disturbances Identification Based on SVMs Classifier. In: 2005 International Conference on Neural Networks and Brain, pp 222–226. https://doi.org/10.1109/ICNNB.2005.1614602
41. Janik P, Lobos T (2006) Automated classification of power-quality disturbances using SVM and RBF networks. IEEE Trans Power Delivery 21:1663–1669. https://doi.org/10.1109/TPWRD.2006.874114
42. Vinayagam A, Othman ML, Veerasamy V (2022) A random subspace ensemble classification model for discrimination of power quality events in solar PV microgrid power network. PLoS ONE 17:e0262570. https://doi.org/10.1371/journal.pone.0262570

43. Ray PK, Mohanty A, Panigrahi T (2019) Power quality analysis in solar PV integrated microgrid using independent component analysis and support vector machine. Optik 180:691–698. https://doi.org/10.1016/j.ijleo.2018.11.041
44. Kumar A, Sharma R (2017) Fuzzy Lyapunov reinforcement learning for non linear systems. ISA Trans 67:151–159. https://doi.org/10.1016/j.isatra.2017.01.026
45. Zhu TX, Tso SK, Lo KL (2004) Wavelet-based fuzzy reasoning approach to power-quality disturbance recognition. IEEE Trans Power Delivery 19:1928–1935. https://doi.org/10.1109/TPWRD.2004.832382
46. Chakravorti T, Patnaik RK, Dash PK (2018) Detection and classification of islanding and power quality disturbances in microgrid using hybrid signal processing and data mining techniques. IET Signal Proc 12:82–94. https://doi.org/10.1049/iet-spr.2016.0352
47. Krishna BV, Baskaran K (2007) Hybrid learning using multi-objective genetic algorithms and decision trees for power quality disturbance pattern recognition. In: International Conference on Computational Intelligence and Multimedia Applications (ICCIMA 2007), pp 276–280. https://doi.org/10.1109/ICCIMA.2007.339
48. Ray PK, Mohanty SR, Kishor N, Catalão JPS (2014) Optimal feature and decision tree-based classification of power quality disturbances in distributed generation systems. IEEE Trans Sustain Energy 5:200–208. https://doi.org/10.1109/TSTE.2013.2278865
49. Biswal B, Dash PK, Panigrahi BK (2009) Power quality disturbance classification using fuzzy C-means algorithm and adaptive particle swarm optimization. IEEE Trans Industr Electron 56:212–220. https://doi.org/10.1109/TIE.2008.928111
50. Huang N, Zhang S, Cai G, Xu D (2015) Power quality disturbances recognition based on a multiresolution generalized s-transform and a PSO-improved decision tree. Energies 8:549–572. https://doi.org/10.3390/en8010549
51. Singh U, Singh SN (2019) A new optimal feature selection scheme for classification of power quality disturbances based on ant colony framework. Appl Soft Comput 74:216–225. https://doi.org/10.1016/j.asoc.2018.10.017
52. Al-Saedi W, Lachowicz SW, Habibi D, Bass O (2013) Power flow control in grid-connected microgrid operation using Particle Swarm Optimization under variable load conditions. Int J Electr Power Energy Syst 49:76–85. https://doi.org/10.1016/j.ijepes.2012.12.017
53. Morsi WG, El-Hawary ME (2010) Novel power quality indices based on wavelet packet transform for non-stationary sinusoidal and non-sinusoidal disturbances. Electric Power Syst Res 80:753–759. https://doi.org/10.1016/j.epsr.2009.11.005
54. Meegahapola L, Pea S (2012) Impact of wind generator control strategies on flicker emission in distribution networks. In: 2012 IEEE 15th International Conference on Harmonics and Quality of Power, pp 612–617. https://doi.org/10.1109/ICHQP.2012.6381178
55. Bansal RC, Bhatti TS, Kothari DP (2004) Automatic reactive power control of isolated wind-diesel hybrid power systems for variable wind speed/slip. Electric Power Compon Syst 32:901–912. https://doi.org/10.1080/15325000490253542
56. Ray P, Salkuti SR (2020) Smart branch and droop controller based power quality improvement in microgrids. Int J Emerg Electr Power Syst 21(6):20200094. https://doi.org/10.1515/ijeeps-2020-0094
57. Das B, Panigrahi PK, Das SR, Mishra DP, Salkuti SR (2022) Power quality improvement in a photovoltaic based microgrid integrated network using multilevel inverter. Int J Emerg Electr Power Syst 23(2):197–209. https://doi.org/10.1515/ijeeps-2021-0040
58. Salkuti SR (2022) Emerging and advanced green energy technologies for sustainable and resilient future grid. Energies 15(18):6667. https://doi.org/10.3390/en15186667
59. Kow KW, Wong YW, Rajkumar RK, Rajkumar RK (2016) A review on performance of artificial intelligence and conventional method in mitigating PV grid-tied related power quality events. Renew Sustain Energy Rev 56:334–346. https://doi.org/10.1016/j.rser.2015.11.064
60. Das SR, Ray PK, Sahoo AK (2021) Artificial intelligence based grid connected inverters for power quality improvement in smart grid applications. Comput Electr Eng 93:107208. https://doi.org/10.1016/j.compeleceng.2021.107208

A Comprehensive Power Quality Mitigation Tool: UPQC

Raavi Satish, Balamurali Pydi, Surakasi Balamurali, Surender Reddy Salkuti, Almoataz Y. Abdelaziz, and Solomon Feleke

Abstract Before the advancement of electronic-based applications in electrical appliances, the only major concern was reactive power compensation for power factor improvement. A steep rise in the use of power electronic-based electrical devices at each and every level of power utilization has completely changed the prior equations. These power electronic-based devices served the purpose of precise control and efficient regulation of loads and equipment. However, the intensifying conflict of electronic devices with their electrical environment disturbs the consistency of voltage and current profile supplied to the utility. Also, in an era of automation and computation, sophisticated equipment with sophisticated controllers demands a current profile that is undesirable for the supply lines. Additionally, the usage of renewable energy sources is one of the recent trends in electric power systems. These generation methods are quite different from the conventional generation approaches in terms of their power generation profile and other system parameters. Thus, incorporating and

R. Satish · S. Balamurali
Department of Electrical and Electronics Engineering, ANITS, Visakhapatnam, Andhra Pradesh—531162, India
e-mail: balamurali.eee@anits.edu.in

B. Pydi
Department of Electrical and Electronics Engineering, Aditya Institute of Technology & Management, Tekkali, Srikakulam, Andhra Pradesh—532201, India
e-mail: balu_p4@yahoo.com

S. R. Salkuti
Department of Railroad and Electrical Engineering, Woosong University, Daejeon—34606, Republic of Korea
e-mail: surender@wsu.ac.kr

A. Y. Abdelaziz (✉)
Faculty of Engineering and Technology, Future University in Egypt, Cairo, Egypt
e-mail: almoatazabdelaziz@hotmail.com

S. Feleke
Department of Electrical and Computer Engineering, DebreBerhan University, DebreBerhan, Ethiopia

S. R. Salkuti et al. (eds.), *Power Quality in Microgrids: Issues, Challenges and Mitigation Techniques*, Lecture Notes in Electrical Engineering 1039,
https://doi.org/10.1007/978-981-99-2066-2_3

sustaining the new generation systems with the prevailing system poses certain challenges, and this needs to be dealt with.As switching element-based power electronic devices are mostly responsible for degrading the power quality, a concurrent solution known as CPD (custom power device) consisting of the same switching elements has made a remarkable impression in electrical power systems. One such device, named UPQC (Unified Power Quality Conditioner) with multi-function compensating capability has been recognized as one of the most effective and efficient power quality enhancement devices.

Keywords Custom power device · Power electronics · Power quality · Unified power quality conditioner

Nomenclature

APF	Active Power Filter
CPD	Custom Power Device
DVR	Dynamic Voltage Restorer
FACTS	Flexible AC Transmission System
PCC	Point of Common Coupling
STATCOM	Static Compensator
THD	Total Harmonic Distortion
UPQC	Unified Power Quality Conditioner
UPQC-L	Left Shunt Unified Power Quality Conditioner
UPQC-R	Right Shunt Unified Power Quality Conditioner
UVT	Unit Vector Template
VSI	Voltage Source Inverter

1 Introduction

In an earlier era of the power distribution system, the electric equipment and machinery were much simpler to operate and quite less sophisticated to control. Thus, these loads were quite robust to various voltage disturbances and quite friendly to the supply system in terms of electrical pollution. Before the advancement of electronic-based applications in electrical appliances, the only major concern was reactive power compensation for power factor improvement. A steep rise in the use of power electronic-based electrical devices at each and every level of power utilization has completely changed the prior equations. These power electronic-based devices served the purpose of precise control and efficient regulation of loads and equipment.

Commercial and industrial electrical loads such as computers, lightning ballast, switch mode power supply, motor drive applications, etc. are some of the vast applications of power electronics. Most of these loads possess non-linear characteristics. A nonlinear load consists of switching elements and introduces nonlinearity, i.e., harmonics in the current [1, 2]. Also, in an era of automation and computation, sophisticated equipment with sophisticated controllers demands a current profile that is undesirable for the supply lines.

The industrial and commercial consumers of electricity are becoming self-aware day by day, thus demanding electricity of high quality and reliability. The modern equipment and devices used in such places are highly sensitive to different types of disturbances. The protection setup to be used individually for each device and each disturbance will increase the complexity and is not a viable option from an economic point of concern.

1.1 Power Quality

The term power quality has gained substantial popularity in electric power-related areas. Since the level of electric pollution in supply systems is more severe at the utilization level, it is important to assess the details of power quality at the terminals of end users in distribution systems. According to international standards and specifications, power quality relates to the quality of the supply being made available at the consumer end that does not cause any harm to its equipment or devices [3]. However, the power quality degradation can be observed either from the supply end or from the load end.

1.2 Major Power Quality Issues

A number of power quality issues have been recognized in an electrical system. These issues can be broadly classified as natural or man-made type. Some of the natural factors for poor power quality are faults. Man-made power quality degrading factors are mostly concerned with load and system operations. Some of the major power quality issues are discussed below.

1.2.1 Voltage Sag

It is concerned with a dip or a reduction in the supply voltage from its nominal rated value for a small duration. This duration can vary between five cycles to a minute. Voltage sag may be caused due to several reasons such as heavy load demand, fault in system operation, starting of large motors, etc.

1.2.2 Voltage Swell

This is related to an increase in supply voltage magnitude from its nominal rated value for a short duration of time. The duration is more or less similar to that of the voltage sag condition. Its occurrence is due to a system fault, immediate turning on and off of large loads, and transfer of the load from one supply to another.

1.2.3 Transient

These are sub-cycle disturbances, existing for not more than one cycle. Detection and measurement of transients are very challenging due to inadequate frequency response or sample rate. Transients are sometimes also termed as spikes, surges, power pulses, etc. This kind of problem arises due to atmospheric disturbances like lightning and solar flashes, interruptions due to fault currents, switching of loads, capacitor bank switching, power line switching, etc.

1.2.4 Voltage Flicker

These are referred to as rapid variations in the supply voltage magnitude with regular random patterns. One of the major causes of its occurrence is the use of arc furnaces and arc lamps.

1.2.5 Distortion in Waveform

Due to several distortion factors such as harmonics, inter-harmonics, DC offset, notching, noise, etc., the waveform of voltage or current may get deviated from its ideal sinusoidal shape.

1.2.6 Harmonics

Any deviation from the original or pure sine waveform of voltage or current signal relates to the presence of harmonics in that particular signal. Harmonic frequencies are integral multiples of the fundamental frequency and are very common in electric power systems. The harmonics can be differentiated as even (2, 4, 6, 8, 10) and odd types (3, 5, 7, 9, 11) depending on their order. Odd harmonics are usually introduced by nonlinear loads and even harmonics are produced due to irregular operations of electrical devices like transformer magnetizing current containing the even harmonic components [4]. The measurement of the harmonic level in a particular signal is usually termed as total harmonic distortion (THD). The adverse effects of high harmonic content include equipment malfunctioning, power loss, overheating of equipment, overloading of the distribution system, etc. It can also cause electromagnetic interference with other communication lines, reduction in efficiency of the motor drive system, and may be a reason for auditory noise in motors and other appliances.

1.2.7 Unbalance

This condition arises when the voltage or current signals of three phases of the supply are either unequal in magnitude or unequally displaced with each other. One of the major reasons for the occurrence of this type of condition is single-phase loading on three-phase networks. Also, line fault conditions give rise to such unbalancing in voltage or current. Unbalanced systems may result in the introduction of a negative sequence component that affects adversely all three-phase loads such as induction motors.

1.2.8 Frequency Variation

Variation in the frequency of an electrical power system arises due to a mismatch between generation and load. These variations are typically caused by sudden fluctuations in the load connected to the system. The deviation in system frequency from its nominal value should be within limits.

1.2.9 Interruptions

An interruption in the supply may be due to unavoidable weather conditions or due to failure in equipment or equipment's protection device. This results in disruption of power supply to the loads.

1.3 Consequences of Power Quality Issues

These power quality related problems may lead to some serious harmful effects on utility as well as on electrical devices and loads. Some of these consequences are listed below:

- Voltage-related disturbances such as harmonics, sag, or swell may be a root cause for the continuous tripping of sensitive devices which in turn may lead to catastrophic outcomes in industries or plants. The frequent occurrence of such events may result in severe financial loss.
- Nowadays, power electronic-based devices are heavy usage, all around the world for electric power-based applications. Especially, an uncontrolled rectifier-based

converter system is quite common in industrial and commercial applications. These devices are the main source of introducing harmonic distortion in the source current. This in turn results in disturbing the voltage profile across the load.
- The unbalancing and harmonics in the current eventually result in unbalanced and distorted voltages. These disturbances in the voltage effects other neighboring loads with regular deterioration in the power quality.

In order to regulate this kind of contamination in a distribution system, some strict regulations are in place by competent authorities. In order to abide by the guidelines and to protect their sensitive equipment, the consumers need to check their power quality degrading factors by taking suitable measures and thus avoiding hefty penalties. These power quality affecting issues can be handled at the consumer end by adapting mitigation devices appropriately.

1.4 Power Electronics Contribution to Power Quality Management

The acceleration equation of electronic devices with their electrical environment disturbs the regularity of voltage and current profile being supplied in a distribution system. In this era of automation and computation, sophisticated equipment with their sophisticated controllers demands a current profile that is undesirable for the supply lines. This significantly impairs the power quality supplied for other loads.

In reference to a popular phrase, "Iron sharpens Iron", a similar approach is in practice nowadays for power quality improvement in the electrical distribution system. As switching element-based power electronic devices are mostly responsible for degrading the power quality, a concurrent solution known as a custom power device consisting of the same switching elements has made a remarkable impression in electrical power systems. Initially, a concept of FACTS was introduced in the late 1980s to provide support for smooth power transfer in a transmission system. These FACTS devices incorporated power electronic-based configurations for required compensation. STATCOM (Static Compensator) and DVR (Dynamic Voltage Restorer) are popular FACTS devices that utilize a controlled VSI (Voltage Source Inverter). STATCOM is popularly used for smooth power flow by providing reactive power compensation whereas DVR is adapted for providing voltage compensation.

2 Research Background on UPQC

The UPQC with multi-function compensating capability has been recognized as one of the most effective and efficient power quality enhancement devices. It has received great attention from various researchers worldwide in the past two decades.

2.1 Initial Study of UPQC and Its Brief Description

In 1989, Moran [5] proposed a combined configuration of series and shunt APF, connected with each other through a common reactor, and termed it as a line voltage conditioner. However, with the practical implementation of such back-to-back inverter configuration in 1998 [6], Akagi brought this device into the limelight and named it as UPQC. A few other nomenclatures adapted for such configuration are series–parallel converter [7], unified APF (UAPF) [8], universal active power line conditioner [9, 10], universal power quality conditioning system (UPQS) [11], load compensation active conditioner [12, 13], universal active filter [14], and so forth.

UPQC is employed in a distribution system and is found quite analogous to another device used in the transmission system, known as UPFC (Unified Power Flow Conditioner) [15]. The application of UPFC is mainly focused on maintaining a smooth power flow in the line, whereas UPQC is subjected to improving the power quality at the distribution level. UPQC is usually employed to handle multiple power quality disturbing issues simultaneously, which include voltage-related factors (sag, swell, unbalance, and harmonics) and current-related factors (harmonics, reactive current, and unbalance). Figure 1 shows a basic representation of UPQC. It usually consists of two back-to-back connected inverters via a common DC link (Cdc).

With a closed loop control operation, the shunt APF part of UPQC is operated as a current-controlled device for current-related compensation and the series APF is operated as a voltage-controlled device for voltage-related compensation.

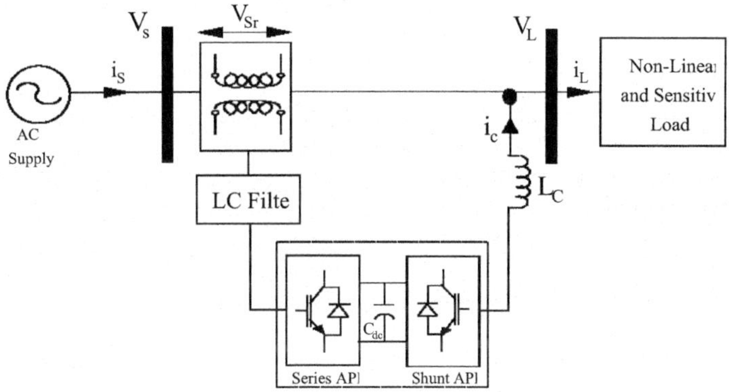

Fig. 1 Single line diagram of UPQC

2.2 Classification of UPQC

UPQC can be classified depending on certain parameters such as converter configuration, type of AC supply, system arrangement, and compensation type as shown in Fig. 2.

2.2.1 Converter Configuration

UPQC can be classified based on two types of converter configurations. The most popular is the VSI (Voltage Source Inverter) based UPQC [16] that utilizes a common DC-link capacitor as shown in Fig. 2. The other configuration is the CSI (Current Source Inverter) based UPQC [17] that implements a common DC-link reactor as shown in Fig. 3. The operational behavior of both configurations is quite similar. However, they differ from each other in certain areas such as in the case of their regulation parameter and in the case of switching device configuration in the inverters of UPQC. The efficient operation of VSI-based UPQC depends on the regulation of voltage across the DC-link capacitor whereas CSI-based UPQC operation depends on the regulation of DC-link inductor current. VSI-based UPQC consists of IGBT modules with anti-parallel diodes connected across each diode [18]. However, CSI-based UPQC consists of discrete IGBTs, and diodes are connected in series with each IGBT.

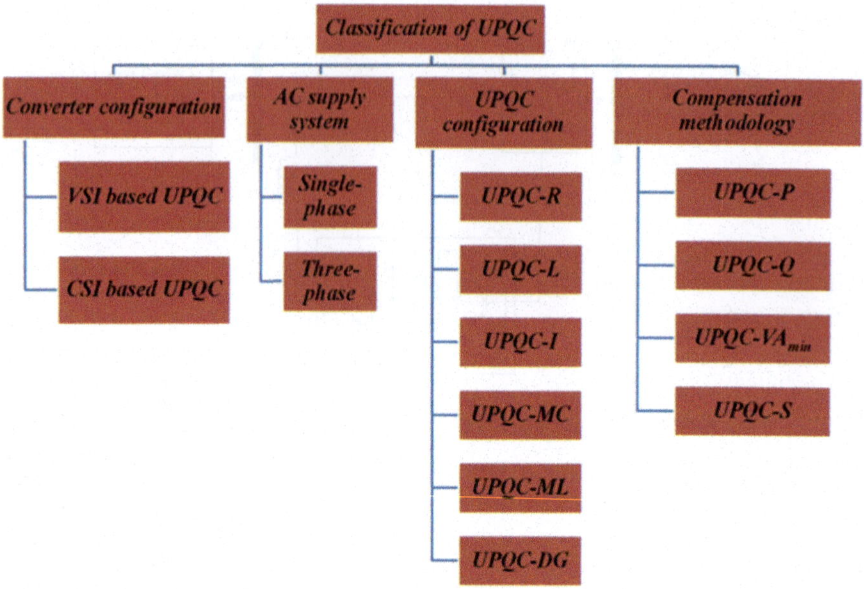

Fig. 2 Classification chart of UPQC

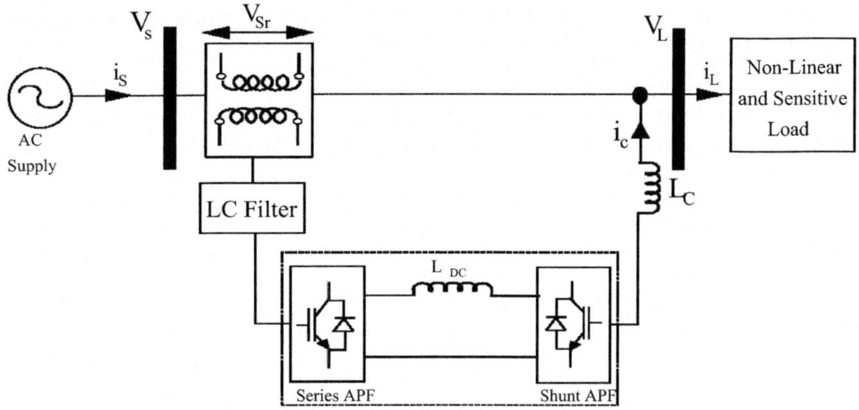

Fig. 3 CSI-based UPQC system

Since IGBT modules with anti-parallel diodes are most popularly adopted for inverter design, VSI-based UPQC is an easier and more feasible configuration to deal with. Also, VSI-based UPQC is more advantageous than the others in terms of cost, design of control function [19], and weight. VSI-based UPQC also finds its applicability and superiority in the case of multi-level inverter configurations, adopted for UPQC design for higher rating applications.

2.2.2 AC Supply System

UPQC system can be implemented with different configurations for single-phase and three-phase AC systems.

Single-Phase System

The single-phase system suffers from high harmonic contamination, reactive power demand, and voltage sag or swell situation [20]. For a single-phase system, the most popular configuration consists of two H-bridge converters with a total of eight semiconductor switches as shown in Fig. 4a. A few other configurations have also been proposed by researchers for a single-phase UPQC system with a reduced number of switches. Two of such single-phase UPQC configurations are shown in Fig. 4b, c. Figure 4b demonstrates a configuration with a six-switch arrangement, i.e. with two switches less than the H bridge type. Here, each of the two series switches is dedicated individually for the shunt and series APF inverter, whereas the third switch leg is commonly used by both APF inverters [21]. Furthermore, in another configuration, one more switch leg can be eliminated, thus conceiving a four-switch arrangement as shown in Fig. 4c. This arrangement utilizes two series switches each for the shunt and series APF inverter. These kinds of configurations with less number of switches help in reducing the overall size of the compensating device [22]. However, they are applicable only for lower rating applications and are less tolerant to system fault conditions. Their compensation performance also gets affected due to the reduced number of switches, thus reducing their degree of usage as compared to the conventional H bridge configuration.

Three-Phase System

The most popularly adapted UPQC configuration is a three-phase three-wire arrangement shown in Fig. 5. Including the problems as stated for a single-phase system, a three-phase system involves more power quality issues like current unbalancing and voltage unbalancing [23]. This UPQC configuration is mostly stressed by researchers for various power quality disturbances and different loading conditions. Some distinct configurations are also proposed for a such system like ten switch topology [24]. However, the application of this configuration is limited to three-phase loads only.

For unbalanced loads, i.e., the combination of single-phase and three-phase loads, the source current gets unbalanced which gives rise to a neutral current [25]. Thus, the neutral current compensation capability becomes an obvious necessity for such compensating devices [26]. The UPQC configurations that have been widely adopted for such unbalanced load conditions (i.e. three-phase four-wire systems) are four-leg, split capacitor, and H-bridge arrangements as shown in Fig. 6. All the reported investigations and conclusions are also valid for similar UPQC configurations. The number of switching devices is highest for the H bridge and lowest for the split capacitor. In the case of a split capacitor, two DC-link capacitors are required while one capacitor is required for another two cases. The minimum DC-link voltage constraint is highest for the split capacitor and lowest for the H bridge. Compensation performance is found most effective in the case of the four-leg system [27]. Thus, the four-leg configuration of UPQC is found to be an appropriate and suitable option for a three-phase four-wire system and has been suitably adopted and explored in this investigation.

2.2.3 UPQC Configuration

UPQC can also be classified depending on the arrangement/connection of inverters within the system or with the external elements. Depending on the placement of the shunt APF inverter, either on the right or left side of the UPQC system [28], UPQC can be categorized as right shunt UPQC (UPQC-R) or left shunt UPQC (UPQC-L).

Fig. 4 Single-phase UPQC system

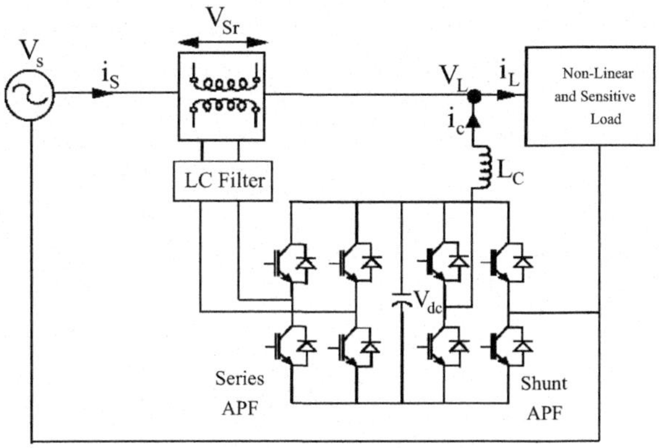

(a) H-bridge arrangement

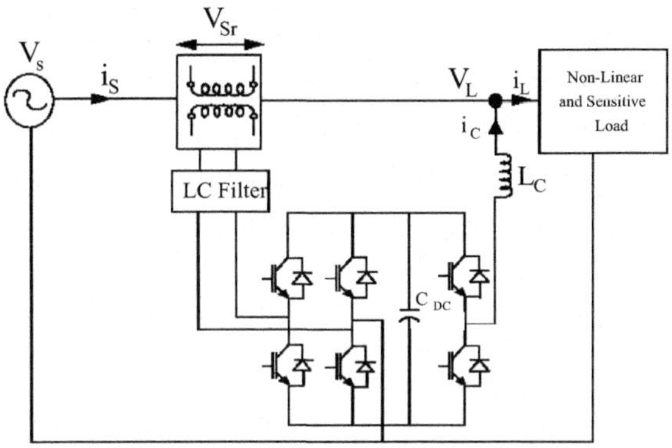

(b) Six-switch arrangement

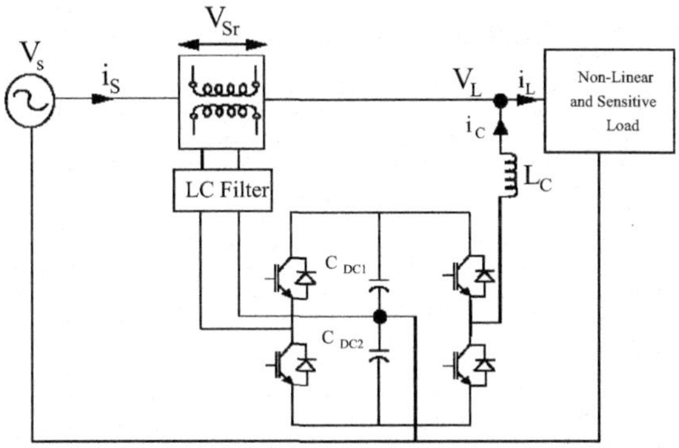

(c) Four-switch arrangement

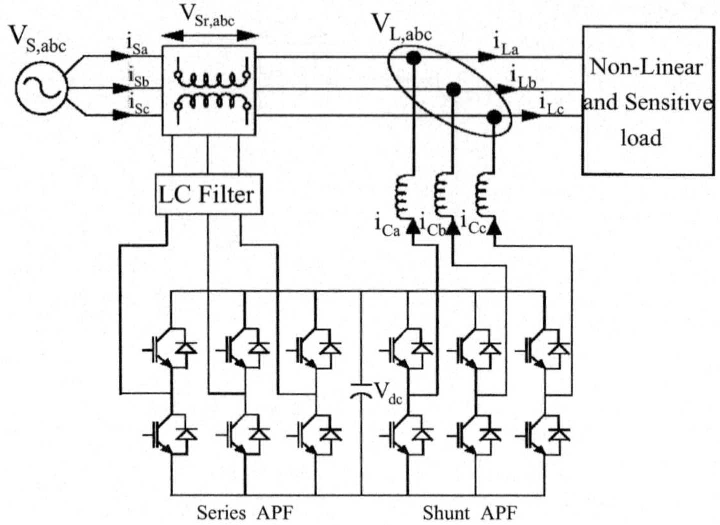

Fig. 5 Three-phase UPQC system

UPQC-R Configuration

The generic term UPQC system mostly signifies UPQC-R configuration. It is the most commonly used arrangement, as it avoids the flow of current harmonics through a series injection transformer [29]. In this thesis also, the UPQC-R configuration is mostly taken under consideration for performance investigation under different conditions.

UPQC-L Configuration

UPQC-L configuration shown in Fig. 7 may be used for special cases such as for voltage-sensitive loads and to avoid interference with passive filters. Also, with the UPQC-L arrangement, the scope of its harmonic mitigation applicability can be increased as reported in this chapter. A comparative analysis of the two popular configurations of UPQC (i.e., UPQC-R and UPQC-L) is performed in this work under different source voltage conditions to confirm their suitability and rating differences [30].

Other UPQC Configurations

Researchers have proposed some other special configurations of UPQC such as interline UPQC (UPQC-I), multilevel UPQC (UPQC-ML), and distributed generators integrated with UPQC (UPQC-DG). Out of these, UPQC-DG is the most popular

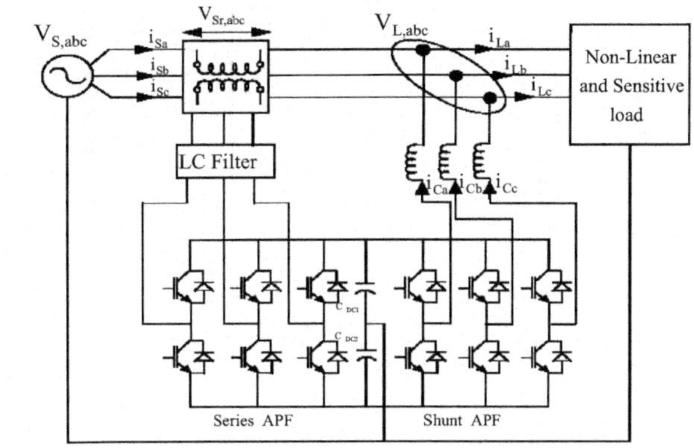

(a) Split-capacitorconfiguration

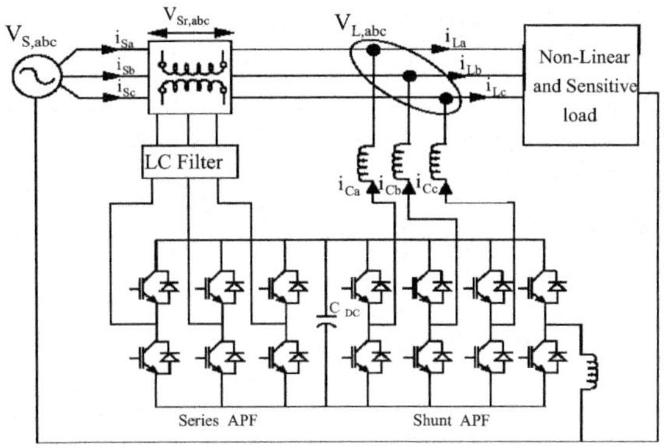

(b) Four-leg configuration

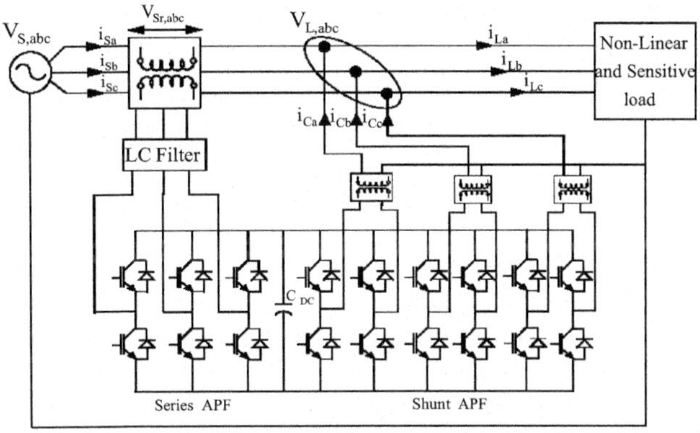

(c) H-bridge configuration

Fig. 6 Three-phase UPQC for three-phase four-wire system

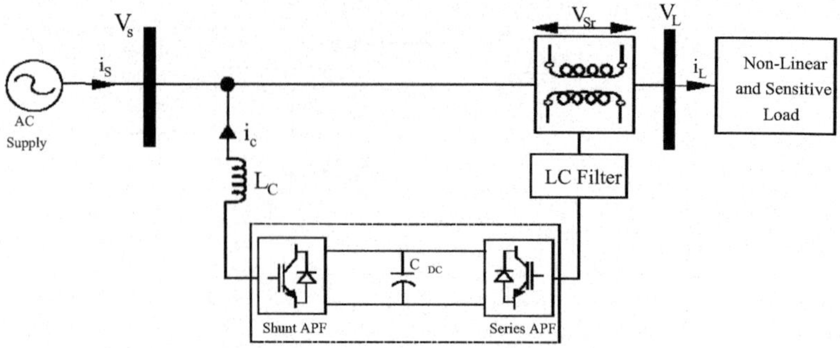

Fig. 7 Single line diagram of UPQC-L system

type of configuration in recent trends [31]. It involves the interconnection of UPQC with an alternative source of energy (solar, wind, fuel cell, etc.). The power flow from DG to the load can be controlled through UPQC along with other power quality improvement factors. It can be operated in two different modes of operation, i.e., interconnected or islanding mode. One such UPQC-DG system with its modified attachment to the main lines is presented in this thesis to enhance the usability of both APF inverters.

2.2.4 Compensation Methodology

Based on different approaches adapted for voltage sag compensation, the UPQC can be classified. In [32], the voltage sag is considered as one of the most common voltage-related issues in the power system. These approaches are the active power control approach (UPQC-P), reactive power control approach (UPQC-Q), minimum Volt-Ampere loading, and simultaneous active-reactive power control approach (UPQC-S).

Active Power Control Method (UPQC-P Approach)

The UPQC-P approach deals with only active power drawn from the source for necessary voltage sag compensation. The UPQC-P approach is also applicable for voltage swell compensation. The voltage phasor diagrams for the UPQC-P approach under both voltage sag and swell conditions are shown in Fig. 8. The voltage component injected by the series APF part of UPQC remains in-phase with the source voltage. The required energy for this voltage injection is drawn by the shunt APF from the source side in the form of active power, thus the magnitude of the source current also changes [33]. The shunt APF in turn transfers the energy to the series APF via a DC-link capacitor. However, this method doesn't involve any reactive power injection from series APF for load reactive power compensation.

Fig. 8 Phasor representation of UPQC-P approach

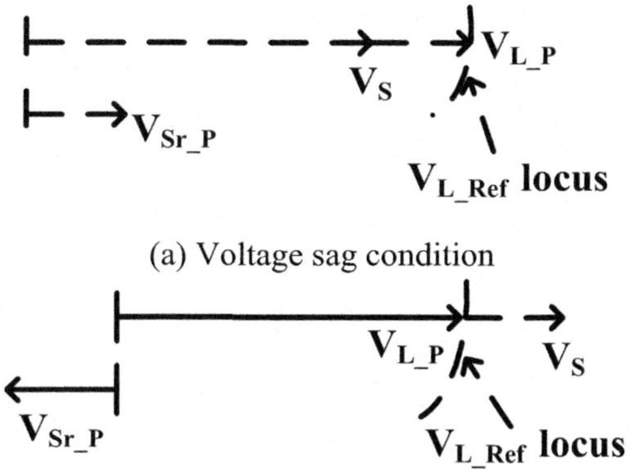

(a) Voltage sag condition

(b) Voltage swell condition

Reactive Power Control Method (UPQC-Q Approach)

The UPQC-Q approach is concerned with the reactive power control approach for voltage sag compensation. This approach is not applicable to voltage swell compensation. The voltage phasor diagrams for the UPQC-Q approach under both voltage sag and swell conditions are shown in Fig. 9. With this method, the voltage component being injected by the series APF remains in quadrature with the source voltage. Thus, the resultant load voltage with the same magnitude as the reference source voltage appears across the load terminals. But the load voltage is slightly shifted in phase as referred to as the source voltage. The UPQC-Q approach is advantageous over UPQC-P in terms of no active power involvement. Also, with this approach, some reactive power support can be provided to the load side through series APF. However, it requires a larger magnitude of injection voltage as compared to the UPQC-P approach for the same amount of sag compensation. This in turn affects the rating of series APF. Most importantly, the UPQC-Q approach is not helpful in the case of voltage swell compensation as can be easily deduced from the phasor diagram shown in Fig. 9. As can be observed, the resultant load voltage will be even higher in magnitude than the source voltage itself under swell conditions. Thus, the UPQC-P approach is preferred to used against the UPQC-Q approach.

Minimum VA (Volt-Ampere) Loading Method (UPQC-VAminApproach)

In contrast to the previously attempted UPQC-P and UPQC-Q approach, another method, known as the minimum VA loading approach came into existence for UPQC systems. This approach differs from the above two mentioned conventional approaches in terms of voltage injection by series APF at a certain optimized angle in reference to the source voltage. The optimized estimation of the injected voltage angle is based on the sole idea of minimizing the rating of the UPQC system during

Fig. 9 Phasor representation of UPQC-Q approach

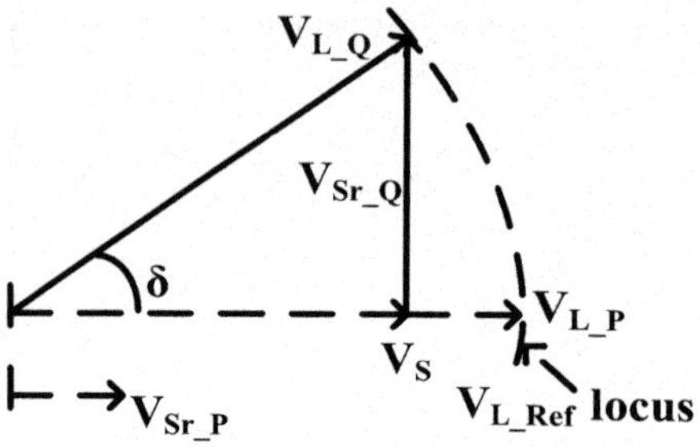

(a) Voltage sag condition

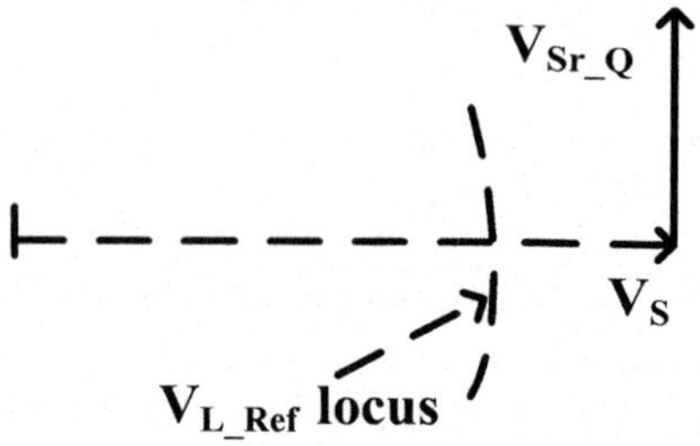

(b) Voltage swell condition

voltage sag compensation. The VA loading minimization of the UPQC system also requires the source current to be taken into consideration along with the injected voltage by series APF. However, this approach doesn't deal with reactive power compensation requirements during voltage disturbance conditions with maximum utilization of UPQC system rating in place of its minimization. The above-stated condition is well resolved with another compensation approach, termed as UPQC-S.

Active-Reactive Power Control Method (UPQC-S Approach)

The UPQC-S approach is found quite similar to UPQC-VAmin approach in terms of dealing with both active and reactive power under voltage sag and swell conditions. It also involves the injection of series voltage at a certain estimated angle in reference to the source voltage. However, the idea of this angle estimation is based on the maximum utilization of both shunt and series APF ratings. This series voltage injection creates a phase difference between the resultant load voltage and source voltage, which is termed as the power angle (δ) as can be seen in the phasor diagram shown in Fig. 10. The estimation of this power angle is based on the amount of reactive power to be shared by the shunt and series APF parts of UPQC. This power angle estimation concept is termed as a power angle control (PAC) approach. Thus, the UPQC-S approach is entirely based on this PAC concept to include reactive power sharing phenomena along with the regular voltage sag and swell compensation. However,

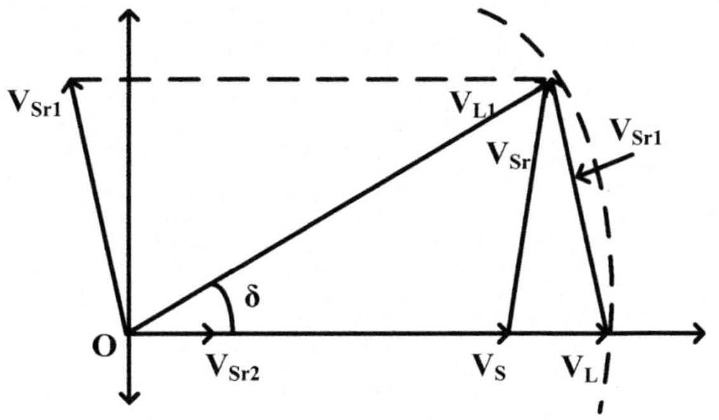

Fig. 10 Phasor representation of UPQC-S approach with voltage sag compensation

it certainly lags in the area of increased complexity which further increases with unbalanced conditions of voltage and load. Also, this approach involves an unequal sharing of reactive power between the shunt and series APF parts of UPQC. With the application of identical inverters of the same rating as the two APFs of UPQC, this unequal sharing approach doesn't prove to be quite efficient for maximum utilization and hence can be replaced with a much simpler equal reactive power sharing approach as presented in this work.

2.3 Different Control Approaches for UPQC

A UPQC device employed for power quality improvement always operates in a controlled closed-loop mode. A closed-loop control strategy implemented for UPQC decides its performance index. It is used to extract the reference signals and generate switching instants, thus providing a controlled operation of shunt and series APF. High-performance closed loop control operation for shunt APF is much more important than for the series APF. This is because, a shunt APF is usually regulated as a current-controlled device, whereas a series APF is controlled as a voltage-controlled device. The desired load voltage is always a known and fixed quantity for a particular system, whereas the current may vary depending on the load applied. Also, the energy required to maintain the DC-link voltage of UPQC is fed from the shunt APF, which necessitates its accountability in the control action of shunt APF. Thus, the control approach for shunt and series APF may differ from each other with a major emphasis on the shunt APF controller. Two of the most broadly used control methods for UPQC (especially for shunt APF) are instantaneous active-reactive power (pq) theory and synchronous reference frame (SRF) method or dq theory. Both these theories are based on transformation techniques and deal with the extraction of DC and fundamental quantities. Both these methods have been well explored by various researchers for the UPQC system.

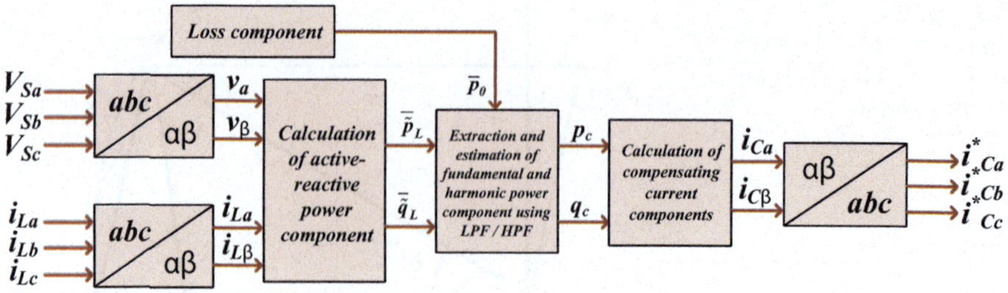

Fig. 11 Current reference estimation technique based on pq theory

2.3.1 Instantaneous Active-Reactive Power (Pq) Theory

During the initial phase of custom power device development for power quality improvement, the control concept of instantaneous pq theory gained worldwide popularity. It was first introduced by H. Akagi for controlling shunt active power filters to successfully mitigate current harmonics along with reactive power compensation. A generalized block diagram for the implementation of pq theory for reference current estimation is shown in Fig. 11. It is based on the transformation techniques and estimation of active and reactive power. The transformed quantities of source voltage and load currents are utilized to estimate the active and reactive power components. These components are fixed or DC quantities as they are derived from fixed orthogonal components of voltage and current. These power signals are then identified as fundamental and harmonic components separately using a low pass filter (LPF) or a high pass filter (HPF). These components are again utilized to generate either the reference compensating current signals or reference source current signals. This pq concept is a quite popular form of reference estimation technique for UPQC and is well explored by researchers. However, the performance of this concept deviates from its desired path under unbalanced or distorted voltage conditions. Hence, this pq method is not found ideally suitable for non-ideal supply voltage conditions as compared to other advanced reference estimation techniques.

2.3.2 Synchronous Reference Frame (SRF) Theory

Another popular reference estimation method used for power quality enhancement devices is the SRF theory. The SRF concept-based reference estimation method is the most popularly adopted approach for UPQC. It is based on a similar platform as that of the pq concept. The difference between the two approaches lies in the context of parameters for extracting fundamental and harmonic components. For pq theory, the controlling parameter is power whereas, in the SRF method, the current is treated as the controlling parameter. A generalized block diagram for the SRF-based current reference estimation technique is shown in Fig. 12. In this approach also, the sensed source current and voltage signals are transformed from abc to αβ frame. The current

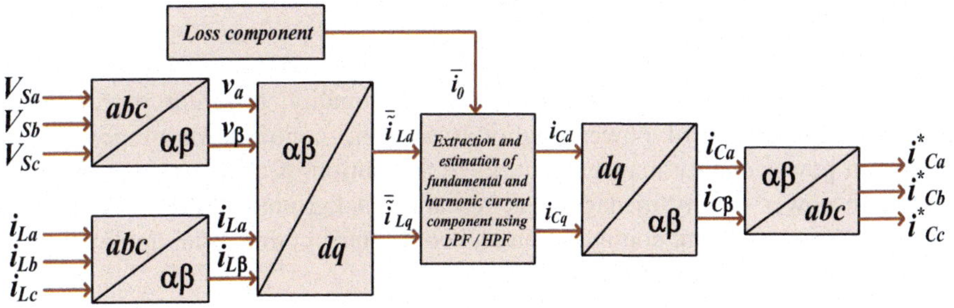

Fig. 12 Current reference estimation technique based on SRF (dq) theory

components so obtained are again transformed from αβ to dq reference frame. These direct and quadrature (dq) components of current are treated as fixed or DC quantities and are distinguished as fundamental and harmonic components with the aid of an LPF or HPF. The filtered or extracted current signals are again utilized to generate either reference compensating current or reference source current with further reverse transformations. Since the fundamental or harmonic extraction process involves only current parameters in this method rather than voltage parameters as in the case of pq theory, it remains quite unaffected due to any kind of voltage disturbances. The utilization level of these current parameters, so obtained by the SRF approach is further enhanced in this work, with their controlled application in the PAC concept.

2.4 Commercial Application and Economical Concern of UPQC

The applicability of power quality improvement devices is always faced with a major challenge of economic concern. STATCOM and only a few other devices have been commercially available till now [8], However, UPQC is not available commercially. A prototype developed at C-DAC, Trivandrum, India, of 250 kVA rating is the most feasible product available. Some other prototypes of higher ratings have also been developed as reported in [33, 34]. All the technical aspects to be known for developing this device have been explored more or less by researchers, but still, some challenges exist in the form of its complexity and cost-effectiveness. Thus, an attempt has been made in this thesis to reduce the control complexity of UPQC and increase the level of its effective and efficient utilization.

3 Conclusions

This chapter deals with an overview of power quality, different power quality issues, consequences of power quality issues, the contribution of power electronics in power quality management, the contribution of a custom power device: UPQC in power quality management, research background, UPQC classification based on supply system, configuration and controllers, commercial application and economical concern of UPQC.

Acknowledgements This research work was supported by "Woosong University's Academic Research Funding–2023".

References

1. Das SR, Mishra DP, Ray PK, Salkuti SR, Sahoo AK (2021) Power quality improvement using fuzzy logic-based compensation in a hybrid power system. Int J Power Electron Drive Syst 12(1): 576–584. https://doi.org/10.11591/ijpeds.v12.i1.pp576-584
2. Das B, Panigrahi PK, Das SP, Mishra DP, Salkuti SR (2022) Power quality improvement in a photovoltaic based microgrid integrated network using multilevel inverter. Int J Emerg Electr Power Syst 23(2): 197–209. https://doi.org/10.1515/ijeeps-2021-0040
3. Ray P, Salkuti SR (2020) Smart branch and droop controller based power quality improvement in microgrids. Int J Emerg Electr Power Syst 21(6). https://doi.org/10.1515/ijeeps-2020-0094
4. Salkuti SR (2022) Emerging and advanced green energy technologies for sustainable and resilient future grid. Energies 15(18):6667. https://doi.org/10.3390/en15186667
5. Patnaik N, Panda AK (2014)Comparative analysis on a shunt active power filter with different control strategies for composite loads. In: TENCON 2014 - 2014 IEEE Region 10 Conference: 1–6. https://doi.org/10.1109/TENCON.2014.7022289
6. Moran S (1989) A line voltage regulator/conditioner for harmonic-sensitive load isolation. In: Conference Record of the IEEE Industry Applications Society Annual Meeting 1: 947–951. https://doi.org/10.1109/IAS.1989.96758
7. Fujita H, Akagi H (1998) The unified power quality conditioner: the integration of series and shunt-active filters. IEEE Trans Power Electron 13: 315–322. https://doi.org/10.1109/63.662847
8. Kamran F, Habetler TG (1998) Combined deadbeat control of a series-parallel converter combination used as a universal power filter. IEEE Trans Power Electron 13:160–168. https://doi.org/10.1109/63.654971
9. Muthu S, Kim JMS (1997) Steady-state operating characteristics of unified active power filters. In: Proceedings of APEC 97—Applied Power Electronics Conference1: 199–205. https://doi.org/10.1109/APEC.1997.581454
10. Aredes M, Heumann K, Watanabe EH (1998) An universal active power line conditioner. IEEE Trans Power Deliv 13:545–551. https://doi.org/10.1109/61.660927
11. Watanabe EH, Aredes M (2002)Power quality considerations on shunt/series current and voltage conditioners. In: 10th International Conference on Harmonics and Quality of Power. Proceedings (Cat.No.02EX630)2: 595–600. https://doi.org/10.1109/ICHQP.2002.1221503
12. Graovac D, Katic V, Rufer A (2000) Power quality compensation using universal power quality conditioning system. IEEE Power Eng Rev 20:58–60. https://doi.org/10.1109/39.890381
13. Prieto J, Salmeron P, Vazquez JR, Alcantara J (2002) A series-parallel configuration of active power filters for var and harmonic compensation. In: IEEE 2002 28th Annual Conference of the

Industrial Electronics Society IECON 02: 2945–2950. https://doi.org/10.1109/IECON.2002.1182864
14. Prieto J, Salmeron P, Herrera RS (2005) A unified power quality conditioner for wide load range: Practical design and experimental results. 2005 IEEE Russia Power Tech: 1–7. https://doi.org/10.1109/PTC.2005.4524583
15. Eakburanawat J, Darapong P, Yangyuen U, Pongam S (2004) A simple control scheme of single phase universal active filter for power quality improvement. In: IEEE Region 10 Conference 3: 248–251. https://doi.org/10.1109/TENCON.2004.1414753
16. Chen G, Chen Y, Smedley KM (2004) Three-phase four-leg active power quality conditioner without references calculation. In: Applied Power Electronics Conference and Exposition 1: 587–593. https://doi.org/10.1109/APEC.2004.1295866
17. Khadkikar V, Chandra A (2009) A novel structure for three-phase four-wire distribution system utilizing unified power quality conditioner (upqc). In: IEEE Transactions on Industry Applications 45: 1897–1902. https://doi.org/10.1109/TIA.2009.2027147
18. Molina CA, Espinoza JR, Espinosa EE, Villarroel FA, Hidalgo JA (2010) Refined control of an unified power quality conditioner under nonlinear and asymmetrical loads. In: 2010 IEEE International Symposium on Industrial Electronics: 2581–2586. https://doi.org/10.1109/ISIE.2010.5636325
19. Viji AJ, Victoire TAA (2014) Enhanced pll based srf control method for upqc with fault protection under unbalanced load conditions. Int J Electr Power Energy Syst 58:319–328. https://doi.org/10.1016/j.ijepes.2014.01.039
20. Pal Y, Swarup A, Singh B (2013) A comparative analysis of different magnetics supported three-phase four-wire unified power quality conditioners—a simulation study. Int J ElectricalPower & Energy Syst 47:436–447. https://doi.org/10.1016/j.ijepes.2012.11.014
21. Ucar M, Ozdemir S (2013) 3-phase 4-leg unified series–parallel active filter system with ultra-capacitorenergy storage for unbalanced voltage sag mitigation. Int J Electr Power Energy Syst 49:149–159. https://doi.org/10.1016/j.ijepes.2013.01.005
22. Modesto RA, O da Silva SA, de Oliveira AA, Bacon VD (2016) A versatile unified power quality conditioner applied to three-phase four-wire distribution systems using a dual control strategy. IEEE Trans Power Electron 31: 5503–5514. https://doi.org/10.1109/TPEL.2015.2487867
23. Aredes M, Heumann K, Watanabe EH (1998) An universal active power line conditioner. IEEE Transactions Power Deliv 13:545–551. https://doi.org/10.1109/61.660927
24. Prieto J, Salmeron P, Vazquez JR, Alcantara J (2002) A series-parallel configuration of active power filters for var and harmonic compensation. In: IEEE 2002 28th Annual Conference of the Industrial Electronics Society. 4: 2945–2950. https://doi.org/10.1109/IECON.2002.1182864
25. Zhili T, Xun L, Jian C, Yong K, Shanxu D (2006) A direct control strategy for upqc in three-phase four-wire system. In: 2006 CES/IEEE 5th International Power Electronics and Motion Control Conference 2: 1–5. https://doi.org/10.1109/IPEMC.2006.4778095
26. Zhili T, Xun L, Jian C, Yong K, Yang Z (2007) A new control strategy of upqc in three-phase four-wire system. In: IEEE Power Electronics Specialists Conference: 1060–1065. https://doi.org/10.1109/PESC.2007.4342139
27. Li X, Zhu G, Duan S, Chen J (2007) Control scheme for three-phase four-wire upqc in a three-phase stationary frame. In: IECON 2007—33rd Annual Conference of the IEEE Industrial Electronics Society: 1732–1736. https://doi.org/10.1109/IECON.2007.4460393
28. Kesler M, Ozdemir E (2010) A novel control method for unified power quality conditioner (upqc) under non-ideal mains voltage and unbalanced load conditions. In: Twenty-Fifth Annual IEEE Applied Power Electronics Conference and Exposition (APEC): 374–379. https://doi.org/10.1109/APEC.2010.5433643
29. Zhili T, Dongjiao Z (2010) A new control strategy for three-phase four-wire upqc when voltage fluctuating on its dc side. In: The 2nd International Symposium on Power Electronics for Distributed Generation Systems: 190–195. https://doi.org/10.1109/PEDG.2010.5545926
30. Rodriguez P, Sainz L, Bergas J (2002) Synchronous double reference frame pll applied to a unified power quality conditioner. In: 10th International Conference on Harmonics and Quality

of Power.Proceedings (Cat. No.02EX630) 2: 614–619. https://doi.org/10.1109/ICHQP.2002.1221506
31. Han B, Bae B, Kim H, Baek S (2006) Combined operation of unified power-quality conditioner with distributed generation. IEEE Trans Power Deliv 21:330–338. https://doi.org/10.1109/TPWRD.2005.852843
32. Basu M, Das SP, Dubey GK (2002) Performance study of upqc-q for load compensation and voltage sag mitigation. In: IEEE 2002 28th Annual Conference of the Industrial Electronics Society 1: 698–703. https://doi.org/10.1109/IECON.2002.1187592
33. Soares V, Verdelho P, Marques GD (2000) An instantaneous active and reactive current component method for active filters. IEEE Trans Power Electron 15: 660–669. https://doi.org/10.1109/63.849036
34. Srinath S, Selvan MP, Vinothkumar K (2010) Comparative evaluation of performance of different control strategies on upqc connected distribution system. In: 5th International Conference on Industrial and Information Systems: 502–507. https://doi.org/10.1109/ICIINFS.2010.5578651

Real-Time Validation of Power Quality Enhancement Techniques in a Distribution Network

Salauddin Ansari, Sameep Sahu, and Om Hari Gupta

Abstract Power quality is an essential element of today's distribution networks as loads become highly sensitive and non-linear loads increase. These loads suffer from reactive power consumption which generated harmonic and injected into the system. The requirement for power quality improvement is unavoidable with the widespread nature of harmonic loads. The traditional filter and capacitor banks eliminate harmonics by supplying reactive power. Despite that, they are bulky and have challenges like resonance. This chapter presents control, modeling, and Real-Time validation for power quality enhancement using a Photovoltaic system with Power Quality (PVSPQ) enhancement. The power quality enhancement techniques are validated in real-time using OPAL-RT (OP4510) simulator. The power quality enhancement techniques are considered parameters to mitigate the power quality issues, such as harmonic distortion compensation, reactive power compensation, reactive power, and harmonic distortion compensation. Results show that the PVSPQ operates successfully with grid support and improved power quality of the system.

Keywords Power quality · PV System · D-STATCOM · Shunt filter · Distribution network

Nomenclature

DVR	Dynamic voltage restorers
APF	Active power filters
CPD	Custom power devices
SVC	Static var compensators
PQI	Power quality improvement

S. Ansari · S. Sahu · O. H. Gupta (✉)
Department of Electrical Engineering, NIT Jamshedpur, Jamshedpur, Jharkhand, India
e-mail: omggupta@gmail.com

S. Sahu
e-mail: 2021pgeeps@nitjsr.ac.in

S. R. Salkuti et al. (eds.), *Power Quality in Microgrids: Issues, Challenges and Mitigation Techniques*, Lecture Notes in Electrical Engineering 1039,
https://doi.org/10.1007/978-981-99-2066-2_4

AVR	Automatic voltage regulator
THD	Total harmonic distortion
MPPT	Maximum power point tracking
CNN	Convolutional neural network
VSC	Voltage source converter
PVSPQ	Photovoltaic system with power quality enhancement
PCC	Point of common coupling
PLL	Phase-locked loop
I_L	Load current
Z_g	Source impedance
Z_L	Load impedance
V_{dc}	DC link voltage
V_{pcc}	PCC voltage
I_{LP}	Active component of load current
I_P	Total active component of current
I_{pv}	PVSPQ current
I_g	Grid current
I_{gref}	Grid reference current
P_L	Load active power
Q_L	Load reactive power
V_g	Grid voltage
P_g	Grid active power
Q_g	Grid reactive power
P_{pv}	PVSPQ active power
Q_{pv}	PVSPQ reactive power

1 Introduction

Renewable energy sources [1] are crucial for the production of electrical power in electrical power system networks. Over the past few years, due to rapid urbanization, there has been a fourfold increase in the usage of sensitive loads including diagnostic equipment in hospitals, schools, detention facilities, etc. This has sparked a debate about the quality power of sensitive loads [2]. Due to the rise in sensitive loads, it is crucial to pay attention to power quality issues. Financial losses will be substantial if these issues are not addressed. In the smart era, almost all of these gadgets are susceptible to interruptions in the electrical supply at any moment and are unable to function effectively. Poor electrical power quality causes equipment failure, data error, loss of memory, software corruption, overheating of equipment, etc.

The usage of non-linear loads injects harmonics into the power supply. This causes the power factor to degrade. Voltage/current abnormalities are the major cause of power quality concerns since they cause frequency variations, which in turn cause consumer equipment to malfunction. Harmonics, transients, voltage sag and swell, flickers, fluctuations, and interruptions are among the most important power quality problems [3]. These issues can harm the working of equipment, for example, computers and medical devices, and can also lead to power losses and increased maintenance costs Solutions to power quality issues include the use of power quality monitors, harmonic filters, and corrective equipment such as dynamic voltage restorers (DVR) [4] and uninterruptible power supplies (UPS).

Several power quality enhancement techniques can be used in a distribution network to enhance the quality of the power delivered to customers. Some of these techniques are power factor correction which involves capacitors to correct the power factor and decrease the amount of reactive power in the system. However, they are bulky and challenging as it results in resonance. Power quality can be enhanced by harmonic filtering [5]. Harmonics [6] are unwanted electrical frequencies that can be generated by non-linear loads. Harmonic filters can be used to reduce or eliminate these harmonic frequencies. Advanced metering infrastructure [7] incorporate advanced meters that provide detailed information on the power consumption and quality of the electrical power supplied to customers, allowing utilities to identify and address power quality issues. The use of smart meters and advanced communication systems can help utilities better monitor and control the power quality in their distribution networks. Similarly active power filters (APF) [8] devices use power electronic converters and control algorithms to cancel out unwanted harmonics in the system. Power quality can be improved by Custom Power Devices (CPD) like DVR, D-STATCOM [9], SVC, etc. These are designed to fulfill the specific power quality needs of the consumer. D-STATCOM is a shunt-connected equipment that uses a power electronic converter (PEC) to control the voltage in the distribution network. Static Var Compensators (SVC) use PEC to dynamically control the voltage level in the system, which can help to mitigate voltage sag and swell. To mitigate power quality problems many power quality improvement (PQI) equipment are used. There are three generations of PQI equipment available.

First generation PQI devices are passive and active filters [10], and hybrid active filters. Inductances and capacitances were combined to create passive power filters, which were designed to remove current harmonics and compensate for reactive power. To solve the limitations of passive filters, APF was developed. APFs can control voltage, reduce harmonics, improve power factor, and correct flicker. There are two types of shunt and series active filter. By injecting a harmonic current 180° out of phase with the harmonic current, the shunt active filter cancels out current harmonics of the same magnitude. Thus, the harmonic current is reduced and the grid current is almost sinusoidal and in phase with the source. By summing a harmonic voltage to the grid in series with the voltage harmonic's opposite phase, the series active power filter cancels out the voltage harmonics and functions as a controlled voltage source. The shunt and series active filter are combined to form a hybrid active power filter.

Second generation [11] PQI devices include DVR, automatic voltage regulator (AVR), DSTATCOM, UPQC, and UPS. The DVR is an electronic device that helps to safe vital loads against voltage imbalances. Another PQI device is an AVR device that adjusts the transformer tap to maintain the critical load voltage at an adequate level. Similarly, DSTATCOM is a device that can absorb and deliver reactive power to the load. It is frequently used in industry and distribution systems to control load voltage.

Next, comes the third generation [11] of PQI devices which are cost-effective and reliable. Smart impedance, electrical springs, and multifunctional DG are a few examples. In a single-phase or three-phase system, the idea of "Smart Impedance" is made by combining a converter, a coupling transformer, a capacitor bank, and an appropriate control scheme. It is capable of regulating harmonic current, harmonic unbalances, enhanced tuning, and displacement power factor. Smart impedance can work as a shunt or series active filter, capacitor bank, or tuned passive filter to eliminate harmonics. In contrast to conventional reactive power controllers like FACTS or SVCs, which solely participate in balancing reactive power, electrical springs regulate input voltage rather than the output voltage and may control both reactive and active power. The electrical spring generates a sinusoidal voltage that is perpendicular to the current and it can control the voltage by 90° leading or lagging to the current.

In [10] the focus is given to shunt active power filters' ability to suppress current harmonics in utility grids that are caused by non-linear loads. Harmonics distortion is reduced by using solar PV energy penetration. An overview of PQI techniques in distribution networks has been discussed in [11]. A Teager energy operator has been proposed in [12] to detect power quality disturbances. This methodology can precisely identify distortions and sudden voltage fluctuations in a wide range of conditions. A matrix pencil approach is investigated for current harmonic and power factor improvement in [13]. APF is utilized to improvement for source current and voltage harmonics. A multi-level STATCOM is used for PQI in distributed networks [14]. Voltage sag, swell, fluctuations, harmonics, and other power quality issues are successfully corrected using STATCOM. The proposed technique employs one-cycle control and it is less prone to error in steady as well as the dynamic state. In [15] to address power quality challenges, this work describes a PV-integrated DVR that uses a rotating reference frame controller. It employs an upgraded incremental conductance Maximum Power Point Tracking (MPPT) technique and an effective PI controller is designed for DVR to boost power quality. The THD of the load voltage was decreased by 3.28% during sag and 3.55% during swell using PV-fed DVR. The voltage profile is improved in [16] by employing hybrid STATCOM using a thyristor-based LC filter. The proposed technique compensates 90% reactive power thus improving the power quality. A double resistive APF is used in [8] to mitigate harmonic voltages and eliminate harmonic propagation. A deep learning-based PQI scheme is proposed in [17]. The main goal is to provide a novel method for classifying a given power signal into the appropriate power quality condition using a convolutional neural network (CNN). The proposed scheme requires a large number of datasets and memory. The overall processing time for the training was 1520 s. After training, the time required for categorization is under 40 s.

Overall, with the above-mentioned power enhancement techniques, real-time validation has not been performed. Moreover, some control strategies can have a high computation burden and are difficult to implement in practical conditions.

1.1 Harmonics in Power System and Its Effect

Harmonics, in this context, mean alternating currents or alternating voltages whose frequencies are integer multiples of the fundamental power frequency. The short pulses result in distorted current/voltage waveforms, and as a result, harmonic currents flow inside power system components. Generally, harmonics are generated by non-linear electronic loads. Some of the loads producing harmonics are rectifiers, variable frequency drives, electronically commutated motors, Electric Vehicle chargers, power electronics converters, arcing devices, switch mode power supplies, and transformers. The power systems generally can handle some amount of harmonic currents, but the issue arises when harmonics start to make up a comparable portion of the total load. The effect that arises with harmonics are discussed below [18]:

- Overheating of electrical equipment like cables, windings of motor, generator, and transformer.
- Low power factor
- False tripping of a circuit breaker during magnetic inrush of transformer
- Overloading of electrical conductors as it causes higher current to flow.
- Harmonics cause skin effects in electrical cables.
- Harmonics lead to an increase in business costs as maintenance is increased.

1.2 Power Factor and Its Effects

The power factor [19] is a metric for evaluating how well an electrical system uses incoming power. It is referred to as the Active Power to Apparent Power Ratio (i.e., kW/kVA). The power factor of the system is impacted by the kind of connected load, whether it is capacitive, inductive, or resistive. Both the power company and the users profit from a high-power factor. Low power factor, on the other hand, denotes inefficient use of electrical power and penalizes users. The impact of low power factor is as follows:

- A low power factor draws more current, and the extra heat produced will decrease the life of the equipment.
- A system with a low power factor will cost more.
- Higher rating of machine required.
- Needed greater size of conductor.
- Result in more copper losses.
- Result in poor voltage regulation

1.3 Sag and Swell in the Power System

The most severe disruptions in distribution networks that affect sensitive loads' voltage stability are known as voltage sag/swell [3]. According to IEEE 1159-2009 standards, voltage sag/swell can last anywhere from a half-period to a minute. Voltage sag [20] is a sharp drop in bus voltage magnitude (rms value) below 0.9 per unit of nominal voltage for an extended period, between a half cycle of a power frequency and one minute. It is mostly caused by abrupt switching caused by momentarily disconnecting the supply, flow of inrush current, and lightning strikes. Voltage swell is a significant rise in the magnitude (rms value) of voltage exceeding 1.1 per unit for an extended time i.e., from a half cycle of a power frequency to one minute. Voltage sag is due to large motor starting, single line to ground fault, inadequate wiring, and inrush current [21]. The cause of voltage swell is capacitor switching, start/stop of heavy loads, and source voltage variation.

This chapter proposes a design and control of a photovoltaic system for power quality enhancement. It focuses on the real-time validation of the power quality enhancement technique [22] of the distribution network using a photovoltaic system with power quality (PVSPQ) enhancement [23, 24]. The rest of this chapter will be structured as follows: Sect. 2 explains a photovoltaic system with power quality enhancement. Section 3 describes the real-time validation of PVSPQ using OPAL-RT. Section 4 summarizes the overall work presented in this chapter.

2 Photovoltaic System with Power Quality Enhancement (PVSPQ)

Energy demand is a result of increased industrialization. Fossil fuels primarily meet the majority of the world's energy needs. Solar energy [25] has become as one of the most promising renewable sources of energy due to its abundance in nature and never-ending supply [26]. Solar Photovoltaic (PV) is used to meet growing power demand. The inverter connects solar power systems to the grid by converting the DC power produced by the PV panel to AC. Solar PV panels generate power by converting sunlight into electricity with the help of PV cells. The PV effect occurs when light photons are absorbed by semiconductor materials, such as silicon, and cause the release of electrons. These electrons can then be captured as a direct current and converted into alternating current for use in homes, businesses, and the grid [27]. The efficiency of a solar panel is measured by its capacity to convert sunlight into electricity, and it can range from 15 to 22%.

In Fig. 1, the one-line diagram of the PV energy conversion system is depicted which is integrated into the grid along with the load. Both types of loads i.e., a linear load having a lagging power factor and a non-linear load have been attached in the system as shown in Fig. 1. The non-linear load is made of a three-phase diode bridge rectifier. The PV system generates DC link voltage i.e., V_{dc}. A capacitor C

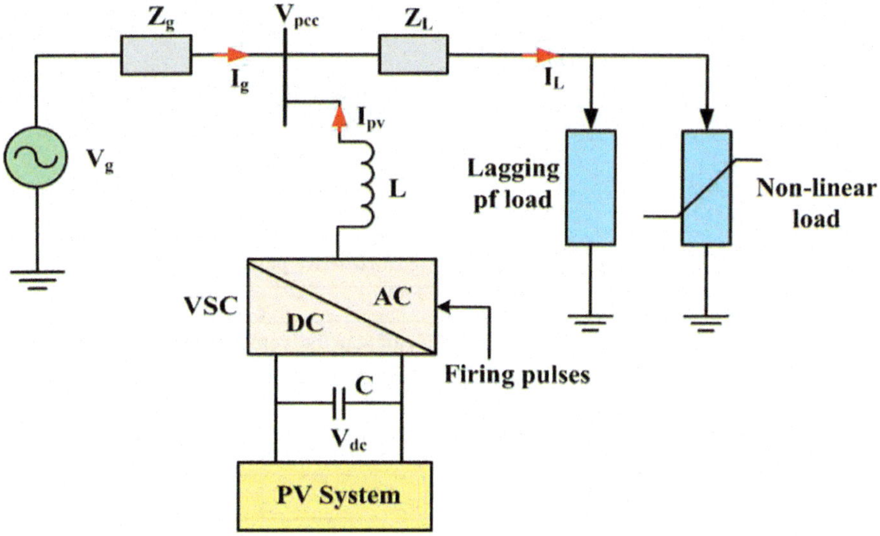

Fig. 1 Single line diagram of PVSPQ

is connected to remove the ripple in the output of the DC link voltage waveform. A voltage source converter (VSC) circuit converts DC to AC side voltage so that it can be integrated into the utility grid. The inductor L removes the ripple in the current waveform which is coming from the PV system. The solar PV energy system supplies power as well as compensates for the harmonics and reactive power.

The algorithm of the PV system with the power quality enhancement (PVSPQ) scheme has been presented in Fig. 2. Initially the voltage at the point of common coupling (PCC), i.e., $V_{pcc,}$ and load current I_L is measured using a measuring device. The instantaneous active power at PCC is obtained which is the product of V_{pcc} and I_L. After that, the average active power is computed which is the average of the instantaneous active power. The active component of load current I_{LP} is calculated as depicted in the flowchart shown in Fig. 2. This active component of current is drawn from the utility grid to maintain the power quality. Additionally, the active component of current called $I_{dcp}*$ must be drawn from the grid and acquired from the DC controller to balance the DC bus voltage V_{dc}. The error signal is generated between V_{dc} and $V_{dc}*$. The error signal is fed to the DC controller which gives $I_{dcp}*$. The phase-locked loop (PLL) circuit extracts the phase from the PCC. The total active component of the current is denoted by I_p which is calculated as shown in Fig. 2. The reference current for the VSC circuit, i.e., $I_{pv}*$ is obtained using the load current I_L and the total active component of current I_p. The firing pulses corresponding to the reference current $I_{pv}*$ and measured PVSPQ current I_{pv} is generated by the hysteresis current controller. The firing gate pulses are applied to the VSC for proper operation.

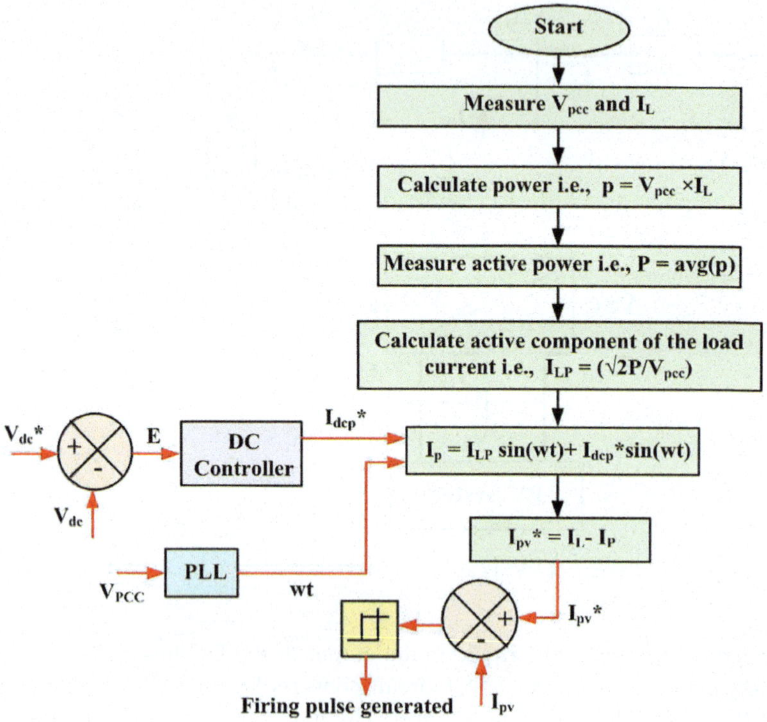

Fig. 2 Flow chart for control scheme of the PVSPQ

3 Real-Time Validations of PVSPQ by OPAL-RT for Power Quality Enhancement

In this section, the real-time validation of PVSPQ by OPAL-RT has been presented. The OPAL-RT simulator (OP4510) depicted in Fig. 3 is comprised of 4 core processor, Kintex-7 FPGA, and a DAC module for communicating with the real-time environment. The test bench of real-time validation of PVSPQ is depicted in Fig. 3. To generate a real-time result, RT-Lab is installed on the host PC for interfacing the Simulink model with the OPAL-RT. The RT-Lab allows the communication between the Simulink model of PVSPQ modeled in the host PC and the OPAL-RT by compiling the model to executable code that can run on the OPAL-RT simulator. Therefore, the central processing unit of the OPAL-RT simulator runs this code and generates real-time results. The real-time result from the OPAL-RT is captured in DSO as depicted in Fig. 3. In this chapter, to make sure the compatibility between the designed model and the OPAL-RT simulator, the sampling time of 10 μs is chosen. The acquired real-time results of PVSPQ for different operating modes are discussed below.

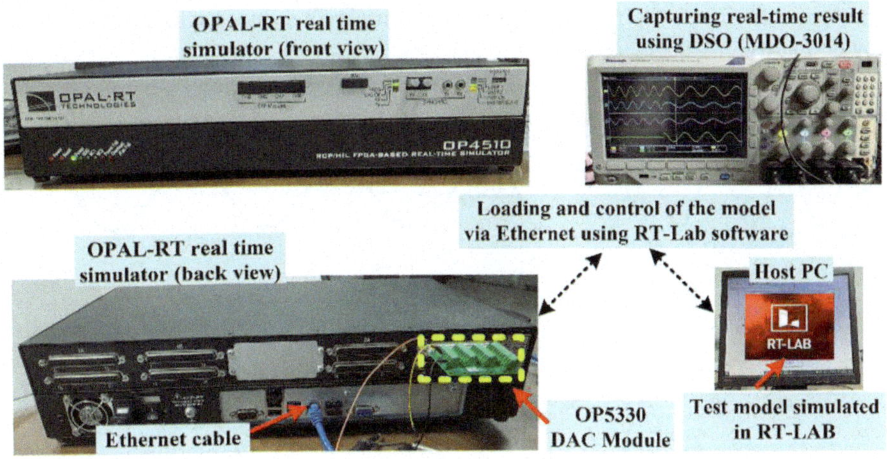

Fig. 3 Test bench of real-time validation of PVSPQ

3.1 Reactive Power Compensation for Power Quality Enhancement

For the reactive power compensation, the PVSPQ is employed to provide the reactive power demand of the load (Q_L). As a result, the load only draws active power from the utility grid. This decreases the voltage drop and current drawn from the grid which leads to a better quality of power. The obtained OPAL-RT result of grid voltage (V_g), grid current (I_g), grid reference current (I_{gref}), and PVSPQ current (I_{pv}) for reactive power compensation is shown in Fig. 4. It can be seen that when PVSPQ is connected then harmonic in V_g and I_g reduces since PVSPQ supplying reactive power to the load and I_{pv} is continued to flow after being connected by PVSPQ as depicted in Fig. 4. It appears that the frequency has changed in instantaneous results. However, the frequency remains the same and it is just due to some unknown reasons related to the interfacing (may be over-run issue). Moreover, OPAL-RT results of the DC bus voltage (V_{dc}), the active power of PVSPQ (P_{pv}), the active power of load (P_L), and the active power of grid (P_g) for reactive power compensation are shown in Fig. 5. Similarly, OPAL-RT results of the DC bus voltage (V_{dc}), reactive power of PVSPQ (Q_{pv}), reactive power of load (Q_L) and reactive power of grid (Q_g) for reactive power compensation are shown in Fig. 6. It is noted from Fig. 5 that before the connection of PVSPQ the active power demand of load is supplied by the grid only. When PVSPQ is connected, surplus active power is delivered into the utility grid as P_g is negative (as shown in Fig. 5) since P_L is less than P_{pv}. Moreover, initially before the connection of PVSPQ, Q_L is supplied by only the utility grid since Q_g is positive as depicted in Fig. 6. However, when PVSPQ is connected then Q_L is supplied by only PVSPQ (i.e., Q_{pv}) since Q_{pv} is positive, and therefore, Q_g becomes zero as shown in Fig. 6.

Therefore, the reactive power burden on the utility grid reduces which enhances the power quality.

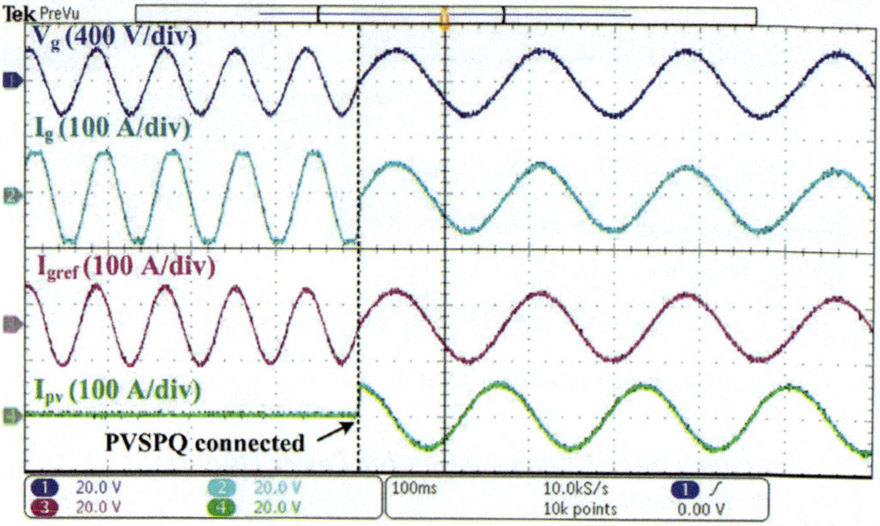

Fig. 4 Grid voltage, grid current, grid reference current, and PVSPQ current for reactive power compensation

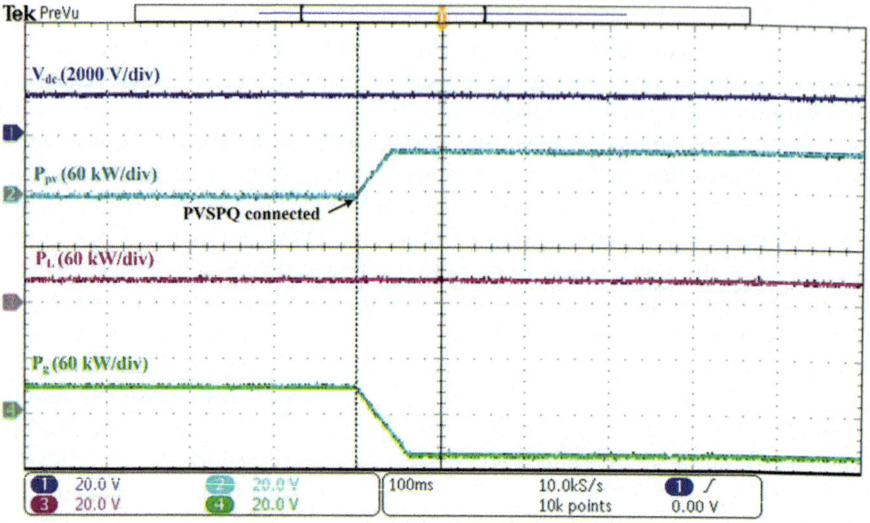

Fig. 5 DC bus voltage, the active power of PVSPQ, the active power of load, and the active power of grid for reactive power compensation

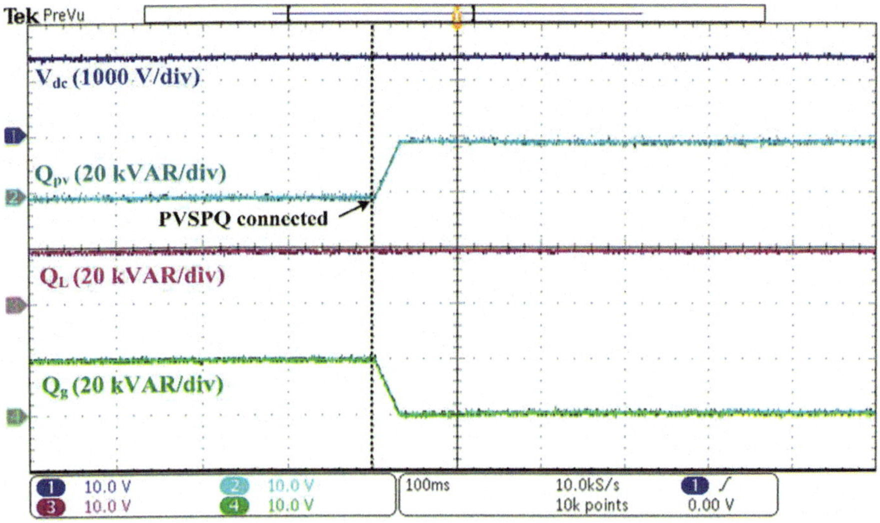

Fig. 6 DC bus voltage, reactive power of PVSPQ, reactive power of load, and reactive power of grid for reactive power compensation

3.2 *Harmonic Distortion Compensation for Power Quality Enhancement*

The PVSPQ is used to compensate for load harmonic distortion for power quality improvement in this section. The PVSPQ removes the harmonic component from the load current and only the fundamental component is drawn from the utility grid. Therefore, it decreases the current coming from the utility grid and decreases the distortion at the bus voltage. Figure 7 depicts the OPAL-RT result of V_g, I_g, I_{gref}, and I_{pv} for harmonic distortion compensation. From Fig. 7, it can be seen that once the PVSPQ is connected harmonic in the I_g improved as the harmonic component of current is supplied by I_{pv}. Moreover, the obtained OPAL-RT result of V_{dc}, P_{pv}, P_L, and P_g are shown in Fig. 8 for the harmonic distortion case. From Fig. 8 it is clear that as the PVSPQ is connected the surplus amount of active power is being delivered by PVSPQ to the utility grid and V_{dc} remains constant. Similarly, the OPAL-RT result for harmonic compensation of V_{dc}, Q_{pv}, Q_L, and Q_g is depicted in Fig. 9. It is clear from Fig. 9 that, the V_{dc} and Q_L are constant and once the PVSPQ is connected, the reactive power demand of load of Q_L is completely supplied by PVSPQ (i.e., Q_{pv}) and hence reactive power supply by utility grid (i.e., Q_g) becomes zero and hence power quality is enhanced.

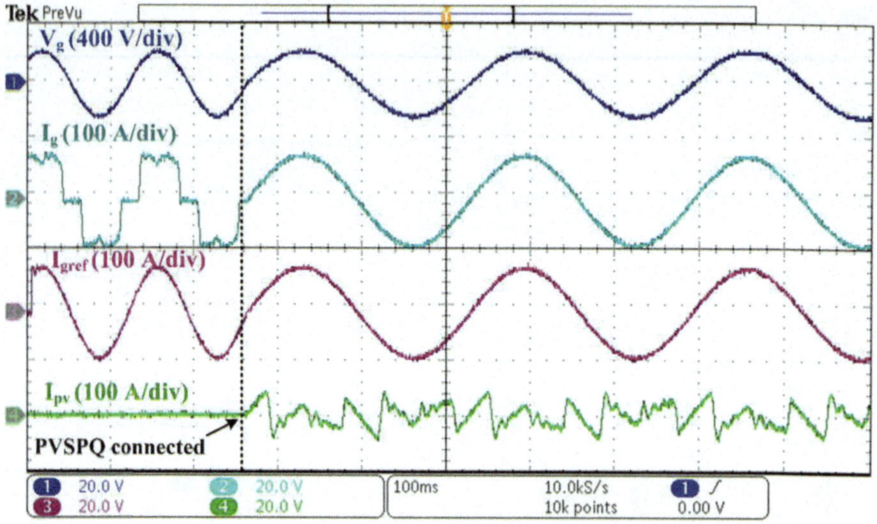

Fig. 7 Grid voltage, grid current, grid reference current, and PVSPQ current for harmonic distortion compensation

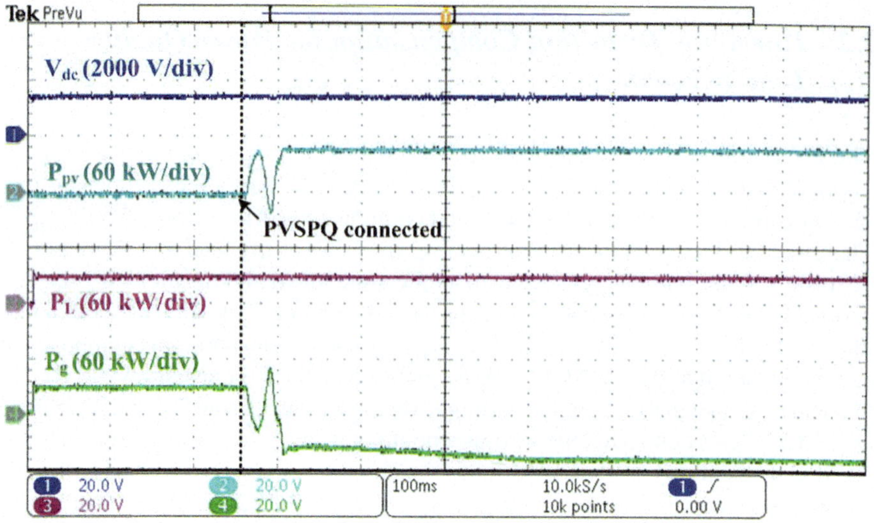

Fig. 8 DC bus voltage, the active power of PVSPQ, the active power of load, and the active power of grid for harmonic distortion compensation

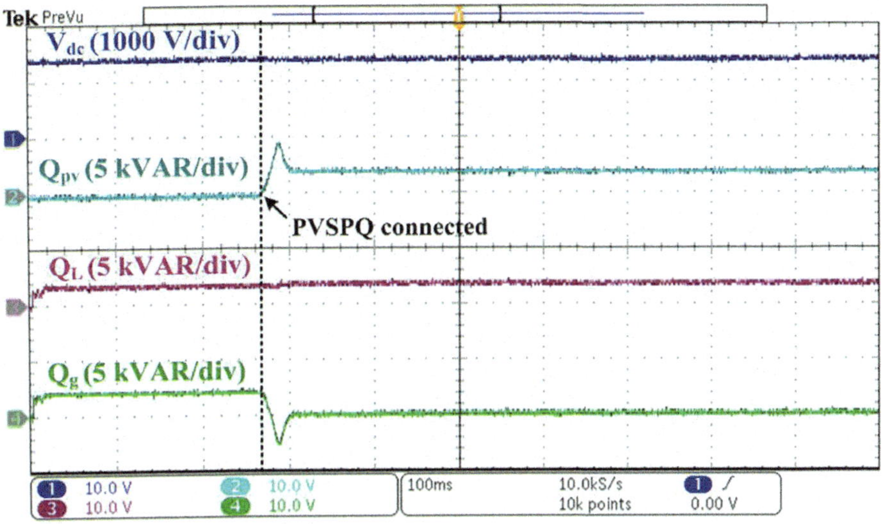

Fig. 9 DC bus voltage, reactive power of PVSPQ, reactive power of load, and reactive power of grid for harmonic distortion compensation

3.3 Reactive Power and Harmonic Distortion Compensation for Power Quality Enhancement

The obtained OPAL-RT result for reactive power and harmonic distortion compensation of V_g, I_g, I_{gref}, and I_{pv} are depicted in Fig. 10. From Fig. 10, it is clear that the I_g is distorted when PVSPQ is not connected, and after the connection of PVSPQ the distortion in I_g is completely removed since a harmonic component of current is being supplied by PVSPQ i.e., I_{pv} and hence power quality is enhanced. Moreover, the obtained OPAL-RT result of V_{dc}, P_{pv}, P_L, and P_g are shown in Fig. 11 for reactive power and harmonic distortion case. From Fig. 11, it is clear that initially, the PVSPQ is not connected therefore, P_L is supplied by the utility grid (i.e., P_g) as shown in Fig. 11 When PVSPQ is connected to the excess amount of active power is supplied by P_{pv} and hence P_g becomes negative as shown in Fig. 11. Similarly, the OPAL-RT result of V_{dc}, Q_{pv}, Q_L, and Q_g is shown in Fig. 12. Before, the connection of PVSPQ the Q_L is supplied by utility grid (i.e., Q_g) and therefore Q_{pv} is zero. Once the PVSPQ is connected, since the load reactive power demand (Q_L) is full fill by PVSPQ (i.e., Q_{pv}) and hence the utility grid is not delivering any reactive power, i.e., $Q_g = 0$ which is seen in Fig. 12 and hence power quality is enhanced.

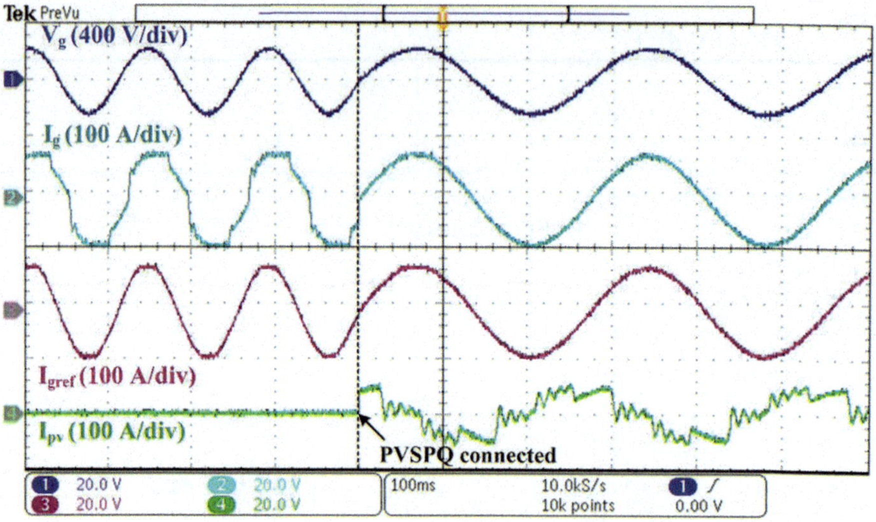

Fig. 10 Grid voltage, grid current, grid reference current, and PVSPQ current for reactive power and harmonic distortion compensation

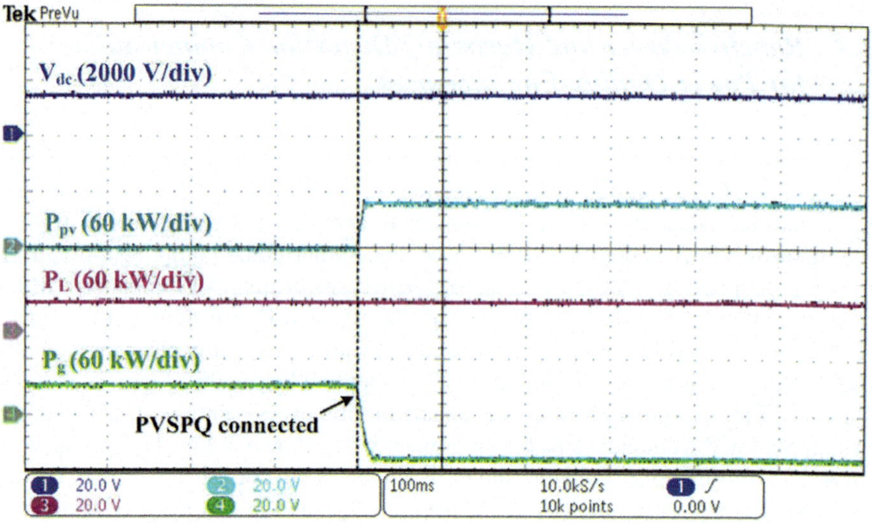

Fig. 11 DC bus voltage, the active power of PVSPQ, the active power of load, and the active power of grid for reactive power and harmonic distortion compensation

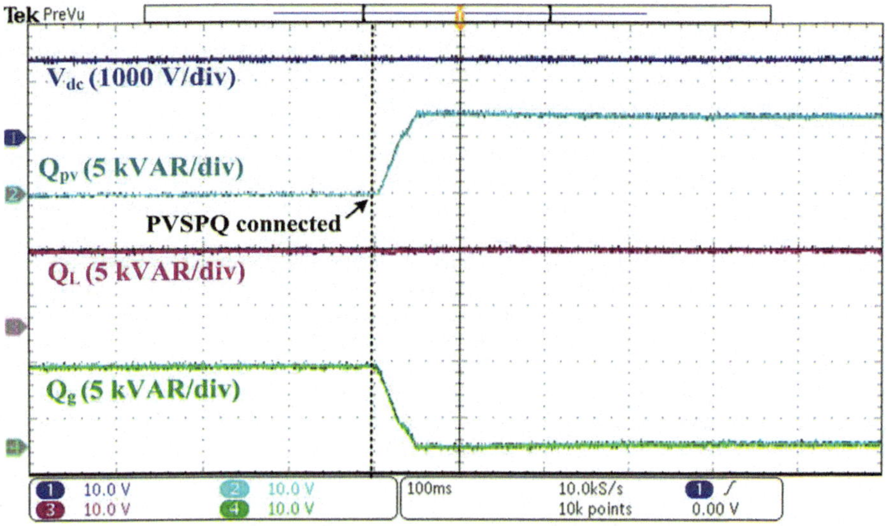

Fig. 12 DC bus voltage, reactive power of PVSPQ, reactive power of load, and reactive power of grid for reactive power and harmonic distortion compensation

4 Conclusion

This chapter discussed the possibility of harmonic and reactive power and harmonic as well as reactive power compensation, using a PVSPQ. The existence of a non-linear load causes changes in the bus voltage where the linear loads are also connected, resulting in non-sinusoidal current flow in the linear loads. As a result, these harmonics should be mitigated to enhance power quality. Similarly, a low power factor increases supply current and, as a result, line losses. It also has an impact on voltage regulation. As a result, reactive compensation is required. Both of these needs can be met by the PVSPQ. OPAL-RT results showed the practicality of the PVSPQ. Apart from the power generated by the PVSPQ, not only reactive power but also harmonics have been adjusted, and successfully validated the results using OPAL-RT Simulator.

References

1. Owusu PA, Asumadu-Sarkodie S (2016) A review of renewable energy sources, sustainability issues and climate change mitigation. Cogent Eng 3. https://doi.org/10.1080/23311916.2016.1167990
2. Thentral TMT, Palanisamy R, Usha S et al (2022) Analysis of power quality issues of different types of household applications. Energy Rep 8:5370–5386. https://doi.org/10.1016/j.egyr.2022.04.010

3. Rao SS, Krishna PSR, Babu S (2017) Mitigation of voltage sag, swell and THD using Dynamic Voltage Restorer with photovoltaic system. In: 2017 International Conference on Algorithms, Methodology, Model and Applications in Emerging Technology ICAMMAET 2017 2017-Jan 1–7. https://doi.org/10.1109/ICAMMAET.2017.8186668
4. Ramya P, Sahithya K, Vamsi T, Pangedaiah B (2022) Improvement in the power quality by using dynamic voltage restorer. MysuruCon 2022—2022 IEEE 2nd Mysore sub section international conference. https://doi.org/10.1109/MysuruCon55714.2022.9972594
5. Kazem HA (2013) Harmonic mitigation techniques applied to power distribution networks. Adv Power Electron 2013. https://doi.org/10.1155/2013/591680
6. Jain SK, Singh SN (2011) Harmonics estimation in emerging power system: Key issues and challenges. Electr Power Syst Res 81:1754–1766. https://doi.org/10.1016/j.epsr.2011.05.004
7. Selvam C, Srinivas K, Ayyappan GS, Venkatachala Sarma M (2012) Advanced metering infrastructure for smart grid applications. Int Conf Recent Trends Inf Technol ICRTIT 2012:145–150. https://doi.org/10.1109/ICRTIT.2012.6206777
8. Sun X, Han R, Shen H et al (2016) A double-resistive active power filter system to attenuate harmonic voltages of a radial power distribution feeder. IEEE Trans Power Electron 31:6203–6216. https://doi.org/10.1109/TPEL.2015.2500913
9. Thirupathaiah M, Venkata Prasad P, Ganesh V (2018) Enhancement of power quality in wind power distribution system by using hybrid PSO-Firefly based DSTATCOM. Int J Renew Energy Res 8:1138–1154. https://doi.org/10.20508/ijrer.v8i2.7002.g7400
10. Pawar MR, Bavdhane VD (2022) Power quality enhancement in utility with photovoltaic renewable energy penetration using active power filter. IEEE Int Conf Data Sci Inf Syst 2022:1–6. https://doi.org/10.1109/icdsis55133.2022.9915995
11. Naderi Y, Hosseini SH, Ghassem Zadeh S et al (2018) An overview of power quality enhancement techniques applied to distributed generation in electrical distribution networks. Renew Sustain Energy Rev 93:201–214. https://doi.org/10.1016/j.rser.2018.05.013
12. Karimian A, Hosseinian SH (2022) Novel method based on Teager Energy Operator for online tracking of power quality disturbances. Electr Power Syst Res 213:108747. https://doi.org/10.1016/j.epsr.2022.108747
13. Chahine K, Tarnini M, Moubayed N, El Ghaly A (2023) Power quality enhancement of grid-connected renewable systems using a matrix-pencil-based active power filter. Sustain 15. https://doi.org/10.3390/su15010887
14. Hasanzadeh S, Shojaeian H, Mohsenzadeh MM et al (2022) Power quality enhancement of the distribution network by multilevel STATCOM-compensated based on improved one-cycle controller. IEEE Access 10:50578–50588. https://doi.org/10.1109/ACCESS.2022.3172144
15. Prasad D, Dhanamjayulu C (2022) Solar PV integrated dynamic voltage restorer for enhancing the power quality under distorted grid conditions. Electr Power Syst Res 213:108746. https://doi.org/10.1016/j.epsr.2022.108746
16. Adware R, Chandrakar V (2022) Power quality enhancement through reactive power compensation using hybrid STATCOM. ICPC2T 2022—2nd international conference on power, control and computing technologies, Proc 8–12. https://doi.org/10.1109/ICPC2T53885.2022.9777006
17. Topaloglu I (2022) Deep learning based a new approach for power quality disturbances classification in power transmission system. J Electr Eng Technol. https://doi.org/10.1007/s42835-022-01177-1
18. Das SR, Mishra DP, Ray PK, Salkuti SR, Sahoo AK (2021) Power quality improvement using fuzzy logic-based compensation in a hybrid power system. Int J Power Electron Drive Syst 12(1):576–584. https://doi.org/10.11591/ijpeds.v12.i1.pp576-584
19. Lin L, Wang J, Gao W (2012) Effect of load power factor on voltage stability of distribution substation. IEEE Power Energy Soc Gen Meet. https://doi.org/10.1109/PESGM.2012.6345690
20. Büyük M, Inci M, Tümay M (2017) Performance comparison of voltage sag/swell detection methods implemented in custom power devices. Rev Roum des Sci Tech Ser Electrotech Energ 62:129–133
21. Kantaria R, Joshi SK (2008) A review on power quality problems and solutions. Power Electron Natl Conf. https://doi.org/10.1109/STPES54845.2022.10006587

22. Balasubrahmanyam CS, Gupta OH, Sood VK (2021) Power quality enhancement and grid support using solar energy conversion system. Microgrid Technol. https://doi.org/10.1002/9781119710905.CH12
23. Salkuti SR (2022) Emerging and advanced green energy technologies for sustainable and resilient future grid. Energies 15(18):6667. https://doi.org/10.3390/en15186667
24. Islavatu S, Kumar P, Kumar A, Salkuti SR (2022) Power quality analysis by H-bridge DSTATCOM control by Icosθ and ESRF SOGI-FLL methods for different industrial loads. Smart Cities 5(4):1590–1610. https://doi.org/10.3390/smartcities5040081
25. Poongothai S, Srinath S (2020) Power quality enhancement in solar power with grid connected system using UPQC. Microprocess Microsyst 79:103300. https://doi.org/10.1016/j.micpro.2020.103300
26. Naga Lakshmi GV, Jaya Laxmi A, Veeramsetty V, Salkuti SR (2022) Optimal placement of distributed generation based on power quality improvement using self-adaptive Lévy flight Jaya algorithm. Clean Technol 4(4):1242–1254. https://doi.org/10.3390/cleantechnol4040076
27. Thangaraj K, Gopalasamy S (2016) Power quality analysis and enhancement of grid connected solar energy system. Circ Syst 07:1954–1961. https://doi.org/10.4236/cs.2016.78170

Harmonic Distortion Assessment in Three-Phase Distribution Networks with the Combined Penetration of Renewable Energy and D-STATCOM

Raavi Satish, B. V. V. L. Kala Bharathi, Dhananjaya Mudadla, Surender Reddy Salkuti, Balamurali Pydi, and Almoataz Y. Abdelaziz

Abstract The harmonic assessment in distribution networks (DNs) is essentially important because of the increasing usage of power electronic equipment due to the penetrations of renewable energy sources (RES) and distribution static synchronous compensator (D-STATCOM). So, the DNs are ceaselessly moving towards the worst situation with the loss of power quality in terms of total rms voltages, total harmonic distortion (THD), and harmonic power loss (HPL). This chapter proposes a harmonic power flow method (HPFM) for three-phase distribution networks (TPDNs) with the penetrations of synchronous based-RES, inverter based-RES, and D-STATCOM and the presence of non-linear loads. The accuracy of this method is tested on IEEE-13 TPDN. Several test studies are conducted on IEEE-13 and IEEE-34 bus TPDNs. The results of the study examples show that THD% and harmonic power loss are reduced

R. Satish
Department of Electrical and Electronics Engineering, Anil Neerukonda Institute of Technologyand Sciences (A), Visakhapatnam, India

B. V. V. L. K. Bharathi
Department of EEE, Aditya Engineering College, Surampalem, Andhra Pradesh, India
e-mail: kalabharathi.bvvl@aec.edu.in

D. Mudadla
Department of Electrical and Electronics Engineering, Anil Neerukonda Institute of Technologyand Sciences (A), Visakhapatnam, India

S. R. Salkuti
Department of Railroad and Electrical Engineering, Woosong University, Daejeon, Republic of Korea

B. Pydi
Department of EEE, Aditya Institute of Technology & Management (A), Andhra Pradesh, Tekkali, Srikakulam, India

A. Y. Abdelaziz (✉)
Faculty of Engineering and Technology, Future University in Egypt, Cairo, Egypt
e-mail: almoataz.abdelaziz@fue.edu.eg

S. R. Salkuti et al. (eds.), *Power Quality in Microgrids: Issues, Challenges and Mitigation Techniques*, Lecture Notes in Electrical Engineering 1039,
https://doi.org/10.1007/978-981-99-2066-2_5

by the penetration of synchronous-based-RES and D-STATCOM device rather than inverter-based-RES.

Keywords Three-phase distribution network · Renewable energy sources · Distribution static synchronous compensator · THD · Linear loads · Non-linear loads · Power flow

Nomenclature

DN	Distribution Network
RES	Renewable Energy Sources
THD	Total Harmonic Distortion
HPL	Harmonic Power Loss
HPFM	Harmonic power Flow Method
TPDN	Three-phase Distribution Network
PEC	Power Electronic Converter
BNM	Bus Number Matrix
BRNM	Branch Number Matrix

1 Introduction

In modern distribution networks (DNs), the trend towards the penetration of RES is enhanced by its environmental and economic advantages. Based on the interfacing device used to integrate the renewable energy sources (RES) into the three-phase distribution networks (TPDN), the RES can be divided into two types. If the interfacing has happened through the only synchronous generator (SG)/induction generator (IG), then the RES is called synchronous-based-RES. If the interfacing has happened through the combination of IG and power electronic converters (PECs) or only PECs, then the RES is called inverter-based-RES. The increasing use of non-linear loads and integration of inverter-based RES can inject harmonic pollution into the TPDN. This has arisen the need for developing the harmonic power flow method (HPFM) for TPDN. The authors of [1] suggest a new method for finding the level of penetration of utility-owned synchronous-based and inverter-based distribution generation (DG) to accomplish maximum penetration of DGs considering the standard limitations of harmonics and constraints for protection coordination. In [2], proposes a power flow method based on backward/forward sweep for analyzing the THD with DG units. In [3], proposes an assessment method for harmonic distortion by considering the injection of DC components from PV inverters. Authors of [4] developed a harmonic mitigation technique to improve power quality in DNs. The Authors of [5] developed an analytical technique to solve harmonic load flows with

PV uncertainties in secondary DNs. In [6], proposed an iterative harmonic power flow method using the point-estimated method for DNs with PV uncertainties. Authors of [7] investigate the use of a D-STATCOM with capacitor-less to compensate for power quality in modern DNs. In [8], presents a technique to allocate the D-STATCOM optimally, taking into consideration the correlation among the uncertain variables. In [9], presents an optimal step least mean square algorithm to control the three-phase D-STATCOM. It can provide reactive power compensation as well as compensation for harmonics in DNs. In [10], proposes an algorithm for the power flow analysis of three-phase DNs with the DGs and D-STATCOM penetrations by utilizing the developed matrices. In [11–13] proposes a power flow algorithm for harmonic evaluation in three-phase DNs with the integration of the D-STATCOM device. Authors of [14] propose a flower pollination algorithm to allocate and size the capacitors optimally in different distribution networks. Authors of [15] propose a novel algorithm to size and place the RES optimally in distribution networks. In [16], various algorithms for optimization are developed by modifying the big bang big crunch method for the realization of the virtual power plant. These algorithms are designed to handle the energy in distribution networks to minimize the amount of energy purchased on the grid. In [17], an optimization algorithm is discussed for the planning, allocation, and sizing of DGs in distribution networks.

This chapter proposes an HPFM for harmonics assessment in TPDNs with the combined penetration of synchronous based-RES, inverter based-RES, and D-STATCOM and the presence of non-linear loads. The modeling of synchronous-based-RES and inverter-based-RES are addressed in this chapter for developing HPFM. The algorithm proposed in [18] is used to find the fundamental solution with penetrations of RES and D-STATCOM. With the fundamental solution obtained, the modeling of linear and no-linear loads is done. This chapter also exploits the bus number matrix (BNM) and branch number matrix (BRNM) proposed in [18] for developing HPFM.

The rest of the chapter is organized in the following order. In Sect. 2, materials for modeling network components for harmonic analysis are discussed. HPFM with penetration of RES and D-STATCOM device is presented in Sect. 3. Results and discussions are presented in Sect. 4. The conclusions of the chapter are outlined in Sect. 5.

2 Materials for Modeling of Network Components for Harmonic Analysis

The modeling of TPDN components like overhead/underground lines, loads, and capacitor banks for fundamental analysis is presented in [19]. The modeling of different interfacing devices such as SGs, IGs, and PECs for penetration of RES is presented in [20–23]. The modeling of D-STATCOM is presented in [24–25].

However, this section presents brief modeling of network components for harmonic analysis.

2.1 Lines

For HPFM, the reactance of each line is considered to be proportionate to the order of harmonics. Say, for harmonic frequency of order-h, the self-impedance for a phase is evaluated with Eq. (1).

$$z^h = R + j.h.X \tag{1}$$

2.2 Capacitor Banks

For HPFM, the susceptance of the capacitance for harmonic frequency of order-h is obtained by multiplying with 'h'.

2.3 Loads

As far as harmonics are concerned, loads are of two types. The first one is linear load and the second one is non-linear load. The general load representation for HPFM is reported in [26].

2.3.1 Linear Loads

The linear loads can produce only a fundamental sinusoidal response when supplied by an AC source. In [27], presents the various methods in which linear loads are modeled for harmonic analysis. Each model will have a different effect on HPFM. The series model presented in [27] is taken in this work.

2.3.2 Non-Linear Loads

These loads are modeled as sources of constant current input [28]. The magnitude of this current source is calculated with the typical harmonic spectrum and load current attained from the fundamental solution as given in Eq. (2). The phase angle of these current sources is obtained with Eq. (3).

$$I_h = I_{rated}\frac{I_{h\_spectrum}}{I_{1\_spectrum}} \tag{2}$$

$$\theta_h = \theta_{h\_spectrum} + h.\left(\theta_1 - \theta_{h\_spectrum}\right) \tag{3}$$

2.4 RES and D-STATCOM

For HPFM the inverter-based-RES is considered the source of harmonics and for synchronous-based-RES, the harmonic current injection is considered to be zero. The D-STATCOM has a low harmonic content, therefore the injection of harmonic current from the D-STATCOM device is regarded as nil for HPFM.

3 Harmonic Power Flow Method for Harmonic Distortion Assessment

The iterative procedure to assess harmonics with penetration of RES and D-STATCOM device is exemplified in the below steps.

1. Read the system data and give the numbering to buses and branches as presented in [29].
2. Construct the matrices BNM and BRNM for the given TPDN as presented in [18].
3. Solve the fundamental power flow solution of the network using the algorithm presented in [18].
4. Select the initial harmonic order, h = ho.
5. Design the impedance of linear loads with the help of converged fundamental bus voltages and specified load.
6. Find the harmonic current injections at a bus due to non-linear loads, synchronous based-RES, inverter based-RES, and D-STATCOM device.
7. Set iteration Count = 1.
8. On the assumption of the harmonic pollution at the substation is zero, assign zero harmonic voltage at every bus as given in Eq. (4) for the initial start of HPFM.

$$[V_{abc}]^h = [0]p.u \tag{4}$$

9. Start at the last section's last bus of BNM and collect its harmonic current. And proceed toward the head bus of this section by applying KCL at every bus and finding the harmonic bus currents and harmonic branch currents. The process of harmonic current calculations at buses and branches is given in Eq. (5) and Eq. (6) for the sample section shown in Fig. 1.

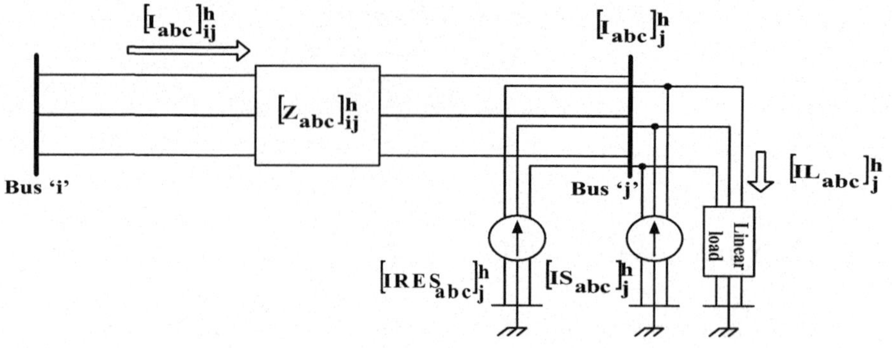

Fig. 1 A sample of two buses in the network for harmonic order-h with RES, linear and non-linear loads at bus-j

$$[I_{abc}]_j^h = -[IS_{abc}]_j^h - [IRES_{abc}]_j^h + [IL_{abc}]_j^h \quad (5)$$

$$[I_{abc}]_{ij}^h = [I_{abc}]_j^h \quad (6)$$

10. Likewise, Find the harmonic bus currents and harmonic Branch currents by moving Towards the Head bus in All Sections.
11. Now, to find out harmonic bus voltages, start with the head bus in section one of BNM and proceed towards the tail bus of this section using Eq. (7).

$$[V_{abc}]_j^h = [V_{abc}]_i^h - [Z_{abc}]_{ij}^h.[I_{abc}]_{ij}^h \quad (7)$$

12. Likewise, Move Toward the Tail bus in the Last Section in Finding harmonic voltages at All buses.
13. Set itr = itr + 1 and repeat the steps from 8 to 12 until the harmonic voltage magnitude mismatches at all buses in two successive iterations are below the tolerance limit.
14. Find the HPL in Branches for harmonic Order-H Using Eq. (8).

$$\begin{bmatrix} HSLoss_a \\ HSLoss_b \\ HSLoss_c \end{bmatrix}_{ij}^h = \begin{bmatrix} (V_a)_i.conj(I_a)_{ij} \\ (V_b)_i.conj(I_b)_{ij} \\ (V_c)_i.conj(I_c)_{ij} \end{bmatrix}^h - \begin{bmatrix} (V_a)_j.conj(I_a)_{ji} \\ (V_b)_j.conj(I_b)_{ji} \\ (V_c)_j.conj(I_c)_{ji} \end{bmatrix}^h \quad (8)$$

15. Now, go to the next harmonic order and repeat steps 5 to 14.
16. Repeat the same procedure as presented in steps 5 to 15 for every harmonic order in the selected range of frequencies.

17. Evaluate the total HPL by using Eq. (9).

$$[Total\_loss] = \sum_{h_o}^{h_m} \sum_{br=1}^{Nbr} [[HSLoss_{abc}]_{br}^{h}] \tag{9}$$

18. Evaluate the Total Rms voltage at Every bus for Every Phase Using Eq. (10).

$$(V_a)_i = \sqrt{\left|(V_a)_i^1\right|^2 + \sum_{h_o}^{h_m} \left|(V_a)_i^h\right|^2} \tag{10}$$

19. Find THD at a bus for a Phase Using Eq. (11).

$$(THD)_i^a = \frac{\sqrt{\sum_{h_o}^{h_m} \left|(V_a)_i^h\right|^2}}{\left|(V_a)_i^1\right|} \tag{11}$$

The flowchart for this HPFM is shown in Fig. 2.

4 Results and Discussion

4.1 Harmonic Solution of IEEE-13 Bus TPDN Without RES and D-STATCOM

Initially, the benchmarks are set for r.m.s voltage profiles, THD%, and total HPL by solving the HPFM without penetration of RES and D-STATCOM device and the presence of linear and no-liner loads. The line data and load data for the IEEE-13 bus TPDN are presented in [30]. The load composition of linear and non-linear loads and current spectra of non-linear loads is taken from [31]. The current spectra of inverter-based-RES are taken from [1]. Table 1 presents the solution of harmonic voltage profiles for the different harmonics of order (h = 3, 5, 7, 9, 11, 13, and 15). The HPL of the network is presented in Table 2. Table 3 presents the total rms voltage solution and THD%. From Table 3, the maximum value of THD% is found to be 5.2263 for 611-bus and in [32] it was reported as 5.23 at the same bus [33]. This value is more than the limit specified by IEEE standard 519–1992 [26]. Therefore, the proposed HPFM is accurate with the literature terms of accuracy. From Table 2, it is found that the total active HPL is 0.90 kW and the total reactive HPL is 7.95 kVAR.

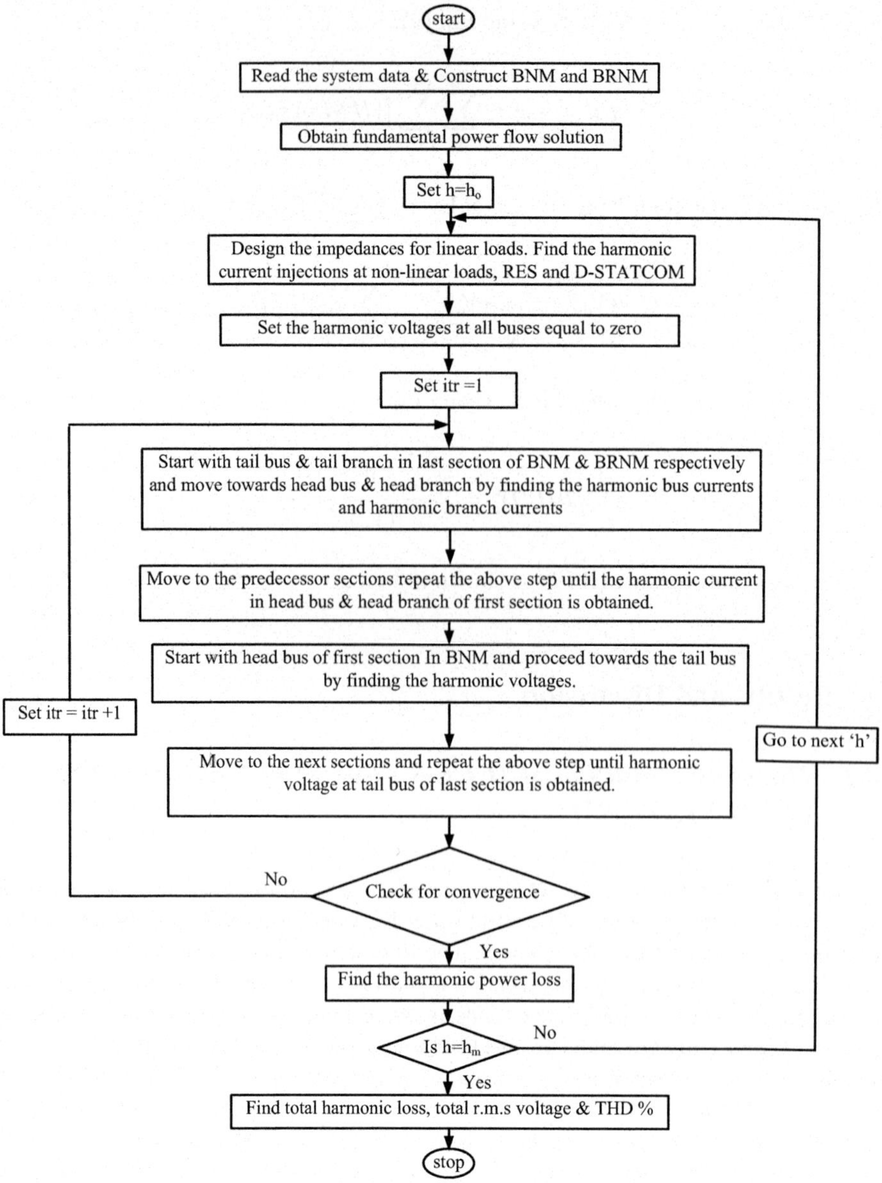

Fig. 2 Flowchart for HPFM for TPDN

Table 1 Harmonic voltage solutions of different harmonic orders for IEEE-13 bus TPDN

Bus	S. no	Phase	Harmonic voltages in p.u						
			Harmonic order						
			3	5	7	9	11	13	15
650	1	a	0	0	0	0	0	0	0
	2	b	0	0	0	0	0	0	0
	3	c	0	0	0	0	0	0	0
632	4	a	0.0139	0.0093	0.0039	0.0043	0.0021	0.0020	0.0034
	5	b	0.0013	0.0012	0.0028	0.0018	0.0013	0.0020	0.0021
	6	c	0.0179	0.0092	0.0039	0.0035	0.0011	0.0022	0.0028
671	7	a	0.0281	0.0188	0.0081	0.0087	0.0043	0.0041	0.0070
	8	b	0.0031	0.0026	0.0058	0.0037	0.0027	0.0040	0.0042
	9	c	0.0365	0.0188	0.0079	0.0071	0.0023	0.0046	0.0057
680	10	a	0.0281	0.0188	0.0081	0.0087	0.0043	0.0041	0.0070
	11	b	0.0031	0.0026	0.0058	0.0037	0.0027	0.0040	0.0042
	12	c	0.0365	0.0188	0.0079	0.0071	0.0023	0.0046	0.0057
633	13	a	0.0138	0.0092	0.0039	0.0042	0.0021	0.0020	0.0034
	14	b	0.0013	0.0012	0.0028	0.0018	0.0013	0.0019	0.0020
	15	c	0.0178	0.0091	0.0038	0.0035	0.0011	0.0022	0.0028
634	16	a	0.0132	0.0088	0.0037	0.0040	0.0020	0.0019	0.0032
	17	b	0.0013	0.0011	0.0027	0.0017	0.0013	0.0019	0.0020
	18	c	0.0172	0.0088	0.0037	0.0033	0.0011	0.0022	0.0027
645	19	b	0.0013	0.0012	0.0028	0.0018	0.0013	0.0020	0.0020
	20	c	0.0179	0.0092	0.0039	0.0035	0.0011	0.0022	0.0028
646	21	b	0.0013	0.0012	0.0028	0.0018	0.0013	0.0020	0.0020
	22	c	0.0179	0.0092	0.0039	0.0035	0.0011	0.0022	0.0028
692	23	a	0.0281	0.0188	0.0081	0.0087	0.0043	0.0041	0.0070
	24	b	0.0031	0.0026	0.0058	0.0037	0.0027	0.0040	0.0042
	25	c	0.0365	0.0188	0.0079	0.0071	0.0023	0.0046	0.0057
675	26	a	0.0298	0.0194	0.0084	0.0091	0.0046	0.0044	0.0073
	27	b	0.0032	0.0026	0.0058	0.0038	0.0028	0.0041	0.0043
	28	c	0.0372	0.0190	0.0082	0.0074	0.0025	0.0048	0.0059
684	29	a	0.0283	0.0188	0.0080	0.0086	0.0043	0.0040	0.0069
	30	c	0.0374	0.0192	0.0080	0.0072	0.0023	0.0045	0.0056
611	31	c	0.0385	0.0197	0.0081	0.0072	0.0023	0.0043	0.0055
652	32	a	0.0283	0.0187	0.0080	0.0086	0.0042	0.0040	0.0068

Table 2 HPL on IEEE-13 bus TPDN

Harmonic order	HPL	
	Active (kW)	Reactive (kVAR)
3	0.80	6.52
5	0.09	1.15
7	0.01	0.12
9	0	0.09
11	0	0.02
13	0	0.02
15	0	0.03
Total HPL	0.90	7.95

4.2 *Case Studies on IEEE-13 Bus TPDN*

The ratings, locations, and modeling of three-phase RES and Three-phase D-STATCOM devices for different example studies are presented in Table 4. These example studies are further divided based on the type of RES. In Case-A of Example-1 and Case-C of Example-2, the RES is considered an inverter based. In Case-B of Example-2 and Case-D of Example-2, the RES is considered synchronous based. The harmonic voltage solution and THD % for Example-1 and Example-2 are presented in Table 5 and Table 6, respectively.

4.2.1 Discussions on IEEE-13 Bus TPDN

From Table 5, in Case-A of Example-1, the maximum THD % is observed to be 6.2161 at bus 634 for the c-phase. Whereas the maximum THD% in Case-B of Example-1 is found to be reduced to 5.2442 by considering that both RES are synchronous based. For Case-A of Example-1, the total active and reactive HPL are obtained to be 1.1074 kW and 12.5531 kVAR respectively. For Case-B of Example-1, these losses are reduced to 0.9299 kW and 8.2585 kVAR respectively. From Table 6, in Case-C of Example-2, the maximum THD % is observed to be 6.1815 at bus 634 for the c-phase. Whereas the maximum THD% in Case-D of Example-2 is found to be reduced to 5.1892 by considering that both RES are synchronous based. For Case-C of Example-2, the total active and reactive harmonic losses are obtained to be 1.0986 kW and 12.4593 kVAR respectively. For Case-D of Example-2, these losses are reduced to 0.9212 kW and 8.1808 kVAR respectively. In Example-2, the penetration of D-STATCOM reduces the harmonic pollution of the network in both types of RES penetrations. Figure 3 compares the total r.m.s voltage profiles in different study examples. Figure 4 presents the comparison of THD% in different study examples. Figure 5 compares the HPL for different study examples on the network.

Table 3 Total harmonic voltage solution and THD% for IEEE-13 bus TPDN

Bus	S. No	Total rms voltage	THD %
650	1	1	0
	2	1	0
	3	1	0
632	4	0.9500	1.9173
	5	0.9839	0.4974
	6	0.9302	2.2737
671	7	0.9117	4.0623
	8	0.9875	1.0363
	9	0.8728	4.9409
680	10	0.9117	4.0623
	11	0.9875	1.0363
	12	0.8728	4.9409
633	13	0.9468	1.9098
	14	0.9819	0.4919
	15	0.9273	2.2648
634	16	0.9209	1.8801
	17	0.9624	0.4873
	18	0.9066	2.2406
645	19	0.9745	0.4991
	20	0.9286	2.2769
646	21	0.9729	0.5000
	22	0.9267	2.2815
692	23	0.9117	4.0623
	24	0.9875	1.0363
	25	0.8728	4.9409
675	26	0.9034	4.3128
	27	0.9887	1.0491
	28	0.8689	5.0687
684	29	0.9100	4.0765
	30	0.8695	5.0741
611	31	0.8663	5.2263
652	32	0.9049	4.0900

4.3 Case Studies and Discussions on IEEE-34 Bus TPDN

The line date and load data for IEEE-34 bus TPDN are taken from [30]. Table 7 presents the linear and non-linear load composition of spot loads. The current spectra

Table 4 Description of Example studies on IEEE-13 bus TPDN

Study Example	Type	Bus location	Active power (kW/ph)	Reactive power (kVAR/ph)	RES description
Example-1	RES	634	300	197	Case-A: Inverter based
	RES	675	260	$100 \leq Q \leq 650$	Case-B: Synchronous based
Example-2	RES	634	300	197	Case-C: Inverter based
	RES	675	260	$100 \leq Q \leq 650$	Case-D: Synchronous based
	D-STATCOM	680	0	$100 \leq Q \leq 500$	

of the non-linear loads are taken from [31]. The current spectra of inverter-based-RES are taken from [1]. The study examples for rating, location, and modeling of RES and D-STATCOM on the network are presented in Table 8.

The summary of the results on the network is presented in Table 9. The test results for Example-1 are taken as benchmarks to analyze the harmonic impacts of RES and D-STATCOM devices. In Case-A of Example-2, which has penetrations of one inverter based-RES and one D-STATCOM device, the THD% maximum value is decreased to 5.4201 and the number of buses with THD % more than 5 is found to be the same as in Example-1. The total power loss on the network including HPL is 127.01 kW and 102.24 kVAR. In Case-B of Example-2, which has synchronous based-RES and D-STATCOM device, the THD% maximum value is reduced to 4.8505 and the number of buses with THD % less than 5 is found to be zero. The total power loss on the network including HPL is 126.67 kW and 100.96 kVAR. In Case-C of Example-3, which has integrations of two inverter-based-RES and one D-STATCOM device, the THD% maximum value is reduced to 6.7344 which is more than in Example-1. The number of buses with THD% more than 5 is found to be 15. The total power loss on the network including HPL is 60.09 kW and 46.91 kVAR. In Case-D of Example-3, which has synchronous based-RES and D-STATCOM device, the THD% maximum value is found to be 4.2408 and the number of buses with THD% more than 5 is found to be zero. The total power loss on the network including HPL is 59.42 kW and 43.84 kVAR.

5 Conclusions

This chapter proposed an HPFM for harmonic assessment with the combined penetrations of RES and D-STATCOM in TPDN by utilizing the BNM and BRNM for its implementation. The proposed method is effectively handling the combined penetrations of both types (synchronous-based and inverter based) and both models (PV

Table 5 Total harmonic voltage solution and THD % for study Example-1

Bus	S. No	Total rms voltage for Case-A of Example-1	THD %	Total r.m.s voltage for Case-B of Example-1	THD%
650	1	1	0	1	0
	2	1	0	1	0
	3	1	0	1	0
632	4	0.9733	2.5118	0.9732	1.9959
	5	0.9974	1.7760	0.9973	0.6219
	6	0.9545	3.1977	0.9543	2.3845
671	7	0.9439	4.6067	0.9437	4.0844
	8	1.0048	2.5902	1.0046	1.2144
	9	0.9072	5.9663	0.9067	4.9782
680	10	0.9439	4.6067	0.9437	4.0844
	11	1.0048	2.5902	1.0046	1.2144
	12	0.9072	5.9663	0.9067	4.9782
633	13	0.9753	2.7230	0.9751	2.0059
	14	1.0008	2.0850	1.0006	0.6309
	15	0.9579	3.4443	0.9577	2.3982
634	16	0.9965	5.3900	0.9952	2.0365
	17	1.0262	4.9795	1.0249	0.6456
	18	0.9847	6.2161	0.9831	2.4403
645	19	0.9881	1.7865	0.9880	0.6240
	20	0.9528	3.2039	0.9526	2.3878
646	21	0.9865	1.7895	0.9863	0.6250
	22	0.9508	3.2105	0.9506	2.3927
692	23	0.9439	4.6067	0.9437	4.0844
	24	1.0048	2.5902	1.0046	1.2144
	25	0.9072	5.9663	0.9067	4.9782
675	26	0.9390	4.8928	0.9388	4.3096
	27	1.0084	2.6999	1.0081	1.2267
	28	0.9064	6.1741	0.9058	5.0896
684	29	0.9421	4.6216	0.9419	4.0965
	30	0.9040	6.0703	0.9035	5.1023
611	31	0.9007	6.1897	0.9003	5.2442
652	32	0.9368	4.6433	0.9366	4.1047

Table 6 Total harmonic voltage solution and THD % for study Example-2

Bus	S. No	Total rms voltage for Case-C of Example-2	THD%	Total rms voltage for Case-D of Example-2	THD%
650	1	0	0	0	0
	2	0	0	0	0
	3	0	0	0	0
632	4	0.9755	2.4988	0.9754	1.9824
	5	0.9996	1.7680	0.9994	0.6181
	6	0.9571	3.1763	0.9569	2.3666
671	7	0.9484	4.5693	0.9481	4.0470
	8	1.0091	2.5722	1.0089	1.2045
	9	0.9125	5.9065	0.9120	4.9260
680	10	0.9495	4.5638	0.9493	4.0422
	11	1.0102	2.5695	1.0099	1.2032
	12	0.9137	5.8984	0.9132	4.9192
633	13	0.9775	2.7095	0.9773	1.9923
	14	1.0029	2.0758	1.0027	0.6269
	15	0.9605	3.4217	0.9602	2.3802
634	16	0.9986	5.3676	0.9974	2.0225
	17	1.0283	4.9591	1.0270	0.6415
	18	0.9872	6.1815	0.9856	2.4218
645	19	0.9903	1.7784	0.9901	0.6201
	20	0.9554	3.1825	0.9551	2.3700
646	21	0.9886	1.7814	0.9885	0.6212
	22	0.9534	3.1890	0.9532	2.3748
692	23	0.9484	4.5693	0.9481	4.0470
	24	1.0091	2.5722	1.0089	1.2045
	25	0.9125	5.9065	0.9120	4.9260
675	26	0.9435	4.8526	0.9433	4.2697
	27	1.0127	2.6811	1.0124	1.2167
	28	0.9116	6.1118	0.9111	5.0360
684	29	0.9466	4.5841	0.9464	4.0590
	30	0.9092	6.0094	0.9087	5.0487
611	31	0.9060	6.1276	0.9055	5.1892
652	32	0.9413	4.6056	0.9411	4.0672

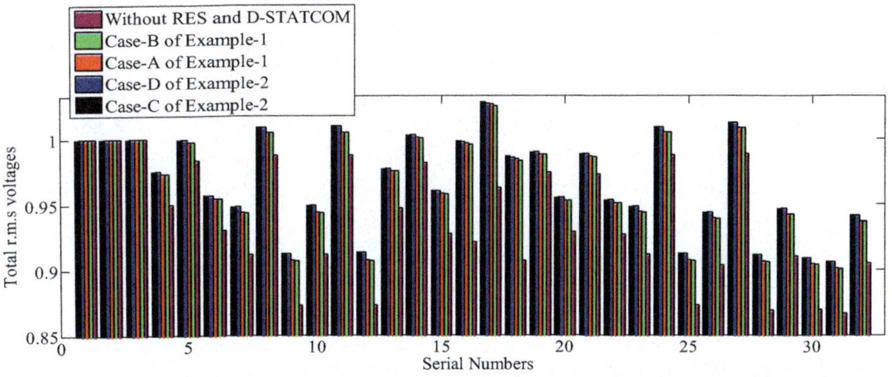

Fig. 3 Total r.m.s voltage profiles comparison for different example studies on IEEE-13 bus TPDN

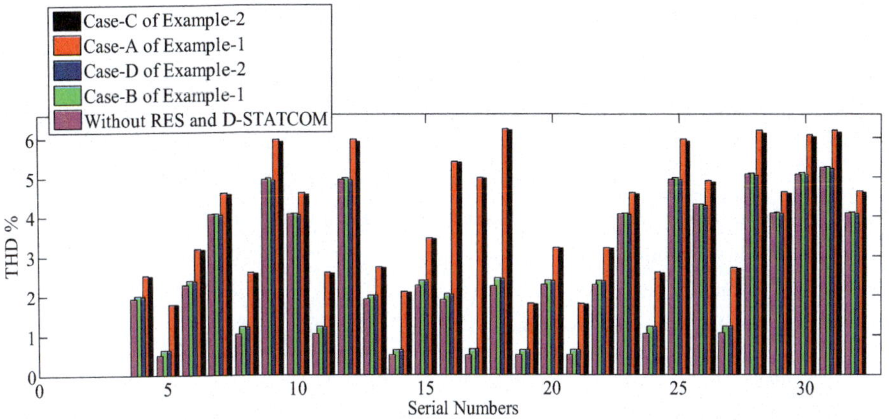

Fig. 4 THD % comparison for different example studies on IEEE-13 bus TPDN

and PQ) of RES and D-STATCOM devices. The test results show that the combined penetration of inverter-based RES and D-STATCOM as in Case-A of Example-3 on IEEE-34 bus TPDN the maximum THD % is observed to be 6.7344 at bus 890 for a-phase and the number of buses with more than 5% is found to be 15. Whereas, the combined penetration of synchronous-basedRES and D-STATCOM as in Case-B of Example-3 on IEEE-34 bus TPDN reduced the THD% at all buses to below 5% which is desirable according to IEEE standard 519–1992.

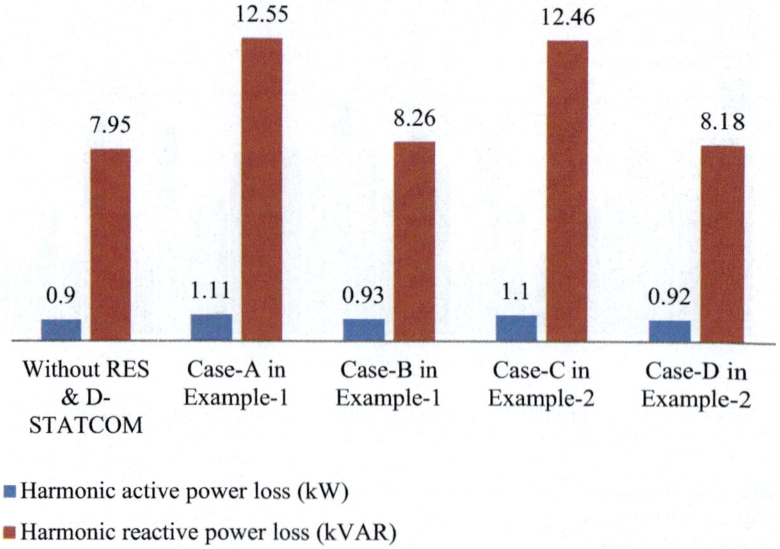

Fig. 5 Comparison of total harmonic power loss for different example studies in IEEE-13 bus TPDN

Table 7 Composition of linear and non-linear loads on IEEE-34 bus TPDN

Bus No	Load composition			
	Non-linear loads			Percentage of Linear loads
	Percentage of Fluorescent light banks	Percentage of Adjustable speed drives	Percentage of Composite residential loads	
830	None	None	80	20
844	30	30	30	10
848	30	30	30	10
890	30	None	60	10
860	30	30	30	10
840	30	30	30	10

Table 8 Description of Example studies on IEEE-34 bus TPDN

Study Example	Type	Bus location	Active power (kW/ph)	Reactive power (kVAR/ph)	RES description
Example-1	Without voltage regulators and capacitor banks				
Example-2	RES	848	150	99	Case-A: Inverter based
	D-STATCOM	650	0	$50 \leq Q \leq 250$	Case-B: Synchronous based
Example-3	RES	848	150	99	Case-C: Inverter based
	RES	890	130	$50 \leq Q \leq 325$	Case-D: Synchronous based
	D-STATCOM	650	0	$50 \leq Q \leq 250$	

Table 9 Concluding results on IEEE-34 bus TPDN

Case study	Minimum fundamental voltage, p.u		Min. total rms voltage, p.u	Maximum THD%	Number of phases of buses (THD) > 5%)
Example-1	0.7837		0.7853	5.5791	4
Example-2	Case-A	0.8386	0.8398	5.4201	4
	Case-B		0.8396	4.8505	0
Example-3	Case-C	0.9030	0.9052	6.7344	15
	Case-B		0.9038	4.2408	0
Case study	Total FPL			Total power loss including total HPL	
	Active (kW)		Reactive (kVAR)	Active (kW)	Reactive (kVAR)
Example-1	260.89		180.49	264.56	188.15
Example-2	Case-A	123.46	94.23	127.01	102.24
	Case-B			126.67	100.96
Example-3	Case-C	56.48	37.73	60.09	46.91
	Case-D			59.42	43.84

Acknowledgements This research work was supported by "Woosong University's Academic Research Funding—2023".

References

1. Pandi VR, Zeineldin HH, Xiao W (2013) Determining optimal location and size of distributed generation resources considering harmonic and protection coordination limits. IEEE Trans Power Syst 28(2): 1245–1254. https://doi.org/10.1109/TPWRS.2012.2209687
2. Milovanovic M, Radosavljevic J, Perovic B (2019) A backward/forward sweep power flow method for harmonic polluted radial distribution systems with distributed generation units. Int Trans Elect Energy Syst 30(5). https://doi.org/10.1002/2050-7038.12310
3. Fortes RRA, Buzo RF, de Oliveira LCO (2020) Harmonic distortion assessment in power distribution networks considering DC component injection from PV inverters. Electr Power Syst Res 188. https://doi.org/10.1016/j.epsr.2020.106521
4. Abbas AS, El-Sehiemy RA, El-Ela AA, Ali ES (2021) Optimal harmonic mitigation in distribution systems with inverter based distributed generation. Appl Sci 11. https://doi.org/10.3390/app11020774
5. Hernandez JC, Rodriguez FJR, Jurado F, Sutil ES (2020) Tracing harmonic distortion and voltage unbalance in secondary radial distribution networks with photovoltaic uncertainties by an iterative multiphase harmonic load flow. Electr Power Syst Res 185. https://doi.org/10.1016/j.epsr.2020.106342
6. Rodriguez FJR, Hernandez JC, Jurado F (2020) Iterative harmonic load flow by using the point-estimate method and complex affine arithmetic for radial distribution systems with photovoltaic uncertainties. Elect Power Energy Syst 118: 1–16. http://dx.doi.org/10.1016/j.ijepes.2019.105765
7. Rohouma W, Balog RS, Peerzada AA, Begovic MM (2020) D-STATCOM for harmonic mitigation in low voltage distribution network with high penetration of nonlinear loads. Renew Ener 145: 1449–1464. https://doi.org/10.1016/j.renene.2019.05.134
8. Marjani SR, Galvani S, Talavat V, Kangarlu MF (2020) Optimal allocation of D-STATCOM in distribution networks including correlated renewable energy sources. Electr Power Energy Syst 122: 1–14. https://doi.org/10.1016/j.ijepes.2020.106178
9. Patel SK, Arya SR, Maurya R (2019) Optimal step LMS-based control algorithm for DSTATCOM in distribution system. Elect Power Comp Syst 47(8): 1–17. https://doi.org/10.1080/15325008.2019.1602797
10. Satish R, Kantarao P, Vaisakh K (2021) A new algorithm for impacts of multiple DGs and D-STATCOM in unbalanced radial distribution networks. Int J Renew Energy Technol 12: 221–242. https://doi.org/10.1504/IJRET.2021.116609
11. Salkuti SR (2021) Optimal allocation of DG and D-STATCOM in a distribution system using evolutionary based bat algorithm. Intern J Adv Comp Sci Appl 12(4): 360–365. https://doi.org/10.14569/IJACSA.2021.0120445
12. Satish R, Kantarao P, Vaisakh K (2022) A new algorithm for harmonic impacts with renewable DG and non-linear loads in smart distribution networks. Technol Eco Smart Grids Sustain Energy 7: 1–19. https://doi.org/10.1007/s40866-022-00134-1
13. Satish R, Vaisakh K, Abdelaziz AY (2022) Evaluation of algorithms for fundamental and harmonic impacts of integration of renewable energy sources in smart power distribution networks. In Salkuti SR, Ray P (eds) Next generation smart grids: modeling, control and optimization. Lecture Notes in Electrical Engineering, vol 824. Springer, Singapore. https://doi.org/10.1007/978-981-16-7794-6_6
14. Abdelaziz AY, Ali ES, Elazim SMA (2016) Flower pollination algorithm for optimal capacitor placement and sizing in distribution systems. Elect Power Comp Syst 44(5): 544–555. https://doi.org/10.1080/15325008.2015.1117540
15. Nowdeh SA, Davoodkhani IF, Moghaddam MJH, Najmi ES, Abdelaziz AY, Ahmadi A, Razavi SE, Gandoman FH (2019) Fuzzy multi-objective placement of renewable energy sources in distribution system with objective of loss reduction and reliability improvement using a novel hybrid method. Appl Soft Comput J 77: 761–779. https://doi.org/10.1016/j.asoc.2019.02.003

16. Othman MM, Hegazy YG, Abdelaziz AY (2017) Electrical energy management in unbalanced distribution networks using virtual power plant concept. Elect Power Syst Res 145: 157–165. https://doi.org/10.1016/j.epsr.2017.01.004
17. Salkuti SR (2022) Emerging and advanced green energy technologies for sustainable and resilient future grid. Energies 15(18): 6667. https://doi.org/10.3390/en15186667
18. Satish R, Vaisakh K, Abdelaziz AY, Shahat AE (2021) A novel three-phase power flow algorithm for evaluation of impact of renewable energy sources and D-STATCOM device in unbalanced radial distribution networks. Energies 14. https://doi.org/10.3390/en14196152
19. Kersting WH (2017) Distribution system modeling and analysis, Fourth edition, CRC Press
20. Losi A, Russo M (2005) Dispersed generation modeling for object-oriented distribution load flow. IEEE Trans Power Deliv 20(2): 1532–1540. https://doi.org/10.1109/TPWRD.2004.838634
21. Chen H, Chen J, Shi D, Duan XZ (2006) Power flow study and voltage stability analysis for distribution systems with distributed generation. IEEE Power Eng Soc General Meet 1-8. https://doi.org/10.1109/PES.2006.1709198
22. Naka S, Genji T, Fukuyama Y (2001) Practical equipment models for fast distribution power flow considering interconnection of distributed generators. Power Eng Soc Summer Meeting Conf Proc 1007–1012. https://doi.org/10.1109/PESS.2001.970196
23. Nehrir H, Wang C, Shaw SR (2006) Fuel cells: promising devices for distributed generation. IEEE Power Energy Magaz 4(1): 47–53. https://doi.org/10.1109/MPAE.2006.1578531
24. Yang Z, Shen C, Crow ML, Zhang L (2000) An improved StatCom model for power flow analyses. Power Eng Soc Summer Meet, 1121–1126. https://doi.org/10.1109/PESS.2000.867536
25. Jazebi S, Hosseinian SH, Vahidi B (2011) DSTATCOM allocation in distribution networks considering reconfiguration using differential evolution algorithm. Energy Conv Manag 52(7): 2777–2783. https://doi.org/10.1016/j.enconman.2011.01.006
26. IEEE Recommended Practices and Requirements for Harmonic Control in Electrical Power Systems (1993) IEEE Standard 519–1992. https://doi.org/10.1109/IEEESTD.2014.6826459
27. Burch R, Cheng G, Hatziadoniu C (2003) Impact of aggregate linear load modeling on harmonic analysis: a common practice and analytical models. IEEE Trans Power Del 18(2): 625–630. https://doi.org/10.1109/TPWRD.2003.810492
28. Task Force on Harmonic Modeling and Simulation (1996) Modeling and simulation of the propagation of harmonics in electric power networks Part I: concepts, models and simulation techniques. IEEE Trans Power Del 11(1): 452–465. https://doi.org/10.1109/61.484130
29. Das D, Nagi HS, Kothari DP (1994) Novel method for solving radial distribution networks. IEE Proc Gener Trans Distrib 141(4): 291–298. https://doi.org/10.1049/ip-gtd:19949966
30. Radial Distribution Test Feeders, http://sites.ieee.org/pes-testfeeders/resources.
31. Hashim RA, Burch R, Chang G (1999) Test systems for harmonic modeling and simulation. IEEE Trans Power Del 14(2): 579–587. https://doi.org/10.1109/61.754106
32. Yang NC, Le MD (2016) Three-phase harmonic power flow by direct ZBUS method for unbalanced radial distribution systems with passive power filters. IET Gener Trans Distrib 10(13): 3211–3219. https://doi.org/10.1049/iet-gtd.2015.1368
33. IEEE Standard (1993) IEEE recommended practices and requirements for harmonic control in electrical power systems. IEEE Std 519–1992, 1–112. https://doi.org/10.1109/IEEESTD.2014.6826459

An Enhanced SRF Theory-Based Multifunctional Control Approach for Power Quality Improvement in Grid-Tied Photovoltaic Systems

Fossy Mary Chacko, M. V. Jayan, A. Prince, and Vidhya Viswambaran

Abstract Renewable energy integration, especially that of photovoltaic (PV) energy resources, into the utility grid presents an immense challenge. The incorporation of PV resources into low-voltage distribution networks results in numerous power quality concerns. The ever-rising number of power electronic loads reduces the power quality further. Traditional methods of power quality augmentation include the use of passive, hybrid, and active filters as well as distribution static compensators (DSTATCOMs). However, in the present-day situation of enhanced renewable resource penetration, the development of ground-breaking approaches for augmenting power quality is deemed vital. The control strategy utilized for the PV inverter plays a key part in ensuring the proper grid integration of solar photovoltaic resources. State-of-the-art controllers for grid-tied photovoltaic (GPV) systems reported in the literature are, in general, found to be computationally intensive, and complicated in structure and real-time implementation. For low-voltage distribution networks, the inclusion of auxiliary capabilities in GPV systems is highly significant. Nevertheless, very limited structures providing both active and reactive power support via the PV inverter are dealt with. This chapter throws light on an enhanced synchronous reference frame (SRF) theory-based multifunctional control approach for the PV inverter. The control configuration is developed with a view to realizing the photovoltaic inverter operation as a practical DSTATCOM. The ability to operate as an active filter and also inject

F. M. Chacko (✉) · A. Prince
Department of Electrical Engineering, Rajiv Gandhi Institute of Technology, (Affiliated to APJ Abdul Kalam Technological University), Kottayam, Kerala, India
e-mail: fossy.mary@rit.ac.in

A. Prince
e-mail: prince@rit.ac.in

M. V. Jayan
Department of Electrical Engineering, Government Engineering College, (Affiliated to APJ Abdul Kalam Technological University), Thrissur, Kerala, India

V. Viswambaran
Department of Electrical and Electronic Engineering, University of Bolton, RAK Campus, Ras Al-Khaima, United Arab Emirates
e-mail: v.ragesh@bolton.ac.uk

S. R. Salkuti et al. (eds.), *Power Quality in Microgrids: Issues, Challenges and Mitigation Techniques*, Lecture Notes in Electrical Engineering 1039,
https://doi.org/10.1007/978-981-99-2066-2_6

renewable energy, are salient characteristics of the GPV system. The performance of the GPV system with the enhanced multifunctional control strategy is assessed using MATLAB/Simulink. The simulation results establish the superiority of the improved multifunctional configuration for the mitigation of power quality issues alongside the control of active and reactive power in GPV systems.

Keywords Power quality · Reactive power · Solar photovoltaic system · Synchronous reference frame · Total harmonic distortion

1 Introduction

Power quality is an important aspect that can be interpreted in many ways. The IEEE 1100 standard [1] articulates power quality as "the concept of powering and grounding electronic equipment in a manner that is suitable to the operation of that equipment and compatible with the premise wiring system and other connected equipment". The outlook of an equipment manufacturer or designer may be that power quality represents a perfectly sinusoidal waveform, with no variations at all in the voltage, and with the absence of noise on the grounding system. However, the perspective of an electric utility engineer may be that power quality is basically the service availability or the outage minutes. Lastly, from an end-user point of view, 'quality power' signifies the power that works for whatsoever equipment the end-user is applying [2]. The term 'power quality' may also be construed as service quality, comprising the aspects of reliability of supply and quality of power delivered [3]. Power quality is thus a significant attribute of electric power systems which is associated with equipment manufacturers, network service providers, and customers [4].

The term 'power quality issue' is interpreted as "any problem manifested in voltage, current or frequency deviation that results in the failure or misoperation of customer equipment" [5]. Some of the foremost power quality concerns include harmonics, supply voltage sag and swell; supply voltage flicker, and voltage notching. So as to maintain the supply quality and provide clean electricity, international agencies like IEEE and IEC have drafted standards regarding several power quality issues that are of mandatory interest to the utility and consumers.

The escalating number of power electronics-based devices has significantly impacted the electric power supply quality. Both high-power industries, as well as domestic loads, create harmonics in the grid voltages. Nonetheless, most of the equipment producing the disturbances is, in fact, sensitive to variations from the ideal sinusoidal network voltages. Thus, power quality issues may either arise in the network or be induced on the consumer side. The introduction and widespread use of high-power, high-frequency semiconductor devices have given the power quality problem a new dimension. The non-linear switching actions produce current harmonics along with the consumption of large VAR. Hence, non-linear currents in the utility have increased at alarming levels, which pollute both the grid current and voltage.

Thyristor-controlled reactors for static VAR compensation, large adjustable speed motor drives, and various non-linear loads are widely used in industries. This equipment contributes to current harmonics and poor power factor. Supply voltage flicker is produced due to large capacity arc furnaces, arc welders, and frequent motor starts. Supply voltage sag may be due to several factors like transformer energization, transmission line faults, and motor starting. On the contrary, supply voltage swell is caused due to single line-to-ground faults, removal of large loads, or addition of capacitor banks. Adjustable speed drives, solid-state rectifiers, etc., cause voltage notching.

Among all these factors, the degeneration in power quality owing to the widespread application of power electronics-based circuits is a problem of the utmost concern since it produces undesirable impacts such as excessive overheating and loss of life of equipment, poor power factor, increased neutral currents, interference in communication systems, increased harmonic distortion leading to inaccuracies in metering, capacitor bank failures, and transformer heating. Nowadays, on account of worldwide energy scarcity and the growing necessity for sustainable energy, there exists a paradigm shift toward the deployment of power supply structures that rely predominantly on renewable energy sources (RES) [6]. The provision of quality power attains increased importance in this current scenario.

Distributed generation (DG) is widespread in today's power systems owing to the intense rivalry in the power sector together with the ever-increasing need to produce electricity from eco-friendly sources like wind, solar, tidal, geothermal, etc. Dispersed generation may be defined as "the integrated or standalone use of small modular electricity generation resources by utilities, utility customers and/or third parties in applications that benefit the electric system, specific end-user customers or both". Distributed generation is also termed as "electric power generation within the distribution network or on the customer side of the network" [7, 8].

Traditional power systems are changing the world over, due to the reason that an enormous number of DG systems based on renewable as well as non-renewable energy resources, such as steam or gas-powered combined heat and power stations, photovoltaic (PV) generators, small hydro, fuel cells, wind turbines, wave generators, etc., are getting integrated into power systems at the distribution side. Due to the inherent variation in the output of the DG units and the load or system demand, an intermediate stage is essential. Power electronic devices that control and effectively convert electric power, play a critical part in the integration of DG units for improved efficiency along with the reliable performance of the power system. Power electronics is thus the facilitating technology for a distributed generation [9].

To overcome the problems associated with the intermittency of renewable energy sources, hybrid combinations of two or more power production technologies together with storage can augment the system's performance [10]. For instance, wind and solar energy resources in a particular region are to some extent complementary on a daily or seasonal basis. Generally, hybrid systems transform all the resources into the same form (usually electrical) and store the energy in a suitable way, which may be the mechanical flywheel, chemical, thermal, compressed air, etc. The combined output can be utilized to supply various dc or ac loads. Although hybridization results in enhanced reliability, appropriate technology selection and sizing of generation units are vital for better operational performance along with dispatch control.

In a distributed power generation system (DPGS) with multiple input power sources, the input power from RES is converted into electricity through a power conversion block that utilizes a structure that is strongly linked with the nature of the power input. The power generated can either be supplied to the utility grid or to the local loads, based on the location where the DPGS is integrated. The most significant aspect of the DPGS is the control system which consists of:

- The input side control scheme extracts maximum power from the input and also plays a crucial role in the protection of the source side power electronic converter.
- The grid side control scheme controls the real power generated to the grid, performs dc bus voltage regulation, and furthermore, controls the reactive power transfer amid the grid and the DPGS. It ensures synchronization with the grid as well as superior injected power quality [11]. Additionally, the grid operator may request auxiliary services such as local voltage and frequency control, active filtering, or voltage harmonics mitigation.

A microgrid may be regarded as a low-scale grid comprising DG units, various electrical energy storage equipment, and loads that are electrically interconnected and controlled in a hierarchical manner. The microgrid has the capacity for either grid-tied or intentionally islanded operation. Power converters play a vital role in microgrids. Based on the operation, these can be categorized as grid-forming, grid-supporting, and grid-feeding types [12]. The grid-forming converter can be depicted as a perfect ac voltage source having low impedance at the output. The local grid voltage magnitude and frequency are set utilizing a control loop. This is applicable in the case of microgrids with one central inverter or small island grids. Conversely, the grid-feeding power converter is essentially designed for power delivery to an energized network. These may be depicted as a perfect current source integrated into the system in a shunt with a large impedance. This current source must be seamlessly in step with the grid voltage at the point of common coupling (PCC) for the effective regulation of the real and reactive power exchange. Lastly, the grid-supporting converter may be depicted as a perfect ac voltage source in series with a link impedance or as a perfect ac-controlled current source in parallel with a shunt impedance. The grid-supporting converter regulates its output voltage or current so that the network frequency and voltage amplitude are maintained near the rated values. The grid-supporting units are controlled to provide auxiliary services for power quality enhancement in addition to extracting maximum real power from the primary energy source.

The ever-rising number of RES and distributed generators being integrated into the electricity grid necessitates the development of new strategies for grid operation and management. These strategies serve to preserve or enhance the power supply quality and reliability. Due to the comparatively greater investment cost of renewable power generation systems, it is crucial to operate the systems at or near their maximum power output point, particularly for the solar PV and wind generation systems. Consequently, maximum power point tracking methods are indispensable. Furthermore, owing to the intermittency of solar PV and wind power resources, accurate predictions and modeling of solar insolation and wind speed are necessary, although difficult.

On the generation side, a rise in the connection of independent power producers with inadequately controlled synchronization will render the maintenance of power quality more challenging. The increase in an embedded generation will create more variations in the voltage magnitude and also result in additional voltage magnitude steps. Wind power is found to cause an intensification in the severity of flicker. Solar power as well as sophisticated techniques of integrating wind power will result in increased harmonic distortion levels [3, 7].

Thus, the intermittent characteristic of RES deems the grid integration of renewable power an uphill task. This situation combined with the proliferation of power electronic loads severely deteriorates the quality of power [13]. In the literature, several methods have been proposed for the alleviation of power quality concerns. Conventionally, harmonic distortion levels have been decreased by utilizing passive filters. Though passive filters offer the advantages of simplicity, reliability, and low cost; they have drawbacks like a slow response, the possibility of series or parallel resonance, bulkiness, limited flexibility for the dynamic compensation of multiple frequency harmonic components, etc. The intensified deterioration of power quality in electric networks has inspired power engineers to develop dynamic as well as adjustable solutions to the power quality concerns, namely active power filters. With the advent of modern self-commutating switching devices, voltage-sourced and current-sourced converters have been put to use as active power filters. These active filters are capable of mitigation of voltage as well as current harmonics, line voltage, reactive power control, and flicker suppression [14]. When active power filters are applied to power distribution networks, they are described as custom power devices. The term 'custom power' [2] refers to "the value-added power that electric utilities and other service providers will offer their customers in the future". Its enhanced level of reliability, with regard to minimal interruptions and variations, will result from the implementation of power electronic controllers in utility distribution networks and at the grid side of commercial and industrial consumers. A custom power device essentially comprises power inverter circuits and energy storage elements like dc capacitors or inductors. The inverter injects voltage, current, or both into the distribution network so as to provide compensation [15].

The major compensating custom power devices include a distribution static compensator (DSTATCOM), dynamic voltage restorer, and unified power quality conditioner. Among all the methods discussed above, the development and applications of DSTATCOM have gained tremendous focus during the past decades on account of the growing concerns about power quality at the distribution as well as consumer ends. Voltage sourced-converter-based DSTATCOMs have become prevalent for the reduction of the various issues affecting the ac distribution system. The DSTATCOM has the capability to perform harmonic current mitigation, load balancing, and reactive power compensation. Though the fundamental DSTATCOM concept is far from novel, current advancements in the field of semiconductors and signal processing have rendered their real-world implementation easier. Thus, the DSTATCOM finds widespread application in the compensation of DPGS including control of asynchronous generators in small hydro plants; voltage and frequency control in wind energy systems; and reactive power compensation, harmonics elimination, and voltage regulation in microgrids.

2 Issues Prevalent in Grid-Tied Photovoltaic (GPV) Systems

Currently, there exists intensified utilization of RES such as hydropower, wind, and solar. Solar photovoltaic systems have gained more popularity than any other technique of power generation owing to benefits such as easy installation, compactness, the absence of moving parts, no fuel cost, durability, less maintenance, and increased power capacity per unit of weight. The IEA 'Renewables 2020' forecast report [16] predicts that during 2023–25, the average annual PV installations are anticipated to vary from 130 to 165 GW, contributing toward nearly 60% of the total renewable energy growth globally and from 13 to 18 GW in India. Hence, this chapter focuses on renewable energy systems based on solar power.

Alongside all the advantages of RES-based power generation are several power quality concerns. Power quality is a significant facet that must be given attention to, as it affects the reliability of distribution grids. For tapping the power of RES, the generating components must be tied to the distribution networks. The operation of such grid-tied systems is critical because of their intermittent nature and also owing to the technological dissimilarities from fossil fuel-based units which are primarily steady power sources. Of late, the extensive incorporation of power electronic inverter-based DG units has created serious issues in distribution power networks including harmonics distortion and frequency instability due to reduced overall inertia.

When solar PV configurations are grid-tied, it produces problems such as voltage, frequency oscillations, and harmonics distortion. The power quality issues prevalent in such renewable energy-based DG systems are predominantly due to the usage of power electronics technologies and the intermittent nature of RES. Variable weather patterns bring about varying RES power output, which leads to voltage and frequency instabilities at the grid [6]. For instance, varying solar power generation under fluctuating climatic conditions can create oscillations in system voltage as well as frequency. Nevertheless, severe intermittency in the available solar power can worsen the grid power quality. Another major power quality concern is the harmonics. Harmonics are produced as the DG inverter injects high-frequency current components into the grid together with the fundamental component. At the PCC, there are several local non-linear loads that add to the harmonics and further deteriorate the power quality. Harmonics distortion results in several issues like resonance, spurious tripping of protection equipment, and overheating in transformers, cables, and lines.

The IEEE Standard 929-2000 [17] identifies voltage, frequency, current harmonic distortion, and voltage flicker as the major factors, indices of which may be utilized to ascertain the power quality performance of DGs. Augmented renewable energy penetration level in the near future, will further aggravate the situation if power quality improvement strategies are not adopted.

3 Power Quality Enhancement Strategies for GPV Systems: State-of-the-Art

In literature, several power quality issue mitigation techniques have been proposed for the seamless grid integration of renewable power systems. These techniques may be generally classified into current quality improvement (CQI) and voltage quality improvement (VQI) techniques. CQI techniques focus on the mitigation of current harmonics owing to the DG system alone or the grid system and load bus. On the other hand, VQI techniques focus on the alleviation of voltage and frequency fluctuations in DGs. Nonetheless, associated with each mitigation technique are certain inadequacies, which need to be overcome by improved strategies.

Smooth grid integration and stable operation of renewable energy systems are deemed essential to pave the way for future smart grids. In the literature, extensive studies have been described that address the reduction of the undesirable effects produced by renewable energy-based DG systems; nevertheless, each one has a few limitations. Hence, this chapter aims at the alleviation of grid integration issues in renewable power systems with a special focus on photovoltaics as a distributed resource.

For the proper functioning of such grid-tied photovoltaic (GPV) systems, three aspects play a major role. The first aspect is the selection of the voltage-sourced inverter (VSI) topology which depends on the type of distribution system where it is to be installed and the nature of the loads to be compensated [18]. For low-voltage distribution systems, the VSI may be directly coupled to the grid without any transformer. Whereas in high voltage distribution systems, the utility grid voltage must be stepped down before the VSI can be connected to the system. The control strategy applied for PV inverter switching is the second key aspect that characterizes the performance and behavior of the GPV system [19]. In the control structure design, the detection algorithm developed for compensating voltage and current reference template generation determines the compensation efficacy of the PV-VSI. The detection algorithm makes use of the instantaneous values of the grid, load voltages, and currents to determine the reference quantities which should be injected into the distribution network for providing accurate compensation. The third aspect involves the current control of the VSI switches such that the actual injected quantities track the reference values generated by the control scheme. This chapter mainly focuses on the second aspect mentioned above, i.e., the control approach developed for PV inverter switching.

Numerous control methods have been suggested in the literature, however, in general, these are complex with limited versatility. Furthermore, in modern-day distribution systems, reactive loads like fans, pumps, etc., account for a major share of the power consumption. These draw lagging power factor currents which increase the distribution system's reactive power burden. This leads to increased feeder losses and diminishes the system's active power flow capacity [20]. Hence, it will be more economical if the GPV systems are equipped with reactive power compensation and power quality enhancement capability rather than using filters or installing separate custom power devices like DSTATCOMs. Consequently, the method of estimation of harmonic currents and real, reactive powers as well as the strategy for compensating current reference template generation play a crucial part in ensuring the superior dynamic performance of the GPV inverter.

Different state-of-the-art control strategies have been elucidated in the literature for the grid connection of PV configurations like instantaneous symmetrical components theory, synchronous reference frame (SRF) theory, and instantaneous reactive power (IRP) theory [21–23]. A multi-level control strategy for a three-phase grid-tied PV system is proposed in [22] comprising external, intermediate, and internal control stages. The control structure derives from the instantaneous power theory in the d-q reference frame. The PV array coupled with VSI provides real power. Additionally, the VSI functions as a static compensator for reactive power control. The external level control determines the real and the reactive power exchanged between the grid and the PV system. The intermediate level control performs the function of ensuring that the expected outputs precisely track set the reference values. The internal control stage generates the gating signals for the switches of the VSI. In [24], a control strategy utilizing IRP theory is proposed, that adequately compensates for different types of loads. Nevertheless, this technique has the disadvantage of indirect inverter current control using gradually changing grid currents. In [25], the authors have designed a synchronous reference frame-based control configuration for grid-integrated PV systems. Although the developed approach shows superior performance under variable environmental conditions; satisfactory power quality enhancement remains unattained because the network current encompasses high-frequency components with oscillatory dc link voltage and grid powers. In [26], the authors have proposed a control strategy using adaptive notch filters (ANF) that possesses the characteristics of robustness and fast response. Nonetheless, the control approach necessitates manifold ANF stages with the drawback of huge computational intensiveness. The authors in [27] have proposed a control scheme utilizing character triangular function and low pass filters for harmonic mitigation through the extraction of the fundamental load current components. However, the proposed method utilizing six low pass filters suffers from the complexity that makes practical realization problematic. Other innovative control methods presented in the literature include those based on neural networks [28], dc bus voltage regulation [29], least mean fourth algorithm [30], and adaptive neuron detection [31]. Nevertheless, all these techniques require complicated computations compared to the more simple three-phase-based SRF or IRP theories.

Keeping in mind the aforesaid disadvantages of the existing control schemes for GPV systems, there is a need to develop an improved control and reference template generation strategy with a reduced computational burden which not only ensures accurate tracking of the dynamically varying operating conditions, but also provides reactive power compensation and power quality enhancement features without any additional hardware. Furthermore, the approach should also achieve accurate active and reactive power regulation.

4 Enhanced SRF Theory-Based Multifunctional Control Approach

The enhanced SRF multifunctional control approach is based on the d-q or synchronous reference frame theory [32]. The classical time domain-based SRF algorithm for reference signal estimation is modified by incorporating the following salient features in the control structure [33]. The GPV system is controlled for operating in two modes, namely, active filter (AF) and renewable power injection (RPI). The insolation *'Irr'* and the total harmonic distortion (THD) in load current i_l. are monitored continuously. The RPI stage is triggered for insolation in excess of 200 W/m^2 or load current THD within the range recommended by IEEE Standard 519. Otherwise, the AF mode is activated. The control approach of the GPV system is developed to provide virtual DSTATCOM capability to the PV inverter. The PV inverter thus performs mitigation of load reactive power and current harmonics; real power injection, and dc bus voltage control. Hence, on the grid side, the GPV system mitigates the current related issues like low power factor and distortion. The incorporation of a load power tracking block in the control approach enables accurate tracking of the fluctuating load requirements. Thus, the inverter is effectively regulated to provide active and reactive power in accordance with the load conditions.

This multifunctional control strategy necessitates the d-q frame transformation of grid voltages v_{gryb}, virtual DSTATCOM currents i_{sryb}, and load currents i_{lryb} [32] using the reference angle 'θ' produced by the phase-locked loop. The required instantaneous real power P and reactive power Q are calculated as in [34] by utilizing the d-q constituents of network voltage v_{gd}, $v_{gq,}$ and load currents i_{ld}, i_{lq}.

The current references, i_{sdCref} and i_{sqCref}, are produced in order to realize dual-stage operation.

4.1 AF Mode

The AF operating mode references are:

$$\begin{bmatrix} i_{sdref} \\ i_{sqref} \end{bmatrix} = \frac{2}{3}\frac{1}{\left(v_{gd}{}^2 + v_{gq}{}^2\right)}\begin{bmatrix} v_{gd} & v_{gq} \\ v_{gq} & -v_{gd} \end{bmatrix}\begin{bmatrix} \tilde{P} \\ Q \end{bmatrix} \tag{1}$$

Thus, the inverter functions like an active filter for compensating the load reactive power Q and mitigating load current distortion. However, the network grid must supply the fundamental part of the total load active power requisite P.

4.2 RPI Mode

The RPI operating mode references are:

$$\begin{bmatrix} i_{sdref} \\ i_{sqref} \end{bmatrix} = \frac{2}{3}\frac{1}{\left(v_{gd}{}^2 + v_{gq}{}^2\right)}\begin{bmatrix} v_{gd} & v_{gq} \\ v_{gq} & -v_{gd} \end{bmatrix}\begin{bmatrix} P \\ Q \end{bmatrix} \tag{2}$$

In this operating mode, the VSI provides the solar power produced during the prevailing insolation and temperature conditions. Consequently, the network must meet the balance of load real power necessity alone [35]. In addition, the PV inverter provides the whole reactive power requirement Q of the load. Thus, in RPI mode, there is a lessening in utility network active and reactive power burden.

For guaranteeing dc bus voltage control, the reference current i_{sdref} obtained above is modified by adding the d-axis current constituent required to retain the dc bus voltage at the preferred value V_{dcref}.

$$i_{sdCref} = i_{sdref} + \left(K_{1p'} + \frac{K_{1i'}}{s}\right)\left(V_{dcref} - V_{dc}\right) \tag{3}$$

$$i_{sqCref} = i_{sqref} \tag{4}$$

A comparison of the inverter current references and the real values is performed. Thus, the active and reactive powers are regulated with the help of the current mode control of the PV inverter.

5 Simulation Results and Discussions

Performance assessment is done using simulation studies under MATLAB/Simulink platform by considering an uncompensated system and a three-phase GPV system [34, 35] having non-linear and linear loads. The simulated results comprise the 'R' phase parameters, namely, the grid current I_{gr}, load current I_{lr}; grid, load, and VSI injected active and reactive power. The superiority of the enhanced SRF theory-based controller in achieving harmonics compensation as well as power flow control is investigated under different scenarios. The results obtained are compared with the uncompensated system as well as with the GPV system with a modified IRP theory-based control approach in α-β frame [34] for the following cases:

5.1 Case 1: Steady-State System Performance in Active Filter Mode of Operation

The uncompensated system [36] comprises the three-phase utility grid with a linear 6 kVA reactive load and 14 kVA non-linear diode rectifier load. Figure 1 depicts the uncompensated system response during these conditions. It can be observed that the load current distortion also reflects in the system current, and the THD value mentioned in Table 1 is 21.87%. Moreover, the utility grid supplies the whole load's active and reactive power requirements.

The GPV system simulation is carried out for insolation of 180 W/m^2 with the same loads at the PCC [37]. During such low irradiance situations, the GPV system functions in the AF mode. Figure 2 depicts the load current waveforms with the enhanced SRF theory-based control approach. It can be perceived that the grid current remains sinusoidal with a THD of 4.5% as indicated in Table 1, in spite of the existence of extremely non-linear loads.

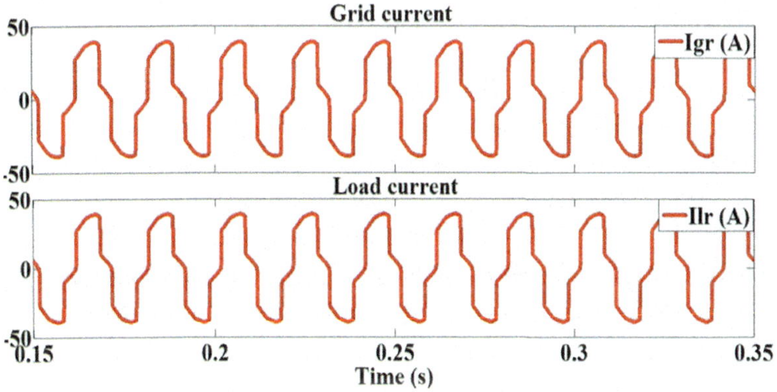

Fig. 1 Uncompensated system response during constant load conditions

Table 1 THD comparison in AF mode of operation

Parameter	Uncompensated system THD (%)	GPV system with enhanced SRF theory-based Control THD (%)
Grid current	21.87	4.5
Load voltage	1.12	1.12
Load current	21.87	22.14

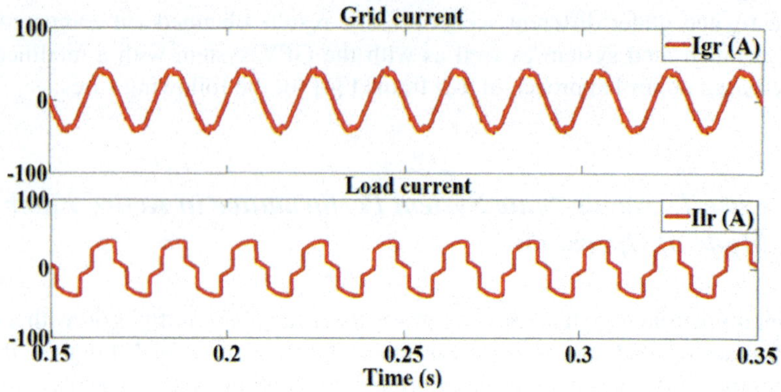

Fig. 2 Response of GPV system in AF mode of operation

Figure 3a, b demonstrate that in the AF mode, the GPV system injects the load reactive power requirement alone and does not inject any active power. The network has to supply the entire load active power requirement as depicted in Fig. 4a. Figure 4b demonstrates that both control strategies totally decrease the grid reactive power burden to zero; yet, the enhanced SRF theory-based controller has faster performance [38]. The comparison of the power-sharing and grid power burden reduction for the GPV system with both control schemes is tabulated as per unit basis in Table 2 as Case 1. The corresponding percentage reduction in grid power burden relative to the uncompensated system is specified in Table 3 (Case 1).

It can be concluded that in the AF operating stage, the enhanced SRF theory-based control approach compensates for load current harmonics and reactive power; also entirely diminishes the grid reactive power requirement. Thus, the system currents remain sinusoidal with THD in the ranges stipulated by the IEEE 519 standard.

5.2 *Case 2: Dynamic System Performance in Renewable Power Injection Mode of Operation Under Varying Insolation Conditions*

Simulation of the GPV system is performed for dynamically varying insolation conditions with the same loads. The insolation is 400 W/m^2 during 0–0.25 s. During 0.25–0.5 s, the insolation abruptly changes to 600 W/m^2. Under such varying insolation, the GPV configuration functions in the RPI state.

Figure 5a, b depict the load, VSI active and reactive power, respectively, of the GPV configuration using both controllers. It is evident that though both approaches provide load reactive power compensation, the enhanced SRF method possesses the superior capacity to inject maximum active power from the PV array during varying insolation.

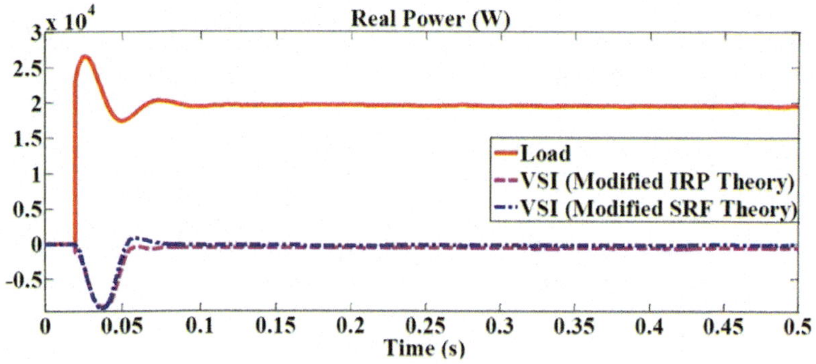

(a)

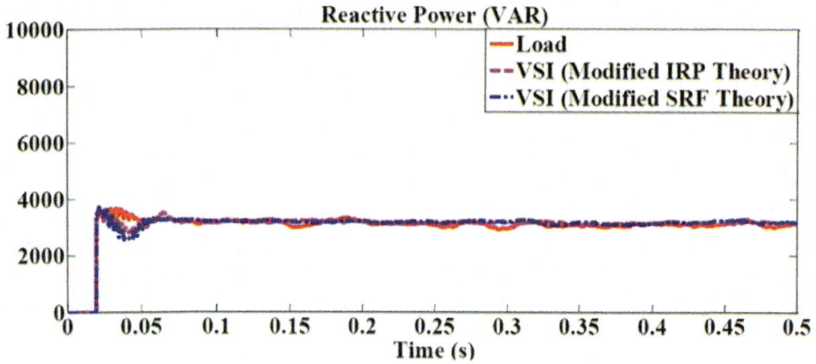

(b)

Fig. 3 GPV system load and VSI power in AF mode of operation: **a** Real power, **b** Reactive power

Thus, the utility grid active power burden decreases more for the enhanced SRF theory-based controller compared with the modified IRP theory-based controller and the uncompensated system as depicted in Fig. 6a. Figure 6b demonstrates that both strategies totally diminish the grid reactive power burden on comparison with the uncompensated network. Nevertheless, the enhanced SRF controller has the benefit of a quicker response. The aforementioned Table 2 (Case 2) summarizes the comparison of the GPV system power sharing and grid power burden reduction using both methods under changing insolation conditions. The corresponding percentage reduction in grid power burden relative to the uncompensated system is specified in the aforementioned Table 3 (Case 2).

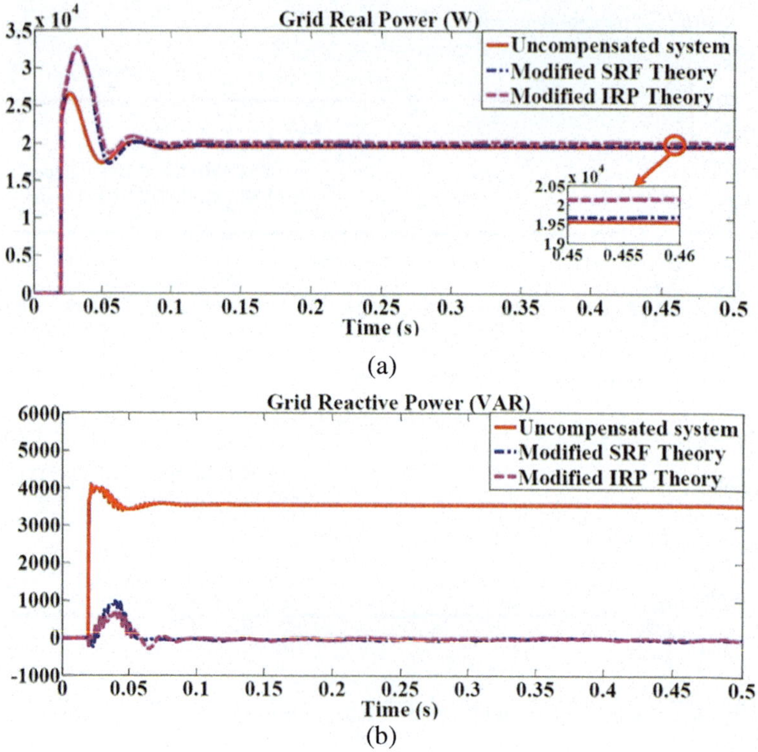

Fig. 4 Comparison of uncompensated and GPV system grid power in AF mode of operation: **a** Grid real power, **b** Grid reactive power

Table 2 Comparison of power sharing and reduction in grid power burden for GPV system

Case	Insolation (W/m^2)	Real power sharing				Reactive power sharing			
		Modified IRP theory		Modified SRF theory		Modified IRP theory		Modified SRF theory	
		Grid	PV-VSI	Grid	PV-VSI	Grid	PV-VSI	Grid	PV-VSI
1	180	1.03	−0.03	1	0	0	1	0	1
2	400	0.63	0.37	0.61	0.39	0	1	0	1
	600	0.43	0.57	0.41	0.59	0	1	0	1

Table 3 Percentage reduction in grid power burden relative to the uncompensated network

Case	Insolation (W/m^2)	% Reduction in power burden of grid relative to uncompensated system			
		Modified IRP theory		Modified SRF theory	
		Real	Reactive	Real	Reactive
1	180	0	100	0	100
2	400	36.73	100	38.77	100
	600	57.08	100	58.88	100

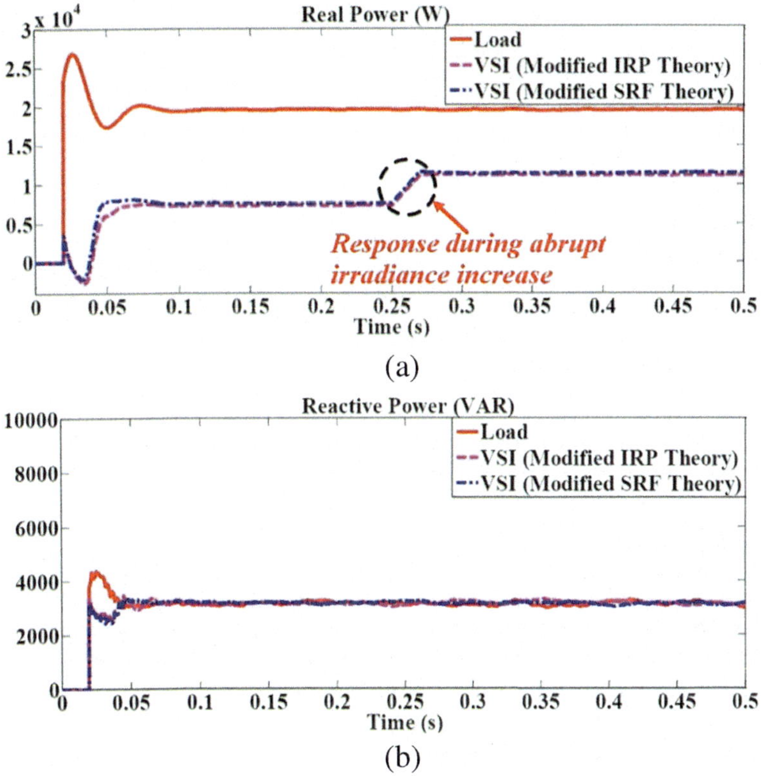

Fig. 5 GPV system load and VSI power in RPI mode of operation: **a** Real power, **b** Reactive power

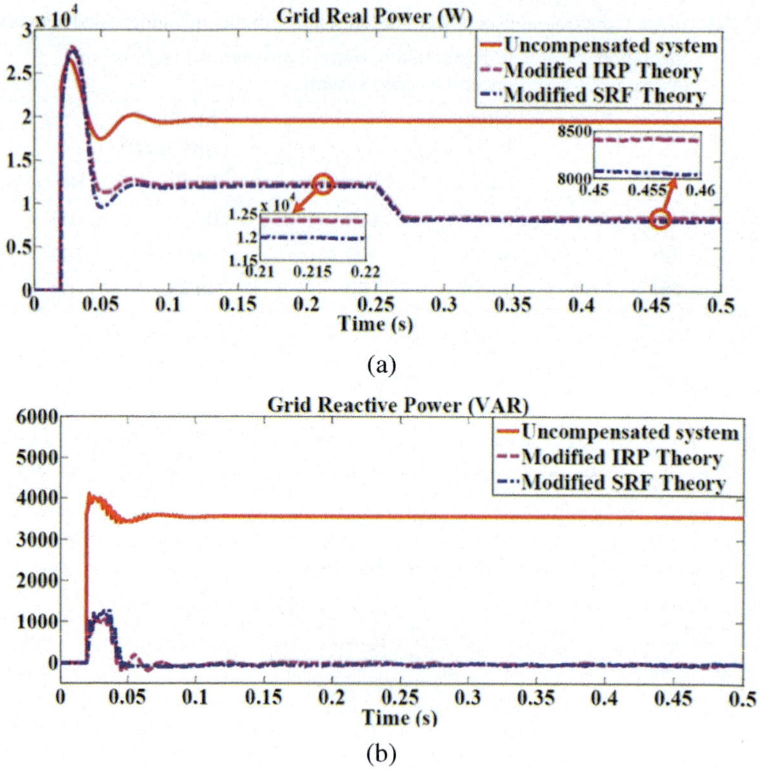

Fig. 6 Comparison of uncompensated and GPV system grid power in RPI mode of operation: **a** Grid real power, **b** Grid reactive power

6 Conclusion

An enhanced SRF theory-based multifunctional control scheme is presented in this chapter, which fulfills the objectives of power quality augmentation along with active and reactive power regulation in GPV systems with fast tracking of the changing insolation conditions. The operation encompassing AF and RPI stages imparts improved functionality for the GPV configuration sans any extra hardware requirement. The simulation results validate the exceptional working of the control approach. The enhanced controller possesses the important benefits of simplicity, easy realization, and fast response, which makes it suitable for extending applications to other DG units, even hybrid ones.

References

1. IEEE Standard 1100-2005, IEEE recommended practice for powering and grounding electronic equipment 1–703. https://doi.org/10.1109/IEEESTD.2006.216391
2. Hingorani NG (1995) Introducing custom power. Proc IEEE Spectr 32:41–48. https://doi.org/10.1109/6.387140
3. Arrillaga J, Bollen MHJ, Watson NR (2000) Power quality following deregulation. Proc IEEE 88:246–261. https://doi.org/10.1109/5.824002
4. Stones J, Collinson A (2001) Power quality. Power Eng J 15:58–64. https://doi.org/10.1049/pe:20010201
5. Dugan RC, McGranaghan MF, Beaty HW (2006) Electric power systems quality, 2nd ed. McGraw Hill
6. Hu R, Wang W, Wu X, Chen Z, Jing L, Ma W, Zeng G (2022) Coordinated active and reactive power control for distribution networks with high penetrations of photovoltaic systems. Sol Energy 231:809–827. https://doi.org/10.1016/j.solener.2021.12.025
7. Macken KJP (2004) Control of inverter–based distributed generation used to provide premium power quality. In: IEEE 35th annual PESC conference, vol 4. Aachen, Germany, pp 3188–3194. https://doi.org/10.1109/PESC.2004.1355345
8. Ackermann T, Andersson G, Soder L (2001) Distributed generation: a definition. Electr Power Syst Res 57:195–204. https://doi.org/10.1016/S0378-7796(01)00101-8
9. Blaabjerg F, Chen Z, Kjaer SB (2004) Power electronics as efficient interfacc in dispersed power generation systems. IEEE Trans Power Electron 19:1184–1194. https://doi.org/10.1109/TPEL.2004.833453
10. Bansal AK (2022) Sizing and forecasting techniques in photovoltaic-wind based hybrid renewable energy system: a review. J Clean Prod 369:133376. https://doi.org/10.1016/j.jclepro.2022.133376
11. Gawhade P, Ojha A (2021) Recent advances in synchronization techniques for grid-tied PV system: a review. Energy Rep 7:6581–6599. https://doi.org/10.1016/j.egyr.2021.09.006
12. Teng Y, Deng W, Pei W, Li Y, Dingv L, Ye H (2022) Review on grid-forming converter control methods in high-proportion renewable energy power systems. Global Energy Interconnect 5:328–342. https://doi.org/10.1016/j.gloei.2022.06.010
13. Bajaj M, Singh AK (2020) Grid integrated renewable DG systems: a review of power quality challenges and state-of-the-art mitigation techniques. Int J Energy Res 44:26–69. https://doi.org/10.1002/er.4847
14. Li D, Wang T, Pan W, Ding X, Gong J (2021) A comprehensive review of improving power quality using active power filters. Electric Power Syst Res 199:107389. https://doi.org/10.1016/j.epsr.2021.107389
15. Ghosh A, Ledwich G (2002) Power quality enhancement using custom power devices. Kluwer Academic Publishers
16. International Energy Agency, Renewables 2020, IEA, Paris, France. https://www.iea.org/reports/renewables-2020
17. IEEE Standard 929-2000, IEEE Recommended Practice for Utility Interface of Photovoltaic (PV) Systems. https://doi.org/10.1109/IEEESTD.2000.91304
18. Iyer S, Ghosh A, Joshi A (2005) Inverter topologies for DSTATCOM applications—a simulation study. Electr Power Syst Res 75:161–170. https://doi.org/10.1016/j.epsr.2005.02.003
19. Chacko FM, Jayan MV, Prince A (2018) Load harmonics extraction based decoupled control of grid connected solar photovoltaic system. IOP Conf Ser: Mater Sci Eng 396:1–8. https://doi.org/10.1088/1757-899X/396/1/012049
20. Singh B, Solanki J (2009) A comparison of control algorithms for DSTATCOM. IEEE Trans Ind Electron 56:2738–2745. https://doi.org/10.1109/TIE.2009.2021596
21. Timbus A, Liserre M, Teodorescu R, Rodriguez P, Blaabjerg F (2009) Evaluation of current controllers for distributed power generation systems. IEEE Trans Power Electron 24:654–664. https://doi.org/10.1109/TPEL.2009.2012527

22. Molina MG, Mercado PE (2008) Modeling and control of grid-connected photovoltaic energy conversion system used as a dispersed generator. IEEE/PES Trans Distrib Conf Exposition 1–8 https://doi.org/10.1109/TDC-LA.2008.4641871
23. Tummuru NR, Mishra MK, Srinivas S (2014) Multifunctional VSC controlled microgrid using instantaneous symmetrical components theory. IEEE Trans Sustain Energy 5:313–322. https://doi.org/10.1109/TSTE.2013.2283739
24. Singh B, Jain C, Goel S, Gogia R, Subramaniam U (2017) A sustainable solar photovoltaic energy system interfaced with grid-tied voltage source converter for power quality improvement. Electric Power Compon Syst 45:171–183. https://doi.org/10.1080/15325008.2016.1233298
25. Hamrouni N, Jraidi M, Dhouib A, Cherif A (2017) Design of a command scheme for grid connected PV systems using classical controllers. Electr Power Syst Res 143:503–512. https://doi.org/10.1016/j.epsr.2016.10.064
26. Singh B, Jain C, Goel S, Chandra A, Al-Haddad K (2016) A multifunctional grid-tied solar energy conversion system with ANF-based control approach. IEEE Trans Ind Appl 52:3663–3672. https://doi.org/10.1109/TIA.2016.2582141
27. Verma AK, Singh B (2018) Harmonics and reactive current detection of a grid interfaced PV generation in a distribution system. IEEE Trans Ind Appl 54:4786–4794. https://doi.org/10.1109/TIA.2018.2830752
28. Singh B, Shahani DT, Verma AK (2014) Neural network controlled grid interfaced solar photovoltaic power generation. IET Power Electron 7:614–626. https://doi.org/10.1049/iet-pel.2013.0166
29. Jain C, Singh B (2017) An adjustable dc link voltage based control of multifunctional grid interfaced solar PV system. IEEE Trans Emerg Sel Top Power Electron 5:651–660. https://doi.org/10.1109/JESTPE.2016.2627533
30. Agarwal RK, Hussain I, Singh B (2016) LMF-based control algorithm for single stage three-phase grid integrated solar PV system. IEEE Trans Sustain Energy 7:1379–1387. https://doi.org/10.1109/TSTE.2016.2553181
31. Singh Y, Hussain I, Mishra S, Singh B (2017) Adaptive neuron detection-based control of single-phase SPV grid integrated system with active filtering. IET Power Electron 10:657–666. https://doi.org/10.1049/iet-pel.2016.0613
32. Naga Lakshmi GV, Jaya Laxmi A, Veeramsetty V, Salkuti SR (2022) Optimal placement of distributed generation based on power quality improvement using self-adaptive Lévy flight Jaya algorithm. Clean Technol 4(4):1242–1254. https://doi.org/10.3390/cleantechnol4040076
33. Schauder C, Mehta H (1993) Vector analysis and control of advanced static VAR compensators. IEE Proc C (Gener, Trans Distrib) 140:299–306. https://doi.org/10.1049/ip-c.1993.0044
34. Chacko FM, George GA, Jayan MV, Prince A (2020) Improved multifunctional controller for power quality enhancement in grid integrated solar photovoltaic systems. World J Eng 17:585–598. https://doi.org/10.1108/WJE-12-2019-0356
35. Subhash B, Rajagopal V, Salkuti SR (2022) Optimization of controller gains to enhance power quality of standalone wind energy conversion system. Int J Emerg Electr Power Syst 23(1):89–104. https://doi.org/10.1515/ijeeps-2021-0024
36. Yazdani A, Iravani R (2010) Voltage-sourced converters in power systems: modeling, control and applications. Wiley-IEEE Press
37. Das SR, Mishra AK, Ray PK, Salkuti SR, Kim SC (2022) Application of artificial intelligent techniques for power quality improvement in hybrid microgrid system. Electronics 11(22):3826. https://doi.org/10.3390/electronics11223826
38. Chacko FM, Jayan MV, Prince A (2021) Voltage harmonics-based islanding detection for grid-tied photovoltaic systems. In: IEEE fourth international conference on electrical, computer and communication technologies (ICECCT), pp 1–5. https://doi.org/10.1109/ICECCT52121.2021.9616663

Adaptive Filtering for Power Quality Features with Optimized PI Gains in Four Wires UPQC

Sabha Raj Arya, Sayed Javed Alam, Rajasekhara Reddy Chilipi, and Papia Ray

Abstract The Unified Power Quality Conditioner (UPQC) is the utmost competitive apparatus. It can simultaneously mitigate the power quality (PQ) disturbances related to voltage and current on both ends of the line. This chapter offers a unique solution for PQ improvement by using Recursive Least Square (RLS) algorithm in the UPQC control strategy. The RLS algorithms have broad applications in many fields, such as processing, controlling, and communicating real-time signals. So, this chapter deals with the noise cancellation property of the RLS technique in extracting the Fundamental Components (FC) to develop a reference signal for the UPQC controller. The result depicts RLS algorithm in the UPQC system performs better in canceling several types of PQ issues. To shorten the tuning time of the Proportional Integral (PI) controller gains, a nature-inspired Spider Monkey Optimization (SMO) algorithm is implemented. A better exploration and exploitation of SMO leads to the prevention of local minima. The action takes place inside each group during the convergence process. The UPQC system, combined with the RLS control algorithm and SMO-optimized PI gains, is developed using a MATLAB environment. It is examined in simulation and hardware for different PQ perturbations in the power distribution line.

Keywords Voltage source converter · Recursive Least Square · Reference signal generation · Active noise cancellation · Optimized PI controller gains · SMO · UPQC

S. R. Arya · S. J. Alam · R. R. Chilipi
Department of Electrical Engineering, Sardar Vallabhai National Institute of Technology, Surat 395007, India

P. Ray (✉)
Department of Electrical Engineering, Veer Surendra Sai University of Technology, Burla, Sambalpur, Odisha, India
e-mail: papiaray_ee@vssut.ac.in

S. R. Salkuti et al. (eds.), *Power Quality in Microgrids: Issues, Challenges and Mitigation Techniques*, Lecture Notes in Electrical Engineering 1039,
https://doi.org/10.1007/978-981-99-2066-2_7

Nomenclature

UPQC	Unified Power Quality Conditioner
RLS	Recursive Least Square
FC	Fundamental Components
PI	Proportional Integral
SMO	Spider Monkey Optimization
PQ	Power Quality
DSTATCOM	Distribution Static Compensator
DVR	Dynamic Voltage Restores
LMS	Least Mean Squares
GA	Genetic Algorithm
SI	Swarm Intelligence
PSO	Particle Swarm Optimization
AVR	Automatic Voltage Regulator
SPWM	Sinusoidal Pulse Width Modulation Controller
PCC	Point of Common Coupling
PID	Proportional Integral Derivative
MG	Maximum number of Groups
PR	Perturbation Rate
ITAE	Integral Time Absolute Error
LLP	Local Leader Phase
GLP	Global Leader Phase
LLL	Local Leader Learning
GLL	Global Leader Learning
LLD	Local Leader Decision
GLD	Global Leader Decision
IGBT	Insulated Gate Bipolar Transistor
THD	Total Harmonic Distortion
DIO	Discrete Input Output
CH	Channel
RMS	Root Mean Square
t_r	Rise time
M_p	Peak overshoot

1 Introduction

The recent broad use of power electronic equipment in the distribution power grid has created a substantial surge in harmonic distortion, voltage fluctuations, and uncontrolled reactive power flow [1]. These disturbances cause problems at the consumer ends like capacitor blowing, neutral currents in excess, overheating, and low-value power factors [2]. To work correctly, the consumer ends equipment requires clean electricity. A detailed survey has been carried out to estimate the power line problems with non-linear loads [3]. Few active and passive filtration methods are developed to overcome the harmonic distortion in the distribution system [4]. However, due to cost concerns, installing passive filters is not very beneficial. Some issues remain with passive harmonic filters, such as aging components, compensation for a fixed quantity, bulky size, and problems regarding resonance conditions, etc. Active power filters with self-supporting dc buses appear to be a viable solution for addressing PQ issues generated by connected load [5].

Several techniques have been implemented to alleviate PQ problems, which can be done using Distribution Static Compensator (DSTATCOM) [6], Dynamic Voltage Restores (DVR) [7] and UPQC [8]. The DSTATCOM is placed in parallel with nonlinear loads, employed to compensate for current harmonics, imbalance in loads, and compensation of reactive load power [9]. Similarly, DVRs, placed in series with the utility and load, can compensate for harmonic voltages, load unbalanced, and voltage fluctuations while the load voltages are regulated at a specified level [10]. On the other hand, the UPQC system can operate DSTATCOM and DVR simultaneously to mitigate utility and load-side PQ problems [11]. The proper function of UPQC depends on the reference load voltage and source current generation followed by the respective gates' pulse sequences. Different control strategies are available in the literature for producing the FC through noisy signals [12]. The authors have presented an overview of various adaptive filtering algorithms for noise cancellation [13].

Adaptive filtering algorithms have abundant applications in biomedical engineering, telecommunications, sonar, radar, and speech processing; they are used for adaptive noise cancellation [14]. The adaptive filtering algorithm has two main classes: Recursive Least Square (RLS) and Least Mean Squares (LMS). LMS algorithm is commonly known to be small computational and vulnerable to variations in the exclusive distribution of the input vector cost correlation matrix [15]. But the RLS algorithms tentatively offer the least square most favorable output at every step/iteration with a low convergence rate option [16]. There are algorithms other than these, such as Affine Projection Algorithm [17] and Fast Newton Transversal Filter Algorithm [18], which lie in between LMS and RLS algorithms regarding the convergence properties and effective computational cost [19]. The coefficients in RLS filters are arranged continuously and directly step-wise during the filtration process [20]. It has an adaptation algorithm that allows it to change parameters for better results. Its purpose is to keep an eye on the surroundings and change the filter switch. It recursively locates filter coefficients that lower the weighted linear least square cost function corresponding to the input signals inside the input data using inverse correlation matrix data. The key properties of the RLS algorithm relevant to the single-channel adaptive filtering problem of the Nth order are used here to mitigate PQ disturbances. Another consideration for RLS is algorithmic stability. Any algorithm is unacceptable if it diverges unexpectedly from the real solution. Still, the study shows that the RLS is not divided and maintains stability for active noise cancellation problems by generating the adaptive filter weights [21]. Based on low computational cost, RLS's rapid convergence and active noise cancellation property are selected in the UPQC controller to estimate fundamental components.

Because of its straightforward construction, user-friendly functionality, and robustness, the proportional integral derivative (PID), the regulator is frequently employed in the design of controller strategies. The controller's primary function is to execute an input-based algorithm and maintain output at a level so that the process variable and the set point do not differ [22]. However, the PID controller's disadvantage is properly selecting its gains. Output such as rise time (t_r), peak overshoot (M_p), time settlement (t_s), and steady-state error (e_{ss}) depend on the controller's gains [23]. Hit and trial or manual calibration of the PID controller's gains does not boost system performance. Ziegler-Nichols, Astrom, Cohen-Coons and Hagglund tuning methods were some of the initial tuning techniques for PID control [24]. However, these techniques had to apply a linear analytical procedure, resulting in larger phase lag, higher overshoot, and incomplete tuning performance [25].

On the other hand, the evolutionary tuning techniques provide better outcomes in PID tuning than primitive methods [26], the Genetic Algorithm (GA), which contributes to an efficient solution to these problems. But later, Swarm Intelligence (SI) based techniques were introduced for designing the PID parameters to produce promising results [27]. These SI techniques provide efficient, positive feedback characteristics, hierarchical search mechanism, and constructive greedy heuristics for more accessible, more accurate, and computationally speedy tuning than conventional and evolutionary methods. A positive feedback search produces good results, as in [28], where PID controller design for the sun tracking system using SI is presented. Similarly distributed search mechanism used to avoid premature convergence is shown in [29], where the simplified Particle Swarm Optimization (PSO) method is applied to plan a PID controller for an Automatic Voltage Regulator (AVR). These algorithms, however, are very successful for any design problem but have various limitations, such as dependency on initial parameters, slow convergence, weak communication between each community member, poor exploitation mechanism, and getting stuck in local minima. To overcome these limitations, a relatively new SI algorithm has been offered by [30] called Spider Monkey Optimization (SMO), whose fission-fusion social behaviour develops the SMO algorithm, and lesser control parameters are required to be tuned [31]. The group's self-organization and division are essential features of SMO to avoid local solutions with better exploration and exploitation inside each group during the convergence process [32]. The greedy heuristic behaviour of spider monkeys towards food has shown results helpful in finding the solution at the early stages of the exploration process. This activity is well used in [33] for tuning the PI controller gains in Perturb and Observe (P&O) method via the SMO algorithm for the Photovoltaic generation system. A similar approach has been applied here to optimize PI controller gains in the UPQC system using the SMO algorithm. A performance like t_s, M_p, t_r and e_{ss} is improved with optimized PI controller gains.

This chapter aims to implement the mitigation strategy using a four-wire UPQC topology for a 3-phase 4-wire distribution system with a nonlinear load. The UPQC control strategy will regulate power quality, such as regulating output voltage, suppressing harmonic load currents, compensating reactive load power, compensating load unbalances, suppressing utility harmonic voltage, and compensating grid voltage unbalances and sag distribution system. After a brief review of the adaptive filter, the RLS method is used for the UPQC controller to estimate FC and investigate different PQ problems. The RLS algorithm is based on both interpolation and prediction. Recursively find the internal filter coefficients that minimize the most minor square cost function weighted linear. It maps the input signals within the input dataset using inverse correlation matrix data. The literature shows that it converges faster than LMS, NLMS, and affine projection algorithms. Its noise cancellation capacity is the best one and also more stable. The difference between the final and initial Signal Noise Ratio (SNR) is higher in RLS than in other algorithms mentioned [34]. It also cancels maximum noise by minimizing error with the rapid step size rate. The UPQC control strategy with the RLS algorithm will provide input currents harmonic free and in phase with supply voltage. So, there is no need for a phase lock loop to grid synchronization and no need for the transformation required to estimate fundamental components. This chapter also uses of SMO technique in the offline approach to demonstrate SMO capacity in obtaining the PI controller gains on the UPQC control strategy to mitigate power quality. Compared with Hit and Trail tuning techniques, the consequence is provided at different time-domain specifications such as M_p, t_s, t_r of ac and dc PI controller gains at various PQ disturbances. This improves both the steady-state and dynamic performance of the system. The RLS control algorithm and SMO technique are efficient enough to mitigate the PQ problems and provide smooth steady-state and dynamic responses for the UPQC system. Performance-based simulation through MATLAB and hardware through d-SPACE Micro Lab Box demonstrates the developed control's efficiency under non-linear load.

2 Description of UPQC

The UPQC is an electronic-based custom power device used to mitigate current and voltage harmonic simultaneously at the common coupling point with an injection of equal and opposite compensation signals. It eliminates harmonic from the polluted line and mitigates other voltage and current-related power quality (PQ) disturbances. Three-phase four-wire UPQC depicted in Fig. 1 is connected with 3 different 1-phase nonlinear loads of the same rating. It is implemented with two voltage source converters (VSC) energy storage capacitors (C_{dc}) connected back-to-back. Each VSC contains six IGBT with anti-paralleling diodes. The VSC, coupled with the supply line, acts as a VSC Series voltage-controlled device.

Similarly, the VSC connected in the shunt with the load acts as a current-controlled device called VSC Shunt. VSC Series mainly generates the compensating voltage to compensate voltages via a series-connected transformer to the main line. Likewise, the VSC Shunt aims to develop compensating current to mitigate harmonics for PQ improvement and reactive power. The L_{se} and L_{sh} interfacing inductors connect series

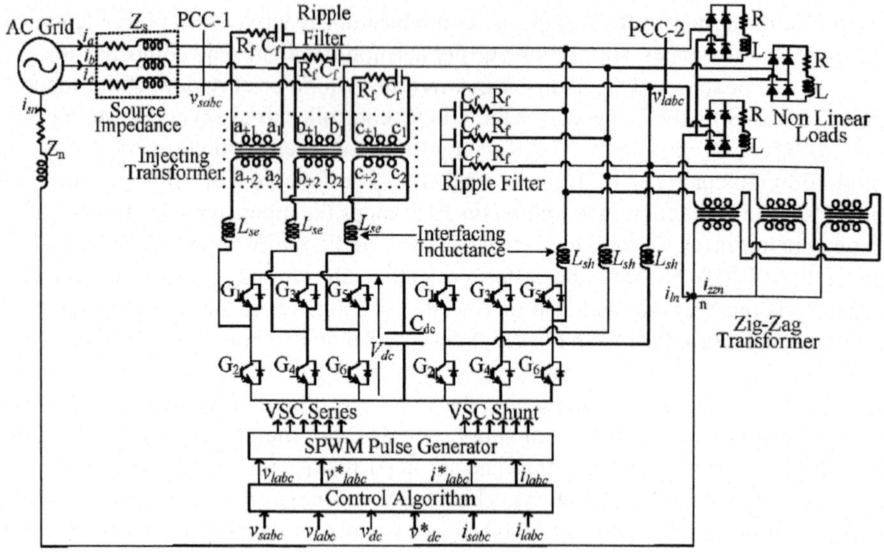

Fig. 1 UPQC layout

and shunt VSC, respectively, to the power line. The ripple filter (R_f, C_f) eliminates switching noise generated by VSCs. Zs stands for the source impedance of a 3-phase line, while Zn stands for the neutral line impedance in a 3-phase and 4-wire system. Using a neutral current, a transformer (zig-zag) is available to compensate for any ground-related defects.

3 Description of Control Algorithm

The control algorithm applied to the designed system here has two vital components. To produce reference load voltage and source current signal, one must extract FC using the RLS method under various PQ conditions. Another is to estimate the optimized PI controller gains value using the SMO technique to enhance the UPQC controller so that the DC link and terminal link voltage at the load side show stable operation [35]. The control algorithm layout is depicted in Fig. 2. Complete control strategy of the proposed system involves Fundamental Magnitude extraction using RLS Algorithm, Reference signal generation using the RLS algorithm for UPQC VSCs, and PI-controller gains estimation using SMO. Each block of the control strategy is addressed in the following sub-section. It is noted that the line-to-line signal is sensed and transformed into line to ground signal for the control process.

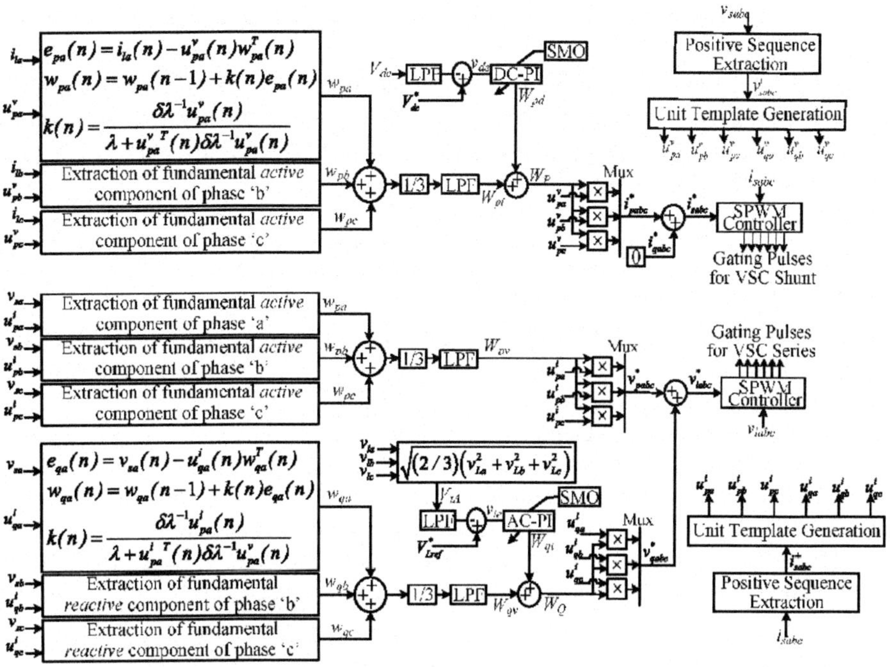

Fig. 2 Layout of reference signal generation with RLS algorithm for UPQC VSCs

3.1 Fundamental Magnitude Extraction Using RLS Algorithm

The RLS adaptive filter is the most effective algorithm for system identification. The RLS adaptive filter can change its coefficients using customizable variables and never requires prior noise information. The RLS algorithm generating adaptive filter weights needs iteration square operations per time step for active noise cancellation. It is fundamentally much more robust than transverse filters and operates consistently with single-precision arithmetic over several millions of iterations [36]. The signal between desired signal and output after filtration is known as the error signal. It acts as a feedback to reduce the error through some adaptation criterion. The algorithm is comparatively much more adaptive for signals those are not stationary and it filters convergence through a computationally complex method. To evaluate weights $w(n)$ of the filter, the algorithm is used to analyze reference data $u(n)$ and primary signal $x(n)$ for every iteration. The approximated weights are chosen so that the RLS filters real output $y(n)$, similar to the interference components of primary signals. Likewise, output components of adaptive noise removal filter error $e(n) = x(n) - y(n)$ which is minimized in estimating the desired signals. Where $y(n) = w^H(n)u(n)$ and

$$w(n) = w(n-1) + k(n)\big(x(n) - w^T(n-1)u(n)\big) \tag{1}$$

where $w^T(n)$ is the transpose position of weight vector (1) premeditated for current iteration n, $u(n)$ is an nth column of reference data. It adapts the weight $w(n)$ adaptively to diminish error, the steps of the RLS algorithm with filter coefficient vector gain $k(n)$ are specified by,

$$k(n) = \frac{p(n-1)\lambda^{-1}u(n)}{\lambda + u^T(n)p(n-1)\lambda^{-1}u(n)} \tag{2}$$

where λ denotes the forgetting factor and indicates the memory of the algorithm. The matrix $P(n)$ determines the inverse correlation matrix of the supply signal, which is given at the starting point as $p\ (0) = I\delta$, where I determined the identity matrix and δ is the regularization coefficient.

Three parameters must be initialized in the RLS algorithm to get $w(n)$: the input correlation matrix $P(0)$, the regularization coefficient δ, and the forgetting factor λ used to remove the older data, and maximum importance will be given to the new data to increase monitoring of time-varying system data. The λ set between 0.95–0.999 is based on the principle of asymptotic sample length [20, 21].

3.2 Reference Signal Generation using RLS Algorithm for UPQC

The RLS noise cancellation principle for the reduction of harmonics is used here. Basic signal processing operations can obtain the required signal from any noisy signal. The primary task is to estimate the fundamental components (FC) or weight from the harmonic polluted power line signals (current and voltages). These filters' output will provide FC termed $w(n)$ as a weight vector. Similarly, the general steps of the RLS algorithm are written in Eqs. (1) and (2) are used here for the estimation of in-phase components (active weight components) of "phase a" from load current as follows.

$$e_{pa}(n) = i_{la}(n) - u^v_{pa}(n)w^T_{pa}(n) \tag{3}$$

$$w_{pa}(n) = w_{pa}(n-1) + k(n)e_{pa}(n) \tag{4}$$

$$k(n) = \frac{\delta\lambda^{-1}u^v_{pa}(n)}{\lambda + \left(u^v_{pa}(n)\right)^T\delta\lambda^{-1}u^v_{pa}(n)} \tag{5}$$

Likewise, the FC for quadrature components (reactive weight components) of phase a from load current is calculated as follows,

$$e_{qa}(n) = i_{la}(n) - u_{qa}^{v}(n)w_{qa}^{T}(n) \tag{6}$$

$$w_{qa}(n) = w_{qa}(n-1) + k(n)e_{qa}(n) \tag{7}$$

$$k(n) = \frac{\delta\lambda^{-1}u_{qa}^{v}(n)}{\lambda + \left(u_{qa}^{v}(n)\right)^{T}\delta\lambda^{-1}u_{qa}^{v}(n)} \tag{8}$$

Similarly, the fundamental components of in-phase and quadrature weights of the sensed signals are calculated for the remaining phases (i.e., phases b and c), as shown in Fig. 2. However, average equivalent in-phase and quadrature components are needed for the reference creation of any signal, whether it be a voltage or current signal [37]. So, average weight components of each phase are averaged and pass through a low pass filter to obtain net in-phase (W_p) and quadrature (W_q) components as,

$$\begin{aligned} W_p &= \frac{w_{pa} + w_{pb} + w_{pc}}{3} \\ W_q &= \frac{w_{qa} + w_{qb} + w_{qc}}{3} \end{aligned} \tag{9}$$

3.2.1 Computation of Positive Sequence Estimation

The positive sequence grid voltage is required to estimate synchronized unit templates correctly during voltage harmonics conditions. Therefore, the use of the symmetrical components method is likely to acquire the symmetrical phasor components (V_+, V_-, V_0) of three-phase grid voltage signals (V_a, V_b, *and* V_c) from as,

$$\begin{bmatrix} V_+ \\ V_- \\ V_0 \end{bmatrix} = \frac{1}{3}\begin{bmatrix} 1 & 1\angle 120^\circ & 1\angle -120^\circ \\ 1 & 1\angle -120^\circ & 1\angle 120^\circ \\ 1 & 1 & 1 \end{bmatrix}\begin{bmatrix} V_a \\ V_b \\ V_c \end{bmatrix} \tag{10}$$

The positive sequence of voltage can be extracted in *abc* signals from the above Eq. (10) as,

$$\begin{bmatrix} V_a^+ \\ V_b^+ \\ V_c^+ \end{bmatrix} = -\frac{1}{3}\begin{bmatrix} -1 & 1\angle -60^\circ & 1\angle 60^\circ \\ 1\angle 60^\circ & -1 & 1\angle -60^\circ \\ 1\angle -60^\circ & 1\angle 60^\circ & -1 \end{bmatrix}\begin{bmatrix} V_a \\ V_b \\ V_c \end{bmatrix} \tag{11}$$

Operations Eq. (11) can be interpreted in various ways to achieve the same response if only fundamental frequency signal components are used in the phase signals. However, each operation modifies the harmonics provided in the original signals differently to get positive sequence components [38]. Therefore, the transformations used here to obtain the positive sequence components of the voltage [34] are as follows,

$$\begin{bmatrix} V_a^+ \\ V_b^+ \\ V_c^+ \end{bmatrix} = \frac{1}{3}\left[-1 \begin{bmatrix} -1 & \frac{1}{2} & \frac{1}{2} \\ \frac{1}{2} & -1 & \frac{1}{2} \\ \frac{1}{2} & \frac{1}{2} & -1 \end{bmatrix} \begin{bmatrix} V_a \\ V_b \\ V_c \end{bmatrix} + \begin{bmatrix} 0 & \frac{\sqrt{3}}{2} & -\frac{\sqrt{3}}{2} \\ -\frac{\sqrt{3}}{2} & 0 & \frac{\sqrt{3}}{2} \\ \frac{\sqrt{3}}{2} & -\frac{\sqrt{3}}{2} & 0 \end{bmatrix} \begin{bmatrix} V_{a90} \\ V_{b90} \\ V_{c90} \end{bmatrix} \right] \tag{12}$$

Instantaneous signals that are 90 degrees after the original ones are denoted by the subscript 90. To create synchronized unit templates, the measured grid voltage for the Shunt VSC is used to determine the positive order of supply voltages. Similarly, Series VSC's synchronized unit templates are created by bringing the supply current's positive sequence, as seen in Fig. 2.

3.2.2 Computation of Positive Sequence Estimation

The amplitude of the point of common coupling voltage (Vs) is needed to minimize the AC loss components under unbalanced circumstances, and it is calculated as,

$$V_s = \sqrt{\frac{2}{3}\left((v_{sa}^+)^2 + (v_{sb}^+)^2 + (v_{sc}^+)^2\right)} \tag{13}$$

Unit templates in-phase $(u_{pa}^v, u_{pb}^v, u_{pc}^v)$ and quadrature $(u_{qa}^v, u_{qb}^v, u_{qc}^v)$ are required to produce symmetrical components from point of common coupling (PCC); voltages of each phase are depicted in (14) and (15) respectively,

$$u_{pa}^v = \frac{v_{sa}}{V_s},\ u_{pb}^v = \frac{v_{sb}}{V_s},\ u_{pc}^v = \frac{v_{sc}}{V_s} \tag{14}$$

$$u_{qa}^v = \frac{-u_{pb}^v + u_{pc}^v}{\sqrt{3}},\ u_{qb}^v = \frac{3u_{pa}^v + u_{pb}^v - u_{pc}^v}{2\sqrt{3}},\ u_{qc}^v = \frac{-3u_{pa}^v + u_{pb}^v - u_{pc}^v}{2\sqrt{3}} \tag{15}$$

Similarly, the unit templates of the in-phase signals $(u_{pa}^i, u_{pb}^i, u_{pc}^i)$ and quadrature signals $(u_{qa}^i, u_{qb}^i, u_{qc}^i)$ from the supply current are generated with the help of peak magnitude current (I_s) obtained at the PCC for the Series VSC.

3.2.3 Computation of Loss Components

The VSC requires less power from the source to feed losses inside the converter. This keeps the terminal voltage and dc-link voltage constant. Its main goal was to increase the speed with which reference signals were generated. To sustain a good response, the VSC supplies energy during transient conditions. For proper reference signal generation to control the VSCs, the DC link voltage must be kept constant at a given reference voltage level. However, the switching losses and control speed of the currents of the VSCs are directly proportional to the difference in voltage between peak magnitude and dc-link voltage. Thus, error (v_{te}) in actual dc-link voltage (V_{dc}) and set reference dc-link voltage (V_{dc}^{*}) are compared. Generated v_{de} error is fed to the PI controller having gains k_{pi} and k_{ii} to maintain dc-link voltage as steady as possible. The yield of PI controller is referred to as DC loss components (W_{pd}) for both VSC and is measured for an nth sample as shown in Eq. (16).

$$W_{pd}(n+1) = W_{pd}(n) + k_{pi}\{v_{de}(n+1) - v_{de}(n)\} + k_{ii}v_{de}(n+1) \tag{16}$$

Similarly, the amplitude of PCC terminal voltage (V_{tA}) at the load side is intended for generating the AC loss components. It is necessary to sustain the PCC voltage at the load side so that the load voltages (V_{labc}) are steadily available at the load. The terminal voltage (V_{tA}) is calculated as,

$$V_{tA} = \sqrt{\frac{2}{3}\left(v_{la}^2 + v_{lb}^2 + v_{lc}^2\right)} \tag{17}$$

Therefore, the calculated amplitude terminal voltage (V_{tA}) is compared with the considered value of reference voltages (V_{Lref}^{*}) and the calculated error (v_{te}) is fed to the PI controller, which includes gains k_{pv} and k_{iv} to maintain the PCC terminal voltage as steady as possible. The generated v_{te} error at any nth sample is presented in Eq. (18).

$$v_{te}(n) = V_{Lref}^{*}(n) - V_{tA}(n) \tag{18}$$

and the PI controller output in reference to AC loss components (W_{qt}) is shown in Eq. (19). The W_{qt} is injected voltage by series VSC via injection transformer and evaluated by output of PI controller at AC bus which is as follows:

$$W_{qt}(n+1) = W_{qt}(n) + k_{pv}\{v_{te}(n+1) - v_{te}(n)\} + k_{iv}v_{te}(n+1)\} \tag{19}$$

where power W_{qt} is considered the reactive part of load voltage at time instant *n*, k_{pv} and k_{iv} are PI-Controller gains respectively at AC bus. The v_{te} is an error between the actual (V_{tA}) terminal voltage and reference (V_{Lref}^{*}) at load side PCC.

3.2.4 Computation of Loss Components

The reference source current formation comprises the input of only active components. Therefore, reference 3-phase in-phase active currents (i_{pabc}^{*}) is calculated as,

$$i_{pa}^{*} = W_{P} \times u_{pa}^{v},\ i_{pb}^{*} = W_{P} \times u_{pb}^{v},\ i_{pc}^{*} = W_{P} \times u_{pc}^{v} \quad (20)$$

where $i_{pa}^{*}, i_{pb}^{*} \& i_{pc}^{*}$ is the reference corresponding active components to fundamental source current estimated from in-phase unit templates. The W_p is the cumulative magnitude of the active elements of the supply reference current, given by $W_P = W_{pi} + W_{pd}$. For the supply current to be in phase with the supply voltage while maintaining zero sequence components in the system, reactive components of supply current are set to zero. Furthermore, reference source currents are measured by the addition of active and reactive components for each phase as,

$$i_{sa}^{*} = i_{pa}^{*} + 0,\ i_{sb}^{*} = i_{pb}^{*} + 0,\ i_{sc}^{*} = i_{pc}^{*} + 0 \quad (21)$$

Finally, the generated reference current and source current from PCC are compared to measure errors. The SPWM controller takes the error signal for generating the switching pulses for Shunt-connected VSC switches.

3.2.5 Computation of Reference Load Voltages Generation

The generation of reference load voltage needs the contribution of both active and reactive components. So, 3-phase in-phase active and quadrature reference load voltage are calculated as,

$$\begin{aligned} v_{pa}^{*} &= W_{pv} \times u_{pa}^{i},\ v_{pb}^{*} = W_{pv} \times u_{pb}^{i},\ v_{pc}^{*} = W_{pv} \times u_{pc}^{i} \\ v_{qa}^{*} &= W_{Q} \times u_{qa}^{i},\ v_{qb}^{*} = W_{Q} \times u_{qb}^{i},\ v_{qc}^{*} = W_{Q} \times u_{qc}^{i} \end{aligned} \quad (22)$$

where the W_{pv} is the total scale of reference load voltage active components $W_{pv} = (w_{pa} + w_{pb} + w_{pc})/3$. Likewise, the W_Q is the total scale of the active components of reference load voltage $W_Q = W_{qv} + W_{qt}$. Furthermore, the reference load voltage is considered by the addition of active and reactive components of individual phases as,

$$v_{la}^{*} = v_{pa}^{*} + v_{qa}^{*},\ v_{lb}^{*} = v_{pb}^{*} + v_{qb}^{*},\ v_{lc}^{*} = v_{pc}^{*} + v_{qc}^{*} \quad (23)$$

Finally, PCC, the three-phase load voltages and the generated reference current are compared to calculate the errors. The SPWM controller takes these error signals, as depicted in Fig. 2, for generating the switching pulses of Series VSC switches.

3.3 *PI Controller Gains Estimation Using SMO*

The SMO is employed to tune the parameters of the PI controller to enhance UPQC dynamic behaviour. The SMO algorithm is a meta-heuristic population-based approach encouraged by nature that is influenced by the social welfare structure of fission-fusion. Spider monkeys stay socially in a group of around 40–50 individuals. By separating themselves into subgroups, they drill at different sites in different directions and recombine to exchange the collected food. This monkey behaviour is used here to collect the optimized gains of PI controller. Premature convergence, stagnation problems, and balancing a healthy search space between exploitation and exploration are some of this algorithm's key characteristics. The Local Leader Limit (LLlim), Global Leader Limit (GLLlim), Maximum number of Groups (MG), and Perturbation Rate (PR) are 4 control parameters for SMO. The SMO algorithm flow map is seen in Fig. 3 and its separate operating phases are briefly discussed below to obtain the PI gains.

Optimal PI controller gains (k_{pi}, k_{ii}, k_{pv}, and k_{iv}) are evaluated with objective function (f_o) defined in Eq. (24) as a minimizing aim of the optimization function.

$$f_o = w_1 * ITAE_1 + w_2 * ITAE_2 \tag{24}$$

where w_1 and w_2 are the weights of $ITAE_1$ and $ITAE_2$ and are taken as 0.5. The ITAE denotes the integral time absolute error data taken from the input of both PI controllers and saved in the workspace setting. The $ITAE_1$ for the DC PI controller and $ITAE_2$ for the AC PI controller were allocated unknown variables as k_{pi}, k_{ii} and k_{pv}, k_{iv}, respectively. The Simulink model's workspace data (ITAE1 and ITAE2) are extracted during each iteration process and given to the algorithm for optimizing PI controller parameters. From Eqs. (16), (19) and (24), the SMO algorithm can get minimize the objective function.

$$f_o = \min\left[0.5\left\{W_{pd}(n+1) + W_{qt}(n+1)\right\}\right] \tag{25}$$

3.3.1 Initialization of Population

The SMO determines population initially with a dimension of ($n \times d$), where 'n' is SM_x spider monkey population size ($x = 1, 2, 3 \ldots n$) and 'd' is several unknown variables that need to be optimized. Each spider monkey is a possible solution to the problem defined under the objective function. The SM_x (xth spider monkey) in the population is initialized by Eq. (26).

$$SM_{xy} = SM_{\min y} + \left(SM_{\max y} - SM_{\min y}\right) \times U(0, 1) \tag{26}$$

SM_{miny} and SM_{maxy} are the xth spider monkey's lower and upper limits in the yth-direction. U is a random number between 0 to 1 and is also uniformly distributed in a given range.

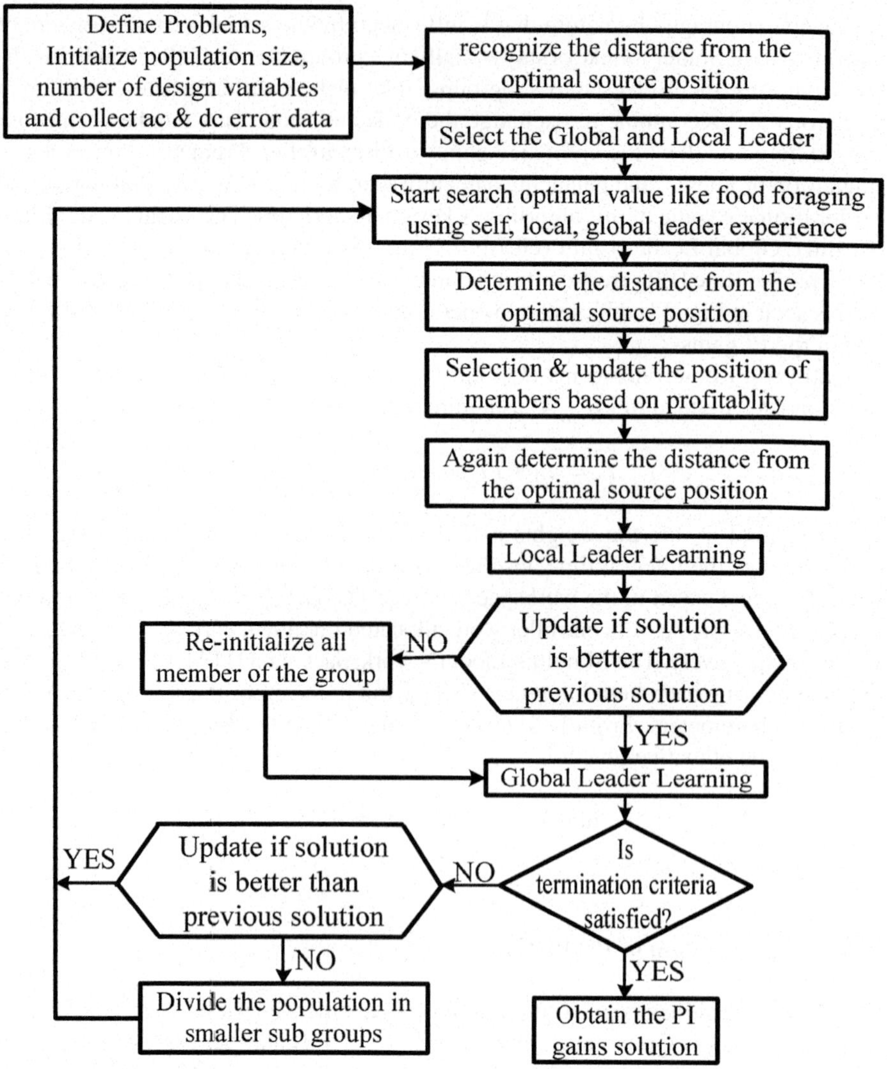

Fig. 3 Flow chart of SMO

3.3.2 Local Leader Phase (LLP)

The function of this stage in the SMO approach is to discover the search space, i.e., exploration, where single spider monkeys change their current status based on an understanding of local leaders and assemble members. New position fitness is again estimated as per the objective function. Here the new position indicates the currently obtained value of all unknown variables. When the fitness value for the older position is greater than the new one, the spider monkey immediately updates to the new position. Equation (27) analyses the fitness value of the new position corresponding to xth monkey (associated with the local pth group).

$$SM_{newxy} = SM_{xy} + \left(LL_{py} - SM_{xy}\right) \times U(0, 1) + \left(SM_{qy} - SM_{xy}\right) \times U(-1, 1) \quad (27)$$

where SM_{xy} is yth-dimension of xth spider monkey, and LL_{py} represents yth-dimension of pth local group leader Position. SM_{qy} is yth-dimension of a qth spider monkey randomly picked within the pth group such that $q \neq x$ U is a random number between -1 to 1 uniformly distributed in a given range.

3.3.3 Global Leader Phase (GLP)

Primary function of this stage in the SMO approach is for exploitation, where all individuals keep posted on their current location using the knowledge of local members and global leaders. The position update of the GLP member is as follows,

$$SM_{newxy} = SM_{xy} + \left(GL_{y} - SM_{xy}\right) \times U(0, 1) + \left(SM_{qy} - SM_{xy}\right) \times U(-1, 1) \quad (28)$$

where GL_y is a global leadership position in yth-dimension. The position of the xth monkey is modified with Eq. (26) with probabilities (PR_x) factor and it determines by,

$$PR_x = 0.1 + \left(\frac{fit_x}{\max\_fit}\right) \times 0.9 \quad (29)$$

where fit_x is the fitness value of the xth monkey and max_fit is the maximum fitness in a set of groups. In this phase, the previous and current position of the spider monkey is compared and a better position is adapted accordingly. In this way, the best monkey has a chance to update and increase their experience.

3.3.4 Local Leader Learning (LLL) Phase

Greedy selection modifies local leader location in this phase. Greedy choice means the monkey's position with the fittest value in any particular group. When the local leader status does not change, the local limit count is incremented by unity.

3.3.5 Global Leader Learning (GLL) Phase

The greedy selection modifies the global leader location in this phase. Greedy choice means the monkey's position with the fittest value in any particular group. When the global leader status does not change, then the local limit count is incremented by unity.

3.3.6 Local Leader Decision (LLD) Phase

Under the local leader limit, if the status of the local leader does not change, then a re-initialization process is initiated for every group member. There are two possible ways to reinitialize each spider monkey in a group. One can be done randomly and another one is through a collection of information from global and local leaders on a PR base using Eq. (30).

$$SM_{newxy} = SM_{xy} + \left(GL_y - SM_{xy}\right) \times U(0, 1) + \left(SM_{xy} - LL_{py}\right) \times U(0, 1) \tag{30}$$

3.3.7 Global Leader Decision (GLD) Phase

Monkey population is divided into several smaller groups if the global position remains unchanged. The process continues over some time. In this way, the SMO algorithm is thus inspired by the social system of a fission-fusion structure. Exploitation and exploration are two significant components of any swarm intelligence-based algorithm and SMO exquisitely depicts the balance between them.

This algorithm estimates the optimal PI controller gains (k_{pi}, k_{ii}, k_{pv}, and k_{iv}) with Eq. (29) as a minimizing optimization objective function. The SMO algorithm uses fifty spider monkeys for four unknown design variables in the speculated time of fifty iterations. Figure 4 shows the performance analysis of the SMO algorithm on RLS control for finding the PI controller parameters at the DC and AC bus of the UPQC system. The fitness value of the objective function given in Eq. (30) is depicted in Fig. 4a concerning iterations; the curve is settled at 222.89 after the 27th iteration. The k_{pi} and k_{ii} variation for iterations for the PI controller's gains at the DC bus terminal is also shown in Fig. 4b, concluded at 1.2801 and 0.103, respectively. Similar to this, Fig. 4c shows the fluctuation in k_{pv} and k_{iv} across iterations for the PI controller's gains at the AC bus terminal, which are fixed at 27.68 and 6.76, respectively. Optimal values of k_{pi}, k_{ii}, k_{pv}, and $\mathrm{k_{iv}}$ are saved from the SMO algorithm and will be used further in the UPQC control.

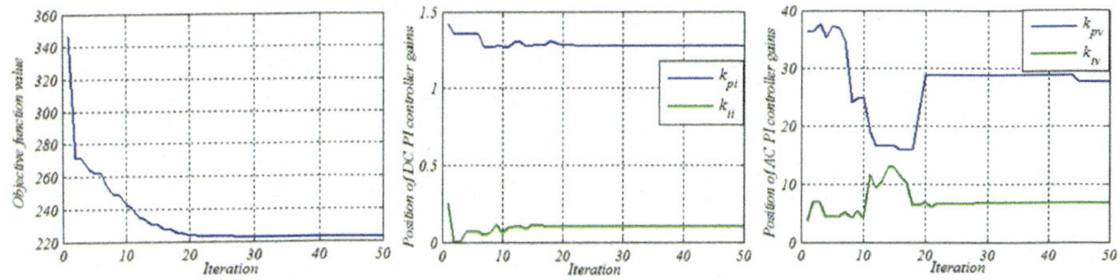

(a) Objective Function value in context with iterations

(b) DC PI controller parameters variation in context with iterations

(c)AC PI controller parameters variation in context with iterations

Fig. 4 Performance analysis of SMO algorithm on RLS control for finding the PI controller parameters

Responses and usefulness of UPQC's PI controllers with SMO algorithm at DC and AC with the PQ above disturbances are expressed in Fig. 5a, b. In its zoomed view, the SMO algorithm's performance behaviour tuning DC bus voltage is shown in Fig. 5c. Likewise, the tuning of AC bus voltage is also clearly shown here in Fig. 5d in its zoomed view. The zoomed view of the Figure depicted a clear picture of the variation of t_r, $t_{s,}$ and maximum M_p for both tuning processes. The t_r is determined at 100% of the final value (i.e., 700 V) and a tolerance band of 3% (i.e., 679–721) is specified for the under-damped system.

Table 1 provides descriptions of the time response parameter on both the PI controller buses. Various time-domain terms like M_p, t_r and t_s of both AC and DC PI controller gains at different PQ perturbations in Table 1 examine the comparison result of SMO with hit and trail tuning. The M_P is lesser in the case of SMO tuning; apart from that, it has a quicker t_r and improved t_s within the 3% tolerance band slightly for both PI controller buses. By hit and trial tuning process, it was noticed from Table1 and Fig. 5 that more stable and faster DC bus voltage and AC bus terminal voltage is done by SMO algorithm. The optimal value of PI gains acquired from SMO will be further considered in control of UPQC, whose performance is addressed in the subsequent section.

4 Simulation Results

The reference current and voltage generation using RLS algorithms for UPQC with optimized PI controller gain is analyzed by simulation using MATLAB/SIMULINK environment under various PQ conditions. Five different PQ perturbations, such as voltage unbalance, harmonics, sag, load unbalances, and current harmonics, are considered for testing proposed RLS algorithms on the UPQC system. All internal signals in the generation of the reference signal are shown with a line to the ground, while the steady-state performance of UPQC is illustrated with a line-to-line signal.

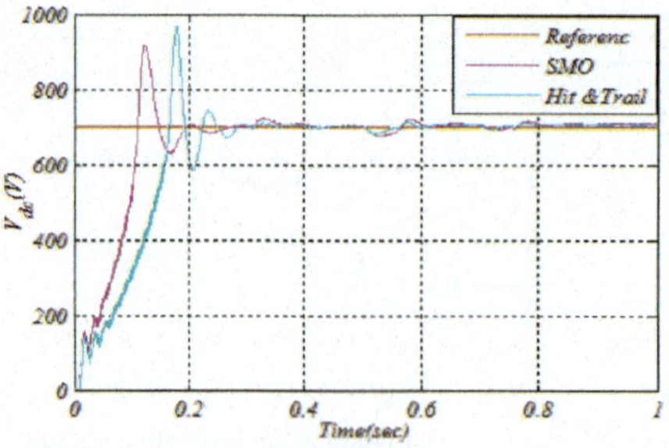

(a) Performance of SMO algorithm in tuning DC bus

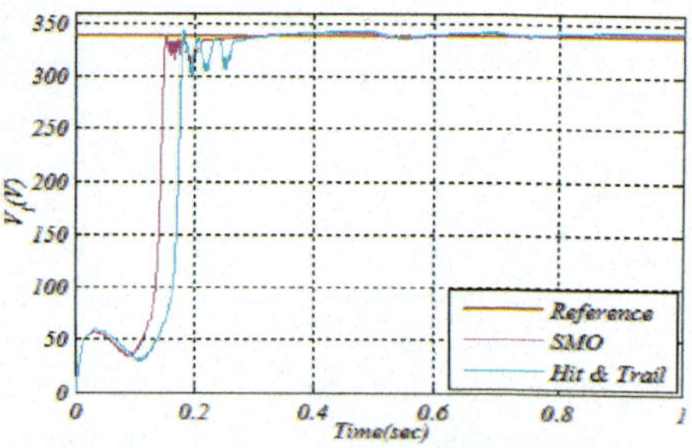

(b) Performance of SMO algorithm in tuning AC bus

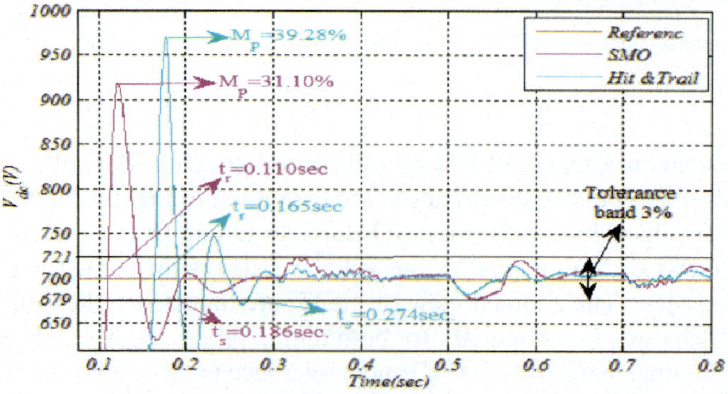

(c) Zoomed view of figure(a)

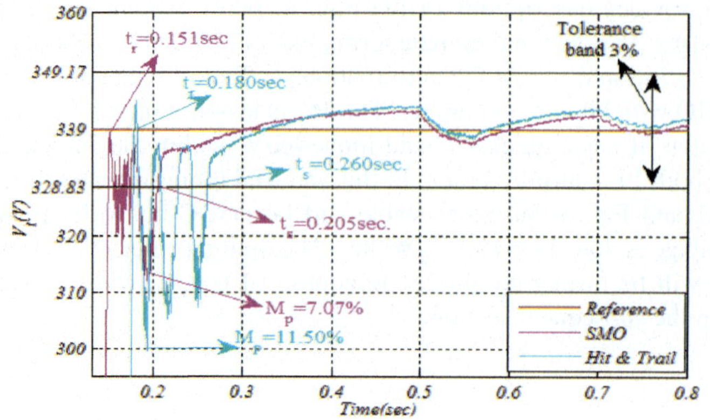

(d) Zoomed view of figure(b)

Fig. 5 a–d Zoomed version of performance behaviour of SMO algorithm on RLS control in tuning DC and AC bus

Table 1 Time response of DC and AC Bus through tuning method

S. No	Tuning METHOD	Rise time (t_r in sec)	Settling time (t_s in sec)	Maximum peak overshoot (M_p in %)
DC bus	SMO	0.110	0.186	31.10
	Hit and trail	0.165	0.274	39.28
AC bus	SMO	0.151	0.205	7.07
	Hit and trail	0.180	0.260	11.50

A three-phase wire with four 415 V, 50 Hz ac sources is considered for feeding a nonlinear bridge rectifier with RL load for each phase. Appendix 1 includes other simulation parameters and their estimated values.

4.1 Reference Load Voltage Generation using RLS Algorithm

The internal signals of RLS control algorithm have been depicted for generation of reference load voltage to the IGBT of the UPQC system in Fig. 6. Reference load voltage (v^{*}_{labc}) signals are derived from supply current (i_{sabc}), positive sequence supply voltage (i^{+}_{sabc}), in-phase unit template (u^{v}_{pabc}), quadrature-phase unit template (u^{v}_{qabc}) obtain from supply current, supply voltage (v_{sabc}), weight of fundamental components (in-phase) of phase a voltage (w_{pa}), total active fundamental component (W_{pv}), reference active (in-phase) components (v_{pabc}), the weight of fundamental components (quadrature) of phase a voltage (w_{qa}), total reactive fundamental component (W_{qv}), AC loss components (W_{qt}), total reference magnitude of the reactive components (W_Q), reference reactive (quadrature) components (v_{qabc}). Terminal voltage (Vt) is shown in Fig. 6. As noticed from subplots, positive sequence components (i^{+}_{sabc}) of supply current (i_{sabc}) are depicted to find in-phase (u^{i}_{pabc}) and quadrature-phase (u^{i}_{qabc}) unit temples. Waveforms are more sinusoidal irrespective of disturbance in line. The subplots of the supply voltage (v_{sabc}) having the disturbances named voltage sag, harmonics and unbalanced in the supply line at time instant 0.5–0.56, 0.6–0.66 and 0.7–0.76 s, respectively. A voltage sag of 0.70 p.u magnitude; voltage twisted through −5th and +7th harmonics order with the scale of 1/10th and 1/15th of its fundamental voltage is set up in supply voltage (v_s). In the same way, a voltage unbalance of 0.60 p.u amount is introduced in "phase a" supply voltage. Subsequently, an unbalanced system voltage appears. Rest instant of v_{sabc} subplots are considered balanced. The subplot w_{pa} is a fundamental in-phase component of phase a voltage extracted from RLS control. The RLS controller correctly estimates the fundamental active components required from the supply voltage during disturbed voltages. Similarly, the other phases' active fundamental components (w_{pb}, w_{pc}) of supply voltage are extracted. After passing through the low pass filter, its averaged value is obtained as W_{pv}, as depicted in Fig. 7. The required three-phase reference active (in-phase) components (v_{pabc}) are extracted by multiplying W_{pv} with u^{i}_{pabc}.

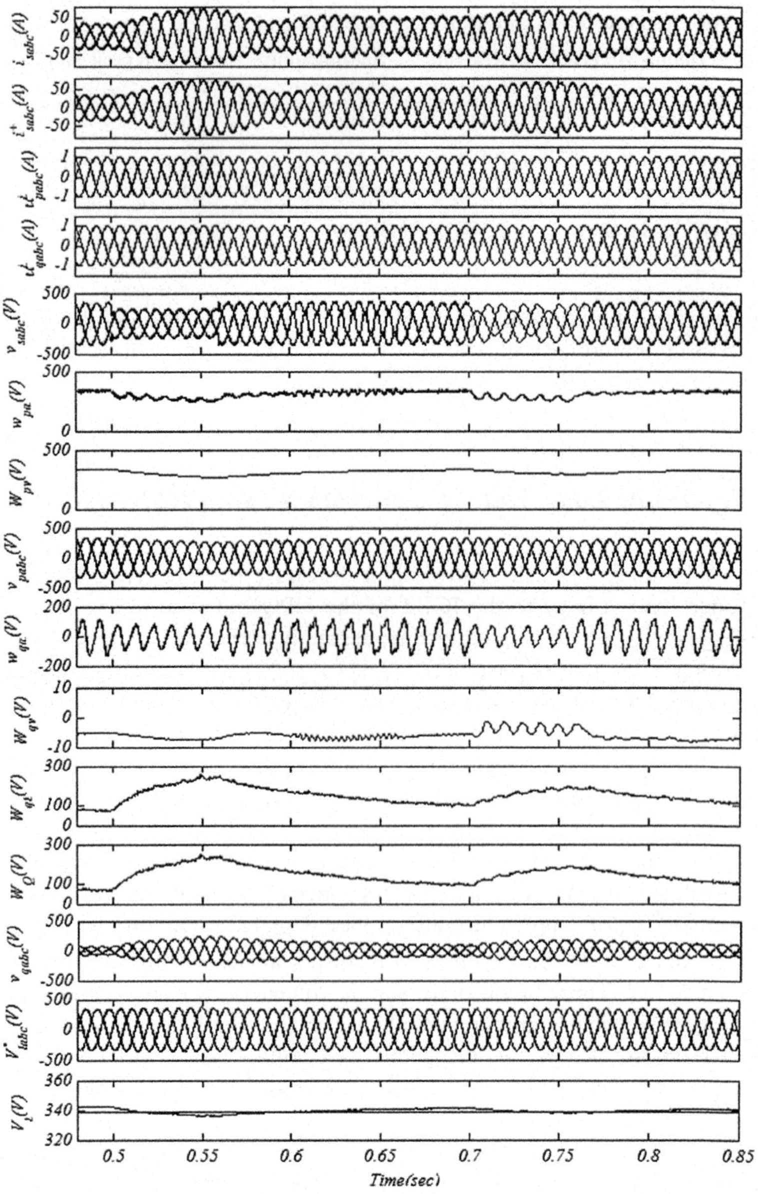

Fig. 6 Using RLS control, various internal control signals are used to generate reference load voltage

Similarly, the subplot named w_{qa} is a fundamental quadrature component of phase a voltage extracted from RLS control. Again, the other phase's reactive fundamental components (w_{qb}, w_{qc}) of supply voltage are extracted and passed through a low pass filter to obtain W_{qv}, its averaged value. The W_{qv} is negative due to the lagging nature of the load, as exhibited in Fig. 6. The instantaneous AC loss components of voltage (W_{qt}) needed to stabilize the terminal voltage under transient situations are likewise depicted in the Fig. 6. Total reference magnitude of reactive components (WQ) is obtained when W_{qv} is added to W_{qt}, as exhibited in Fig. 6, to stabilize the terminal voltage. In addition, the dynamics in W_{pv}, W_{qv}, W_{qt}, and WQ are stable and within a limit, irrespective of disturbances in supply voltages. Voltage's 3-phase reference quadrature components (v_{qabc}) are extracted by multiplying WQ u^{v}_{qabc}. Now, as shown in Fig. 6, the necessary balanced 3-phase reference load voltages (v^*l_{abc}) are acquired by adding v_{pabc} to v_{qabc}. The SPWM controller is used to produce gate pulses for Series VSC using sensed load voltages and generated reference load voltages. This Figure also shows the terminal voltage (V_t) curve, which stabilizes and falls after an overshoot of 3 V from its regulated value of 339 V.

4.2 Reference Load Voltage Generation using RLS Algorithm

Figure 7 illustrates the dynamic analysis of the three-phase four-wire IGBT-based UPQC for PCC load voltage control, voltage unbalance, voltage harmonics, and voltage sag as neutral current compensation and load unbalancing of a three-phase four-wire nonlinear loading. Supply voltage (v_{sabc}), injection voltage by a series transformer ($v_{inj}a$, $v_{inj}b$, $v_{inj}c$), load voltage (v_{labc}), source current (i_{sabc}), load current (i_{labc}), compensator current ($v_{com}a$, $v_{com}b$, $v_{com}c$), load neutral current (i_{Ln}), zig-zag neutral current(i_{zzn}), source neutral current (i_{sn}), dc bus voltage (V_{dc}) and terminal voltage at load PCC voltage (V_t) are exhibited in Fig. 7 under different PQ conditions. The three-phase load is disconnected from the system at times 0.4–0.5 s. Thus, it causes unbalanced loading conditions in the system. Similarly, from time 0.5–0.56 s, a 0.70 pu of voltage sag; and 0.6 to 0.66 s, waveform deformed through -5th and + 7th harmonics with a scale of 1/10th and 1/15th of fundamental supply voltage are set up in v_{sabc}. Again, at 0.7 to 0.76 s, a voltage unbalance of 0.60 p.u amount is introduced in source voltages (v_s) of "phase a." Subsequently, it brings an unequal voltage condition in the system. The UPQC part of Series VSC provides compensation for voltage harmonics, sags, voltage unbalanced & causes load voltages (v_{labc}) distortions free, as illustrated in Fig. 7. It maintains load voltage at a decent sinusoidal required level, regardless of fluctuations in supply voltages. In the v_{inja}, v_{injb}, and v_{injc} subplots of the corresponding phases, voltage injected by series VSC via the series transformer is shown. It should be noted that the required injecting voltage is applied to line via a series transformer, regulating terminal voltage to reference required voltage.

Similar to this, the shunt VSC of UPQC simultaneously compensates for load current distortions (i_{labc}) and load unbalancing situations, and, as depicted in Fig. 7, makes the source currents (i_{sabc}) sinusoidal in a wave shape regardless of the current

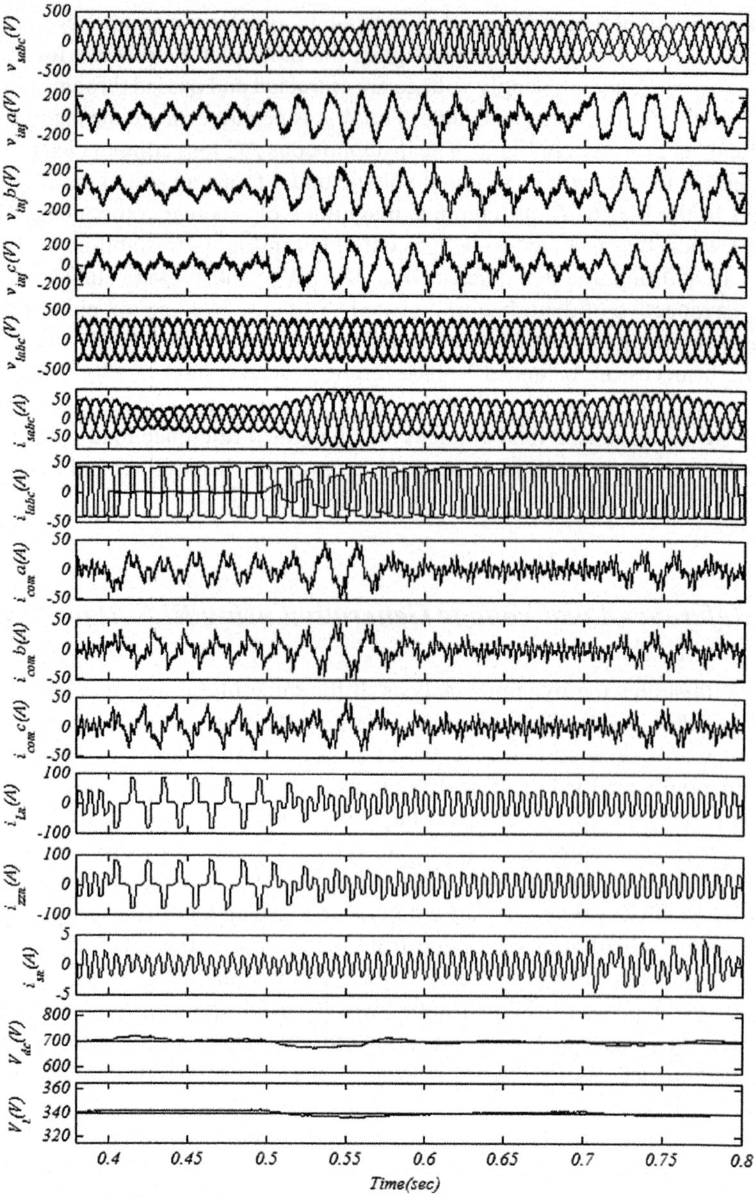

Fig. 7 UPQC's dynamic performances under RLS-based control

Table 2 Steady-State harmonic compensation performance analysis on UPQC using RLS-based control

Waveform distortion	Parameters	Percentage of harmonics and magnitude
Harmonics level	Source voltage (v_{sa})	14.91%, 581.8 V
	Supply current (i_{sa})	3.19%, 52.43A
	Load voltage (v_{la})	3.98%, 590.8 V
	Load current (i_{la})	38.59%, 52.49A

load wave-shape. As a result, the power factor is held at unity, and the source current is balanced. The source current magnitude increases or decreases because the load demand increases or reduces during different PQ problems. The current profile of the source is perfectly balanced and sinusoidal as Shunt VSC provides the compensating current ($i_{com}a$, $i_{com}b$ and $i_{com}c$) in all phases. The compensating current injects the harmonic current of equal and opposite load current in the network. However, the shape of compensating current during unbalancing in "phase a" is different from the other phases because of trying to inject a sinusoidal supply to the source. The zig-zag transformer circulates zero sequences fundamental current of load neutral current (i_{Ln}) produced by the unbalanced load current. Thus, the supply neutral current (i_{sn}) is maintained at nearly zero, as revealed in subplot i_{sn}. The proposed controller maintains self-supported DC-link voltage (Vdc) between two VSC near the reference level under all PQ disturbances. Alike V_{dc}, the terminal voltage (V_t) shown here are also maintained at their reference voltage at the PCC terminal. In just a few cycles, the DC link voltage regulator PI controller can take action and restore the dc voltage.

The Total Harmonic Distortion (THD) is evaluated for source voltage, source current, load voltage, and load current before and after correction on line-to-line quantities to evaluate the steady-state performance of UPQC, as illustrated in Table 2. It demonstrates that the THD improves from 38.59 to 3.98% in the source current and 14.91–3.19% in the load voltage for the RLS controller implemented on the UPQC after compensation. The source current and load voltage are better results using the RLS algorithm in the UPQC controller. Their THDs are below the requirements in the IEEE 519–2014 standard recommendations on harmonic levels.

5 Hardware Results

The RLS control algorithm is implemented per response and features for the UPQC system under nonlinear load. The fixed sampling time of 50 μsec is taken for implementation through the Micro Lab box processor. The dynamic performance of a UPQC has been examined with voltage sag, harmonics, voltage unbalance and load removal-related PQ problems. The dynamic nature of UPQC is demonstrated in Figs. 8, 9, 10, 11 and 12 to mitigate voltage sag and harmonic issues using an

Fig. 8 Experimental setup of UPQC with 3-phase four wires VSC-based topology

RLS-based control technique. The dynamic disturbances are seen in half of the DSO screen, while the other half shows standard working conditions. Signals are provided here when the disturbances begin to return to normal. All internal signals of the control algorithm from Figs. 8, 9, 10, 11 and 12 are measured in "line to line" values. All the waveforms represented in Figs. 8, 9, 10, 11 and 12 concerning phase 'ab'. Similarly, a PQ spectrometer FLUKE-4B is used to analyze steady-state results during harmonic distortions. Detailed system parameters and control algorithms for hardware are given in Appendix 2.

5.1 PV Cell Modeling

As seen in Fig. 8, the UPQC prototype was developed using a d-SPACE Micro Lab Box-based processor with a 40 μsec sampling time. The Micro Lab Box is a real-time processor made by NXP, QorlQ P5020, as depicted in the photograph of the experimental setup. The voltages and currents were measured with LEM-made LV-25P and LA-55P, respectively, for the controller. The A/D channel was used to send the sensed singles (voltages and currents) to the processor to control the operation. After the proposed control algorithms have been processed, gate signals are acquired from processor through DIO channel. The dynamics were recorded under all PQ scenarios using a 4-channel Digital Storage Oscilloscope DSO-X-2004A. The Fluke-43B PQ analyzer is used to capture steady-state response as per availability in our laboratory.

5.2 Observation of Dynamic Performance

The behaviour of the source voltage (v_{sab}) in CH1, the load voltage (v_{lab}) in CH2, the source current (isa) in CH3, and the load injection, respectively, in phase "ab," is shown in Fig. 9a–d. These Figures show a voltage sag of 10%; harmonic distortions of order 5th plus 7th are superimposed on source voltages as depicted in Fig. 9a, b in CH1, respectively. According to changes in supply voltages, supply current magnitude increases during voltage sag excursion, as shown in Fig. 9. (in CH3). In CH2, the load voltage is simultaneously kept at the proper sinusoidal level regardless of disruptions. As in Fig. 9a, b in CH2, UPQC mitigates harmonic distortions and minimizes load voltage distortion. Figure 9c, d depicts the response of neutral current compensation and unbalanced loading condition on the load side in CH4. Here in Fig. 9c, source voltage (v_{sab}) in CH1, zig-zag transformer neutral current (i_{zzn}) in CH2, load neutral current (i_{Ln}) in CH3 and load current (i_{la}) in CH4 are recorded in DSO waveforms. The sinusoidal feature of the source current is maintained while considering the impacts of load disturbance. The neutral current from the load compensates for transformer's neural current, and output load voltage is also kept constant. According to waveforms, the zigzag transformer delivers zero sequences current in precise phase opposition to neutral current. Therefore, as a result, the source neutral current dropped to nearly zero, allowing the VSC to source harmonic current other than zero-sequence current. From Fig. 9d, it can be seen that the source current (i_{sa}) in CH2 is maintained sinusoidal even when the load is unbalanced and during this compensator current (i_{ca}) in CH3 is injecting a sinusoidal current. In all Fig. 9a–d, the load current appears to be very non-sinusoidal and the proposed adaptive controller of UPQC maintains the source current sinusoidal waveform. During disruptions and normal conditions, the load voltage (v_{lab} in CH2) and source current (i_{sa} in CH3) are in phase, as shown in Fig. 9a–d, which demonstrates how UPQC maintains regulated voltage at the unity power factor.

Figure 10a, b illustrates supply voltage (v_{sab}) in CH1, load voltage (v_{lab}) in CH2, compensating voltage of phase a (v_{ca}) in CH3, and DC-link voltage (V_{dc}) in CH4 at the voltage sag and harmonics disturbance. This is acknowledged in Fig. 10a, b; throughout voltage sag and harmonic distortions excursion, essential compensated voltage injected via series transformer is in-quadrature. During the dynamic disturbance, the variations of DC-link voltage can be realized in all Figures. It settles at the desired level within 2 cycles after a slight variation of 2–3 V in the DC bus voltage.

Figure 11a, b illustrates source voltage (v_{sab}) in CH1, load voltage (v_{lab}) in CH2, compensating voltage of phase a (v_{ca}) in CH3, and terminal voltage (V_t) in CH4 at the above-said PQ disturbance. This is recognized in Fig. 11a–c. Required compensated voltage injected through the series transformer is in-quadrature during disturbances. During the dynamic disturbance, the variations of terminal voltage can be observed in all figures. All Figures from 10 to 12 concluded that the 3-phase 4-wires UPQC dynamic performance with an RLS-based control algorithm successfully compensates for voltage sag and harmonics problems.

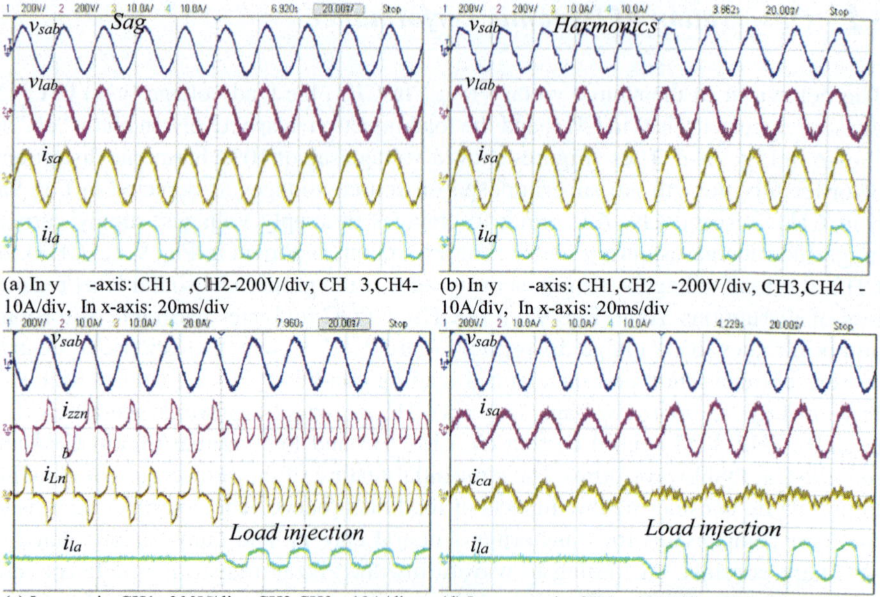

(a) In y -axis: CH1 ,CH2-200V/div, CH 3,CH4-10A/div, In x-axis: 20ms/div

(b) In y -axis: CH1,CH2 -200V/div, CH3,CH4 -10A/div, In x-axis: 20ms/div

(c) In y -axis: CH1 -200V/div, CH2,CH3 -10A/div, CH4-20A/div, In x-axis: 20ms/div

(d) In y -axis: CH1 -200V/div, CH2,CH3,CH4 -10A/div, In x-axis: 20ms/div

Fig. 9 a–d Dynamic response of UPQC during **a** sag, **b** harmonics disturbances, **c** neutral current compensation, **d** load injected disturbances

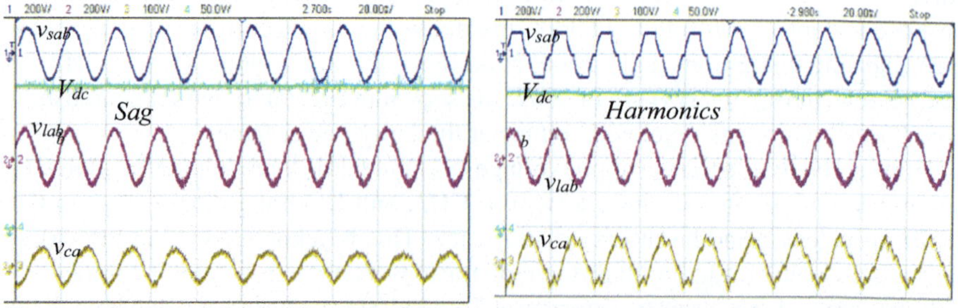

(a) In y -axis: CH1,CH2-200V/div, CH3-100V/div, CH4-50V/div; In x-axis: 20ms/div

(b) In y -axis: CH1,CH2 -200V/div, CH3 -100V/div, CH4-50V/div; In x-axis: 20ms/div

Fig. 10 a-b Dynamic response of UPQC during **a** sag and **b** harmonics disturb- ances with dc-link variations

5.3 Steady-State Performance Analysis of UPQC with RLS Control Under Voltage Disturbances

The steady-state performance of the 3-phase four-wire UPQC with RLS control algorithm has been illustrated in Fig. 12a–l under sag disturbances with non-linear loads. All waveforms indicate each signal's steady-state Root Mean Square (RMS)

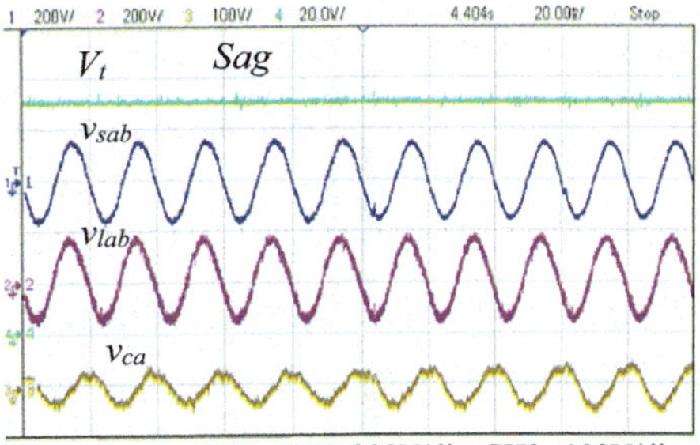

(a) In y -axis: CH1 ,CH2-200V/div, CH3 -100V/div, CH4-20V/div; In x-axis: 20ms/div

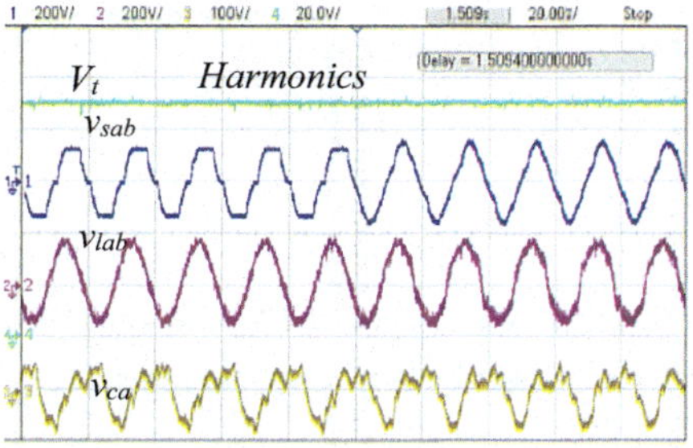

(b) In y -axis: CH1,CH2 -200V/div, CH3 -100V/div, CH4-20V/div; In x-axis: 20ms/div

Fig. 11 **a, b** Dynamic response of UPQC during **a** sag and **b** harmonics disturbances with terminal voltage link variations

value in the respected current and voltage signal. Figure 12a–c depicts the waveforms of supply voltages (v_{sab}), (v_{sbc}), and (v_{sca}) having 10% voltage sag present in the line and supply current (i_{sa}) after mitigation, respectively. Similarly, Fig. 12d expresses an image of mitigated load voltage (v_{lab}) after compensation and distortional load current (i_{la}), respectively. All three load voltages (v_{lab}), (v_{lbc}) and (v_{sca}) are up to 110 V RMS values without having a voltage dip. In Fig. 12e-h, the THD for supply voltage ab (v_{sab}), supply current an (isa), load voltage ab (v_{lab}), and load current a (ila) during sag disturbances are, respectively, 3, 4, 4.8, and 25.5%. During voltage sag, as indicated in Fig. 12e, the supply voltage has an RMS value of 100.4 V with a THD of 3.0%. After that mitigation, the source current has an RMS value of 5.88A current with a THD of 4.0%, which is depicted in Fig. 12f. The mitigated load voltage is seen in Fig. 12g from the sag disturbances and has a 109 V RMS value and 4.8% THD. The nonlinear load measured has an RMS value of 4.55 A and 25.50% THD as revealed in Fig. 12h. The required compensating voltage of phase a (v_{ca}) injected during sag disturbances is shown in Fig. 12i concerning supply current (i_{sa}). Similarly, Fig. 12j describes the image of load voltage ab (v_{lab}) after mitigation with the required compensator current of phase an (i_{ca}). Figure 12k–l illustrates the role of obtaining neutral current compensation using a zigzag transformer. The zig-zag transformer neutral current (i_{zzn}) appears to provide zero sequences current in exact phase opposition to neutral current (i_{Ln}); as seen by these images and the source neutral current became almost zero. After compensation, the supply current and load voltage have a sinusoidal shape with the desired amplitude. Figure 12a–d shows that the i_{sa} and v_{la} come under the IEEE-519 standards limit.

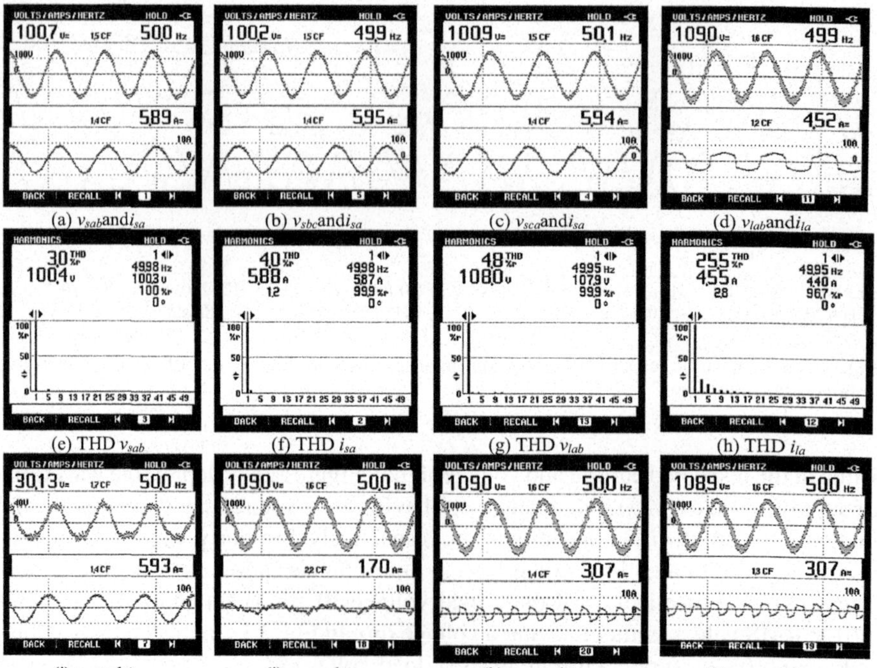

(a) v_{sab}andi_{sa} (b) v_{sbc}andi_{sa} (c) v_{sca}andi_{sa} (d) v_{lab}andi_{la}

(e) THD v_{sab} (f) THD i_{sa} (g) THD v_{lab} (h) THD i_{la}

(i) $v_{ca}$and i_{sa} (j) $v_{lab}$and i_{ca} (k) v_{lab} and i_{zzn} (l) v_{lab} and i_{ln}

Fig. 12 a–l Voltage Sag compensation through UPQC using RLS-based control

6 Conclusion

The UPQC is assembled to compensate for the various PQ issues through transformer-based topology. The RLS technique is a control algorithm for estimating reference signals. The RLS algorithm control strategy to extract fundamental components is vital in generating balanced reference load voltage and source current. This RLS algorithm is based on the adaptive noise cancellation theory, which emphasizes the enhancement of PQ by providing fundamental components to the UPQC controller. The SMO helps tune the PI controller parameters for the above-said PQ disturbances. It is noted that the performance of the PI controller with the SMO algorithm is better in peak overshoot, settling time, and rise time compared to hit and trail tuning for the UPQC system. Also, Table 1 justifies the improved response of SMO in comparison with hit and trail tuning. The SMO algorithm also helps to stabilize DC and AC link voltage levels. After 20 iteration, PI-controller proportional gain (k_{pi}, k_{ii}, k_{pv}, *and* k_{iv}) are obtained at 1.2801, 0.103, 27.68, and 6.76, respectively, which maintains DC and AC link voltage levels to the required magnitude. The performance study of the four-wire UPQC system is appreciably raised by using RLS and SMO algorithms in terms of fundamental components extraction and tuning of PI gains, respectively.

Appendix

1. RLS Control Parameter on UPQC

AC line voltage (v_{sabc}) = 415 V, 50 Hz; Nonlinear load: 3 single phase bridge rectifier with $R = 5\ \Omega$, $L = 250$ mH; Source impedance (Z_s) R = 0.060 Ω, L = 2 mH; AC bus voltage (V_t) = 339 V DC link voltage (V_{dc}) = 700 V; DC bus Capacitor (C_{dc}) = 7000 μF; zig-zag transformer 7KVA, 120/120 V; Injecting interfacing transformer 5 KVA, 120/120 V; neutral line impedance: $R = 0.02\ \Omega$, $L = 30$ mH; Shunt side Interfacing inductance (L_{sh}) = 2.5 mH; Series side Interfacing inductance (L_{se}) = 1 mH; Low pass filter $R_f = 4\ \Omega$, $C_f = 20\ \mu$F; PI Controller at DC bus $k_{pi} = 1.280$, $k_{ii} = 0.103$; PI Controller at AC bus $k_{pv} = 27.68$, $k_{iv} = 6.76$; Low pass filter = 2 * pi * 10; RLS gains: $\lambda = 0.95$ and $\delta = 0.0001$; SMO control parameter: Population size = 50, number of iterations = 20, and number of runs = 10. $LL_{lim} = 25$, $GL_{lim} = 250$, *MG.* = 10, *PR* = 0.1–0.9, *increasing linearly.*

2. RLS Control Parameter on UPQC for Hardware Setup

AC line voltage (v_{sabc}) = 110 V, 50 Hz; Connected nonlinear load: 3 single phase bridge rectifier with R = 5Ω, L = 250mH; AC bus voltage (V_t) = 89 V Injecting transformer 4KVA, 125/125 V; zig-zag transformer 1KVA, 120/120 V; DC link voltage (V_{dc}) = 200 V; DC bus Capacitor (C_{dc}) = 3500 μF; Shunt side Interfacing inductance (L_{sh}) = 4 mH; DC PI Controller at DC bus $k_{pi} = 0.80$, $k_{ii} = 0.01$; AC PI Controller at AC bus $k_{pv} = 20.5$, $k_{iv} = 0.01$; Low pass filter = 2 * pi * 12; RLS gains: $\lambda = 0.95$ and $\delta = 0.0001$.

References

1. Bose BK (2017) Power electronics and motor drives: advances and trends. Academic Press, 2020. ISBN 13: 978-0-12-088405-6
2. Emadi A, Nasiri A, Bekiarov SB (2017) Uninterruptible power supplies and active filters. CRC Press. ISBN 0-8493-3035-1
3. Acha E, Madrigal M (2001) Power system harmonics: computer modelling and analysis. John Wiley. ISBN: 978-0-471-52175-4
4. Zobaa AF, Abdel Aleem SHE (2014) A new approach for harmonic distortion minimization in power systems supplying nonlinear loads. IEEE Trans Ind Inf 10(2):1401–1412. https://doi.org/10.1109/TII.2014.2307196
5. de Araujo RL, de Azevedo CC, de Sousa RM (2012) A robust adaptive control strategy of active power filters for power-factor correction, harmonic compensation, and balancing of nonlinear loads. IEEE Trans Power Electron 27(2):718–730. https://doi.org/10.1109/TPEL.2011.2161334

6. Badoni M, Singh A, Singh B (2015) Variable forgetting factor recursive least square control algorithm for DSTATCOM. IEEE Trans Power Deliv 30(5):2353–2361. https://doi.org/10.1109/TPWRD.2015.2422139
7. Naidu TA, Arya SR (2021) Maurya R and Padmanaban S (2021) Performance of DVR using optimized PI controller based gradient adaptive variable step LMS control algorithm. IEEE J Emerg Select Top Ind Electron 2(2):155–163. https://doi.org/10.1109/JESTIE.2021.3051553
8. Karanki SB, Geddada N, Mishra MK, Kumar B (2013) K (2013) A Modified three-phase four-wire UPQC topology with reduced DC-link voltage rating. IEEE Trans Ind Electron 60(9):3555–3566. https://doi.org/10.1109/TIE.2012.2206333
9. Kumar M, Swarnkar A, Gupta N (2017) Niazi KR (2017) Design and operation of DSTATCOM for power quality improvement in distribution systems. J Eng 2017(13):2328–2333. https://doi.org/10.1049/joe.2017.0747
10. Abas N, Dilshad S, Khalid A, Saleem MS, Khan N (2020) Power quality improvement using dynamic voltage restorer. J IEEE Access 8:164325–164339. https://doi.org/10.1109/ACCESS.2020.3022477
11. Paramanik S, Sarker K, Chatterjee D, Goswami SK (2019) Smart grid power quality improvement using modified UPQC. In: Proceedings of 2019 IEEE devices for integrated circuit (DevIC), Kalyani, India, pp 356–360. https://doi.org/10.1109/DEVIC.2019.8783704
12. Borbely AM, Kreider JF (2001) Distributed generation: the power paradigm for the New Millennium, Boca Raton. CRC Press. ISBN: 9780367397197
13. Poulo SRD (2008) Adaptive filtering, algorithms and practical implementations, 4th ed. Springer US. ISBN: 978-1-4614-4106-9
14. Dixit S, Nagaria D (2017) LMS adaptive filters for noise cancellation: a review. Int J Electr Comput Eng 7(5):2520–2529. https://doi.org/10.11591/ijece.v7i5.pp2520-2529
15. Marei MI, El-Saadany EF, Salama MMA (2012) A flexible DG interface based on a new RLS algorithm for power quality improvement. IEEE Syst J 6(1):8–75. https://doi.org/10.1109/JSYST.2011.2162930
16. Skidmore ID, Proudler IK (2003) The KaGE RLS algorithm. IEEE Trans Sig Process 51(12):3094–3104. https://doi.org/10.1109/TSP.2003.818997
17. Badoni M (2018) Singh A and Singh B (2018) Power quality improvement using DSTATCOM with affine projection algorithm. IET Gener Transm Distrib 12(13):3261–3269. https://doi.org/10.1049/iet-gtd.2017.0841
18. Djendi M, Benallal A, Guessoum A, Berkani D (2003) Three new versions for the Newton type adaptive filtering algorithm. In: Proceedings of seventh international symposium on signal processing and its applications, vol 2, Paris, France, pp 559–562. https://doi.org/10.1109/ISSPA.2003.1224938
19. Martinez R, del Cacho A, de Haro, Calvo M (2002) Comparative study of LMS and RLS adaptive algorithms in the optimum combining of uplink W-CDMA. In: Proceedings 56th IEEE vehicular technology conference, vol 4, Vancouver, BC, Canada, pp 2258–2262. https://doi.org/10.1109/VETECF.2002.1040622
20. Abdullah AH, Yusof MI, Baki SRM (2002) Adaptive noise cancellation: a practical study of the least-mean square (LMS) over recursive least-square (RLS) algorithm. In: Proceedings of student conference on research and development, Shah Alam, Malaysia, pp 448–452. https://doi.org/10.1109/SCORED.2002.1033154
21. Yu F, Bouchard M (2001) Recursive least-squares algorithms with good numerical stability for multi-channel active noise control. In: Proceedings of 2001 IEEE international conference on acoustics, speech, and signal processing. Proceedings (Cat. No.01CH37221), vol 5, Salt Lake City, UT, USA, pp 3221–3224. https://doi.org/10.1109/ICASSP.2001.940344
22. Araki M (2010) Control systems, robotics and automation-Vol 2-PID control. Kyoto University-Japan. ISBN: 978-1-84826–591-2
23. Astrom K (2001) J and Hagglund T (2001) The future of PID control. J Control Eng Pract 9(11):1163–1175. https://doi.org/10.1016/S0967-0661(01)00062-4
24. Ghosal S, Darbar R, Neogi B, Das A, Tibarewala DN (2012) Application of swarm intelligence computation techniques in PID controller tuning: a review. In: Proceedings of international

conference on information systems design and intelligent applications 2012 held in Visakhapatnam, India. Springer, Berlin, Heidelberg, pp 195–208. https://doi.org/10.1007/978-3-642-27443-5_23
25. Meshram PM, Kanojiya RG (2012) Tuning of PID controller using Ziegler-Nichols method for speed control of DC motor. In: Proceedings of IEEE-international conference on advances in engineering, science and management (ICAESM-2012), Nagapattinam, India, pp 117–122. INSPEC Accession Number: 12818603
26. Krohling RA, Rey JP (2001) Design of optimal disturbance rejection PID controllers using genetic algorithms. IEEE Trans Evol Comput 5(1):78–82. https://doi.org/10.1109/4235.910467
27. Ghosal S, Darbar R, Neogi B, Das A, Tibarewala DN (2012) Application of swarm intelligence computation techniques in PID controller tuning: A review. In: Proceedings of the international conference on information systems design and intelligent applications 2012 (INDIA 2012), Visakhapatnam, India. Springer, Berlin, Heidelberg, pp 195–208. https://doi.org/10.1007/978-3-642-27443-5_23
28. Sabir MM, Ali T (2016) Optimal PID controller design through swarm intelligence algorithms for sun tracking system. J Appl Math Comput 274:690–699. https://doi.org/10.1016/j.amc.2015.11.036
29. Panda S, Sahu BK, Mohanty PK (2012) Design and performance analysis of PID controller for an automatic voltage regulator system using simplified particle swarm optimization. J Franklin Inst 349(8):2609–2625. https://doi.org/10.1016/j.jfranklin.2012.06.008
30. Das B, Panigrahi PK, Das SR, Mishra DP, Salkuti SR (2022) Power quality improvement in a photovoltaic based microgrid integrated network using multilevel inverter. Int J Emerg Electr Power Syst 23(2):197–209. https://doi.org/10.1515/ijeeps-2021-0040
31. Ray P, Salkuti SR (2020) Smart branch and droop controller based power quality improvement in microgrids. Int J Emerg Electr Power Syst 21(6):20200094. https://doi.org/10.1515/ijeeps-2020-0094
32. Bansal JC, Sharma H, Jadon SS, Clerc M (2014) Spider monkey optimization algorithm for numerical optimization. J Memetic Comput 6(1):31–47. https://doi.org/10.1007/s12293-013-0128-0
33. Gupta K, Deep K, Bansal JC (2017) Spider monkey optimization algorithm for constrained optimization problems. J Soft Comput 21(23):6933–6962. https://doi.org/10.1007/s00500-016-2419-0
34. Sharma H, Hazrati G, Bansal JC (2019) Spider monkey optimization algorithm. In: proceedings of evolutionary and swarm intelligence algorithms. Springer, Cham, pp 43–59. https://doi.org/10.1007/978-3-319-91341-4_4
35. Salkuti SR (2022) Emerging and advanced green energy technologies for sustainable and resilient future grid. Energies 15(18):6667. https://doi.org/10.3390/en15186667
36. Mishra DP, Nayak AS, Tripathy T, Salkuti SR, Mishra S (2021) A novel artificial neural network for power quality improvement in AC microgrid. Int J Power Electron Drive Syst 12(4):2151–2159. https://doi.org/10.11591/ijpeds.v12.i4.pp2151-2159
37. Behera TK, Behera MK, Nayak N (2018) Spider monkey based improve P&O MPPT controller for photovoltaic generation system. In: Proceedings of technologies for smart-city energy security and power (ICSESP), Bhubaneswar, India, 1–6. https://doi.org/10.1109/ICSESP.2018.8376735
38. de Souza HEP, Bradaschia F, Neves FAS, Cavalcanti MC, Azevedo GMS, de Arruda JP (2009) A method for extracting the fundamental-frequency positive-sequence voltage vector based on simple mathematical transformations. IEEE Trans Ind Electron 56(5):1539–1547. https://doi.org/10.1109/TIE.2008.2009525

Fractional Order High Pass Filter Based Extremum Seeking Control for Grid Connected PV System

Laxman Bhukya, Narender Reddy Kedika, Rambabu Motamarri, Surender Reddy Salkuti, and Srinivas Punna

Abstract The solar-based renewable energy became the leading alternate energy injection into the local utility grid to limit the usage of fossil fuels. This power injection must be of high quality and with the maximum possible, hence the utilization of the Maximum Power Point Tracking (MPPT) technique is essential to stabilize and maximize the power supplied by the grid-connected PV source. In this work, an MPPT algorithm called fractional order High-Pass Filter (HPF) based Extremum Seeking Control (ESC) for grid-connected solar photovoltaic (PV) systems is proposed. The DC link voltage and the power injected into the grid are regulated with voltage source inverter (VSI) control using d-q components of the grid currents. The proposed is a benchmarked algorithm against conventional Perturb & Observe (P&O), Integer Order extremum seeking control (IOESC) methods. The efficiency of the proposed method is illustrated using MATLAB simulations under uniform and variable irradiances. Utilizing the Fractional order operators in an ESC scheme enhances the robustness, convergence speed, efficiency, and performance of the solar PV system.

L. Bhukya
Department of Electrical and Electronics Engineering, Methodist College of Engineering and Technology, Hyderabad, India

N. R. Kedika
Department of Electrical and Electronics Engineering, Institute of Aeronautical Engineering, Hyderabad, India
e-mail: k.narenderreddy@iare.ac.in

R. Motamarri
Department of Electrical and Electronics Engineering, Vignan's Institute of Information Technology, Vishakhapatnam, India

S. R. Salkuti (✉)
Department of Railroad and Electrical Engineering, Woosong University, Daejeon 34606, Republic of Korea
e-mail: surender@wsu.ac.kr

S. Punna
Division of Motor Controller Design, Mahindra and Mahindra, Bengaluru, India
e-mail: 25016896@mahindra.com

S. R. Salkuti et al. (eds.), *Power Quality in Microgrids: Issues, Challenges and Mitigation Techniques*, Lecture Notes in Electrical Engineering 1039,
https://doi.org/10.1007/978-981-99-2066-2_8

Keywords Maximum power point tracking · Inverter control · Extremum seeking control · Solar Photovoltaic system · High pass filter

Nomenclature

MPPT	Maximum power point tracking
HPF	High-Pass Filter
PV	Photovoltaic
VSI	Voltage source inverter
P&O	Perturb & Observe
IOESC	Integer Order extremum seeking control
OC	Open Circuit
SC	Short Circuit
FLC	Fuzzy Logic Control
GL	Grunwald-Letnikov
MPP	Maximum power point
RL	Riemann-Liouville
SP	Series-Parallel
ESC	Extremum seeking control
PWM	Pulse width modulation
PLL	Phase locked loop
SISO	Single input-single output

1 Introduction

In past years, however, the photovoltaic (PV) powered loads connected to the native utility grid have increased significantly. These grid-connected solar PV systems have PV panels that provide some or total power required during the day time, while still being connected to the native grid during the night time. The sun-irradiated solar PV system can sometimes produce more power than needed, especially during the summer months. This surplus or extra power is either stored in batteries or directly fed back to the utility grid. In this grid-connected PV source, the loads can consume all or part of the energy they require from the solar system while continuing to draw power from the utility grid at night, on overcast days, and during rainy weather. A DC-DC converter, which maximizes the compatibility between the utility grid and solar panels, PV panels to produce energy from solar irradiance, and an inverter to convert the DC to AC and inject it into the utility grid are all parts of a grid-connected PV system [1].

Nowadays industries are increasing solar-based renewable energy injection into the local utility grid to limit the usage of fossil fuels. This power injection must be of high quality (zero reactive power with low harmonics) and with the maximum possible, hence the utilization of MPPT plays a key role to stabilize and maximize the power supplied by the grid-connected PV source [2]. Owing to the photovoltaic nature of the PV panels, the I-V curves depend on insolation levels, the type of load connected, and temperature. The direct connection of the load to the PV source also effects the overall efficiency [3]. Nevertheless, maximum power is extracted with an intermediate DC-DC converter connection between the PV system and the load, numerous MPPT methods have been investigated which vary in some aspects of tracking speed, complexity, and cost [4]. Among all the MPPT techniques, P&O has gained huge interest due to its simplicity in implementation and acceptable results [5]. But it produces fluctuations around the maximum power point (MPP) and loses this point during the changing insolation. Hence, to make sure good quality power flow between the grid and the PV source needs a better MPPT technique. The authors in [6, 7] proposed fractional Open Circuit (OC) voltage and fractional Short Circuit (SC) current MPPT methods, which have a linear relationship between the operating current (voltage) and SC current (OC voltage). However, these techniques have power loss, due to the SC current, and the OC voltage is measured by short-circuiting or shutting down the PV panels periodically. Moreover, the MPP of FOCV and FSCC is not an exact MPPT. To overcome the drawbacks discussed above Fuzzy Logic Control (FLC) [8] and neural network [9] based MPPTs are proposed. But these techniques need complex computational structures to handle different stages and huge data storage for training. Some control techniques have been proposed in grid-connected PV systems [10–12]. Over the last decade, the authors introduced different converter topologies with P&O MPPT in a grid-connected PV system. However, there is less improvement owing to the disadvantage of P&O MPPT, which produces fluctuations around MPP and slow convergence during varying irradiance.

To overcome the aforementioned problem, the extremum seeking control (ESC) is used for MPPT in a grid-connected PV source [13]. Although the ESC-based MPPT is faster than P&O, the efficiency and tracking speed is still challenging issue. For this reason, Fractional order HPF-based extremum seeking control (FOESC) is proposed in this chapter for MPPT in a grid-connected PV source. The supremacy of the proposed MPPT is debated in Sect. 6. The key contributions of this chapter are summarized below:

- The FOESC is proposed by replacing the HPF with fractional order HPF from the IOESC controller.
- The proposed FOESC presents better results concerning the control of the grid, and load and is also well adapted for PV panels for fast-changing insolation.
- Utilizing the fractional order operators in an ESC scheme enhances the robustness, convergence speed, performance, and efficiency of the PV system.

The organization of this chapter is as follows: The fundamental definitions of fractional calculus are covered in Sect. 2. MPPT is used in Sect. 3 to control the grid-connected system. Section 4 provides an explanation of the proposed fractional order HPF-based ESC. Part 5 goes into detail about the simulation results and debates. Finally, Part 6 provides conclusions.

2 Basic Definitions of Fractional Order

The basic operator for non-integer-order differentiation and integration is ${}_aD_t^k$, where $k \epsilon \mathbb{R}$ denotes the order of the integration or differentiation and, a and t stand for the operation's boundaries. It is defined as,

$$ {}_aD_t^k = \begin{cases} \frac{d^k}{dt^k}, & k > 0 \\ 1, & k = 0 \\ \int_a^t (d\tau)^k, & k < 0 \end{cases} \tag{1} $$

There exist various definitions for the fractional order differ-integral [14, 15]. The two widely used definitions are Grunwald Letnikov (GL) and Riemann Liouville (RL) [16, 17]. The GL is defined as,

$$ {}_aD_t^k f(t) = \lim_{h \to 0} h^{-k} \sum_{i=0}^{\left[\frac{t-a}{h}\right]} (-1)^i \binom{k}{i} f(t - ih) \tag{2} $$

where [.] represents the integer part. The RL is defined as,

$$ {}_aD_t^k f(t) = \frac{1}{\Gamma(n-k)} \frac{d^n}{dt^n} \int_a^t \frac{f(\tau)}{(t-\tau)^{k-n+1}} d\tau \tag{3} $$

where $n - 1 < k < n$. For resolving engineering complications, the Laplace transform is used. Using Laplace transform method the RL fractional derivative is given by,

$$ \int_0^\infty e^{-st} {}_0D_t^k f(t) dt = s^k F(s) - \sum_{x=0}^{n-1} s^k {}_0D_t^{k-x-1} f(t)|_{t=0} \tag{4} $$

For $n - 1 < k \leq n$, 's' is the Laplace operator. The additional significant uses and characteristics of fractional integrals and derivatives can be found in [18, 19].

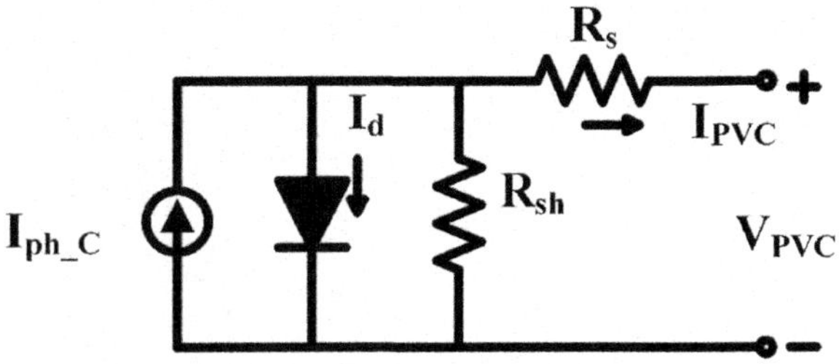

Fig. 1 One diode Photovoltaic cell

3 Solar Photovoltaic System

3.1 Solar Cell Model

The photovoltaic cell's single-diode type has a photocurrent with a parallel diode, one parallel resistor, and a series resistor across the load depicted in Fig. 1. Using KCL, The output current of solar Photovoltaic cell, I_{PVC} obtained as,

$$I_{PVC} = I_{ph\_C} - I_d - \frac{V_{PVC} + R_s * I_{PVC}}{R_{sh}} \tag{5}$$

where I_{ph_C} is photocurrent, I_d denotes diode current. The Photocurrent, I_{ph_C} of a Photovoltaic cell, which is affected by temperature and irradiation is stated as [20–22],

$$I_{ph\_C} = \frac{G}{G_{\text{ref}}}\left[I_{ph\_C\text{ref}} + \mu_{sc}(T_{\text{cell}} - T_{\text{ref}})\right] \tag{6}$$

And the diode current (I_d) is

$$I_d = I_o\left(e^{\frac{V_{PVC}+R_s*I_{PVC}}{V_t}} - 1\right) \tag{7}$$

The I_o, diode saturation dark current can be obtained as,

$$I_o = I_{o,\text{ref}}\left(\frac{T_{\text{cell}}}{T_{\text{ref}}}\right)^3 \exp\left[\left(\frac{q\varepsilon_G}{A \bullet K}\right)\left(\frac{1}{T_{\text{ref}}} - \frac{1}{T_{\text{cell}}}\right)\right] \tag{8}$$

3.2 PV Module Model

The PV module is a collection of photovoltaic cells arranged in series–parallel (SP) configurations. The PV module generates a smaller amount of power. To enhance

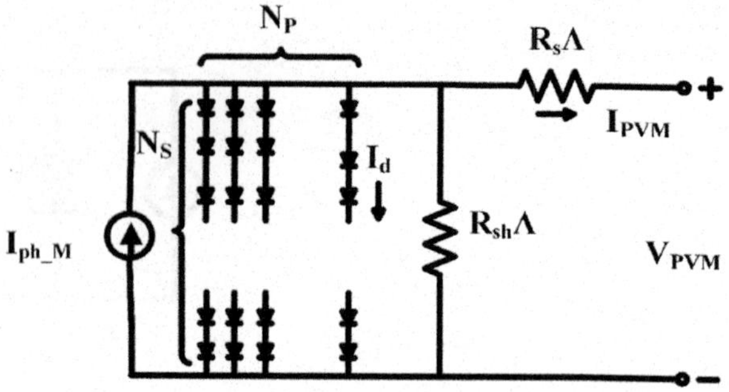

Fig. 2 PV array

the required amount of energy generation, the PV modules are constructed as an SP arrangement. The SP arrangement of PV modules is termed a PV array. As shown in Fig. 2, the PV array is made up of N_p number of parallel cells and N_s number of series cells. As a result, the output current of the PV module, I_{PVM} is calculated as,

$$I_{PVM} = N_p I_{ph\_M} - N_p I_o \left[\exp\left(\frac{V_{PVM} + I_{PVM} * R_s \Lambda}{N_s V_t} \right) - 1 \right] - \frac{V_{PVC} + \Lambda R_s * I_{PVC}}{\Lambda R_{sh}} \tag{9}$$

where

$$\Lambda = \frac{N_s}{N_p} \tag{10}$$

4 Control of Solar PV System Connected to Grid

As solar power is generating a DC at the terminals of PV panels, hence in a grid-connected PV system, the PV source is connected to the grid via a DC-DC converter and a VSI [23]. Typically, 3-phase inverters are used in larger power plants. The complete design of the solar PV source, which is connected to the utility grid is presented in Fig. 3. A 3-phase VSI is used as a power conditioning unit, which transfers the generated DC energy into the utility grid. For solar PV grid-connected system, the purpose is power flow control between the solar PV system and the grid, also power factor of inverter with good quality of power; the power flow is attained with voltage controller permitting the suitable dc link voltage based on MPPT, meanwhile, the power factor is regulated using a current regulator [24, 25].

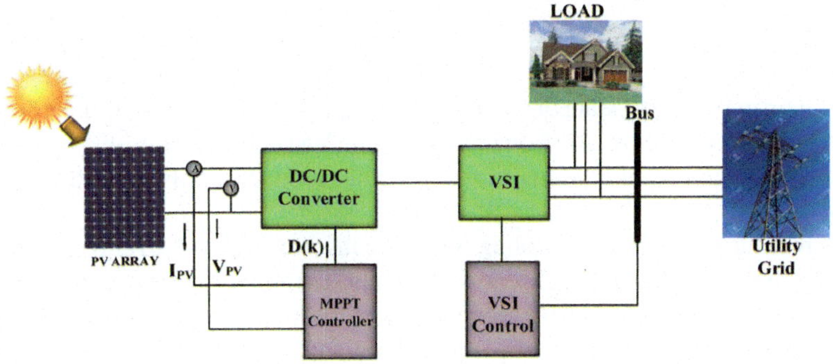

Fig. 3 Grid-connected PV system

4.1 Voltage Source Inverter Control

In this, the 3-phase system (currents and voltages) must be analyzed in a synchronous reference frame to control the real and reactive power injected into the grid, which uses the d-q components obtained by park transformation. For a 3-phase system, a park vector in the (d-q) frame is determined using the Clark transformation plus a rotation by an equivalent angle to the grid voltage phase angle, the Clark transformation is given below [26].

$$\begin{bmatrix} V_o \\ V_\alpha \\ V_\beta \end{bmatrix} = \sqrt{\frac{2}{3}} \begin{bmatrix} \frac{1}{\sqrt{2}} & \frac{1}{\sqrt{2}} & \frac{1}{\sqrt{2}} \\ 1 & -\frac{1}{2} & -\frac{1}{2} \\ 0 & \frac{\sqrt{3}}{2} & -\frac{\sqrt{3}}{2} \end{bmatrix} \begin{bmatrix} V_a \\ V_b \\ V_c \end{bmatrix} \tag{11}$$

The park (d-q) components obtained by a rotation are,

$$\begin{bmatrix} V_d \\ V_q \end{bmatrix} = \sqrt{\frac{2}{3}} \begin{bmatrix} \cos\theta & \sin\theta \\ -\sin\theta & \cos\theta \end{bmatrix} \begin{bmatrix} V_\alpha \\ V_\beta \end{bmatrix} \tag{12}$$

The real and reactive power is,

$$P = v_d i_d + v_q i_q \tag{13}$$

$$Q = v_q i_d - v_d i_q \tag{14}$$

where v_d, v_q, i_d, and i_q represent the d-q frame of the 3-phase grid voltages and currents. If vector v_q is equal to zero (i.e. aligned with the d-axis), which ensures a separate control of real and reactive powers.

$$P = v_d i_d \tag{15}$$

$$Q = -v_d i_q \tag{16}$$

To retain zero steady-state error, a PI controller is used because of the continuous nature of d-q components. As shown in Fig. 4, the VSI control is composed of a phase locked loop (PLL) block, an inner loop, an outer loop, and pulse width modulation (PWM).

The measured 3-phase grid voltages and the inverter output currents are presented in the d-q reference frame (synchronous) using parks transformation, which are inputs to the inner loop, where these d-q components of the line currents are controlled using PI controllers, so as to regulate the injected real and reactive powers into the utility grid. In the inner loop, the i_d component is compared with the output of the outer loop component i_{dref} to control the active power, and i_q component is set to zero (i_{qref}) to inject the zero reactive power into the grid. A PI regulator used in the outer loop, which controls the DC voltage to follow the reference voltage, so the energy flow between the grid and the solar PV source is ensured. The PLL calculates the necessary phase angle of the three-phase grid voltage for park transformation and synchronizes the inverter's output voltage with the 3-phase voltage of the grid [27, 28]. The output d-q components of the inner loop are transformed into a 3-phase frame, which are the inputs to the PWM block, which controls the commutation of inverter switches, to follow the reference voltage.

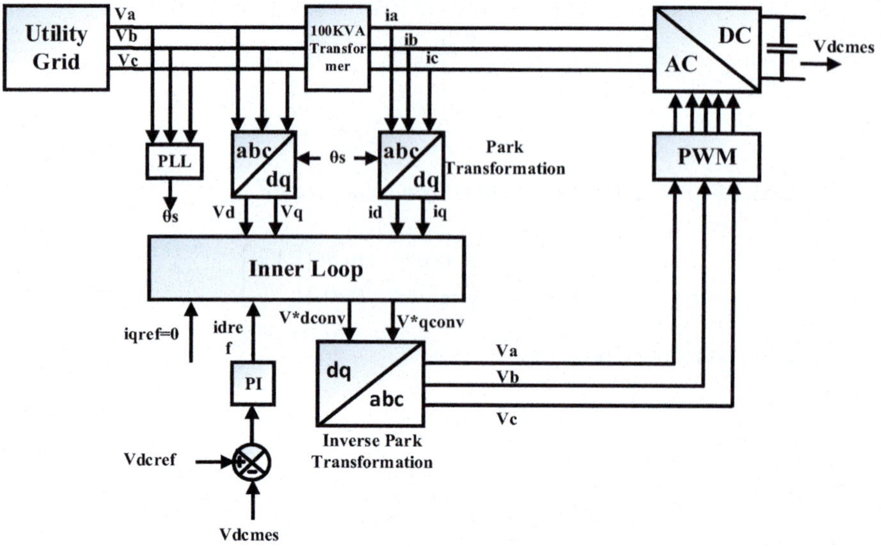

Fig. 4 VSI control

4.2 Extremum Seeking Control

Several ESC configuration types have been explored in the past, and among the numerous structures, sinusoidally perturbed ESC configuration has garnered significant interest from researchers [29, 30]. The overall structure for the sinusoidal perturbed single input-single output [SISO] extremum-seeking algorithm is predicted in Fig. 5. This kind of ESC structure uses a sluggish periodic perturbation signal, $\sin(\omega t)$, and adds it to the assessed signal, $\widehat{\theta}$. Due to the aforementioned sluggish perturbation, the plant seems like a static map, $c = f(\theta)$ to ESC, and its dynamics do not obstruct the ESC scheme. The perturbation signal, $a\sin(\omega t)$ produces a response of 'c', which is either in phase or out of phase with $a\sin(\omega t)$ when the estimated signal, $\widehat{\theta}$ is on both sides of the optimal point $\theta*$. The HPF in the ESC configuration removes the "DC component" of c. Therefore, the output signal, $\left(\left(\frac{s}{s+\omega_h}\right)c\right)$ and the perturbation signal, $a\sin(\omega t)$ are two approximate sinusoids, which are out of phase for $\widehat{\theta} > \theta^*$ and in phase for $\widehat{\theta} < \theta^*$ [31–33]. The operation of ESC is described in Fig. 6, in which the output of ESC is predicted when the operating point is greater, equal, or lesser than the extremum point. As the product of two signals which are in phase provides a positive mean and out of phase provides a negative mean, this concept will be used to extract the optimal point using a gradient detector. In both cases, the product of two sinusoidal signals produces a "DC component" which is extracted with the LPF, $(w_l/s + w_l)$. Thus, the direct component ξ can be claimed to be almost the sensitivity $\left(\frac{a^2}{2}\right)\frac{\partial f}{\partial \theta}\left(\widehat{\theta}\right)$ and the gradient update law $\dot{\widehat{\theta}} = k\left(\frac{a^2}{2}\right)\frac{\partial f}{\partial \theta}\left(\widehat{\theta}\right)$ which is determined by the sensitivity function to tune $\widehat{\theta}$ to θ^* and control goal is achieved. Mathematically, the ESC scheme can be written as,

$$\begin{cases} c = f(\widehat{\theta} + a\sin(\omega t)) \\ \dot{\widehat{\theta}} = -k\xi \\ \xi = v * L^{-1}\left\{\frac{\omega_l}{s+\omega_l}\right\} \\ v = (c * L^{-1}\left\{\frac{s}{s+\omega_h}\right\})\sin(\omega t) \end{cases} \tag{17}$$

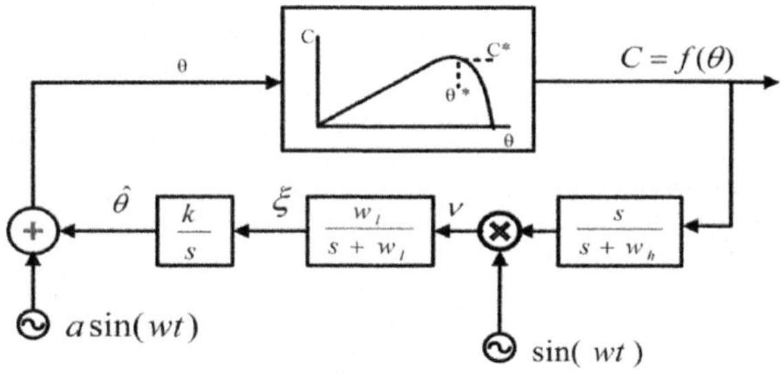

Fig. 5 Sinusoidal perturbed extremum seeking algorithm

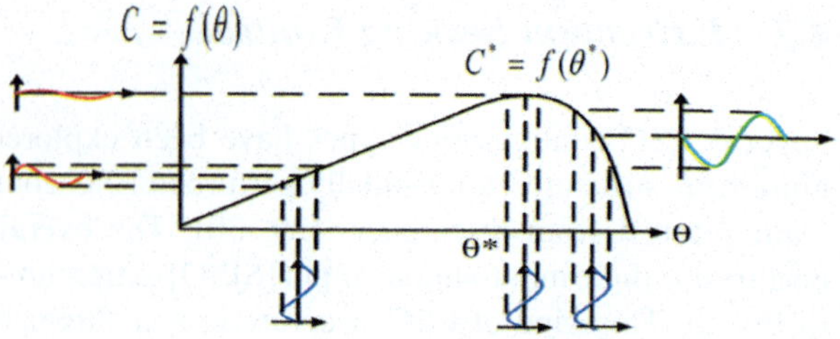

Fig. 6 Operation of ESC algorithm

4.3 Fractional Order Extremum Seeking Algorithm

The linearized model concerning to error signal and the optimized point for IOESC is depicted in Fig. 5.

$$\frac{\tilde{\theta}}{\theta*} = \frac{1}{1 + L(s)} \tag{18}$$

where

$$L(s) = \frac{ka^2}{2s}\left(\frac{s + j\omega}{s + j\omega - \omega_h} + \frac{s - j\omega}{s - j\omega + \omega_h}\right) \tag{19}$$

Hence,

$$\frac{\tilde{\theta}}{\theta*} = \frac{1}{1 + L(s)} = \frac{s(s^2 + 2\omega_h s + \omega_h^2 + \omega^2)}{s^3 + (2\omega_h + ka^2)s^2 + (\omega_h^2 + \omega^2 + ka^2 w_h)s + ka^2\omega^2} \tag{20}$$

The fractional order Extremum seeking control is obtained by substituting the integer order integrator and filters with fractional order integrator and fractional filters. This replacement advances the convergence speed and efficiency of the ESC algorithm. The linearized model for FOESC can be obtained by replacing s with s^λ as shown in Fig. 7.

$$\frac{\tilde{\theta}}{\theta*} = \frac{1}{1 + L(\gamma)} = \frac{\gamma(\gamma^2 + 2\omega_h\gamma + \omega_h^2 + \omega^2)}{\gamma^3 + (2\omega_h + ka^2)\gamma^2 + (\omega_h^2 + \omega^2 + ka^2 w_h)\gamma + ka^2\omega^2} \tag{21}$$

where $\gamma = s^\lambda$. By looking at Eq. (20), when ka^2 is small compared to ω, the IOESC transfer function, $\frac{\tilde{\theta}}{\theta*}$ is asymptotically stable for all $k > 0$.

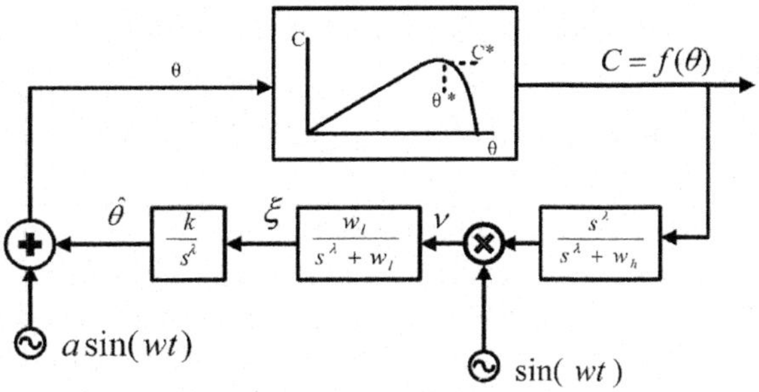

Fig. 7 Fractional order ESC

5 Proposed Fractional Order HPF Based ESC

The integrator and the low-pass filter used in ESC are in series and have no effect on the rate of convergence, and their purpose is to diminish the noise in the error signal. The HPF reduces the DC component of '*c*'. For these reasons, the HPF is replaced by fractional order HPF (FO-HPF), without loss of generality, utilizing the fractional operator called fractional order HPF-based ESC (FO-HPF ESC), shown in Fig. 8. Considering the FO-HPF with an integer order LPF and integrator in ESC, the FO-HPF is given by,

$$G_{FO-HPF} = \frac{s^{\lambda}}{s^{\lambda} + \omega_h}, 0 < \gamma < 1 \tag{22}$$

Therefore, the linearized model of FO-HPF ESC is,

$$\frac{\tilde{\theta}}{\theta *} = \frac{1}{1 + L(s)} = \frac{s(s^{2\gamma} + 2\omega_h s^{2\gamma} + \omega_h^2 + \omega^2)}{s(s^{2\gamma} + 2\omega_h s^{\gamma} + \omega_h^2 + \omega^2) + ka^2(s^{2\gamma} + 2\omega_h s^{\gamma} + \omega_h^2 + \omega^2)} \tag{23}$$

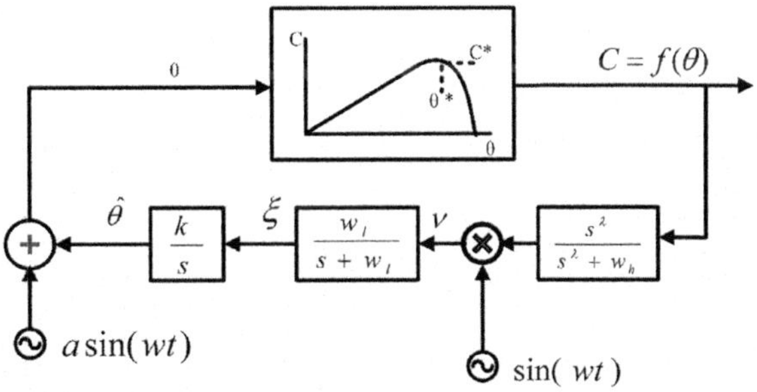

Fig. 8 Fractional order HPF structured ESC

6 Simulation Results and Discussions

6.1 Results Under Uniform Insolation Condition

The time domain simulations for the grid-connected solar photovoltaic system are performed using Matlab/Simulink. The simulated results are contrasted with those from the IOESC and P&O approaches to show the superiority of the proposed Fractional order HPF-based ESC for the MPPT controller. Three MPPT algorithms are evaluated for their effectiveness, robustness, and convergence speed using three different solar irradiances (1.0, 0.75, and 0.5 kW/m^2). Figure 9 exhibits both the V-I and V-P curves.

The simulated results of grid-connected PV array for aforementioned irradiances with various MPPT methods are illustrated in Fig. 10. The proposed FOESC, IOESC, and P&O reach MPP with convergence time 0.065, 0.07, and 0.12 s for 1000 W/m^2, and 0.06, 0.08, and 0.106 s for 750 W/m^2, and 0.05, 0.075, and 0.11 s for 500 W/m^2. The results demonstrate that the proposed FOESC reduces convergence speed by 45.83% for 1.0 kW/m^2, 43.39% for 0.75 kW/m^2, and 54.54% for 0.5 kW/m^2.

A comparative study of various MPPT approaches is listed in Table 1. It demonstrates how the proposed FOESC outperforms the IOESC and P&O in terms of convergence speed and efficiency.

In general, the P&O method is widely used for grid-connected solar PV systems, but its main drawback is fluctuations around MPP, which causes the loss of energy, however, the proposed FOESC technique produces stable PV output power and achieves MPP in short convergence speed and with better accuracy. The 3-phase voltages shown in Fig. 11, and three-phase currents injected into the utility grid illustrated in Fig. 12 have sinusoidal waveforms ensuring low voltage and current harmonic injections.

These grid-injected voltages and currents are in phase shown in Fig. 13, which confirms that zero reactive power is injected into a utility grid. Figure 14 represents

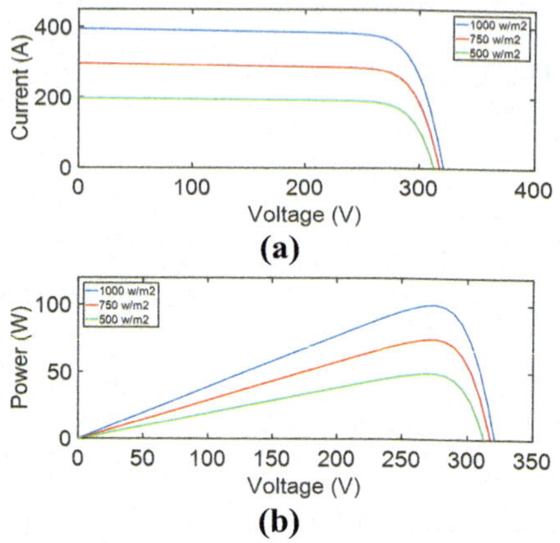

Fig. 9 **a** V-I curves and **b** V-P curves

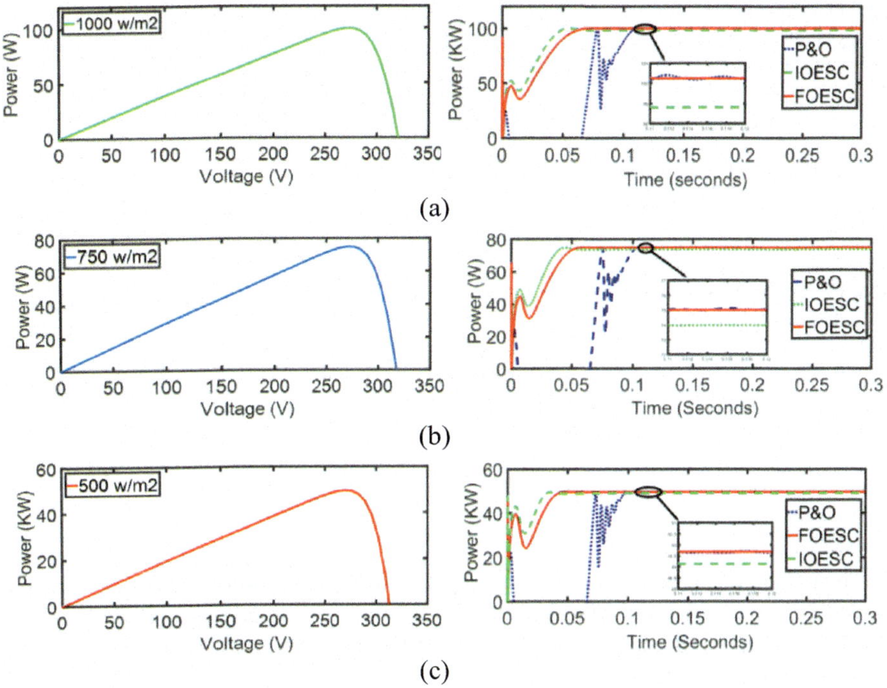

Fig. 10 The power output **a** for insolation 1.0 kW/m^2, **b** for insolation 0.75 kW/m^2, and **c** for insolation 0.5 kW/m^2

Table 1 Comparative study on MPPT algorithms

Insolation	Method	Power at MPP (KW)	Tracking time (Sec)	Maximum power (KW)	Efficiency
1000 W/m^2	P&O	100.23	0.12	100.7	99.53
	IOESC	98.82	0.07		98.13
	FOESC	100.237	0.065		99.54
750 W/m^2	P&O	75.01	0.106	75.2	99.74
	IOESC	75.02	0.08		98.4
	FOESC	75.02	0.06		99.76
500 W/m^2	P&O	49.68	0.11	49.85	99.65
	IOESC	49.14	0.075		98.57
	FOESC	49.7	0.05		99.69

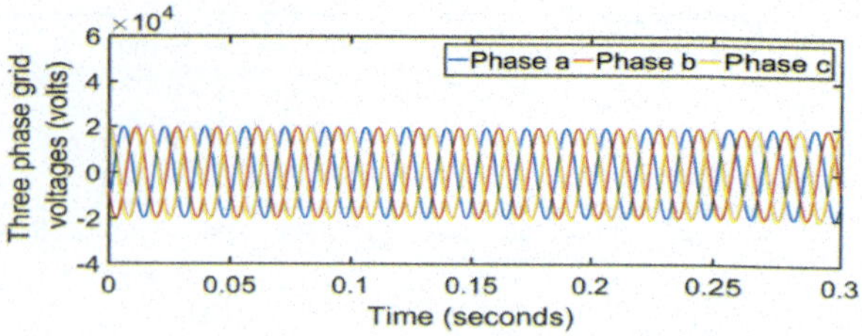

Fig. 11 Three phase voltages injected into the utility grid

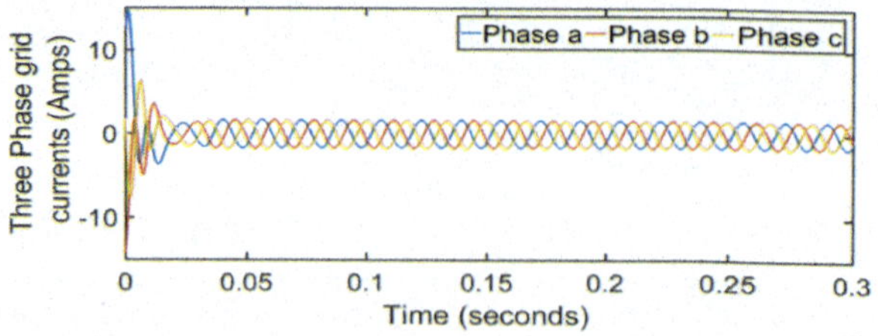

Fig. 12 Three phase currents injected into the utility grid

the instantaneous real and reactive power injected into the utility grid. The reactive power is zero and the real power is 49.10 KW, which is the peak power delivered by solar PV when irradiance is 500 W/m^2. The DC link voltage follows the reference value shown in Fig. 15 exactly.

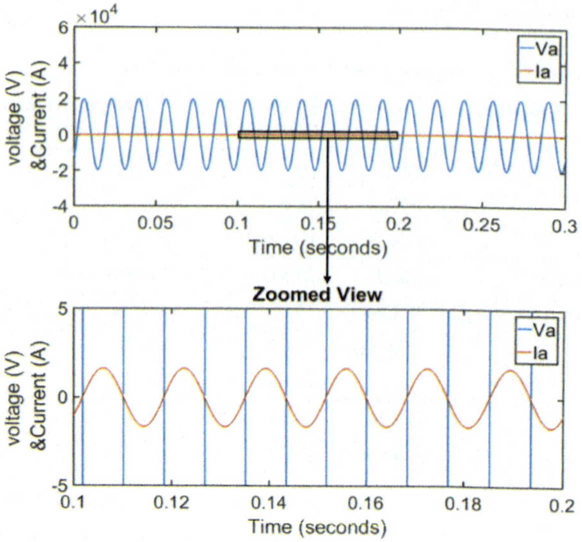

Fig. 13 Phase 'a' grid current and voltage

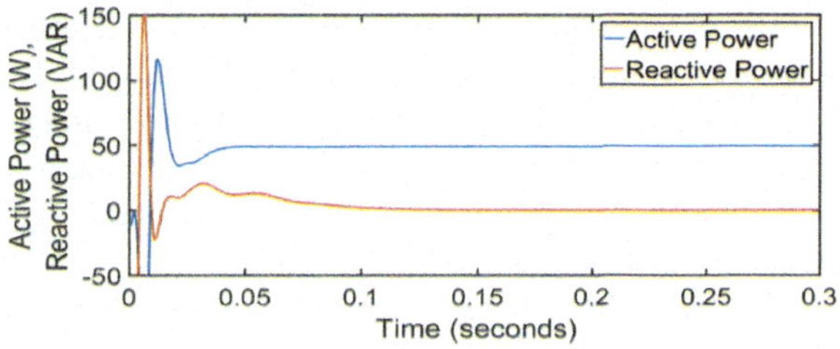

Fig. 14 Real and reactive power injected into the utility grid

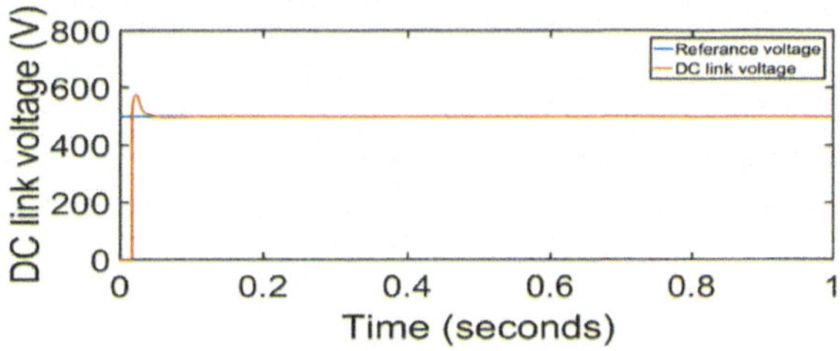

Fig. 15 DC link voltage

6.2 Results Under Variable Insolation Conditions

In general, the sun irradiations are not constant for this reason, the variable insolation profile (assumed climatic conditions) shown in Fig. 16 is used to investigate the performance of the proposed ESC-based MPPT method. Figure 17 shows how effective the impact of the proposed MPPT is on convergence rate and tracking efficiency of PV output power with MPP for each time insolation is changed.

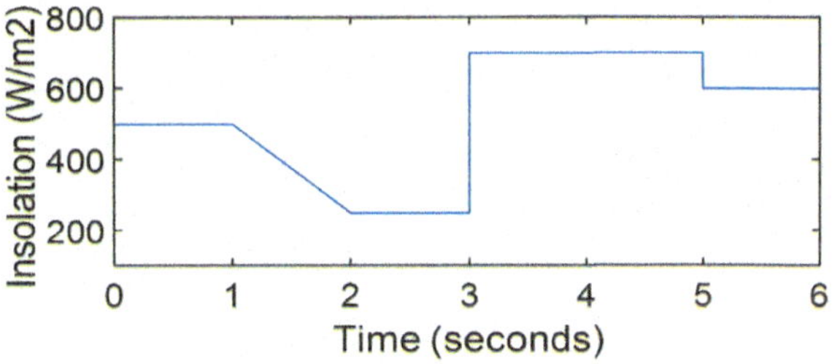

Fig. 16 Insolation profile in W/m^2

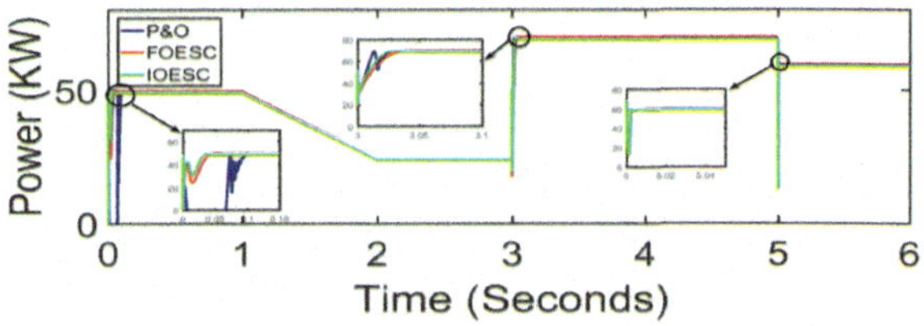

Fig. 17 Solar PV power output under fluctuating climatic conditions

The inverter output voltage changes with the change in input DC voltage, which leads to the flow of reactive power into the grid. Hence, maintaining the DC link voltage constant and equal to the reference voltage is required to supply the grid with a fixed real power. Figure 18 represents the regulated DC link voltage, which continuously follows the reference voltage under variable insolation. The active and reactive power is presented in Fig. 19.

To synchronize the grid voltage with the inverter output voltage the VSI control uses a PLL block, which obtains the phase angle of the grid voltage. Figures 20 and 21 represent the 3-phase grid voltages and currents, which ensures that zero reactive power with a low harmonic rate is injected into the utility grid.

Fig. 18 DC link voltage under variable insolation

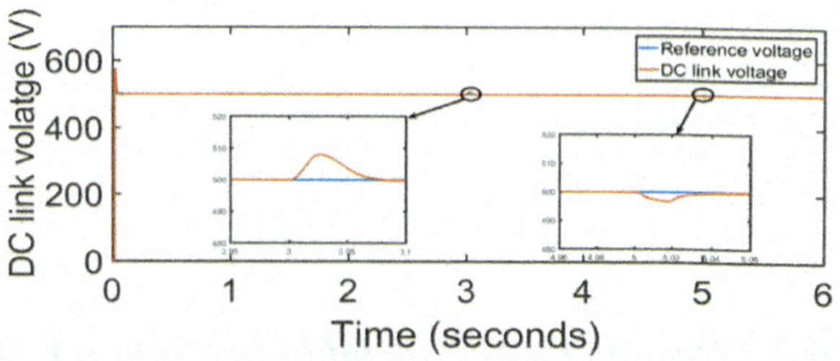

Fig. 19 3-phase active and reactive power injected into the grid

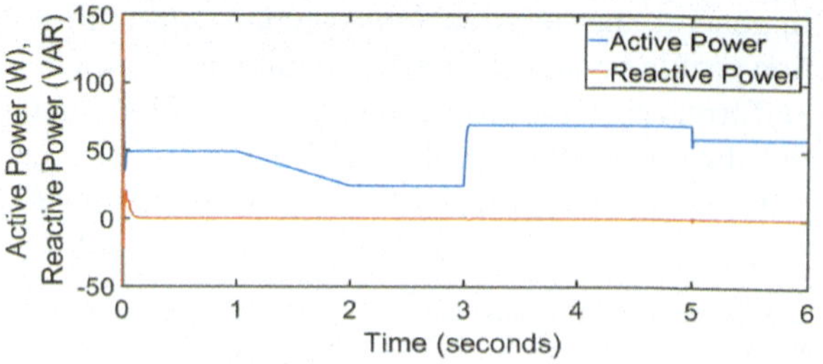

Fig. 20 3-phase grid injected voltages

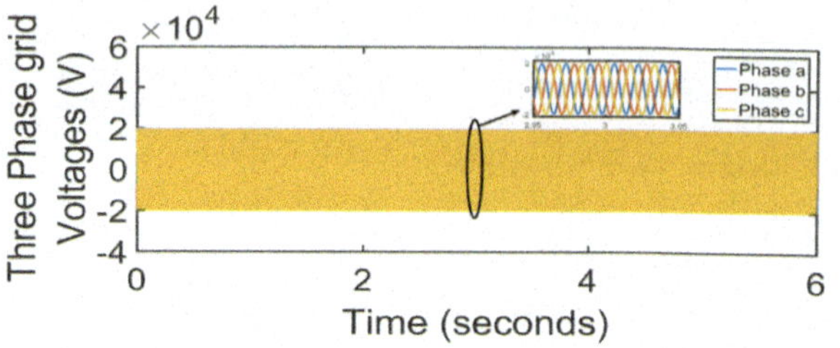

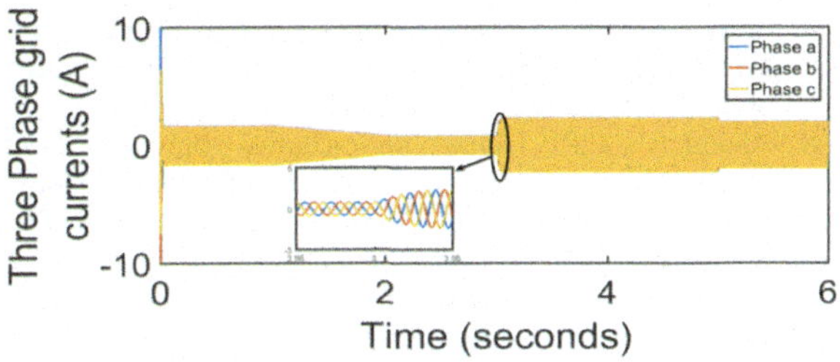

Fig. 21 3-phase grid injected currents

7 Conclusions

In this work, the fractional order HPF-based ESC algorithm called FOESC is proposed to track the MPP in a grid-connected solar PV system. This proposed MPPT algorithm is investigated under uniform and variable insolations, which tracks the maximum PV output power with short convergence speed and with low ripples on the DC side. Also, the AC side 3-phase gird injected currents are in phase with the grid voltages, which ensures that zero reactive power is injected into the grid. The proposed technique is benchmarked against traditional P&O and IOESC methods. The simulation results witness the proposed FOESC technique is better than IOESC and P&O, which enhances convergence speed and efficiency.

Acknowledgements This research work was supported by "Woosong University's Academic Research Funding—2023".

References

1. Trejos A, Gonzalez D, Ramos-Paja CA (2012) Modeling of step-up grid-connected photovoltaic systems for control purposes. Energies 5(6):1900–1926. https://doi.org/10.3390/en5061900
2. Subudhi B, Pradhan R (2013) A comparative study on maximum power point tracking techniques for photovoltaic power systems. IEEE Trans Sustain Energy 4(1):89–98. https://doi.org/10.1109/TSTE.2012.2202294
3. Hernandez J, Gordillo G, Vallejo W (2013) Predicting the behavior of a grid-connected photovoltaic system from measurements of solar radiation and ambient temperature. Appl Energy 104:527–537. https://doi.org/10.1016/j.apenergy.2012.10.022
4. Esram T, Chapman PL (2007) Comparison of photovoltaic array maximum power point tracking techniques. IEEE Trans Energy Convers 22(2):439–449. https://doi.org/10.1109/TEC.2006.874230
5. Nedumgatt JJ (2011) Perturb and observe MPPT algorithm for solar PV systems-modeling and simulation. In: Annual IEEE India conference (INDICON), Hyderabad, India, pp 1–6. https://doi.org/10.1109/INDCON.2011.6139513
6. Kollimalla SK, Mishra MK (2014) A novel adaptive P&O MPPT algorithm considering sudden changes in the irradiance. IEEE Trans Energy Convers 29(3):602–610. https://doi.org/10.1109/TEC.2014.2320930

7. Sivakumar P (2015) Analysis and enhancement of PV efficiency with incremental conductance MPPT technique under non-linear loading conditions. Renew Energy 81:543–550. https://doi.org/10.1016/j.renene.2015.03.062
8. Alajmi BN (2011) Fuzzy-logic-control approach of a modified hill-climbing method for maximum power point in microgrid standalone photovoltaic system. IEEE Trans Power Electron 26(4):1022–1030. https://doi.org/10.1109/TPEL.2010.2090903
9. Rai AK (2011) Simulation model of ANN based maximum power point tracking controller for solar PV system. Sol Energy Mater Sol Cells 95(2):773–778. https://doi.org/10.1016/j.solmat.2010.10.022
10. Wai RJ, Wang WH (2008) Grid-connected photovoltaic generation system. IEEE Trans Circuits Syst I Regul Pap 55(3):953–964. https://doi.org/10.1109/TCSI.2008.919744
11. Castilla M (2009) Control design guidelines for single-phase grid-connected photovoltaic inverters with damped resonant harmonic compensators. IEEE Trans Industr Electron 56(11):4492–4501. https://doi.org/10.1109/TIE.2009.2017820
12. Cacciato M (2010) Soft-switching converter with HF transformer for grid-connected photovoltaic systems. IEEE Trans Industr Electron 57(5):1678–1686. https://doi.org/10.1109/TIE.2009.2032201
13. Krstić M (2000) Performance improvement and limitations in extremum seeking control. Syst Control Lett 39(5):313–326. https://doi.org/10.1016/S0167-6911(99)00111-5
14. Tepljakov A (2017) Fractional-Order Modeling and Control of Dynamic Systems. Springer International Publishing, Cham (Springer Theses). https://doi.org/10.1007/978-3-319-52950-9
15. Luo C, Wang X (2013) Chaos in the fractional-order complex Lorenz system and its synchronization. Nonlinear Dyn 71(1):241–257. https://doi.org/10.1007/s11071-012-0656-z
16. Reddy SS, Sandeep V, Chitti Babu B, Jung CM (2018) Multi-objective based optimal generation scheduling considering wind and solar energy systems. Int J Emerg Electric Power Syst 19(5). https://doi.org/10.1515/ijeeps-2018-0006
17. Cardoso LC, Dos Santos FLP, Camargo RF (2018) Analysis of fractional-order models for hepatitis B. Comput Appl Math 37(4):4570–4586. https://doi.org/10.1007/s40314-018-0588-4
18. Kujur S, Dubey HM, Salkuti SR (2023) Demand response management of a residential microgrid using chaotic aquila optimization. Sustainability 15(2):1484. https://doi.org/10.3390/su15021484
19. Petráš I (2011) Fractional-order nonlinear systems: modeling, analysis and simulation. Springer, New York, NY, USA. https://doi.org/10.1007/978-3-642-18101-6
20. Chenni R (2007) A detailed modeling method for photovoltaic cells. Energy 32(9):1724–1730. https://doi.org/10.1016/j.energy.2006.12.006
21. Bhukya L, Annamraju A, Nandiraju S (2021) A novel maximum power point tracking technique based on Rao-1 algorithm for solar PV system under partial shading conditions. Int Trans Electr Energy Syst 31(9). https://doi.org/10.1002/2050-7038.13028
22. Bhukya L, Nandiraju S (2020) A novel photovoltaic maximum power point tracking technique based on grasshopper optimized fuzzy logic approach. Int J Hydrogen Energy 45(16):9416–9427. https://doi.org/10.1016/j.ijhydene.2020.01.219
23. Ram V, Infantraj SSR (2022) Modelling and simulation of a hydrogen-based hybrid energy storage system with a switching algorithm. World Electric Vehicle J 13(10):188. https://doi.org/10.3390/wevj13100188
24. Rey-Boué AB (2012) An integrative approach to the design methodology for 3-phase power conditioners in photovoltaic grid-connected systems. Energy Convers Manage 56:80–95. https://doi.org/10.1016/j.enconman.2011.11.012
25. Salkuti SR (2022) Emerging and advanced green energy technologies for sustainable and resilient future grid. Energies 15(18):6667. https://doi.org/10.3390/en15186667
26. Gui Y (2017) Direct power control of grid connected voltage source inverters using port-controlled Hamiltonian system. Int J Control Autom Syst 15(5):2053–2062. https://doi.org/10.1007/s12555-016-0521-9
27. Kayisli K (2023) Super twisting sliding mode-type 2 fuzzy MPPT control of solar PV system with parameter optimization under variable irradiance conditions. Ain Shams Eng J 14(1). https://doi.org/10.1016/j.asej.2022.101950

28. Záskalický P (2017) Mathematical model of a five-phase voltage-source PWM-controlled inverter. Electr Eng 99(4):1179–1184. https://doi.org/10.1007/s00202-017-0643-y
29. Xu T, Chen G, Zhou G, Liu Z, Zhang Z, Yuan S (2022) Fast source seeking with obstacle avoidance via extremum seeking control. In 13th Asian control conference (ASCC), Jeju, Republic of Korea, pp 2097–2102. https://doi.org/10.23919/ASCC56756.2022.9828209
30. Liu SJ, Krstic M (2012) Introduction to extremum seeking. In: Stochastic averaging and stochastic extremum seeking. Communications and control engineering. Springer, London. https://doi.org/10.1007/978-1-4471-4087-0_2
31. Krstic M, Wang HH (2000) Stability of extremum seeking feedback for general nonlinear dynamic systems. Automatica 36(4):595–601. https://doi.org/10.1016/S0005-1098(99)00183-1
32. Malek H, Chen Y (2016) Fractional order extremum seeking control: performance and stability analysis. IEEE/ASME Trans Mechatron 21(3):1620–1628. https://doi.org/10.1109/TMECH.2016.2517621
33. Vweza AO, Chong KT, Lee DJ (2015) Gradient-free numerical optimization-based extremum seeking control for multiagent systems. Int J Control Autom Syst 13(4):877–886. https://doi.org/10.1007/s12555-013-0221-7

Design and Analysis of Maximum Power Point Tracking-Based Charging System

Partha Sarathi Panuya, Surender Reddy Salkuti, and Seong-Cheol Kim

Abstract Conventional energy sources cover the most percentage of our requirement, which in turn leads to pollution and global warming. In recent years there is a major focus on the use of non-conventional sources of energy such as solar which is intermittent due to environmental conditions. To extract maximum power from the solar photovoltaic (PV) panel, it is essential to use different types of Maximum Power Point Tracking (MPPT) algorithms. But these algorithms need rigorous testing before they can be applied practically on PV panels which is troublesome due to the higher dependency on solar irradiance, large space requirements if power rating increases, etc. These reasons lead us to develop a PV emulator which costs less than the commercially available PV emulator with acceptable performance. This chapter proposed a PV emulator which consists of a DC–DC buck converter. This power stage is controlled in such a way that its operating points run on the characteristics of a real PV source. The control stage consists of PV modeling and solving the same with the help of a numerical method, this together decides the operating point of the power circuit using the direct referencing method. The theoretical analysis has been validated by simulation results. Finally, the PV emulator is used for battery charging through the MPPT-controlled DC–DC boost converter. The results are compared with the real PV panel and it is found that they are in close agreement. This new testing bed is more flexible in terms of emulating different ratings and brands of solar panel at our disposal, portable so that it can be carried easily, and also provide controlled regulation in terms of irradiance and temperature. Controlled regulation of environmental factors leads to more tunable optimization of MPPT controllers. Testing of the MPPT in the solar PV battery charging scenario is performed in a software-based environment.

Keywords MPPT · Photovoltaic emulator · DC-DC boost converter · Battery charging

P. S. Panuya · S. R. Salkuti (✉) · S.-C. Kim
Department of Railroad and Electrical Engineering, Woosong University, Daejeon, Republic of Korea
e-mail: surender@wsu.ac.kr

S. R. Salkuti et al. (eds.), *Power Quality in Microgrids: Issues, Challenges and Mitigation Techniques*, Lecture Notes in Electrical Engineering 1039,
https://doi.org/10.1007/978-981-99-2066-2_9

Nomenclature

MPPT	Maximum Power Point Tracker
PV	Photovoltaic
STS	Solar Tracking System
CCM	Continuous conduction mode
DCM	Discontinuons conduction mode
PV Emulator	Photovoltaic Emulator
PV Array	Photovoltaic Array
NR Method	Newton–Raphson Method
MOSFET	Metal–oxide–semiconductor field-effect transistor
P&O Algorithm	Perturb and observe algorithm
RMS	Root mean square
MATLAB	Matrix Laboratory

1 Introduction

The era of photovoltaic (PV) offers many benefits in creating a sustainable wellspring of energy as it has no fuel costs, requires less support, does not being polluting, and emanates no noise. The DC–DC boost converter is utilized to amplify the voltage from PV to an appropriate level of voltage acknowledged by the load. The brain of the boost converter will be the MPPT tracker which attempts to settle the voltage delivered by the converter through its controlled algorithm. MPPT control is introduced to tackle maximum power from PV emulators under various sunlight-based irradiance and temperature setting. It is a second-order system comprised of an inductor, a capacitor, and a diode with the load resistance associated in corresponding with the capacitor. The utilization of devices increases PV output voltages, which is fundamental for energizing the battery [1, 2]. As a result, PV is not consistent due to the temperature variation and environmental conditions, the design of such a converter is essential to guarantee standard charging up of the battery [3]. This boost converter is designed based on the information about PV voltages as well as the determined duty cycle for MPPT. MPPT algorithm is utilized primarily in a system where the wellspring of the source is nonlinear, for example, the sun-powered PV modules [4, 5]. Therefore, to extract maximum power from it, a power converter coupled with an MPPT controller [6–8].

In this chapter, an MPPT coordinated boost converter is used in sun-powered PV applications with a battery charger associated with the PV framework. The reason for charging a battery is only for the capacity of electrical energy [9]. This energy if it comes from the sun-based PV framework then quick charging of the battery should be possible with the assistance of MPPT controlled boost converter [10, 11]. Lately, continuous recreation and imitation are now compelling devices for architects to approve and test different MPP calculations before executing them in the equipment

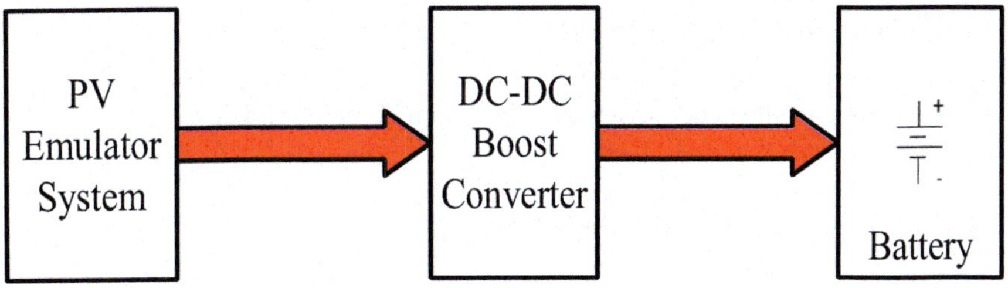

Fig. 1 PV emulator fed system

stage. A common block diagram of a PV emulator-fed system that has a battery load is addressed in Fig. 1.

2 Description of the System

In the next section, we will cover the topic mainly of power, control stages, and the design of MPPT-controlled charging system. Buck converter with emulator control and battery as the load is used here. Two autonomous control cycles are proposed to control the independent PV emulator MPPT system [12, 13]. Emulator control is positioned to control the buck converter to such an extent that it acts like a genuine PV source [14, 15]. Further, emulator control is introduced to provide the control strategy combined with an electrical circuit model of a genuine PV source and tackle something very similar with the assistance of mathematical calculation [16]. An overview of the proposed system has been depicted in Fig. 2.

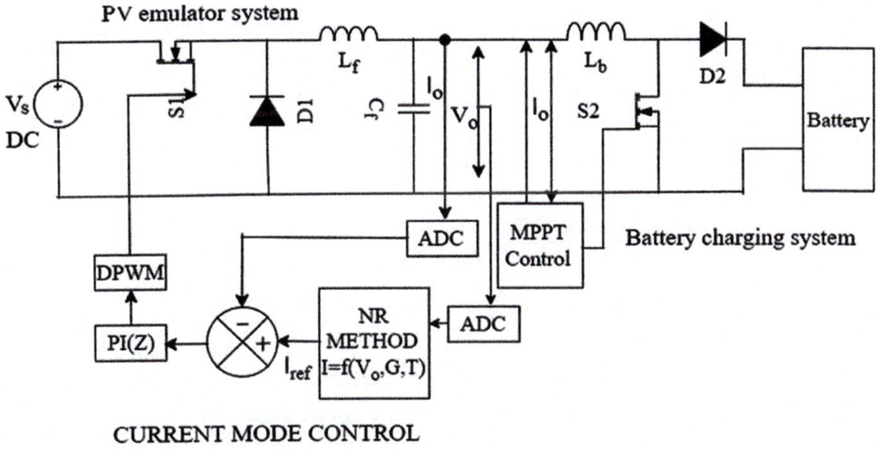

Fig. 2 Battery charging system

2.1 Power Stage Component

The power stage of the proposed system has been depicted in Fig. 3.

2.1.1 PV Emulator (DC–DC Buck Converter)

To accomplish continuous energy flow from source to destination, the two important DC–DC converters [17, 18], applied in this research, were using a pair of switching devices, typically one regulated and one unregulated. One capacitance and one inductance are also used to store and release power from the source to the destination of the converters. Current and voltage are also filtered. CCM and DCM are two different phases. As a result, both types of operations must be considered when designing a converter and its regulation. Whenever the device is turned on for DT seconds, the inductor current is conducted and the diode is reverse biased [19, 20]. This leads to a positive voltage throughout the inductor, $V_L = V_S - V_O$. The circuit diagrams of buck converter and boost converter are depicted in Fig. 4.

The inductor current i_L increases linearly as the voltage rises. Due to the obvious inductor storing energy, i_L keeps flowing even when the switches are switched off. Till the device is turned on again, this current flow and $V_L = -V_O$ for a time length (1-D) T. The integral of the inductive voltage beyond one timespan is now equal to zero [21].

$$\int_0^T V_L dt = \int_0^{t_{on}} V_L dt + \int_0^{t_{off}} V_L dt = 0 \tag{1}$$

$$(V_S - V_O) \times DT + (-V_O) \times (1 - D)T = 0 \tag{2}$$

$$\frac{V_O}{V_S} = D \tag{3}$$

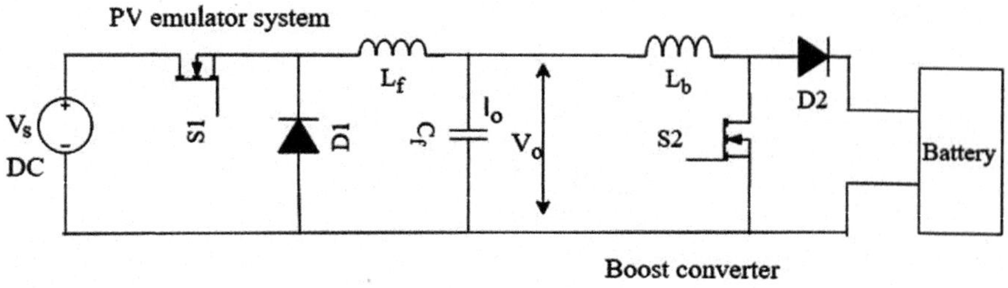

Fig. 3 Power stage of the proposed system

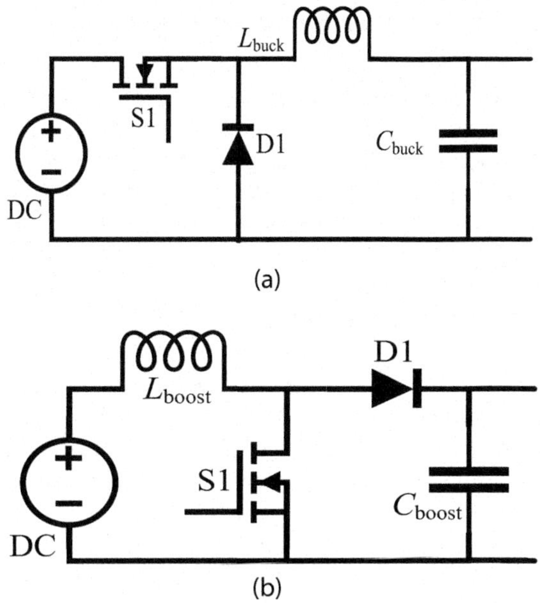

Fig. 4 Circuit Diagram of **a** Buck Converter, **b** Boost Converter

2.1.2 DC–DC Boost Converter

Boost converter generates high output voltage than its input voltage. Once the switched on for DT seconds, the inductor current is conducted and the diode is reverse biased. As an outcome, the inductor receives a positive voltage $V_L = V_S$. The current through the inductor i_L increases linearly as the voltage rises [22, 23]. Due to the obvious inductor storing energy, i_L keeps flowing even when the switch is switched off. Till the switched on again, this current flows through the diode and $V_L = V_S - V_O$ for a time length (1-D) T. The integral of the inductor voltage beyond one timespan is now equal to zero [24].

$$\int_0^T V_L dt = \int_0^{t_{on}} V_L dt + \int_0^{t_{off}} V_L dt = 0 \tag{4}$$

$$(V_S) \times DT + (V_S - V_O) \times (1 - D)T = 0 \tag{5}$$

$$\frac{V_O}{V_S} = \frac{1}{1\text{-}D} \tag{6}$$

In the next section, the control stage has been described. This helps the power stage to operate by using the algorithm introduced in it.

2.2 Controller Stage Component

2.2.1 Emulator Control

The design of the emulator control algorithm has been described in the previous chapter [25]. In this chapter, the modeling of the controller and implementation using the NR method have been specified.

2.2.2 MPPT Controller

PV array has a nonlinear curve and output power depends on environmental factors like sunlight-based illumination and temperature [26, 27]. There is a point on the I–V, P–V curve of the PV array called as Maximum power point, where the PV panel creates its maximum power. Area of MPP changes with changes in environmental conditions. The need for MPPT is to change the solar working voltage near MPP under changing sudden conditions. To ceaselessly accumulate the maximum power from the PV array, it needs to work at its MPPT despite the inhomogeneous change in ecological circumstances [28].

In this chapter, the Hill climbing algorithm is used, and its flow chart is shown in Fig. 5. The Hill climbing technique is simpler contrasted with P&O since no compensator is utilized in the control system. As a result, the problem with the photovoltaic module was isolated. That basic computation illustrated in Fig. 5 underpins the Hill climbing methodology [29, 30]. Once the voltage and current are assessed at the beginning and the signals are sent to the processor. V_O, G and T are used to calculate the I_{pv}. I_{pv} matches with I_O assuming I_{pv} bigger compare to I_O, and I_O increments. To increment I_O, V_O is expanded by enlarging the duty. The duty expanded by D_{step}.

Nonetheless, assuming I_{pv} is more modest than I_O, I_O diminishes. To diminish I_O, V_O is diminished by lessening the D [31]. The duty is diminished as indicated by the consistent D_{step}. Changes of D supplanted with I_{ref}. Figure 6 guides us to surmise that augmenting the voltage builds the power while working on the left of the power point, yet power diminishes while working on the right half of MPP. Assuming that power is expanding, the perturbation remains; however, then again power diminishes, and perturbation reverses [32]. This should be revolving. So, MPP is accomplished but arriving at a similar it oscillates around that point. The recurrences of the oscillation are controlled by the perturbation step size. If the step size decreases oscillation is reduced yet, the controller becomes drowsy.

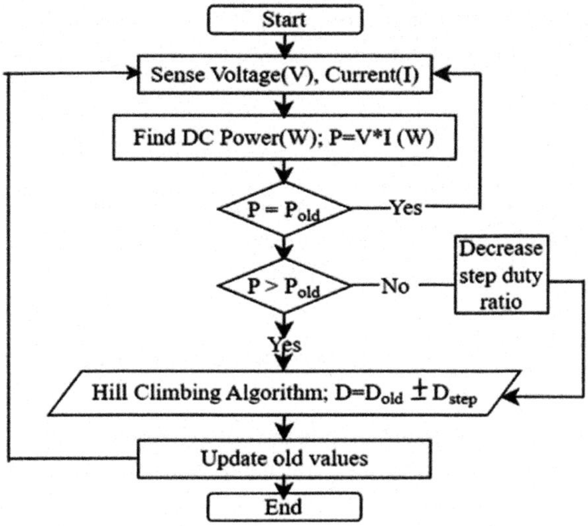

Fig. 5 Flow chart of Hill Climbing Algorithm

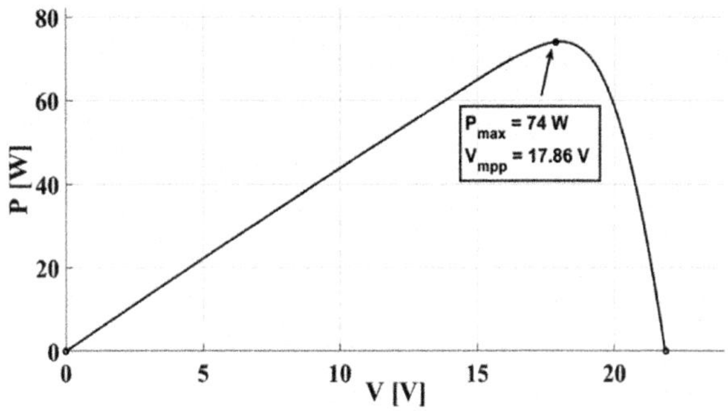

Fig. 6 Power curve

3 Design of Passive Components for Buck Converter

The ideal design of the passive component is a very crucial part of the ordinary activity of the power stage. This part influences the dynamic performances of the converter. These are likewise answerable for the misfortunes at various phases of a converter, further they additionally filter the undesirable signal from the output of the converter.

3.1 *Filter Inductor for Buck Converter* (L_f)

The filter inductor is the exigent element of the output stage of the buck converter. It stores and feeds energy to the load [33]. Designating an inductor straightforwardly impacts how much current ripple is seen on the inductor current, as well as the current capacity of the buck converter itself. The equation is formulated as

$$L_f = \frac{D(1\text{-}D)V_s}{f_{sw}dI} \tag{7}$$

where D is the duty ratio, V_s is the source voltage, f_{sw} is the switching frequency and dI is the inductor ripple current.

3.2 *Filter Capacitor for Buck Converter (C_f)*

Output capacitance affects the output voltage overshoot that happens during changes in load current. Increases the capacitance and diminishes the amount of ripple voltage present. The capacitor additionally has a series resistance that impacts the voltage ripple. Misfortunes because of capacitor series resistance will restrain the capacitor capacity of quick shorting [34]. At the initial stage, high series resistance expands high-frequency noise across the capacitor, diminishing filter adequacy. The equation for the capacitor is formulated as

$$C_f = \frac{D(1\text{-}D)V_s}{8f_{sw}^2 LdV} \tag{8}$$

where D is the duty ratio, V_s is the source voltage, f_{sw} is the switching frequency, and dV is the voltage ripple.

4 Design of Passive Component of Boost Converter

The plan rules are pretty much like buck converter. Yet, the position is different. The inductance of the boost converter is associated with the input stage and the presence of this inductor at the input makes it sturdy, subsequently gravitating toward a steady current.

4.1 *Inductance of Boost Converter*

An inductor in a boost converter is also useful for ripple current. Because of ripple, current RMS builds prompting undesirable warming issues. In a boost converter, it is prescribed all the time to plan so that the inductor current is higher than the average output current. Higher inductor esteem higher resistance so an ideal compromise must be done such that resistance is within proper limits because of inductor resistance we will have misfortunes at the input stage itself. The equation for an inductor is formulated as

$$L_b = \frac{DV_s}{2f_{sw}dI} \tag{9}$$

4.1.1 Battery Specification

Battery selection is a very important part of solar PV application because of the remittent ecological condition.

Steps for the Selection of Battery:

(i) Identify Space constrained of the battery:

It is based on application, emergence, and outlay. If it is space-constrained and the most amount of storage is required from the least amount of space, then lithium-ion nickel manganese cobalt is best.

(ii) DOD identification:

If we need a very low discharge current for a long time then a lead acid tubular battery and, for high discharge current lead acid (SLI) battery is suitable.

(iii) Determine the Watt-Hour requirement.
(iv) Find the strength of the battery:

$$Wh_{battery} = \frac{Wh_{load}}{DOD} \tag{10}$$

$$Ah_{battery} = \frac{Wh_{battery}}{V_{battery\_nominal}} \tag{11}$$

where DOD is the depth of discharge and Ah is the charge in the battery.

(v) Find C-Rate:

Find out the average battery discharge current and choose C from the following equation:

$$I^n t = C \tag{12}$$

where t is the time in an hour, n = 1 for low discharge current, and n = 2 for high discharge current. For high discharge, the current battery capacity rate follows a square law with the discharge current.

5 Specification and Design Parameters

Details are chosen by the accessibility of different parts at our premises and the reason for which they will be utilized. It might shift as indicated by individual necessities and accessibility of parts. The power stage design parameters, i.e., LC values for buck and boost converter are presented in Table 1.

The specification of PV module for 74 W panel PV source is listed in Table 2, along with $C_f = 2\ \mu F$. PV emulator parameters and battery specifications for 74 W panel are presented in Tables 3 and 4, respectively.

The specification of PV module for 300 W panel PV source is listed in Table 5. PV emulator parameters and battery specifications for 300 W panel are presented in Tables 6 and 7, respectively.

Table 1 LC values for buck and boost converter

Parameter	Value
Filter inductor for buck converter (L_f)	1 mH
Filter capacitor for buck converter (C_f)	2 μF
Filter inductor for boost converter (L_b)	1 mH

Table 2 Parameters of PV module for 74 W panel

PV module (Model No. SS74P)	
Maximum Power (P_{max})	74 W
Voltage at MPP (V_{mpp})	17.86 V
Current at MPP (I_{mpp})	4.19 A
Open circuit voltage (V_{oc})	21.89 V
Short circuit current (I_{sc})	4.48 A

Table 3 PV emulator parameters for 74 W panel

Parameter	Value	Parameter	Value
Source Voltage (V_s)	30 V	Filter Inductor (L_f)	1 mH
Switching Frequency (f_{sw})	50 kHz	Filter Capacitor (C_f)	10 μF
Peak–Peak ripple Current (dI)	5% of I_o	Proportional gain (K_p)	0.5
Peak–Peak ripple Voltage	1% of V_o	Integral Gain (K_I)	200 s^{-1}
Load Resistance (R)	0–30 Ω	Sampling Frequency	1 MHz

Table 4 Battery Specifications 74 W panel

Battery type	Lead acid
Nominal voltage	24 V
Capacity	25 Ah
Cut-off Voltage	18 V
Fully charged Voltage	26.13 V
Discharge current	5 A

Table 5 Parameters of PV module for 300W panel

PV module (Model No. SOMERA VSM)	
Maximum Power (P_{max})	300 W
Voltage at MPP,V_{mpp}	32.7 V
Current at MPP,I_{mpp}	9.18 A
Open circuit voltage (V_{oc})	40.3 V
Short circuit current (I_{sc})	9.68 A

Table 6 PV emulator parameters for 300W panel

Parameter	Value	Parameter	Value
Source voltage (V_S)	60 V	Filter Inductor (L_f)	1 mH
Switching Frequency	50 kHz	Filter Capacitor (C_f)	2 μF
Peak–Peak ripple current	5% of I_O	Proportional Gain	0.1
Peak–Peak ripple voltage	1% of V_O	Integral Gain (K_I)	200 s^{-1}
Load Resistance (R)	0–30 Ω	Sampling Frequency	1 MHz

Table 7 Battery specifications for 300W panel

Battery type	Lead acid
Nominal voltage	48 V
Capacity	25 Ah
Cut-off Voltage	36 V
Fully charged Voltage	50.13 V
Discharge current	5 A

6 PV Emulator: Results and Discussions

The PV emulator system-based MPPT system is simulated. This system is simulated for various situations and these situations are mainly caused by diverse environmental conditions. Irradiance (G) and Temperature (T) are those environmental factors that govern the behavior of a real PV source, further to incorporate their behavior onto the emulator system we have used mathematical equations. We are trying to emulate a real PV source with model no. SS74P by Sova Power, followed by software validation of the MPPT system constituting an MPPT controller coupled with a DC–DC Boost converter. This overall system will give the reader an insight into how we can eradicate the need of real PV source in development and testing process of solar PV equipments like MPPT controllers, inverter, charge controller etc. In our system, we are testing for MPPT controller and for every test situation our load is lead acid battery.

6.1 Standard Test Condition (STC) Operation

The STC operation includes the solar irradiance of 1000 W/m^2 and the temperature of 298 K. Figure 7 shows the power, voltage and current curve for 74W after simulation at STC condition and Table 8 corresponds to the values which are obtained from simulation along with error.

Now, the irradiance has been changed at $G = 250\ W/m^2$ to observe the changes in power, voltage and current characteristic of PV emulator is depicted in Fig. 8. The values obtained from PV emulator system (74 W specification) are presented in Table 9.

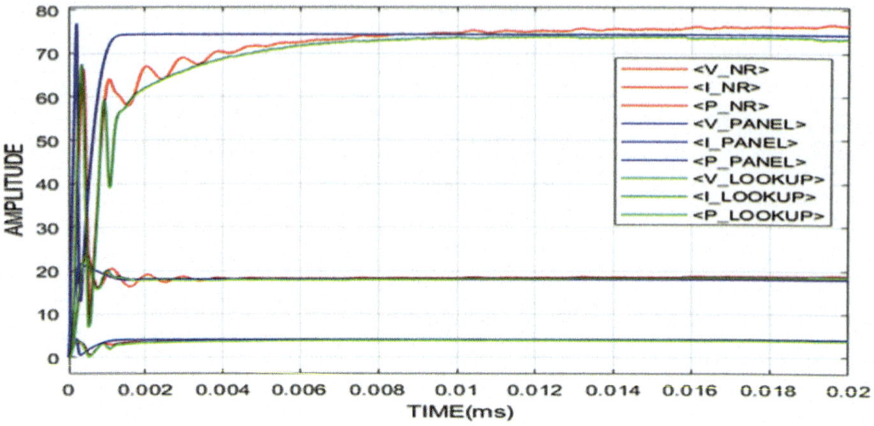

Fig. 7 Power, voltage and current curve 74 W Specification

Table 8 Values obtained from PV Emulator system (74 W Specification)

$G = 1000$ W/m^2	PV module (SS 74P)	PV emulator (NR Method)	PV emulator (Lookup Table)
P_{max_steady}	74 W	73.77 W	73.89 W
V_{mpp}	18 V	18.02 V	18.05 V
I_{mpp}	4.11 A	4.094 A	4.094 A
Error	$P_{max\_NR} = 0.31\%$; $P_{max\_Look\text{-}up\ table} = 0.15\%$		

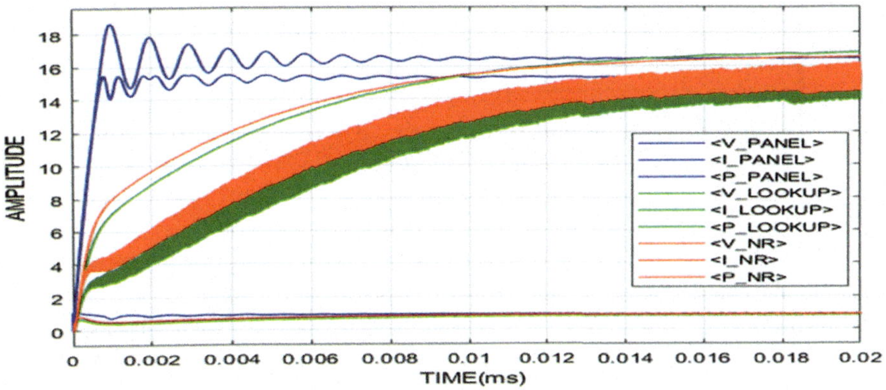

Fig. 8 Power, voltage and current curve for $G = 250$ W/m^2

Table 9 Values obtained from PV Emulator system (74 W Specification at $G = 250$ W/m^2)

PV module (SS 74P)	PV emulator (NR Method)	PV emulator (Lookup Table Method)
$P_{max\_steady} = 15.31$ W	$P_{max\_steady} = 15.68$ W	$P_{max\_steady} = 15.36$ W
$V_{oc} = 21.89$ V	$V_{oc} = 21.89$ V	$V_{oc} = 21.89$ V
$I_{sc} = 4.48$ A	$I_{sc} =$ \, 4.48 A	$I_{sc} = 4.48$ A
$V_{mpp} = 16.44$ V	$V_{mpp} = 16.52$ V	$V_{mpp} = 16.52$ V
$I_{mpp} = 0.931$ A	$I_{mpp} = 0.94$ A	$I_{mpp} = 0.93$ A
Error	$P_{max\_NR} = -2.41\%$; $P_{max\_Look\text{-}up\ table} = -0.32\%$	

6.2 PV Emulator System for 300W Specification

In this section, the PV emulator system is simulated for 300W specification. Figure 9 shows the power, voltage, and current curves for the 300W specification. Table 10 corresponds to the values which are obtained from the simulation at STC condition. Also, the system is simulated under different irradiance which is described further in those sections.

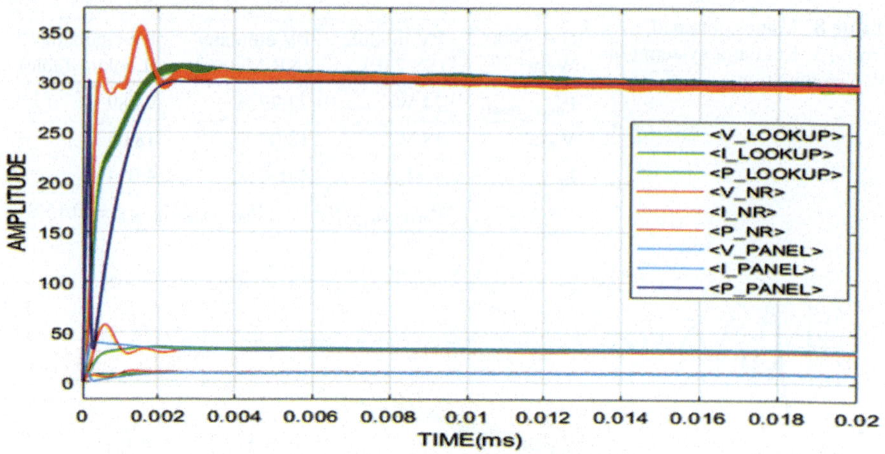

Fig. 9 Power, voltage and current curve for 300W specification

Table 10 Values obtained from PV Emulator system (300 W Specification)

$G = 1000\ W/m^2$	PV Model	PV emulator (NR Method)	PV emulator (Lookup Table Method)
P_{max} at Steady state	300.3 W	301 W	300.2 W
V_{mpp}	32.39 V	30.99 V	32.85 V
I_{mpp}	9.27 A	9.712 A	9.137 A
V_{oc}	40.3 V	40.3 V	40.3 V
I_{sc}	9.68 A	9.68 A	9.68 A
Error	$R_s = 0.3\Omega,\ R_p = 370.3\Omega$ $P_{\max\_NR} = -0.23\%;\ P_{\max\_Look-up\ table} = -0.03\%$		

Now, the irradiance has been changed at $G = 250\ W/m^2$ to observe the changes in power, voltage, and current characteristic of the PV emulator. Figure 10 depicts the power, voltage, and current curve for 300 W specification at $G = 250\ W/m^2$. The values obtained from PV emulator system (300 W specification at $G = 250\ W/m^2$) are presented in Table 11.

In the above cases, one can see that our system is being simulated for changing irradiance and constant temperature conditions. From the simulation results, it is evident that our MPPT controller can churn out maximum power at every respective condition used. According to the PV equations, the decrease in irradiance leads to a decrease in photo-generated or PV current in a real PV source, and the same has been successfully implemented in the system. Therefore, our system is also showing a decrease in PV current with a respective decrease in irradiance values. But the change in PV voltage is minimal and less affected compared to the photo current. Real PV source is also affected by a change in temperature which is very substantial

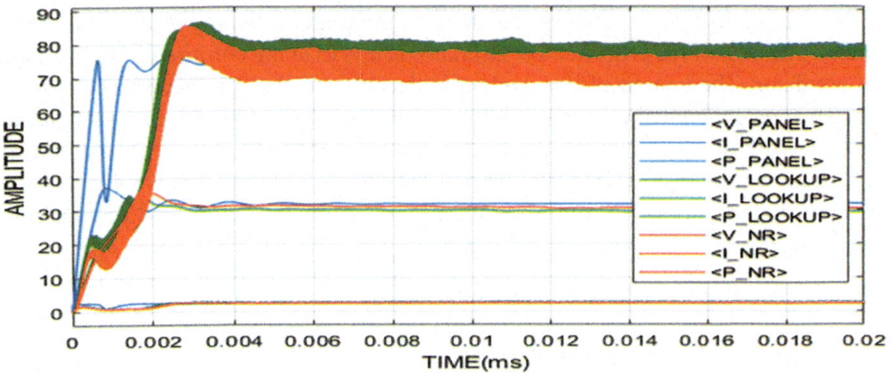

Fig. 10 Power, voltage, and current curve for 300 W Specification at G = 250 W/m^2

Table 11 Values obtained from PV emulator system (300 W specification at G = 250 W/m^2)

G = 250 W/m^2	PV model	PV emulator (NR Method)	PV emulator (Lookup Table Method)
P_{max} at Steady state	75.3 W	75.5 W	75.56 W
V_{mpp}	32.1 V	30.7 V	32.52 V
I_{mpp}	2.345 A	2.459 A	2.323 A
V_{oc}	40.3 V	40.3 V	40.3 V
I_{sc}	9.68 A	9.68 A	9.68 A
Error	$R_s = 0.3\ \Omega,\ R_p = 370.3\ \Omega$ $P_{\max\_NR} = -0.26\%;\ P_{\max\_Look-up\ table} = -0.34\%$		

from the manufacturer's datasheet. PV voltage and current have certain temperature coefficient associated with them, that is with the increase and decrease of temperature PV voltage and current also changes.

7 Conclusions

This chapter proposed a Photovoltaic emulator system which has a smaller number of switches in the power stage. Due to this, we are able to decrease the complexity of control circuit. This all translates to lower making cost and less maintenance. Control circuit uses single diode modeling and implementation using Newton–Raphson method which is more accurate than other methods. MPPT-controlled boost converter battery charging setup acts as constant power load to PV source. We tested our proposed PV emulator system with this constant power load and compared the results when using a PV source. During the testing, we also put the PV Emulator to change in irradiance and temperature. Like PV source PV emulator is also adapting to the change in these variables. All these claims are validated using simulation results. The cost of the proposed PV emulator is less than the commercially available PV emulator. The benefit of this system is that it decreases the necessity of room, constancy on ecological elements, and power utilization, yet has higher adaptability to change the parameter according to the user. Accordingly, this is best suited for development as well as testing custom power devices. Utilizing this we are effective in the testing of the MPPT controller and programming level approval implanted in this controller.

7.1 Future Research Direction

The future research direction will be as follows:

- To validate the theoretical analysis and simulation results using hardware prototypes for the proposed photovoltaic emulator system. Hardware prototype will allow us to analyze the discrepancies between simulation/theoretical results and experimental results due to the presence of non-idealities present in the overall system.
- Further, we will look for much more sophisticated but less complex control approach so as to further improve the dynamic response and reduction of error in tracking of operating points.

Acknowledgements This research work was supported by "Woosong University's Academic Research Funding—2023."

References

1. Fărcaş C, Ciocan I, Tulbure A (2018) Solar emulator for a photovoltaic module. IEEE 24th International Symposium for Design and Technology in Electronic Packaging, pp 314–318. https://doi.org/10.1109/SIITME.2018.8599283
2. Soulatiantork P, Cristaldi L, Faifer M, Laurano C, Ottoboni R, Toscani S (2018) A tool for performance evaluation of MPPT algorithms for photovoltaic systems. Measurement 128:537–544. https://doi.org/10.1016/j.measurement.2018.07.005
3. Grandi G, Ienina A, Bardhi M (2014) Effective low-cost hybrid LED-halogen solar simulator. IEEE Trans Ind Appl 50(5):3055–3064. https://doi.org/10.1109/TIA.2014.2330003
4. Sanaullah A, Khan HA (2015) Design and implementation of a low cost Solar Panel emulator. IEEE 42nd Photovoltaic Specialist Conference, pp 1–5. https://doi.org/10.1109/PVSC.2015.7356274
5. Ünlü M, Çamur S (2017) Solar irradiance emulator based on DC power supply for photovoltaic panels. 9th International Conference on Electronics, Computers and Artificial Intelligence, pp 1–4. https://doi.org/10.1109/ECAI.2017.8166455
6. de Brito MAG, Galotto L, Sampaio LP, Melo GdAe, Canesin CA (2013) Evaluation of the main MPPT techniques for photovoltaic applications. IEEE Trans Ind Elect 60(3): 1156–1167. https://doi.org/10.1109/TIE.2012.2198036

7. Jayawardana A, Ho CNM, Pokharel M (2019) Design and implementation of switch-mode solar photovoltaic emulator using power-hardware-in-the-loop simulations for grid integration studies. IEEE Energy Conv Cong Exp 889–894. https://doi.org/10.1109/ECCE.2019.8912636
8. Adam SM, Wang K, Ma J, Man KL (2021) Soft modeling of the photovoltaic modules based on MATLAB/Simulink. 25th International Conference Electronics, pp 1–6. https://doi.org/10.1109/IEEECONF52705.2021.9467417
9. Villalva MG, Gazoli JR, Filho ER (2009) Comprehensive approach to modeling and simulation of photovoltaic arrays. IEEE Trans Power Electron 24(5):1198–1208. https://doi.org/10.1109/TPEL.2009.2013862
10. Zeng Q, Song P, Chang L (2002) A photovoltaic simulator based on DC chopper. IEEE Canadian Conf Elect Comp Eng 1:257–261. https://doi.org/10.1109/CCECE.2002.1015221
11. Segura GM, Mestre JL, Casas MT, Andreu AS (2007) Development of a photovoltaic array emulator system based on a full-bridge structure. 9th International Conference on Electrical Power Quality and Utilisation, pp 1–6. https://doi.org/10.1109/EPQU.2007.4424165
12. Koutroulis E, Kalaitzakis K, Tzitzilonis V (2006) Development of an FPGA-based system for real-time simulation of photovoltaic modules. Seventeenth IEEE Intern Workshop Rapid Syst Protot 200–208. https://doi.org/10.1109/RSP.2006.14
13. Piazza MCD, Vitale G (2010) Photovoltaic field emulation including dynamic and partial shadow conditions. Appl Energy 87(3):814–823. https://doi.org/10.1016/j.apenergy.2009.09.036
14. Carralero LLO, Barbara da S. e Silva GS, Costa FF, Tahim APN (2019) PV emulator based on a four-switch buck-boost DC-DC converter. IEEE 15th Brazilian Power Electronics Conference and 5th IEEE Southern Power Electronics Conference pp 1–5. https://doi.org/10.1109/COBEP/SPEC44138.2019.9065364
15. Garg P, Priyanshi, Bhuvaneswari G (2018) Power electronic circuit based implementation of a solar PV emulator using a power factor corrected buck converter. IEEMA Engineer Infinite Conference, pp 1-6. https://doi.org/10.1109/ETECHNXT.2018.8385297
16. Dolan D, Durago J, Crowfoot J, Taufik (2010) Simulation of a photovoltaic emulator. North American Power Symposium, pp 1-7. https://doi.org/10.1109/NAPS.2010.5618941
17. Rout KK, Mishra DP, Mishra S, Rishitha GT, Salkuti SR (2023) Application of fuzzy logic technique to track maximum power point in photovoltaic systems. Indonesian J Elect Eng Comp Sci 29(1): 101–109. http://doi.org/https://doi.org/10.11591/ijeecs.v29.i1.pp101-109
18. Muoka PI, Haque ME, Gargoom A, Negnevitsky M (2013) Modeling, simulation and hardware implementation of a PV power plant in a distributed energy generation system. IEEE PES Innov Smart Grid Technol Conf 1–6. https://doi.org/10.1109/ISGT.2013.6497917
19. Bhukya L, Kedika NR, Salkuti SR (2022) Enhanced maximum power point techniques for solar photovoltaic system under uniform insolation and partial shading conditions: a review. Algorithms 15(10):365. https://doi.org/10.3390/a15100365
20. Salkuti SR (2022) Emerging and advanced green energy technologies for sustainable and resilient future grid. Energies 15(18):6667. https://doi.org/10.3390/en15186667
21. Rout KK, Mishra DP, Mishra S, Patra S, Salkuti SR (2023) Perturb and observe maximum power point tracking approach for microgrid linked photovoltaic system. Indonesian J Electr Eng Comp Sci 29(2): 635–643. http://doi.org/https://doi.org/10.11591/ijeecs.v29.i2.pp635-643
22. Balakishan CH, Sandeep N (2014) Development of a Microcontroller Based PV Emulator with Current Controlled DC/DC Buck converter. Intern J Renew Energy Res 4(4): 1049–1055. https://doi.org/10.20508/ijrer.v4i4.1618.g6446
23. Ma CT, Tsai ZY, Ku HH, Hsieh CL (2021) Design and implementation of a flexible photovoltaic emulator using a gan-based synchronous Buck Converter. Micromachines 12(12):1587. https://doi.org/10.3390/mi12121587
24. Abhishek RS, Dhanus KA, Hariharan S, Hariharasudhan S, Saravanan, S (2022) Design of solar PV emulator using raspberry Pi controller. Inter Conf Invent Comput Technol 581–584. https://doi.org/10.1109/ICICT54344.2022.9850727
25. Suwito S, Rivai M, Ashari M, Mustaghfirin MA (2021) Simple and fast response photovoltaic panel emulator using transistor current source. Inter Conf Elect Inform Techn 223–227. https://doi.org/10.1109/IEIT53149.2021.9587441

26. Khawaldeh HA, Al-Soeidat M, Farhangi M, Lu DDC, Li L (2021) Efficiency improvement scheme for PV emulator based on a physical equivalent PV-cell model. IEEE Access 9:83929–83939. https://doi.org/10.1109/ACCESS.2021.3086498
27. Boutabba T (2021) dSPACE real-time implementation of maximum power point tracking based on Kalman filter structure using photovoltaic system emulator. Intern Conf Control Autom Diagn 1–6. https://doi.org/10.1109/ICCAD52417.2021.9638743
28. Alaoui M, Maker H, Mouhsen A (2020) A new PV source emulator based on modified hybrid referencing technique with nonlinear Lyapunov controller. 1st International Conference on Innovative Research in Applied Science, Engineering and Technology, pp 1–6. https://doi.org/10.1109/IRASET48871.2020.9092261
29. Ali AN, Premkumar K, Vishnupriya M, Manikandan BV, Thamizhselvan T (2021) Design and development of realistic PV emulator adaptable to the maximum power point tracking algorithm and battery charging controller. Sol Energy 220:473–490. https://doi.org/10.1016/j.solener.2021.03.077
30. Chavarría J, Biel D, Guinjoan F, Poveda A, Masana F, Alarcón E (2014) Low cost photovoltaic array emulator design for the test of PV grid-connected inverters. IEEE 11th International Multi-Conference on Systems, Signals & Devices, pp 1–6. https://doi.org/10.1109/SSD.2014.6808860
31. Mishra S, Raju S, Kshirsagar A, Mohan N (2018) An extremely low-cost multi-panel PV emulator for research and education. IEEE Energy Conver Cong Expos 4760–4765. https://doi.org/10.1109/ECCE.2018.8557361
32. Estrada L, Vazquez N, Ortega J, López H, Hernández C, Vaquero J (2021) Low cost PV emulator based on programmable DC power supply and LabVIEW. 47th Annual Conference of the IEEE Industrial Electronics Society, pp 1–6. https://doi.org/10.1109/IECON48115.2021.9589184
33. Shinde UK, Kadwane SG, Gawande SP, Keshri R (2016) Solar PV emulator for realizing PV characteristics under rapidly varying environmental conditions. IEEE Inter Conf Power Electr Drives Energy Syst 1–5. https://doi.org/10.1109/PEDES.2016.7914286
34. Chariag D, Sbita L (2017) Design and simulation of photovoltaic emulator. Inter Conf Green Energy Conv Syst 1–6. https://doi.org/10.1109/GECS.2017.8066121

Fault Detection, Classification, and Location in Underground Cables

Smrutisikha Jena, Debani Prasad Mishra, and Surender Reddy Salkuti

Abstract Underground (UG) cables have become widespread in the present scenario as these cables' durability is high and there are fewer environmental issues. With time, the insulation of the UG cables begins to deteriorate. As a result, their conduction efficiency decreases. To achieve the integrity of an underground cable, it is vital to accurately identify a faulty segment to limit the amount of time that the system is unavailable during a fault. Therefore, a fast and accurate fault detection approach is necessary to expedite system restoration, reduce outage duration, minimize economic losses, and considerably increase system reliability. There are several traditional techniques used for fault classification and localization and those are arduous, sluggish, and computationally intensive, they are reliant on mathematical modeling and necessitate specialized skill. To overcome such problems, several artificial intelligence approaches such as machine learning techniques as well as deep learning methods have been deliberated in this chapter. The above-mentioned approaches are considered for fault location, classification, and detection in the UG cable. All the methods are taken from some conference papers, journals, and e-books from 2002 to 2022. The benefits and drawbacks of all fault diagnosis approaches have also been explored.

Keywords Underground cables · Fault classification · Location and detection · Machine learning · Deep learning

---

S. Jena · D. P. Mishra
Department of Electrical and Electronics Engineering, IIIT Bhubaneswar, Bhubaneswar, Odisha, India
e-mail: c121012@iiit-bh.ac.in

D. P. Mishra
e-mail: debani@iiit-bh.ac.in

S. R. Salkuti (✉)
Department of Railroad and Electrical Engineering, Woosong University, Daejeon, Republic of Korea
e-mail: surender@wsu.ac.kr

S. R. Salkuti et al. (eds.), *Power Quality in Microgrids: Issues, Challenges and Mitigation Techniques*, Lecture Notes in Electrical Engineering 1039,
https://doi.org/10.1007/978-981-99-2066-2_10

Nomenclature

ML	Machine learning
DL	Deep learning
ANFIS	Adaptive neuro-fuzzy interference system
ANN	Artificial neural network
FLS	Fuzzy logic system
SVM	Support vector machine
CNN	Convolutional neural network
RNN	Recurrent neural network
LSTM	Long short-term memory
SUGPD	Smart underground power distribution system
ADAGRAD	Adaptive gradient descent
RMSPROP	Root mean square prop
ADAM	Adaptive moment estimation
DBN	Deep belief network
GRNN	General Regression neural network
PT	Potential transformer
CT	Current transformer

1 Introduction

Due to safety regulations, overhead lines become difficult in large cities and densely inhabited areas. In many circumstances, overhead lines are not an option. As a result, insulated conductors are put beneath the earth, resulting in underground (UG) cables or UG cables. Long current-carrying devices with their insulation and an earthed exterior surface are referred to as cables [1]. So basically, UG cable is made up of one or more conductors that are surrounded by a protective cover and are covered with suitable insulation. It is suitable for use in a crowded environment. UG cable is more beneficial where tower construction is challenging [2]. More safety, less interface with amenities, and a better sight are all advantages of using UG cable. So, in a nutshell, the key benefits of an underground cable are damage due to storms and lightning is less likely to occur, low upkeep cost, fewer possibilities of fault, a reduced voltage drop, and a more appealing overall appearance [3, 4].

The insulated cables start to decline with time. Therefore, the efficiency of their conduction diminishes. Excessive moisture in the paper insulation of cables is the most common cause of cable faults. As a result, the lead sheath that covers the cable may be damaged. A lead sheath can be harmed in several ways. The majority of them can have the chemical effect of soil on the lead during installation, as well as mechanical damage and lead crystallization caused by vibration. Some of the most prevalent types of faults in underground cables are listed below [5, 6].

Open circuit: In an underground cable, a discontinuity occurs when one or more cable conductors (cores) fail. Mechanical stress causes the cable to come out of its joint, causing this discontinuity. This is the open circuit fault. Here, the conducting material of the cable ruptures which causes this failure. The most typical sort of fault in an underground cable is this. An instrument known as a megger is used to check for this type of fault. The resistance in an open circuit is infinite. In this sort of fault, the cores are shortened at the far end and then placed on the ground. The resistance between each core and the ground is then determined using the megger. If it reads 0, then the core is not destroyed. In contrast, if the megger reads infinite resistance, then the core is destroyed and must be substituted.

Short circuit: This is the second most prevalent fault in underground cable. It can only be found in multi-cored cables. When an insulator fails, the two wires of a multi-core cable jointly get into an electric contact, indicating short-circuit defect. Visual detection is impossible without dismantling the cable. So, this fault can also be tested with the megger. Zero resistance is the characteristic of short-circuiting. The two terminals of the megger are linked to any two wires in this configuration. When the megger reads 0 between the electrical conductors, a fault has occurred. Taking another two conductors at a time, repeat the operation.

Earth fault: This failure happens when some of the conductors of the cable come in touch with the ground. This happens when the outer sheath is destroyed by chemical reactions with the soil, vibrations, or mechanical crystallization. A megger is also needed in this fault. The two inputs of the megger are attached to any two wires and the reading is recorded. When it reads nearly zero between the wires, it indicates a fault. The same procedure can be followed with the remaining two conductors' one at a time. Due to the inaccessibility of detecting the location of a fault is a difficult task in an underground cable. Because of several drawbacks, traditional fault locating techniques cannot be employed for underground wires. The effectiveness of artificial intelligence approaches owing to a higher ability to learn and generalize (from training patterns) and these strategies have been used by a number of researchers for location and identification of a fault.

The remainder of this work is organized as follows: Sect. 2 presents traditional methods for fault classification, location, and detection; Sect. 3 some signal processing methods; Sect. 4 presents artificial intelligence techniques using machine learning and deep learning Sect. 5 presents features of different techniques and Sect. 6 presents the conclusion.

2 Conventional Methods for Fault Classification, Location, and Detection

2.1 Distance Relay

The operating principle of a distance relay is quite basic. It has a potential transformer (PT) for the voltage supply and a current transformer (CT) for the current element. Both of these are connected in series with the entire circuit. The deflecting torque is caused by the secondary current of the CT, whereas the restoring torque is caused by the potential transformer [7]. The operation of an impedance relay is dependent on the proportion of voltage and current which is the impedance value. Only when the impedance is less than the relay's predetermined impedance value, then the distance relay begin to operate. When there is a risk of a fault then the current rises faster than the voltage. Then the current on the line is inversely proportional to the impedance of the relay. As a result, the relay begins to function in this state [8]. The impedance of the line falls below the predetermined value of the relay, and the relay begins to operate by sending a tripping command to the circuit breaker. If the fault extends beyond the positive condition, the relay contacts will be open.

2.2 Traveling Wave Method

Using traveling wave theory, this technique determines the fault. It reduces complexity, lowers costs, and improves accuracy, although it has certain reliability and maintenance issues. This approach has a 0.5–2% of accurateness. This strategy relies on a study of how long it takes light to travel from one end of a transmission line to the other. It can be seen in two different methods: the double-ended way and the single-ended method [9].

2.2.1 Double-Ended Method

Figure 1 depicts the double-ended method of the traveling wave phenomenon. It is a case of both sides observing the same thing at the same time. In the figure, m is calculated by using $\frac{1}{2}$ $[l + (T_L - T_R)]v$ and it is the distance to fault equation. l is the transmission line's length, T_L is the arrival time of the traveling wave at L, T_R is the arrival time of the traveling wave at R, and v denotes the speed at which a traveling wave propagates. This approach is time-consuming and expensive, and it does not work for complex topologies.

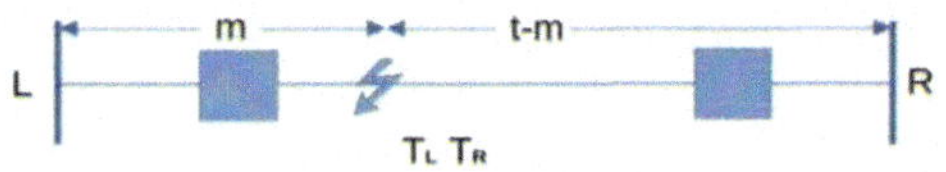

Fig. 1 Double-ended method of traveling wave phenomenon

2.2.2 Single-Ended Method

This is the method of a single point of observation. This strategy works regardless of the network's setup or the devices that are connected to it. These approaches are extremely precise, but they necessitate a high sample rate and are more costly to execute than the impedance method.

2.3 Impedance-Based Method

This technique demonstrates the one and two-ended impedance-based fault location algorithms that are routinely used in power system networks to locate faults. The fundamental frequency current and voltage are the most important factors here. This method is simple and inexpensive, but it produces incorrect results for large values of fault resistance and has restrictions related to fault path resistance, line loading, source characteristics, and other factors [10]. As a result, the fault location accuracy is roughly 2–3% of the whole line length. Only one end of the transmission line is considered in a one-ended impedance-based fault location algorithm. There is no requirement for a communication channel because it just takes data from one end. This method is primarily focused on the estimation of apparent impedance from the monitoring location to the fault using voltage and current during a fault period [11]. Two-ended impedance-driven fault location uses the very same procedures as single-ended impedance-driven fault location, but it uses data from both terminals of the transmission line to estimate the defect in the line. In this case, a communication route is required to transport data from one intelligent electronic device (IED) to another, with data from both IEDs being gathered and analyzed at a central location [12].

2.4 Bridge and Step Voltage Method

The technique depends on the bridge balance principle and is used to determine the degree of resistance deviation. The relationship between the resistance and the cable length can then be used to calculate the location of the defect. The step voltage method uses the deflection and sway of the pointer to estimate the degree of potential variation between distinct places, with the location of the fault.

2.5 Blavier Test

The earth fault identification in a UG cable is determined by using Blavier's Test. When an earth fault occurs in a single cable and no other cables than the faulty one is present, then this test can be used to find the fault. Here, the resistance can be analyzed in two ways. The first one is by isolating the far end from the earth and the second one is to shorten the far end of the cable to the earth. In this type of test, a bridge network is used with a low-voltage supply, an ammeter, and a voltmeter.

The sending end of this cable is open as well as isolated. The resistance between the sending end and the earth point is determined by maintaining the far end segregated from the earth, letting it be R_A. Then, the resistance between the sending end and the earth point is shown in Fig. 2, and it is analyzed by retaining the far end of the damaged cable shorted to the earth and letting it be R_B.

The overall line resistance is represented as R Ω. The resistance between the faulty end and sending end is denoted by r_a, whereas the resistance between the faulty and far ends is represented as r_b. As a result, the sum of the r_a and r_b resistances equals the total resistance R which means $R = r_a + r_b$.

The resistance between the sending end and the ground is determined by maintaining the far end isolated from the ground is equal to the total resistance of r_a and r loop. It is given by

$$R_A = r_b + \mathrm{r} \tag{1}$$

Since R_B is measured at the test end while the far end is earthed, thus r and r_a are parallel. The resistance between the sending end and the ground by retaining the far end of the damaged cable shorted to the ground is equal to the total resistance of the overall loop, and it is given by,

$$R_B = r_b + \frac{r r_a}{r + r_a} \tag{2}$$

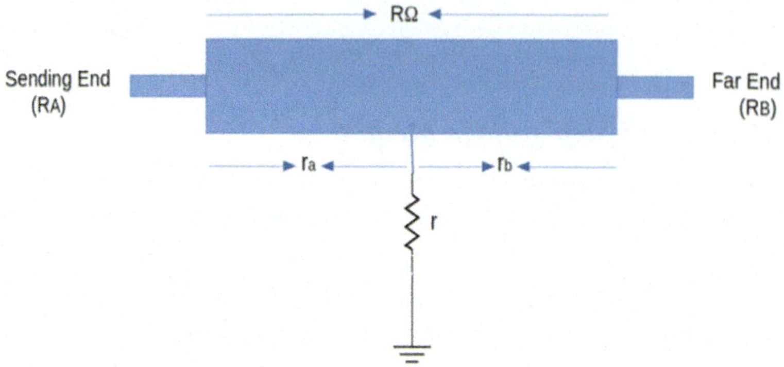

Fig. 2 Circuit connection of Blavier test

From Eq. (1), we can write $r = R_A - r_b$. By substituting the value of r in Eq. (2), we will get

$$R_B = r_b + \frac{(R_A - r_b)r_a}{(R_A - r_b) + r_a} \tag{3}$$

$$\Rightarrow (R_B - r_b)\{(R_A - r_b) + r_a\} = (R_A - r_b)r_a \tag{4}$$

$$\Rightarrow R_A R_B - R_A r_b - r_b R_B + {r_b}^2 + r_a R_B - r_a r_b = r_a R_A - r_a r_b \tag{5}$$

$$\Rightarrow R_A R_B - R_A r_b - r_b R_B + {r_b}^2 + r_a R_B = r_a \tag{6}$$

$$\Rightarrow R_A R_B - r_b R_B - r_b R_B + r_b R_B + {r_b}^2 + r_a R_B = r_a R_A + R_A r_b \tag{7}$$

From (7), we can write

$$R_A R_B - 2r_b R_B + {r_b}^2 + R R_B = R_A R \tag{8}$$

$$\Rightarrow {r_b}^2 - 2r_b R_B + {R_B}^2 - {R_B}^2 + R_A R_B = R_A R - R R_B \tag{9}$$

$$\Rightarrow (r_b - R_B)^2 = R(R_A - R_B) + {R_B}^2 - R_A R_B \tag{10}$$

$$\Rightarrow (r_b - R_B)^2 = (R_A - R_B)(R - R_B) \tag{11}$$

$$\Rightarrow r_b = R_B \pm \sqrt{(R_A - R_B)(R - R_B)} \tag{12}$$

As r_b is less than R_B, a negative sign should be taken. Therefore,

$$r_b = R_B - \sqrt{(R_A - R_B)(R - R_B)} \tag{13}$$

2.6 *Murray Loop Test*

The premise of the Wheatstone bridge is employed in the Murray loop test to locate cable faults in an underground cable. Here, the location of the fault is figured out by comparing the resistance in the bridge. In this experiment, a specified cable length should be used. In the Murray loop test, there are mainly two types of connections to determine the fault location. One is when there is a ground fault and another one is a short circuit failure. Figure 3a, b are of these types of connections [13].

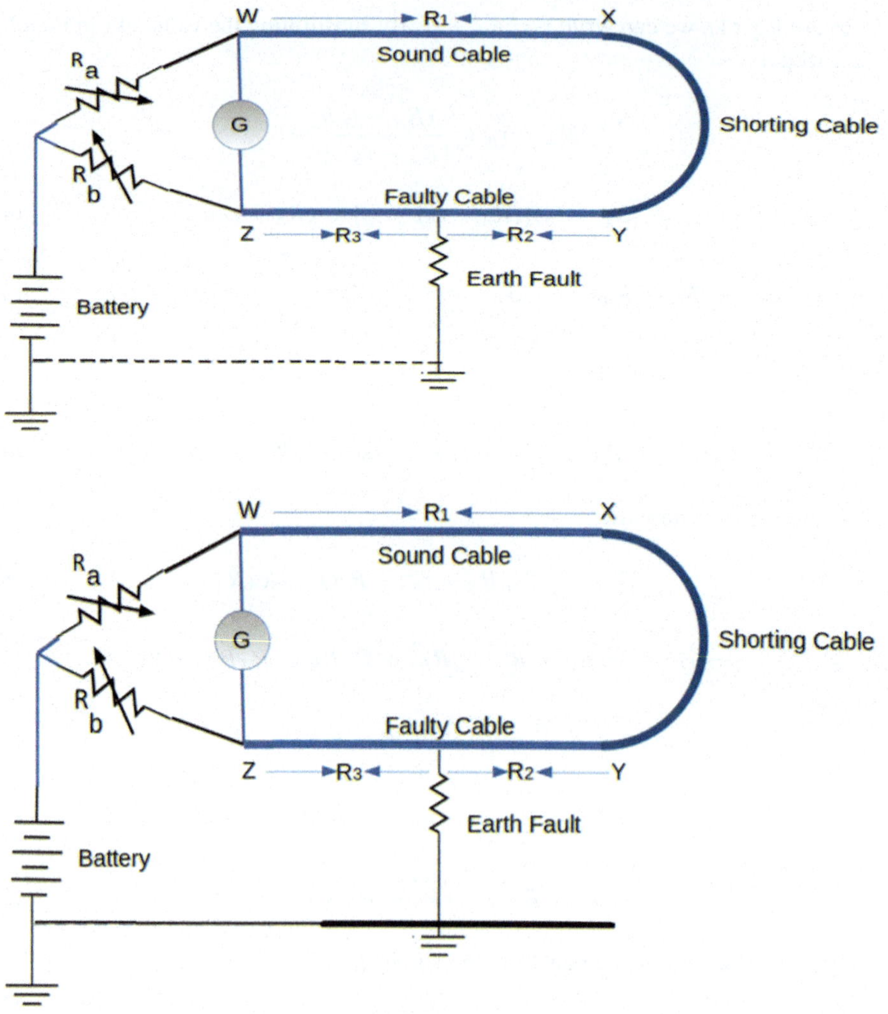

Fig. 3 **a** Circuit connection for the ground fault of Murray loop test. **b** Circuit connection for short circuit fault of Murray loop test

The damaged wire is connected to the sound cable through a low-resistance connection. That resistance must have no effect on the overall resistance of the cable and it can be able to flow loop current towards the bridge circuits without degradation. The bridge is balanced by modifying the resistance of the ratio of arms of the bridge. The Murray loop procedure is done by shorting the adjacent sound cable and the damaged cable with a jumper conductor at the far end. A voltage source is connected to the test side end via a pair of resistors. The galvanometer will read zero deflection when the bridge is balanced. The overall loop resistance created by the sound cable and the damaged cable is $[R_1 + R_2 + R_3]$. At the balanced condition of the wheat stone bridge, we have

$$\frac{R_a}{R_b} = \frac{R_1 + R_2}{R_3} \tag{14}$$

$$\Rightarrow \frac{R_a}{R_b} + 1 = \frac{R_1 + R_2}{R_3} + 1 \tag{15}$$

$$\Rightarrow \frac{\mathrm{R_a} + \mathrm{R_b}}{\mathrm{R_b}} = \frac{R_1 + R_2 + R_3}{R_3} \tag{16}$$

$$\Rightarrow \frac{\mathrm{R_a} + \mathrm{R_b}}{\mathrm{R_b}} = \frac{2R}{R_3} \tag{17}$$

Here, the resistance of one of the cables while it is fault-free is denoted by R. When both the area of the cross section of the sound cable and defective cable are similar, the resistance of the conductors is equivalent to their lengths. If L is the cable's length in meters and R/L is the cable's resistance per meter, then the fault distance from the testing end.

$$\frac{R_x}{R} = \frac{LR_x}{R} = \frac{\mathrm{R_b}}{\mathrm{R_a} + \mathrm{R_b}}2R \times \frac{L}{R} = \frac{\mathrm{R_b}}{\mathrm{R_a} + \mathrm{R_b}}2L \tag{18}$$

Varley loop test, microcontroller-based method, cable thumping, and potential divider network are also some traditional methods to detect fault location in UG cable.

3 Signal Processing Methods

3.1 Wavelet Transform (WT)

The wavelet transform concept relies on signal analysis using several times and frequency scales. Faulted signals in a UG cable are often nonperiodic with high-frequency oscillations and correlated with fast electromagnetic transients [14]. It can extract information from transient signals in both the time and frequency domains at the same time. This skill aids in the analysis of signal properties that can be used to characterize the cause of transient and the state of underground cables after a disturbance. There are multiple kinds of WTs, and one approach is favored over the others depending on the application. These are discussed below.

3.1.1 Continuous Wavelet Transform

When the input signals are continuous the time as well as scale characteristics are normally continuous. So, here the continuous wavelet transform is the logical choice [14, 15]. The continuous wavelet transform (CWT) can be defined as

$$\mathrm{CWT}(\alpha, \beta) = \int_{-\infty}^{+\infty} w(t)\varphi^{*}_{\alpha,\beta}(\mathrm{t})\mathrm{dt}\alpha > 0 \tag{19}$$

Here, $w(t)$ is the signal to be examined. The mother wavelet $\varphi_{\alpha,\beta}$ is shifted with the factor β and scaled with the factor α. The larger and smaller scales correspond to lower and higher frequencies respectively. * denotes the complex conjugation.

$$\varphi_{\alpha,\beta}(\mathrm{t}) = \frac{1}{\sqrt{\alpha}}\varphi\frac{t-\beta}{\alpha}\alpha > 0, -\infty < \beta + \infty \tag{20}$$

3.1.2 Discrete Wavelet Transform

Based on the changes in size and location, CWT creates huge data having the form of wavelet coefficients. This results in a significant computational cost. Discrete wavelet transform (DWT) is utilized to get around this constraint. The DWT is based solely on CWT subsamples. It improves the efficiency of analysis, makes it simple to implement, and reduces computing time. On the other hand, the DWT preserves the original CWT. To put it another way, DWT on discretized samples is used to implement WT in digital computers. The DWT employs dyadic dilations and translations, which are scale and position values based on powers of two [16]. The decomposition of the discrete wavelet transform is depicted in Fig. 4.

With no data loss, the signal can be entirely retrieved from its DWT. Scaling and translation parameters are discretized in the following way, $\alpha = \alpha_0^{\mathrm{p}}$, $\beta = \mathrm{q}\beta_0\alpha_0^{\mathrm{p}}$ where $\alpha_0 > 1$, $\beta_0 > 0$. Where p and q are two integers. The DWT can be defined as

$$\mathrm{DWT}(\mathrm{p}, \mathrm{q}) = \int_{-\infty}^{+\infty} x(t)\varphi_{\mathrm{p,q}}(t)\mathrm{dt} \tag{21}$$

$$\varphi_{\mathrm{p,q}}(\mathrm{t}) = \alpha_0^{\frac{-p}{2}}\varphi\frac{t - \mathrm{q}\beta_0\alpha_0^{\mathrm{p}}}{\alpha_0^{\mathrm{p}}} \tag{22}$$

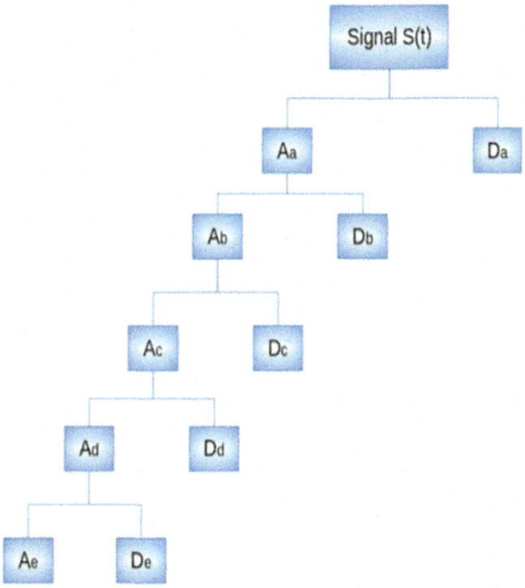

Fig. 4 Decomposition of discrete wavelet transform

3.1.3 Wavelet Packet Transform

The WPT is a modification of wavelet disintegration that provides a larger variety of signal evaluation options. The details and the similarities can be split in WPT, as shown in Fig. 5. As WPT generates a huge number of nodes. As a result, the computational overhead is increased. Only similarities are further decomposed in DWT, reducing the level of decomposition and, as a result, the number of computing attempts. The equations that define a wavelet packet transfer are as follows [17]:

$$W_{2n}^{i}(t) = \sqrt{2}\sum_{x} h(x)W_{n}^{i}(2t - x) \tag{23}$$

$$W_{2n+1}^{i}(t) = \sqrt{2}\sum_{x} g(x)W_{n}^{i}(2t - x) \tag{24}$$

Here, $W_0^0(t)$ represents the scaling function ϕ(t) that is $W_0^0(t) = \phi(t)$, and $W_1^0(t)$ represents the basic wavelet function ψ(t) that is $W_1^0(t) = \psi(t)$ [3]. The ith level wavelet packet basis is indicated by the superscript (i) and this is the decomposition level. At the ith level, there will be 2^i wavelet packet bases. $h(x)$ and $g(x)$ are high-pass filter and low-pass filter respectively.

3.1.4 Stationary Wavelet Transform

In contrast to DWT, stationary wavelet transform SWT is time-invariant due to decimation. It's been utilized for de-noising, singularity detection, and sharp transient detection, among other applications in UG cable. In conclusion, the SWT approach can be characterized as follows: when the high and low-pass filters are applied to

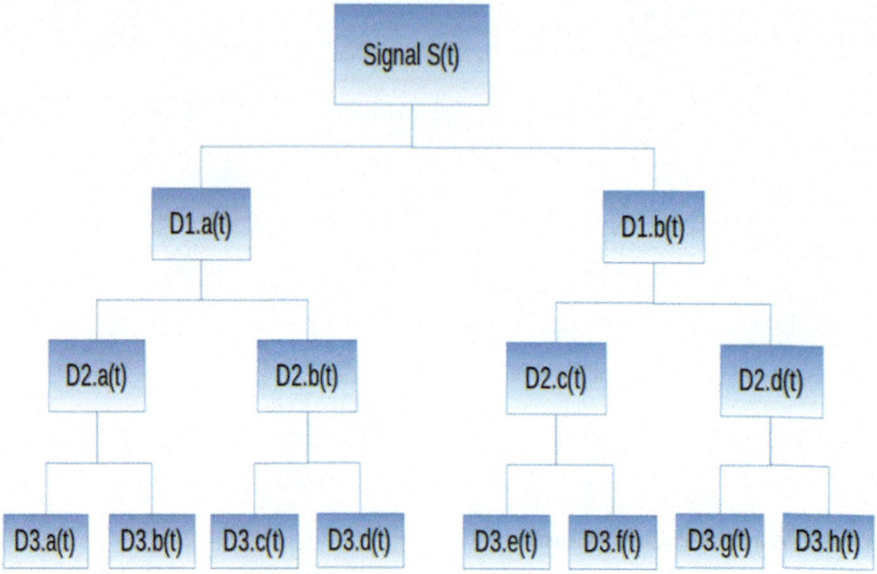

Fig. 5 Decomposition of wavelet packet transform

the data at each level, the two new sequences have the same length as the original sequences. This is accomplished by not decimating the original data. On the other hand, the filters at each level are changed by padding them with zeros.

Assuming that a function $f(x)$ is projected on the subset $U_i(\ldots\ldots \subset U_3 \subset U_2 \subset U_2 \subset U_0)$ at each step *i*. The scalar product $c_{\mathrm{i,j}}$ of $f(x)$ with the dilated and translated scaling function $\phi(x)$ defines this projection:

$$c_{\mathrm{i,j}} = (\mathrm{f(x)}, \phi_{\mathrm{i,j}}(\mathrm{x})) \tag{25}$$

$$\phi_{\mathrm{i,j}}(\mathrm{x}) = 2^{-i}\phi(2^{-i}\mathrm{x} - \mathrm{j}) \tag{26}$$

Here, $\phi(x)$ is a low-pass filter scaling function. $c_{\mathrm{i,j}}$ is known as a discrete approximation at 2^i. The wavelet coefficients of the wavelet function $\psi(\mathrm{x})$ are

$$\psi_{\mathrm{i,j}} = (\mathrm{f(x)}2^{-i}\psi(2^{-i}\mathrm{x} - \mathrm{j})) \tag{27}$$

Here, $\psi_{\mathrm{i,j}}$ is known as a discrete detail signal at 2^i. The scaling function $\phi(x)$ has the property $\frac{1}{2}\phi\frac{\mathrm{x}}{2} = \sum_{\mathrm{k}} \mathrm{h(k)}\phi(\mathrm{x} - \mathrm{k})$. Here, h(k) is the low-pass filter. The wavelet function $\psi(x)$ has the property $\frac{1}{2}\psi\frac{\mathrm{x}}{2} = \sum_{\mathrm{k}} \mathrm{g(k)}\psi(x - \mathrm{k})$. Here, $g(k)$ is the low-pass filter [16, 17].

3.2 S Transform

S-transformed time-frequency analysis is based on continuous wavelet transformation and short-time Fourier transformation. S transform is used for non-stationary signals where the window width varies in inverse proportion to frequency. The S-transform (ST) has the advantage over other signal processing tools in that it offers information on the time, frequency, and phase angle of the signal. The S-transform is noise-resistant. As a result, it is commonly employed in power system fault research [3, 18]. Mathematically, the S-transform can be expressed as

$$\mathrm{S[a, b]} = \sum_{h=0}^{N-1} X[b+h]e^{\frac{-2\pi^2 h^2}{b^2}} e^{\frac{i2\pi ha}{N}} \ \mathrm{b} \neq 0 \tag{28}$$

$$\mathrm{S[a, b]} = \frac{1}{N}\sum_{h=0}^{N-1} X[h]\,\mathrm{b} = 0 \tag{29}$$

where $\mathrm{X[b]} = \frac{1}{N}\sum_{h=0}^{N-1} X[h]e^{\frac{-i2\pi ha}{N}}$.

3.3 Time–Time (TT) Transform

The S-transform is a method of representing a real-time series as a set of complex, time-localized spectra. This method is used to derive the time–time transform representation. The S-transform becomes the Fourier transform of the primary time series when it is integrated across time. Similarly, the TT-transform goes back to the original time series when averaged across the primary time variable. The TT-transform's invertibility indicates the possibility of filtering and signal-to-noise improvements in the time domain, as well as some insight into the S-transform's localized spectra. The TT-transform could be useful in underground cable fault detection methods as here different frequency components of a signal need to be localized.

4 Artificial Intelligence Techniques

Generally, the fault in the power system is handled by a microcontroller based on relays and circuit breakers. To make the system faster we have to make our computer as intelligent as a technician sitting in the substation who can say the type of fault that occurs in the power system just by visualizing the voltage and current. The basic steps that need to be completed in this technique are raw data generation/extraction, data analysis and visualization, feature selection, data normalization and encoding, and

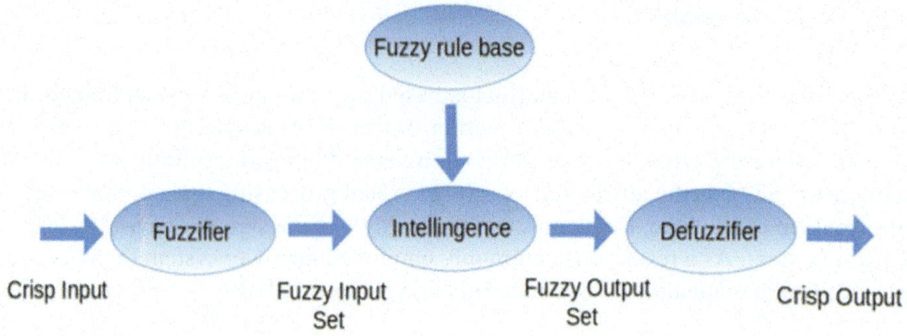

Fig. 6 Fuzzy logic system

algorithm for classification and regression. The following are some of the artificial intelligence methods that are used to locate faults in underground cables.

4.1 Fuzzy Logic-Based System

Fuzzy logic is a type of computation that uses "degrees of truth" instead of the traditional "true or false" (1 or 0) approach. It performs on sloppy datasets, but the algorithms must still be written by humans. Fuzzification, fuzzy inference system, fuzzy rule base, and defuzzification for fault categorization are the basic components of a fuzzy scheme. Crisp numbers are used in the fuzzification stage. The fuzzy set has been mapped. The fuzzy inference system is provided inputs, and then the type of fault is determined by the fuzzy rule basis in the final product. Finally, during the defuzzification state, the fuzzy data is removed. Crisp fault type is assigned to the output set. The functioning of the fuzzy logic system is depicted in Fig. 6.

4.2 ANFIS

Adaptive neuro-fuzzy interference system (ANFIS) is a simple data learning method that transforms a given input into a target output using a fuzzy inference system model. Membership functions, fuzzy logic operators, and conditional rules are all used in this transformation. In a typical ANFIS operation, input fuzzification, application of fuzzy operators, application method, output aggregation, and defuzzification are the five primary processing stages which are shown in Fig. 7.

The term "adaptive network" refers to a multilayer network in which each node performs a specific function based on the dataset being used. The node's function changes from node to node. It functions similarly to a fuzzy inference system and is

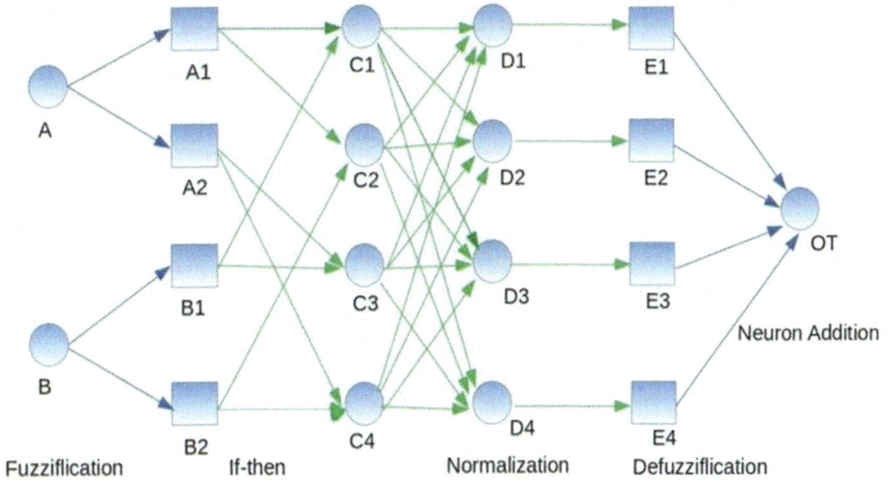

Fig. 7 Adaptive neuro-fuzzy interference system (ANFIS)

similar to a neural network. The ANFIS system can be used to locate and classify faults in underground cables [19, 20].

4.3 CNN

Initially, CNN was applied to reduce the number of preprocessing processes. CNN is characterized by its characteristic invariance. The input layer, convolution layer, pooling layer, full connection layer, and output layer are the five layers in CNN. Convolutional kernels are a set of filters found in each convolutional layer. The filter is a matrix of integers the same size as the kernel that is applied to a subset of the input pixel values. Each pixel is multiplied by the kernel's corresponding value, then the result is summed for a single value that represents a grid cell in the output channel/feature map, similar to a pixel. Each convolution is a sort of affine function, and they are all linear transformations. When the same parameters of the convolution kernel are applied to different places of the preceding layer, new features might be obtained. To get local features, CNN performs a convolution operation on each layer and combines local characteristics within the layer. In terms of classification, identification of faults CNN outperforms [21, 22].

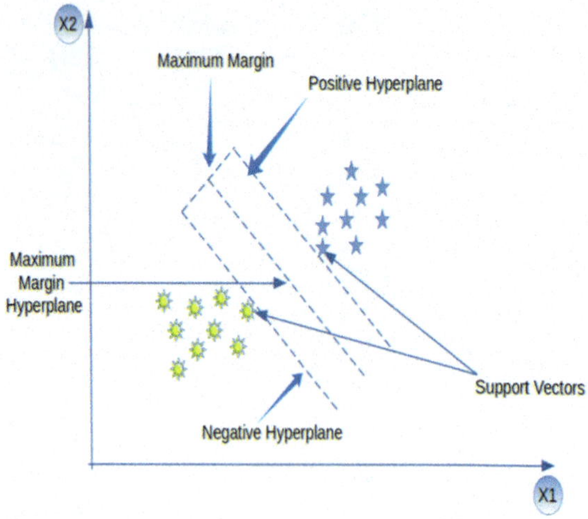

Fig. 8 Support vector machine (SVM)

4.4 SVM

The support vector machine (SVM) divides the samples by establishing a decision boundary known as a hyperplane. This hyperplane distinguishes samples belonging to one class from those belonging to another. The functioning of SVM is depicted in Fig. 8. The loss function is used by the SVM to punish each incorrect sample classification. As the SVM uses Kernel functions to shift non-linearly separable samples into higher dimensional space where they are more likely to be linearly separable, it is the greatest technique to overcome incorrect categorization. Rather than local minima, it always provides a global solution. The cost parameter controls the error bound, while the gamma parameter controls the breadth of the hyper-axis. SVM is used to improve the accuracy of identifying and classifying faults.

4.5 ANN

Artificial neural network (ANN) is well known for their pattern recognition capabilities. It is important in underground cables for identifying and classifying damaged lines. By training the network with prior records, measurements, available data, and experiences, ANN acts on the learning ability of a human from its environment, modifies itself to it, and acts appropriately. It is employed in both real-time and offline applications because of its simple, greater generalization property, and adaptive nature. As an input, a faulty signal is trained, and as an output, a fault state is diagnosed. The ability of ANN is to solve linear and non-linear issues. A full protection mechanism for fault identification, categorization, and localization in transmission lines has been created using ANN [23]. In distribution networks, ANN is also used to analyze faults [24]. However, there have been few studies utilizing ANN to investigate shunt faults in hybrid subterranean and overhead transmission lines.

4.6 LSTM

A timing-cycle neural network is known as an LSTM. It is mostly used to overcome the problem of broad RNN long-term reliance. LSTM is suitable for processing and predicting critical events with very long intervals and delays in time series because of its unique architectural structure. This method can be used to visualize features. By clustering, this approach produces data mining results from visualized features [25]. Vibration signals with varied pressures at the inlet are used to identify problems in this method [26].

4.7 PNN

A probabilistic neural network (PNN) is a type of neural network that is used to solve categorization and pattern identification tasks. There are three levels to it (input level, radial basis level, and competitive level). The first level calculates the length between the input and training input vectors, resulting in a vector. The elements of this vector reflect how similar the input is to that training input. The second level adds up these and produces the second level's net output. Then it selects the greatest of this probability, yielding a 1 for that category and a 0 for the others. The training technique starts with a randomly initialized weight and gradually increases this spreading on a radial basis level from 0.0001 to 0.1, which corresponds to a bias value ($b = 0.8326$ Spread). The number of minimum errors is calculated using a 0.0001 increment step. This loop is repeated until either the maximum value (0.1) or the minimum errors equal zero, at which point the training is terminated.

5 Features of Various Fault Diagnosis Techniques

For underground system fault diagnosis, numerous methods have been discussed so far, including conventional methods, ML, and DL-based methods. This section focuses on the features of these fault diagnosis methods in order to assist readers in better understanding the limitations of these methods (Table 1).

Table 1 Features of various fault diagnosis techniques

S. no	Methods	Features
1	Traveling wave method	This method provides great accuracy because it is unaffected by fault resistance, variations in loads, series capacitor banks, fault type, or fault inception angles. High implementation cost. Mostly used in OV lines. Data synchronization is required to estimate accurate fault location [27]
2	Impedance based method	This method is simple, economical, and only requires volt and current from the measurement node [27]. This procedure is time-consuming and mathematically complex. Certain assumptions are used in the calculations, which lead to mistakes in fault location estimation
3	ANFIS	During the training period, there is a high rate of convergence. It lowers the search space dimensions. The parameters are suitably tuned using the hybrid learning rule. The level of computational complexity is extremely high. The number of inputs is limited. Learned information is difficult to comprehend
4	Neural network	ANN learns on its own and does not require reprogramming. It's simple to use, with only a few parameters to tweak [24, 28]. CNN is well known for its image-based fault categorization. When it comes to fault categorization, DNN has a very high level of accuracy. In PNN, there is no need to study anything and no need to specify the network's starting weights. There is no link between the learning and recalling processes [16]. ANN is slow to converge in the BP algorithm. Here Convergence is determined by the initial value of the weigh parameters that link to the network. CNN and DNN require a significant quantity of data to train which requires a large amount of memory. It takes longer to train a model with a neural network. For good generalization ability, the number of hidden layers is an essential issue
5	SVM	Possibilities of misclassification are rare. Because of the maximum marginal boundary, the chances of error are less. Noise, load angle fluctuation, fault initiation angle, and fault resistance do not affect SVM-based fault localization and classification [29, 30]. Multiclass categorization is highly difficult and necessitates extensive memory. It takes more time to train and test [31]
6	FLS	Fuzzy systems are well suited to uncertain or approximate reasoning. It enables decision-making based on estimated values in the face of incomplete or uncertain data. It is a way of forecasting a future event [24]. It is a time-consuming experiment, and its pattern recognition efficiency is lower than that of a neural network
7	LSTM	Used to visualize features. Then by clustering, it produces data mining results. It can immediately classify raw data as well as learn the information contained within it [32]. It takes longer to train and demands additional memory. Various random weight initializations affect its performance
8	Logistic regression	For binary class categorization, the logistic regression method is simple is desirable [33]. It is unsuitable for data that is unbalanced because it is biased toward the majority class

6 Conclusion

This study provides a complete review of underground cable fault detection, classification, and localization applying several approaches as well as a brief explanation of different types of faults. The phases of development in fault detection have been briefly discussed starting with traditional methods and progressing to artificial intelligence techniques. The use of the system, methodologies, application area, features, and outcomes are compared by taking some selective and essential papers. The traditional method of fault diagnosis was a time-consuming and inefficient procedure that depended on model-based methodologies and domain expertise to monitor the power system for fault problems. Machine learning and deep learning algorithms that rely on historical data can quickly adjust to changes in power system designs over time. Advanced instrumentation has made it simpler to accurately detect and continuously record power system data. The data is currently employed by process history-based methodologies to train the model and operate the power system for defect diagnostics through process history-based methodologies. This study will aid the researcher in their research and development in this area.

Acknowledgements This research work was supported by "Woosong University's Academic Research Funding—2023."

References

1. Orton H (2013) History of underground power cables. IEEE Electr Insul Mag 29:52–57. https://doi.org/10.1109/MEI.2013.6545260
2. Salkuti SR (2022) Emerging and advanced green energy technologies for sustainable and resilient future grid. Energies 15(18). https://doi.org/10.3390/en15186667
3. Swamy BYVNR, Rama Krishna D, Pruna Chandu CH, Venkatesh K, Sasidhar Reddy A (2018) Underground cable fault detection. Int J Res Appl Sci Eng Technol 6(3):3060–3064. https://doi.org/10.22214/ijraset.2018.3663
4. Hans MR, Kor SC, Patil AS (2017) Identification of underground cable fault location and development. In: International conference on data management, analytics, and innovation, ICDMAI 2017. Institute of Electrical and Electronics Engineers Inc., pp 5–8 https://doi.org/10.1109/ICDMAI.2017.8073476
5. Asif RM, Hassan SR, Rehman AU et al (2020) Smart underground wireless cable fault detection and monitoring system. In: International conference on engineering and emerging technologies, ICEET 2020. Institute of Electrical and Electronics Engineers Inc. https://doi.org/10.1109/ICEET48479.2020.9048210
6. Sinha T, Ray P, Salkuti SR (2018) Protection coordination in microgrid using fault current limiters. J Green Eng 8(2):125–150. https://doi.org/10.13052/jge1904-4720.822
7. Chandra S, Chanki R, Prashant P, Rajpoot S, Kumar S (2021) A dynamic SUGPDS model for faults detection and isolation of underground power cable based on detection and isolation algorithm and smart sensors. J Electr Eng Technol 16(4):1799–1819. https://doi.org/10.1007/s42835-021-00715-7
8. Mishra S, Patra S, Swain SC, Routray A (2021) Parameter estimation for underground cable fault using stochastic optimization. IECON proceedings—industrial electronics conference. https://doi.org/10.1109/IECON48115.2021.9589994

9. Jia H (2017) An improved traveling-wave-based fault location method with compensating the dispersion effect of traveling wave in wavelet domain. Math Probl Eng. https://doi.org/10.1155/2017/1019591
10. Das S, Santoso S, Gaikwad A, Patel M (2014) Impedance-based fault location in transmission networks: theory and application. IEEE Access 2:537–557. https://doi.org/10.1109/ACCESS.2014.2323353
11. Zimmerman K, Costello D (2005) Impedance-based fault location experience. In: 58th annual conference for protective relay engineers. College Station, TX, USA, pp 211–226. https://doi.org/10.1109/CPRE.2005.1430435
12. Min SW, Nam SR, Kang SH, Park JK (2007) Fault location algorithm for cross-bonded cables using the singularity of the sheath impedance matrix. Electr Eng 89(7):525–533. https://doi.org/10.1007/s00202-006-0035-1
13. Helles SM, Hvidsten S, Balog G, Furuheim KM (2011) Calculation of water ingress in a HV subsea XLPE cable with a layered water barrier sheath system. J Appl Polym Sci 121(4):2127–2133. https://doi.org/10.1002/app.33568
14. Orlando Oliveira M, Horacio Reversat J, Alberto Reynoso L (2019) Wavelet transform analysis to applications in electric power systems. Wavelet Transform Complex. https://doi.org/10.5772/intechopen.85274
15. Han J, Crossley PA (2013) Fault location on mixed overhead line and cable transmission networks. IEEE Grenoble Conference PowerTech, POWERTECH 2013. https://doi.org/10.1109/PTC.2013.6652311
16. Ngaopitakkul A, Suttisinthong N (2012) Discrete wavelet transform and probabilistic neural network algorithm for classification of fault type in underground cable. In: International conference on machine learning and cybernetics, Xi'an, China, pp 360–366. https://doi.org/10.1109/ICMLC.2012.6358940
17. Jung CK, Lee JB, Wang XH, Song YH (2007) Wavelet based noise cancellation technique for fault location on underground power cables. Electric Power Syst Res 77(10):1349–1362. https://doi.org/10.1016/j.epsr.2006.10.005
18. Shi Q, Troeltzsch U, Kanoun O (2010). Detection and localization of cable faults by time and frequency domain measurements. In: 7th international multi-conference on systems, signals and devices. https://doi.org/10.1109/SSD.2010.5585506
19. Salahshoor K, Kordestani M, Khoshro MS (2010) Fault detection and diagnosis of an industrial steam turbine using fusion of SVM (support vector machine) and ANFIS (adaptive neuro-fuzzy inference system) classifiers. Energy 35:5472–5482. https://doi.org/10.1016/j.energy.2010.06.001
20. Abellan-Nebot JV, Subirón FR (2010) A review of machining monitoring systems based on artificial intelligence process models. Int J Adv Manuf Technol 47:237–257. https://doi.org/10.1007/S00170-009-2191-8
21. Guo MF, Zeng XD, Chen DY, Yang NC (2018) Deep-learning-based earth fault detection using continuous wavelet transform and convolutional neural network in resonant grounding distribution systems. IEEE Sens J 18(3):1291–1300. https://doi.org/10.1109/JSEN.2017.2776238
22. Peng C, Zhe Z, Rui L, Cheng C, Shaokang C (2020) A CNN recognition method for early stage of 10 kV single core cable based on sheath current. Electric Power Syst Res. https://doi.org/10.1016/j.epsr.2020.106292
23. Ekici S, Yildirim S, Poyraz M (2008) Energy and entropy-based feature extraction for locating fault on transmission lines by using neural network and wavelet packet decomposition. Expert Syst Appl 34:2937–2944. https://doi.org/10.1016/j.eswa.2007.05.011
24. Rafinia A, Moshtagh J (2014) A new approach to fault location in three-phase underground distribution system using combination of wavelet analysis with ANN and FLS. Int J Electr Power Energy Syst 55:261–274. https://doi.org/10.1016/j.ijepes.2013.09.011
25. Zheng S, Zhao J (2020) A new unsupervised data mining method based on the stacked auto-encoder for chemical process fault diagnosis. Comput Chem Eng. https://doi.org/10.1016/j.compchemeng.2020.106755

26. Wei X, Chao Q, Tao J, Liu C, Wang L (2021) Cavitation fault diagnosis method for high-speed plunger pumps based on LSTM and CNN. Acta Aeronauticaet Astronautica Sinica 42(3). https://doi.org/10.7527/S1000-6893.2020.23876
27. Gururajapathy SS, Mokhlis H, Illias HA (2017) Fault location and detection techniques in power distribution systems with distributed generation: a review. Renew Sustain Energy Rev 74:949–958. https://doi.org/10.1016/j.rser.2017.03.021
28. Raza A, Benrabah A, Alquthami T, Akmal M (2020) A review of fault diagnosing methods in power transmission systems. Appl Sci. https://doi.org/10.3390/app10041312
29. Gashteroodkhani OA, Majidi M, Etezadi-Amoli M, Nematollahi AF, Vahidi B (2019) A hybrid SVM-TT transform-based method for fault location in hybrid transmission lines with underground cables. Electric Power Syst Res 170:205–214. https://doi.org/10.1016/j.epsr.2019.01.023
30. Livani H, Evrenosoglu CY (2014) A machine learning and wavelet-based fault location method for hybrid transmission lines. IEEE Trans Smart Grid 5(1):51–59. https://doi.org/10.1109/TSG.2013.2260421
31. Mishra DP, Ray P (2018) Fault detection, location and classification of a transmission line. Neural Comput Appl 30:1377–1424. https://doi.org/10.1007/s00521-017-3295-y
32. Swaminathan R, Mishra S, Routray A, Swain SC (2021) A CNN-LSTM-based fault classifier and locator for underground cables. Neural Comput Appl 33:15293–15304. https://doi.org/10.1007/s00521-021-06153-w
33. Cai T, Chow MY, Lu W, Li L (2010) Statistical feature selection from massive data in distribution fault diagnosis. IEEE Trans Power Syst 25(2):642–648. https://doi.org/10.1109/TPWRS.2009.2036924

Performance Analysis of Fuzzy-Based Controller for Wind and Battery Fed UPQC

Koganti Srilakshmi, Sravanthy Gaddameedhi, Uday Kumar Neerati, Surender Reddy Salkuti, Ponamanenni Anoop Rao, Thattiparthi Pavan Kumar, and Machidi Akshith

Abstract Recently, renewable energy sources are integrated into the distribution system to overcome the effect of active and reactive power losses in order to meet the requirement of the load. However, the arrival of the power electronics equipment to control the nonlinear loads has made an impact on the power quality (PQ). The unified power quality conditioner (UPQC) is a FACTS device comprised of two back-to-back connected converters via a DC-Link capacitor. This chapter suggests an intelligent hybrid controller for wind power generation systems (WPGS) with Battery storage (BS) integrated UPQC. The suggested system consists of a hybrid controller which includes both a fuzzy-logic controller (FL-C) and a proportional-integral (PI) controller. Here the self-tuning filter with unit vector generation scheme (STF-UVGS) was designed to maintain the synchronization of phases in series and shunt filters and also to obtain the superior performance of the proposed system under different uncertain conditions like unbalance/distortions in the supply voltage. However, the purpose of using STF is to isolate the fundamental harmonic components without phased locked loop (PLL), low pass filters (LPFs), High pass filters (HPFs) to generate phase synchronization (SYP). The aim of the proposed fuzzy logic plus proportional integral controller (FLPIC) is to stabilize DC link capacitor voltage

K. Srilakshmi · S. Gaddameedhi
Department of Electrical Engineering, Sreenidhi Insitute of Science and Technology, Hyderabad, India

U. K. Neerati
Department of Electrical and Electronics Engineering, Vasavi College of Engineering, Hyderabad, India
e-mail: uday@staff.vce.ac.in

S. R. Salkuti (✉)
Department of Railroad and Electrical Engineering, Woosong University, Daejeon 34606, Republic of Korea
e-mail: surender@wsu.ac.kr

P. A. Rao · T. P. Kumar · M. Akshith
Department of Electrical and Electronics Engineering, Sreenidhi Insitute of Science and Technology, Hyderabad, India

S. R. Salkuti et al. (eds.), *Power Quality in Microgrids: Issues, Challenges and Mitigation Techniques*, Lecture Notes in Electrical Engineering 1039,
https://doi.org/10.1007/978-981-99-2066-2_11

(DCLV) during load and wind velocity variations, diminish the harmonics in the current signals, power factor (PF) enhancement, maximum mitigation of swell, sag and supply voltage distortions. The performance of the proposed hybrid controller is studied with four different test cases for different combinations of loads, supply voltages, and issues and comparison was carried out with other controllers like PI, sliding mode controller (SM-C), and FL. The FLPIC shows extraordinary performance in diminishing THD thereby improving PF and reducing voltage distortions.

Keywords Self-tuning filter · Fuzzy-logic controller · Unified power quality conditioner · Wind power generation system · Battery system

Nomenclature

SUAPF	Shunt active power filter
SEAPF	Series active power filter
FC	Fundamental signal Component
HC	Harmonic signal Component
V_{s_abc}	Supply voltage in abc
i_{s_abc}	Supply current in abc
$V_{s\_\alpha\beta0}$	Supply voltage in αβ0
$V'_{s\_\alpha\beta0}$	FC of supply voltage in αβ0
$V''_{s\_\alpha\beta0}$	HC of supply voltage in αβ0
R_s	Supply side Resistance
L_s	Supply side Inductance
V_{l_abc}	Load voltages in abc
$V_{\mathrm{se\_abc}}$	Series compensated voltage in abc
$V^{ref}{}_{abc}$	Reference voltages in abc
$V^{ref}{}_{se\_dq}$	Series compensated reference voltage in dq
$V^{ref}{}_{se\_\alpha\beta0}$	Reference of compensated voltage in αβ0
$V^{ref}{}_{\alpha\beta0}$	Reference of voltage in αβ0
$V^{ref}{}_{dq}$	Reference of voltage in dq
i_{l_abc}	Load current for abc phase
$i_{l\_\alpha\beta0}$	Load currents in αβ0
$i'_{l\_\alpha\beta0}$	Load current FC in αβ0
$i''_{l\_\alpha\beta0}$	Load current HC in αβ0
i''_{l_dq0}	HC of load current in dq0
$i^{ref}{}_{l\_abc}$	Reference load current for abc phase
$i^{ref}{}_{l\_\alpha\beta0}$	Reference load current in αβ0
$i^{\mathrm{ref}}{}_{\mathrm{l\_dq}}$	Reference load current in dq frame
i_{sh_abc}	SUAPF compensated current in abc
$i^{ref}{}_{sh\_abc}$	Reference SUAPF compensated current in abc
C_{sh}	Capacitance of SUAPF
R_{sh}	Resistance of SUAPF

L_{sh}	Inductance of SUAPF
C_{dc}	Capacitance at DC link
V_{dc}	Voltage at DC link
$V^{ref}{}_{dc}$	Reference voltage at DC link
$i^{ref}{}_{dc}$	Reference current at DC link
Δi_{dc}	Output current error of DC link
P_W	Power from wind system
V_W	Voltage of wind system
I_W	Current of wind system
$P_{DC\text{-}Link}$	DC link power
P_{BS}	Battery power
V_{BS}	Battery voltage
i_{BS}	Battery current
Q	Battery charge
DPD	DC link power demand
MF	Membership function
E	Fuzzy Error
CE	Change of error
SRFT	Synchronous Reference-Frame Theory
B-C	Boost-Converter
B-B-C	Buck Boost Converter
SOCB	State of charge of the battery
PMSG	Permanent magnet synchronous generator

1 Introduction

Nowadays, the power distribution system is facing PQ issues like interruptions, disturbances, flicker, sag/swell, harmonics, PF, etc. due to the integration of inconsistent behavior of wind, tidal, solar, etc., the large non-linear and unbalanced loads with power electronic equipment. Due to the huge demand of heavy nonlinear loads causes a reduction in pf leads to power quality problems. Hence effective maintenance of PQ is the major objective for researchers. The different techniques were adapted to control the 1ϕ and 3ϕ SUAPF for 3-wire and 4-wire systems to eliminate the PQ problems were addressed [1]. The SRFT-based controller was proposed for the effective mitigation of PQ issues with various unbalanced distorted loads [2].

The new hybrid controller with the integration of both FL-C and PI-C for SUAPF with different case studies was introduced to diminish PQ problems and consequently reduce THD [3]. Besides, the PV battery STF for SUAPF was developed in view of regulating the reactive power and minimizing current harmonics efficiently. While generating an appropriate reference current the Maxikalman filter was introduced [4]. Further, the improvement in the intelligent-based controllers like FL, ANN, etc.

for SUAPF was implemented to reduce PQ problems efficiently during the variations in load [5, 6].

To study the overall performance of the proposed system with lesser THD, a solar-fed UPQC with a novel MPPT technique was implemented and analyzed with various combinations of loads [7, 8]. Besides, SRFT-UPQC with PI controller for a fuel cell was implemented for SUAPF to regulate the DLCV effectively and to minimize the distortions in the current waveform [9]. The hybrid optimization algorithm-based controller was proposed for optimal tuning of SUAPF with an aim of minimizing THD and effective maintenance of reactive power [10]. Nevertheless, for effective management of reactive power, a feed-forward ANN controller was designed for wind-solar-based UPQC and also to maintain the DCLV constant [11].

A new adaptive control procedure was applied on the multilevel UPQC with 8 switches with the goal of minimizing current and voltage distortions efficiently [12]. Besides, a multi-level UPQC integrated with a microgrid was implemented with the objective of reducing imperfections in voltage and current waveforms effectively [13]. The UPQC with furnace load was recommended in order to suppress the current/voltage harmonics However, to prove its superior performance different case studies were considered and compared with DSATCOM [14]. The development of an efficient controller by the combination of both the features of ANN, FL was carried out for the solar PV integrated UPQC to address the PQ problems and to show the viability of the proposed controller performance investigation was carried out for different types of loads and supply voltage conditions [15]. P-Q theory-based control technique was suggested for UPQC using an STF and eliminating the usage of LPFs or PLL [16].

However, to eliminate the complexity of the proposed system in terms of LPF and PLL the STF-based SUAPF was illustrated and to prove its viability theoretical and/or practical performance investigation was done [17]. The FL-based UPQC with nonlinear induction motor load was adapted to suppress the imperfections in waveforms[18]. A new Predictive phase dispersion method was suggested for the multilevel UPQC for reducing source voltage fluctuations, and harmonics, and maintaining the stable voltage at DC-Link [19]. The microgrid integrated UPQC based on Fourier transform was suggested with the objective of minimizing the fluctuations in the supply voltage and harmonics in output load current [20]. The different SVPWM control techniques were suggested to address various PQ problems [21].

The IOT-based control unit with ANN controller was designed for SUAPF to identify the PQ issues with the goal of minimizing THD [22]. The intelligent Fuzzy-PI and Fuzzy-PID controllers were suggested for the AC-DC microgrid with a target of addressing PQ problems and thereby enhancing the voltage stability. Moreover, the performance was demonstrated in two test cases with variable loads [23]. Besides, a BBO optimization algorithm based UPQC was implemented to obtain the optimal values of Kp/Ki of the PI controller with an aim of reducing the DCLV oscillations [24].

The adaptive neuro-fuzzy hybrid controller was introduced to eliminate harmonics and in addition to the improvement in the utilization of the network [25]. A novel automatic shifting between grid/ or island was designed for the solar-battery connected

UPQC to reduce PQ issues, in addition, validation was carried out with experimental results [26]. The hybrid optimization technique-based novel controller-based UPQC was implemented with an intention of diminishing the error of power imbalances by appropriate tuning of K_p/K_i values [27].

The reactive power control method with UVGT with an ANN controller for the solar-integrated UPQC was developed to deal with PQ issues effectively and to reduce the stress and ratings on converters in addition, to keeping the stable voltage across DC-Link [28]. The sequence-component detection technique combination with UVGT was projected for the two stages PV incorporated UPQC to attend to PQ-related problems effectively [29]. The FL-based SEAPF was designed to solve the PQ issues like voltage and current signal distortions in addition to regulating DC link capacitor voltage [30].

Although many controllers were developed, still there exists a scope for developing a new hybrid controller and intelligent methods for handling power quality-related issues. In this chapter, a hybrid controller comprised of FL and PI properties was designed for UPQC connected to WPGS and BS systems (U-WBS) with the objective of diminishing the current THD thereby enhancing PF, regulating DCLV during load/wind velocity variation, compensation during with sag/swell and disturbance condition [31]. The STF-UVGS was suggested to generate SYP for controllers instead of PLL and also it abolishes the requirement of LPFs and HPFs for splitting both FC and HC of current. The performance analysis of the projected FLPIC for the U-WBS system was analyzed for 5 different test cases, and to exhibit its supremacy comparative study was done with the conventional PI, SM, and FL controllers. Section 2 gives the structure of the developed U-WBS. Section 3 provides the design of the suggested FLPIC. Section 4 portrays a discussion of the result analysis, and finally, Sect. 5 gives the Conclusion.

2 Structure of Proposed U-WBS

The structure of U-WBS is illustrated in Fig. 1. The WPG and BS are associated with the DC Link of UPQC via a B-C, and B-B-C to regulate DCLV during the wind velocity and dynamic load variations, and reduce the burden of converters. This chapter designs a fuzzy-based hybrid controller FLPIC to exploit both the unique qualities of FL and PI controller [32]. The SEAPF mitigates voltage issues like sags/swell/disturbance, and supplies voltage unbalances by injecting suitable compensating voltage V_{se} via injecting transformer. Similarly, the need for SUAPF is to reduce the imperfections in the current signal by introducing the compensating current i_{sh} and maintaining DCLV. The WPGS and BS specifications are given in Table 1.

The power distribution at the DC link of the proposed U-WBS is given by,

$$P_W + P_{BS} - P_{DC-Link} = 0 \quad (1)$$

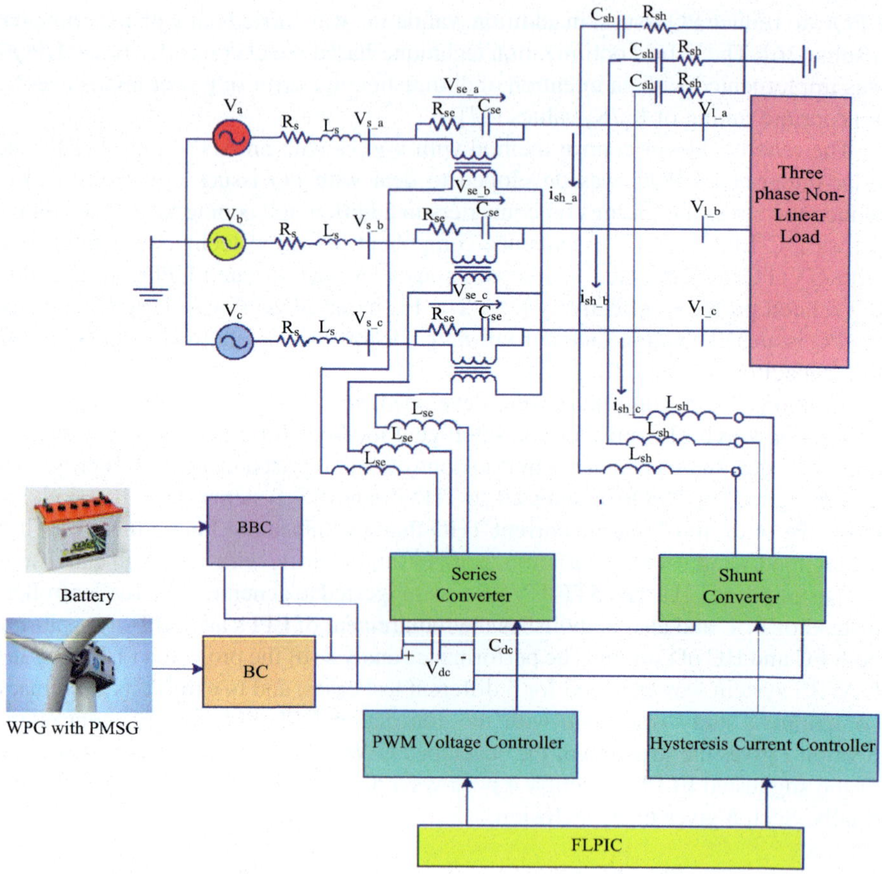

Fig. 1 Components of U-WBS

Table 1 Wind system and Battery Ratings

Device	Parameters	Values
Wind Turbine	Nominal power	3 MW
	Base speed of the wind	11 m/sec
	Integral-gain	5
	Proportionality gain	25
	Highest pitch's angle	45^0
	Rate of the change of the pitch's angle	25 deg./second
Li-ion battery	Rated Capacity of battery	350 Ah
	Maximum capacity of the battery	450 Ah
	Normal Voltage	650 V
	Fully-charge voltage	756 V

2.1 BS System

The lead-acid battery is considered in the work connected to DC link via B-B-C to maintain constant DLCV [33]. The controllers of the battery system are given in Fig. 2. SOCOB plays a vital role in maintaining the life of the battery which is evaluated by,

$$SOCOB = 100\left(1 + \int i_{BS}\, dt\ Q\right) \quad (2)$$

The power produced by WPG will choose the state of working of a battery: charging or discharging by satisfying the upper and lower constraints given by,

$$SOCOB_{max} \leq SOCOB \leq SOCOB_{min} \quad (3)$$

The i_{dc}^{ref} is estimated by minimizing the DLCV error $V_{dc,err}^{ref}$ by a PI controller. The reference error current of the battery $i_{BS,\ err}$ is calculated by PI through a battery's error current $i_{BS,err}$. Where $i_{BS,err}$ is the difference between i_{dc}^{ref} and battery reference current i_{BS}^{ref} received from LPF.

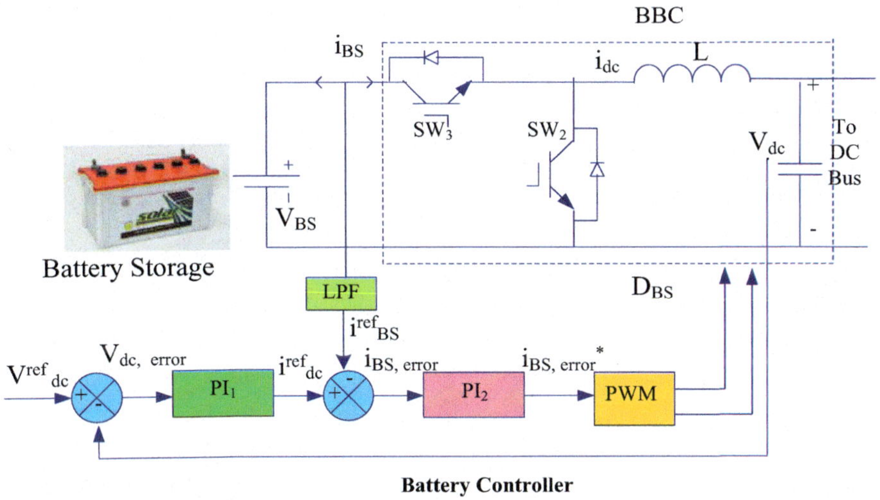

Fig. 2 Battery controller

2.2 Wind Power Generation System (WPGS)

The major parts of the WPGS system are the turbine, a PMSG, and BC illustrated in Fig. 3. The output power of the system is specified by,

$$P_W = 12\delta A v^3 C_P(\mu, \theta) \tag{4}$$

where δ is the air density in kg/m^3, A is the area of rotor blades in m^2, V is the wind velocity in m/s, C_P is the power coefficient which is the function of (TSR,μ), and pitch angle (θ).

PMSG has been selected because of its minimum maintenance expenditure. The PMSG output depends on the wind speed. However, the generator AC output is transformed into DC through the rectifier and then the voltage magnitude is boosted by means of BC as given in Fig. 3. The i_{dc}^{ref} is estimated by minimizing DLCV error $V_{dc,err}$ by a PI controller. The reference error current of wind $i_{W,err}*$ is calculated by PIC through an error current produced by wind $i_{W,err}$. Where $i_{W,err}$ difference value of i_{dc}^{ref} and wind system reference current i_W^{ref} received from the LPF. Table 2 provides the distribution of power at the DC link.

3 Control Strategy of Proposed U-WBS

During large faults or sudden changes in the load of the distribution network, the power varies which in turn changes the DCLCV. However, the necessary action has to be taken in order to bring DCLCV back to the stable in a short duration of time.

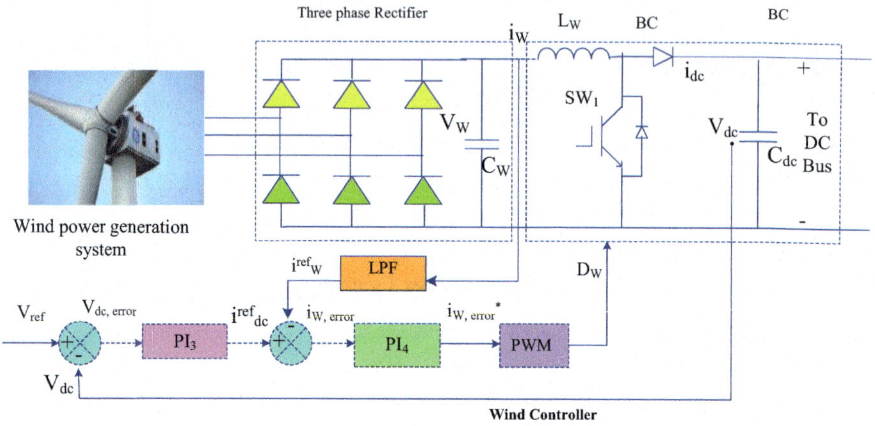

Fig. 3 Wind systemcontroller

Table 2 DC-link power distribution

Condition	Action
$P_W > P_{dc}$	Excess P_W charges Battery till it attains $SOCOB_{max}$
$P_W = P_{dc}$	P_W alone will supply the required power to P_{dc}
$P_W < P_{dc}$	The difference is supplied by Battery until it attains $SOCOB_{min}$
Nil P_W	BS will supply the required power to P_{dc}

Generally in conventional mechanisms, SUAPF, SEAPF, PLL, LPFs, and HPFs are used to generate pulses but in this work STF- UVGS is used to produce SYP from the source voltages in addition to performing the role of PLL, HPFs, and LPFs for isolating the FC/HC of currents. The pulses of series-connected and shunt-connected VSCs are generated from the PWM and hysteresis current controllers with FLPIC. The controlling structure is given below.

3.1 STF-UVGS

The proposed STF-UVGS produces SYP from the V_S as shown in Fig. 4. By using Clarke's frame $V_{S\_\alpha\beta 0}$ is transformed into V_{S_abc} given in Eq. (5).

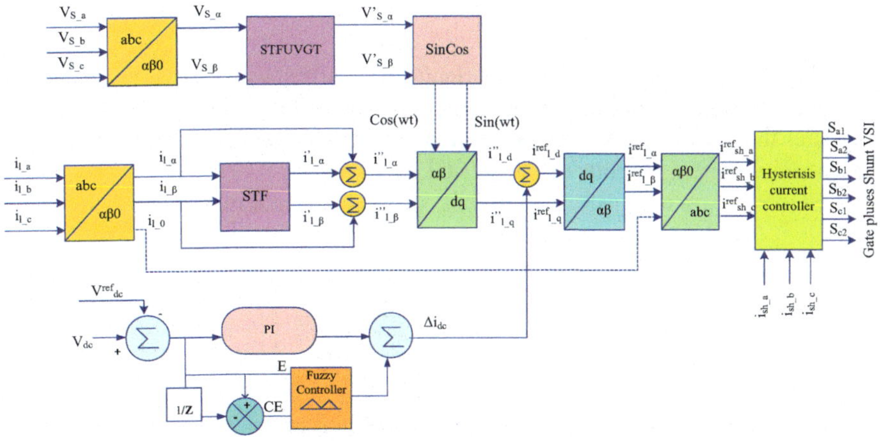

Fig. 4 FLPIC for SUAPF

$$\begin{bmatrix} V_{S\_\alpha} \\ V_{S\_\beta} \\ V_{S\_0} \end{bmatrix} = \sqrt{2/3} \begin{bmatrix} 1 & -1/2 & -1/2 \\ 0 & \sqrt{3/2} & -\sqrt{3/2} \\ 1/2 & 1/2 & 1/2 \end{bmatrix} \begin{bmatrix} V_{S\_a} \\ V_{S\_b} \\ V_{S\_c} \end{bmatrix} \tag{5}$$

However, the FC, and HC were isolated from supply voltage through Eq. (6).

$$\begin{bmatrix} V_{S\_\alpha} \\ V_{S\_\beta} \end{bmatrix} = \begin{bmatrix} V'_{S\_\alpha} + V''_{S\_\alpha} \\ V'_{S\_\beta} + V''_{S\_\beta} \end{bmatrix} \tag{6}$$

By Laplace transformation of STF Eq. (7)is obtained.

$$\begin{bmatrix} V'_{S\_\alpha}(s) \\ V'_{S\_\beta}(s) \end{bmatrix} = \frac{k_1}{s} \begin{bmatrix} V_{S-\alpha}(s) - V'_{S\_\alpha}(s) \\ V_{S-\beta}(s) - V'_{S\_\beta}(s) \end{bmatrix} + \frac{2\prod fkc_1}{S} \begin{bmatrix} V'_{S\_\beta}(s) \\ -V'_{S\_\alpha}(s) \end{bmatrix} \tag{7}$$

However, to omit the PLL, SYP $sin(wt)$, $cos(wt)$ is produced by,

$$\begin{bmatrix} \sin(wt) \\ \cos(wt) \end{bmatrix} = \frac{1}{\sqrt{(V'_{S\_\alpha})2 + (V'_{S\_\beta})2}} \begin{bmatrix} V'_{S\_\alpha}(s) \\ V'_{S\_\beta}(s) \end{bmatrix} \tag{8}$$

3.2 *SUAPF*

The role of SUAPF is to reduce the imperfections in current waveforms by introducing the compensating current and maintaining constant DLCV. The suggested FLPIC compares the actual DLCV with the reference DLCV and supplies the error value to the system. Figure 4 exhibits the suggested controller.

In SUAPF $i_{l\_\alpha\beta 0}$ is transformed to i_{l_abc} using the Clarke method by,

$$\begin{bmatrix} i_{S\_\alpha} \\ i_{S\_\beta} \\ i_{S\_0} \end{bmatrix} = \sqrt{2/3} \begin{bmatrix} 1 & -1/2 & -1/2 \\ 0 & \sqrt{3/2} & -\sqrt{3/2} \\ 1/2 & 1/2 & 1/2 \end{bmatrix} \begin{bmatrix} i_{S\_a} \\ i_{S\_b} \\ i_{S\_c} \end{bmatrix} \tag{9}$$

The STF splits the HC and FC using the Laplace function is given by,

$$\begin{bmatrix} i'_{S\_\alpha}(S) \\ i'_{S\_\beta}(S) \end{bmatrix} = \frac{k_2}{s} \begin{bmatrix} i_{S-\alpha}(S) - i'_{S\_\alpha}(S) \\ i_{S-\beta}(S) - i'_{S\_\beta}(S) \end{bmatrix} + \frac{2\prod fkC_2}{S} \begin{bmatrix} i'_{S\_\beta}(S) \\ -i'_{S\_\alpha}(S) \end{bmatrix} \tag{10}$$

$$i''_{l\_\alpha} = i_{l\_\alpha} + i'_{l\_\alpha} \tag{11}$$

$$i''_{l\_\beta} = i_{l\_\beta} + i'_{l\_\beta} \tag{12}$$

$$i''_{l\_d} = i''_{l\_\alpha}\sin(wt) - i''_{l_\alpha}\cos(wt) \tag{13}$$

$$i''_{l\_q} = i''_{l\_\alpha}\sin(wt) - i''_{l\_\beta}\cos(wt) \tag{14}$$

The DLCV plays a major role in the operation of SUAPF. The proposed FLPIC injects an appropriate suitable error current Δi_{dc} by,

$$\Delta i_{dc} = k_{p,sh}e_1(t) + k_{i,sh}\int_{o}^{t} e_1(t)dt \tag{15}$$

$$e_1(t) = V_{dc}^{ref} - V_{dc}(t) \tag{16}$$

$$i_{l\_d}^{ref} = i''_{l\_d} - \Delta i_{dc} \tag{17}$$

The $i^{ref}{}_{l\_\alpha\beta 0}$ is transformed to $i^{ref}{}_{l\_abc}$ into abc by,

$$\begin{bmatrix} i_{l\_\alpha}^{ref} \\ i_{l\_\beta}^{ref} \end{bmatrix} = \begin{bmatrix} \sin(wt) & \cos(wt) \\ -\cos(wt) & \sin(wt) \end{bmatrix} \begin{bmatrix} i_{l\_d}^{ref} \\ i_{l\_q}^{ref} \end{bmatrix} \tag{18}$$

$$\begin{bmatrix} i_{sh\_a}^{ref} \\ i_{sh\_b}^{ref} \\ i_{sh\_c}^{ref} \end{bmatrix} = \sqrt{2/3} \begin{bmatrix} 1 & 0 \\ -1/2 & \sqrt{3/2} \\ -1/2 & -\sqrt{3/2} \end{bmatrix} \begin{bmatrix} i_{l\_\alpha}^{ref} \\ i_{l\_\beta}^{ref} \end{bmatrix} \tag{19}$$

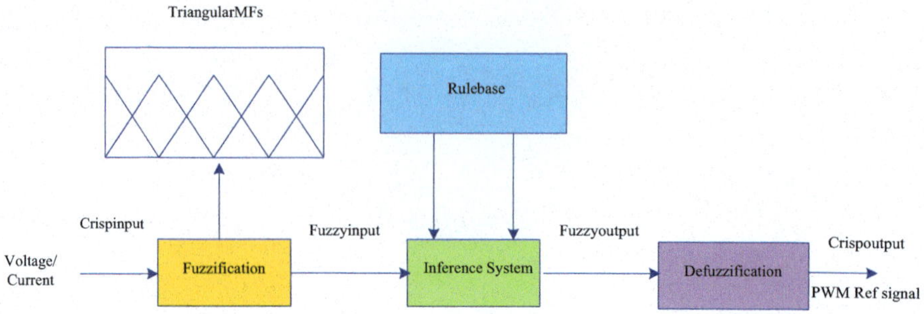

Fig. 5 FLC working process

Fig. 6 MF of "E"

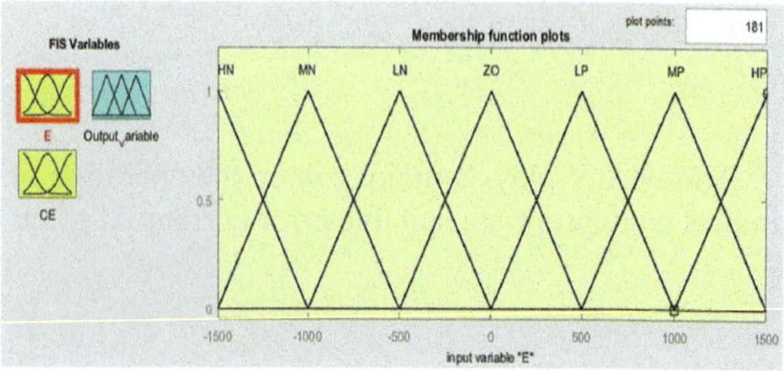

3.2.1 Proposed FLIPC

The FL controller works on the linguistic rules. The working of an FL controller begins with fuzzification, knowledge base, and lastly defuzzification. In the process of fuzzification arithmetical values are transformed into linguistic terms with MF. However, the knowledge base saves the latest information about all input as well as output relations. Lastly, defuzzification transfers linguistic to crisp output. The working model of FL is given in Fig. 5. The triangle's MFs of E, CE (inputs), and output are given in Figs. 6, 7, and 8. The FL variables in triangular MF as shown in Figs. 6 to 8 containing High positive (HP), medium positive (MP), low positive (LP), Zero (ZO), High negative (HN), medium negative (MN), and Low negative (LN). The DLCV is considered in linguistic terms through 49 sets of MFs as given in Table 3. The PI controller's k_p, k_i are chosen randomly as 1.302, 0.010.

3.3 SEAPF

SEAPF function is to reduce PQ problems that are related to the voltage by introducing compensating voltage via a transformer.

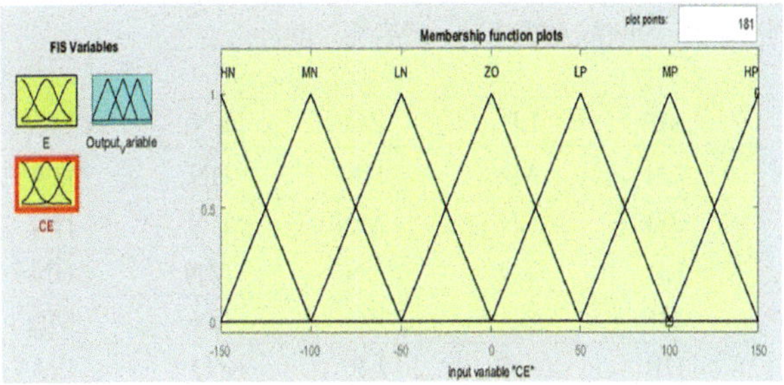

Fig. 7 MF of "CE"

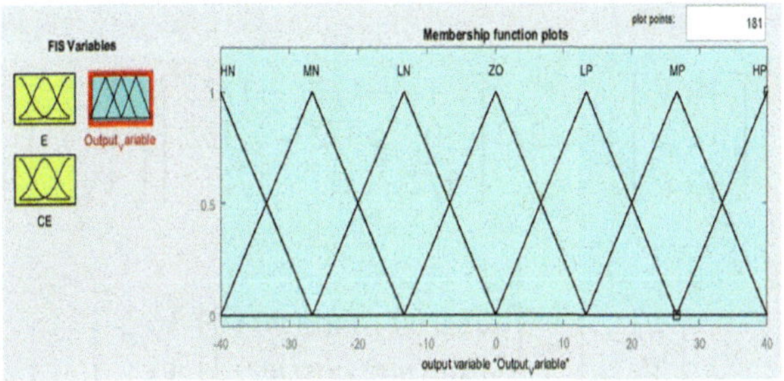

Fig. 8 MF of "Output"

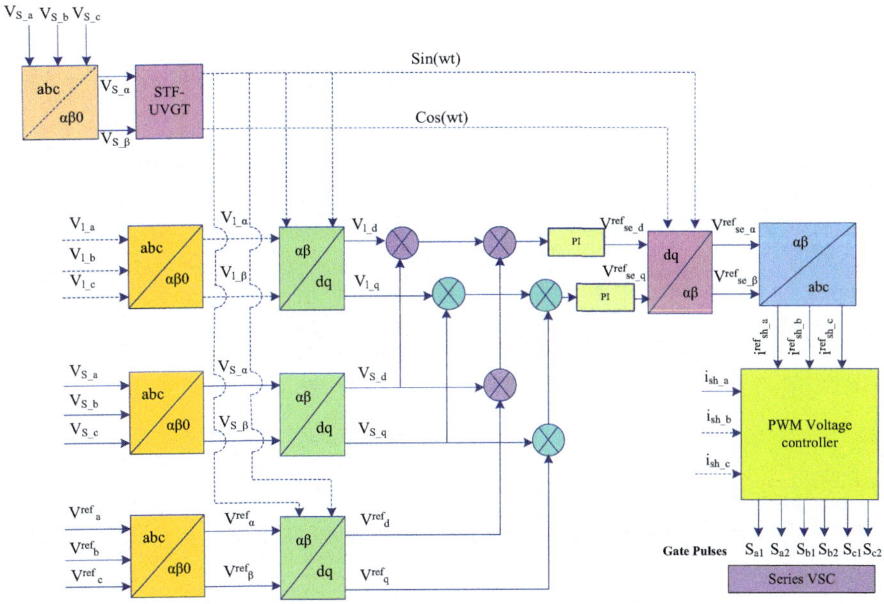

Fig. 9 Series controller

Table 3 MF mapping for voltage across DC-link

Er	CE						
	HP	MP	LP	ZO	LN	MN	HN
HN	ZO	LN	MN	HN	HN	HN	HN
MN	LP	ZO	LN	MN	HN	HN	HN
LN	MP	LP	ZO	LN	MN	HN	HN
ZO	HP	MP	LP	ZO	LN	MN	HN
LP	HP	HP	MP	LP	ZO	LN	MN
MP	HP	HP	HP	MP	LP	ZO	LN
HP	HP	HP	HP	HP	MP	LP	ZO

$$\begin{bmatrix} V_{S\_\alpha} \\ V_{S\_\beta} \\ V_{S\_0} \end{bmatrix} = \sqrt{2/3}\begin{bmatrix} 1 & -1/2 & -1/2 \\ 0 & \sqrt{3/2} & -\sqrt{3/2} \\ 1/2 & 1/2 & 1/2 \end{bmatrix}\begin{bmatrix} V_{S\_a} \\ V_{S\_b} \\ V_{S\_c} \end{bmatrix} \tag{20}$$

$$\begin{bmatrix} V_{S\_d} \\ V_{S\_q} \end{bmatrix} = \begin{bmatrix} \cos(wt) & \sin(wt) \\ -\sin(wt) & \cos(wt) \end{bmatrix}\begin{bmatrix} V_{S\_\alpha} \\ V_{S\_\beta} \end{bmatrix} \tag{21}$$

The voltage is shifted to dq using the park method by,

$$\begin{bmatrix} V_{l\_\alpha} \\ V_{l\_\beta} \\ V_{l\_0} \end{bmatrix} = \sqrt{2/3}\begin{bmatrix} 1 & -1/2 & -1/2 \\ 0 & \sqrt{3/2} & -\sqrt{3/2} \\ 1/2 & 1/2 & 1/2 \end{bmatrix}\begin{bmatrix} V_{l\_a} \\ V_{l\_b} \\ V_{l\_c} \end{bmatrix} \tag{22}$$

By Eq. (22) the V_{l_α} is obtained from to V_{l_abc}.

$$\begin{bmatrix} V_a^{ref} \\ V_b^{ref} \\ V_c^{ref} \end{bmatrix} = V_{max}\begin{bmatrix} \sin(wt) \\ \sin(wt - \frac{2\pi}{3}) \\ \sin(wt + \frac{2\pi}{3}) \end{bmatrix} \tag{23}$$

$$\begin{bmatrix} V_\alpha^{ref} \\ V_\beta^{ref} \\ V_0^{ref} \end{bmatrix} = \sqrt{2/3}\begin{bmatrix} 1 & -1/2 & -1/2 \\ 0 & \sqrt{3/2} & -\sqrt{3/2} \\ 1/2 & 1/2 & 1/2 \end{bmatrix}\begin{bmatrix} V_a^{ref} \\ V_b^{ref} \\ V_c^{ref} \end{bmatrix} \tag{24}$$

$$\begin{bmatrix} V_d^{ref} \\ V_q^{ref} \end{bmatrix} = \begin{bmatrix} \cos(wt) & \sin(wt) \\ -\sin(wt) & \cos(wt) \end{bmatrix}\begin{bmatrix} V_\alpha^{ref} \\ V_\beta^{ref} \end{bmatrix} \tag{25}$$

The V_{dq}^{ref} are obtained from $V_{\alpha\beta 0}^{ref}$ by Eqs. (26)-(29) by implementing Clark and Park shifting.

$$V_{se\_q}^{ref} = (V_d^{ref} - V_{s\_d}) - (V_{l\_d} - V_{s\_d}) \tag{26}$$

$$V_{se\_q}^{ref} = (V_q^{ref} - V_{s\_q}) - (V_{l\_q} - V_{s\_q}) \tag{27}$$

$$\begin{bmatrix} V_{se\_\alpha} \\ V_{se\_\beta} \end{bmatrix} = \begin{bmatrix} \sin(wt) & \cos(wt) \\ -\cos(wt) & \sin(wt) \end{bmatrix} \begin{bmatrix} V_{se\_d}^{ref} \\ V_{se\_q}^{ref} \end{bmatrix} \tag{28}$$

$$\begin{bmatrix} V_{se\_a}^{ref} \\ V_{se\_b}^{ref} \\ V_{se\_c}^{ref} \end{bmatrix} = \sqrt{2/3} \begin{bmatrix} 1 & 0 \\ -1/2 & \sqrt{3/2} \\ -1/2 & -\sqrt{3/2} \end{bmatrix} \begin{bmatrix} V_{se\_\alpha}^{ref} \\ V_{se\_\beta}^{ref} \end{bmatrix} \tag{29}$$

4 Simulation Results

The analysis of the proposed U-WBS with FLPIC controller was designed in MATLAB Simulink environment as shown in Fig. 10. The Vs was considered to be balanced for case studies 1, and 2, and for unbalanced case studies 3, 4, and 5 respectively with various conditions of swell/sag, disturbances in voltage with different values of irradiation, and steady temperature conditions. Besides, the THD was also achieved for i_{sh} of the proposed system for all the case studies and further, it is compared with those of PI/SMC and FLC controllers as illustrated in Fig. 11. The PF is computed by Eq. (30) for all 5 test case studies as given in Fig. 12. The U-WBS specifications are presented in Table 4.

$$PF = \cos\theta * \frac{1}{\sqrt{1 + THD^2}} \tag{30}$$

Five test studies with various types of balanced, un-balanced and non-linear loads with conditions like swell/sag/disturbance at a constant temperature of 25^0C and variable irradiation were considered to analyze the efficient performance of the proposed U-WBS with FLPIC controller as given in Table 5.

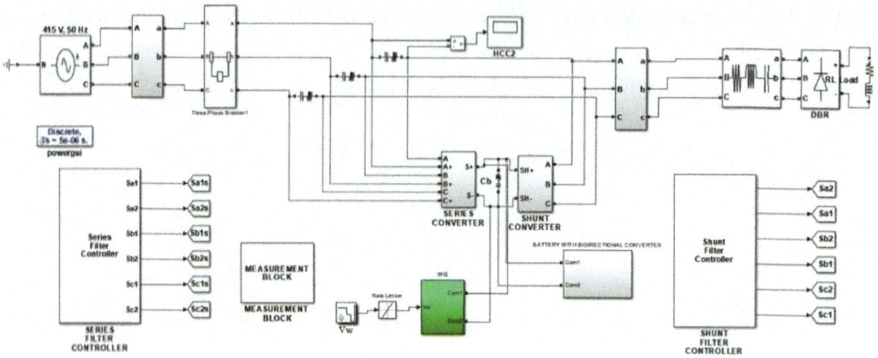

Fig. 10 U-WBS model in Simulink

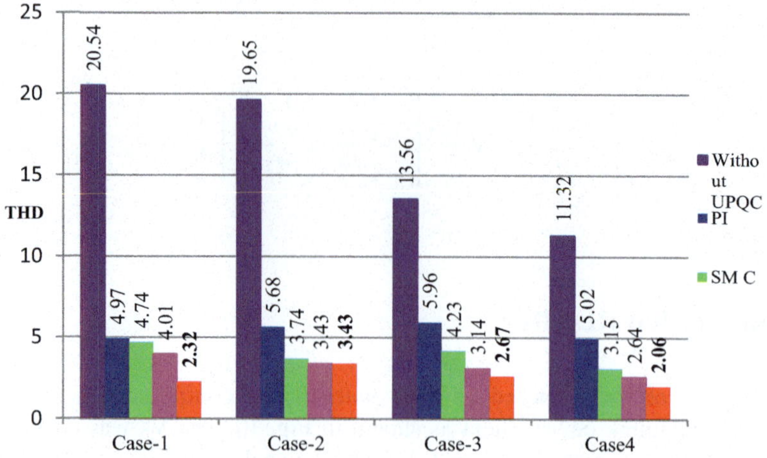

Fig. 11 THD comparison

Fig. 12 PF for test cases

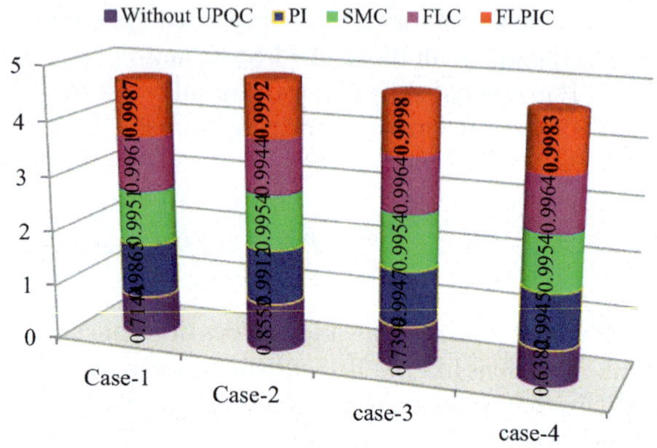

Table 4 Ratings of U-WBS

Source	Grid voltage: 415 V; Frequency: 50 Hz; supply side resistance:0.1Ω; supply side Inductance: 0.151mH
SEAPF	R_{se}: 1Ω; L_{se}: 3.60mH; C_{se}: 60.0 μf
SUAPF	R_{sh}: 0.00101 Ω; L_{sh}: 2.150mh; C_{sh}: 1.0μF; Hysteresis-band: 0.01A
DC-link	load current: 9400μf; V_{dc}: 700 V

Table 5 Description of Cases 1 to 4

Condition and Loads	Case 1	Case 2	Case 3	Case 4
Balanced V_S	✓	✓		
Unbalanced V_S			✓	✓
Swell, sag, &disturbance	✓	✓		
Current under steady state	✓	✓	✓	✓
Voltage under steady state			✓	✓
11 m/s wind speed	✓		✓	
10 m/s wind speed		✓		✓
Load 1: Balanced-rectifier- 30.00Ω & 20.00mH	✓	✓	✓	
Load 2: Unbalanced RL:R: 10, 20 & 15Ω; L: 9.50mH, 10.50mH & 18.50mH			✓	✓
Load 3: Induction-furnace- LC = 400mh, 50 μF, RL = 10Ω,100mH		✓		

In case 1, balanced V_s was considered with 30% sag for the period of 0.2–0.3 s and 30% swell for 0.4–0.5 s respectively. Additionally, the disturbance was also initiated during the time period of 0.60–0.70 s, as illustrated in Fig. 13a. In Fig. 13b, it is visible that the load current waveform is viewed as balanced and non-sinusoidal because of the nonlinear load. On the other hand, it clearly exhibits that the U-WBS can eliminate PQ problems related to voltage and current by introducing the required compensating voltage/current. Such development in the structure of obtained signal also reciprocated in the THD and PF. Hence, the THD of the current signal was diminished from 20.54 to 2.32%, which is lower when compared to previous controllers as given in Table 6 and PF boosts up to 0.9987 by introducing V_{Se} and currents. Moreover, the proposed controller affords a steady DCLV during 11 m/sec velocity as shown in Fig. 13c.

In case 2, balanced V_S was considered with voltage-related issues similar to case 1, but with a wind velocity of 10 m/sec as exhibited in Fig. 14a. From Fig. 14b, it is clearly described that the load current signal is noticed to be balanced and non-sinusoidal because of the nonlinear furnace and rectifier loads. From the waveforms, it is obviously shown that the proposed configuration effectively mitigates the PQ-related issues and reduces the THD value from 19.65 to 3.43%, and improves the PF from 0.8552 to 0.9992 by introducing the suitable compensating voltage and

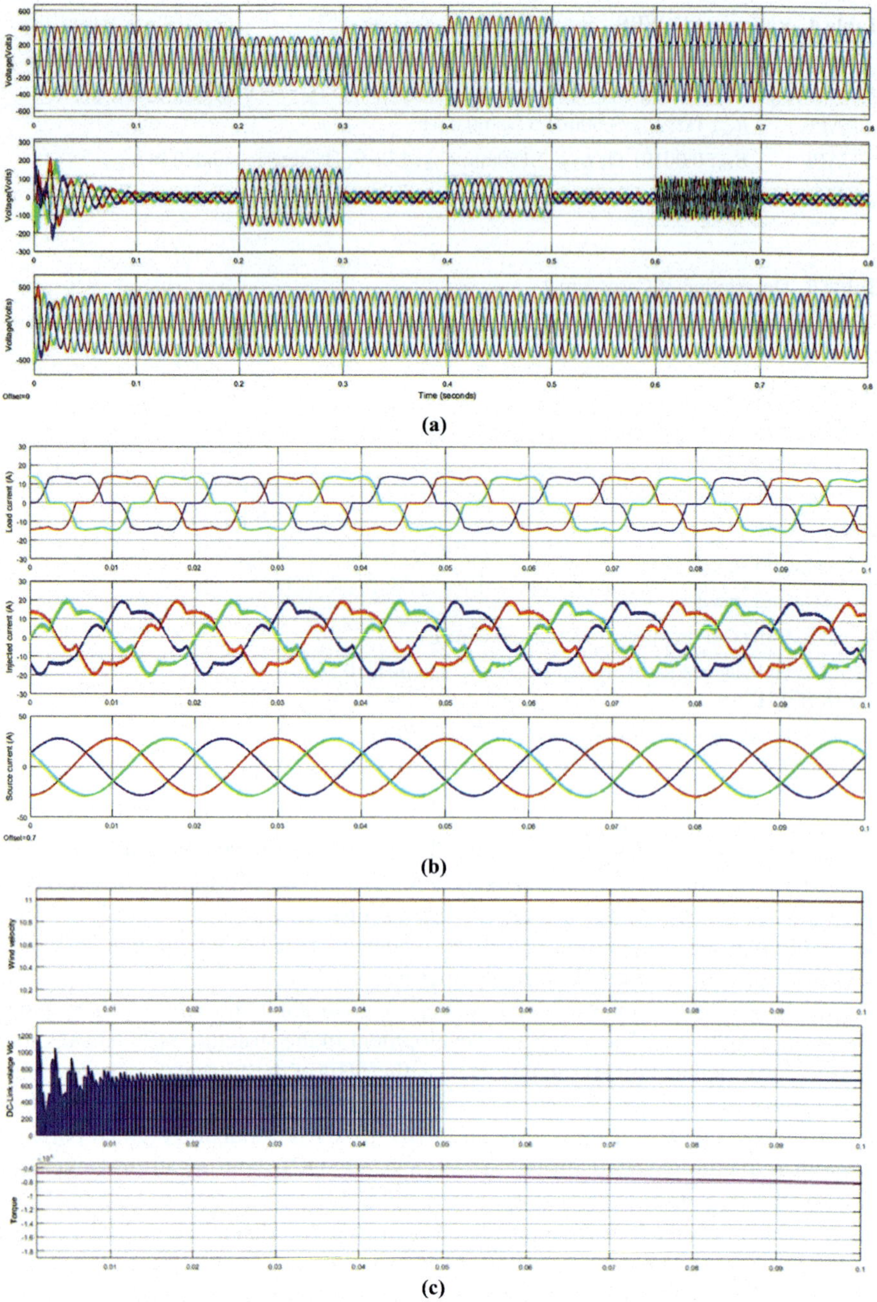

Fig. 13 Waveform of U-WBS for case 1; **a** V_S, V_{Se}, V_l; **b** i_l, i_{sh}, i_s; **c** velocity of wind, VLDC, torque

Table 6 THD (in %) Comparison

Controller	Case 1	Case 2	Case 3	Case 4
Without U-WBS	20.54	15.31	12.63	19.63
PIC	4.97	5.68	4.25	5.02
SMC	4.74	3.74	4.23	4.09
FLC	4.01	3.43	3.14	4.14
PIC [20]	3.27	–	–	–
SMC [20]	2.43	–	–	–
PIC [25]	14.73	–	–	–
FLC [25]	6.12	–	–	–
ANFISC [25]	2.42	–	–	–
PIC [30]	3.64	–	–	–
FLC [30]	2.51	–	–	–
FLPIC	**2.32**	**3.43**	**2.67**	**2.06**

current. As exhibited in Fig. 14b the recommended controller maintains constant DLCV during both load and wind velocity variation simultaneously.

In case 3, unbalanced V_S with 3φ balanced rectifier load was considered for 11 m/sec wind velocity as given in Fig. 15a. Because of the presence unbalanced nonlinear load, the load current was noticed as non-sinusoidal in nature described in Fig. 15b. The proposed FLPIC effectively diminishes the THD from 13.56 to 2.67%, improves PF from 0.7399 to 0.9988and balances load voltage successfully. Additionally, it also makes the VLDC stable as in Fig. 15c.

In case 4, V_S and V_S both were considered to be un-balanced with 10 m/sec e in Fig. 16a. It is noticed from Fig. 16b that the load current is sinusoidal and unbalanced. The FLPIC is able to suppress the current harmonics and decreases THD from 11.32 to 2.06% while boosting up the PF from 0.6382 to 0.9983. Figure 16c exhibits its performance in regulating the voltage at the DC link during load and velocity of wind variation. The frequency spectrum for THD of the proposed cases is exhibited in Fig. 17.

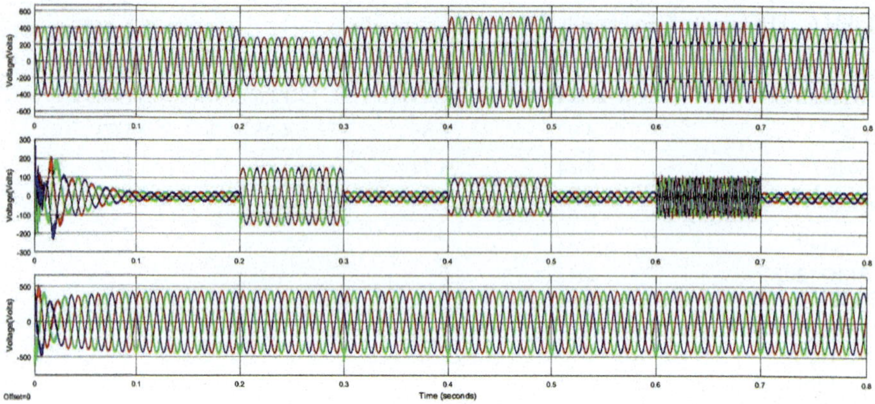

(a)

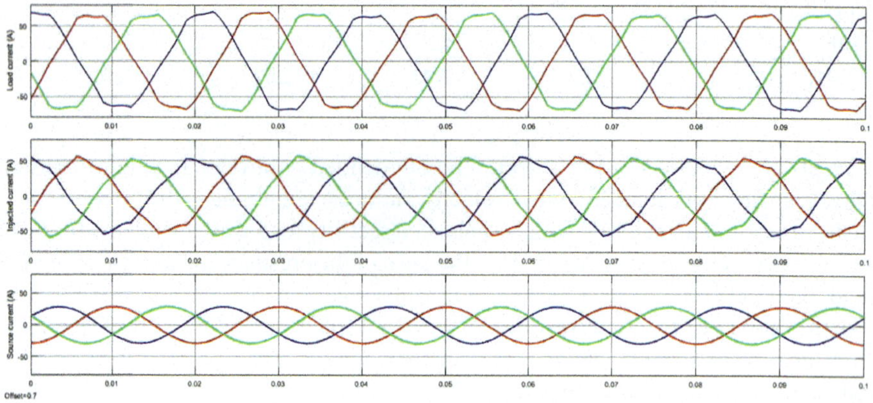

(b)

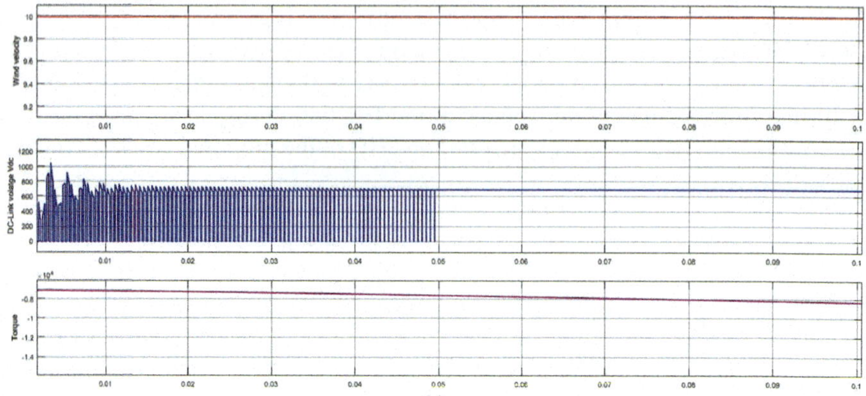

(c)

Fig. 14 Waveform of U-WBS for case 2; **a** V_S, V_{Se}, V_l; **b** i_l, i_{sh}, i_s; **c** velocity of wind, VLDC, torque

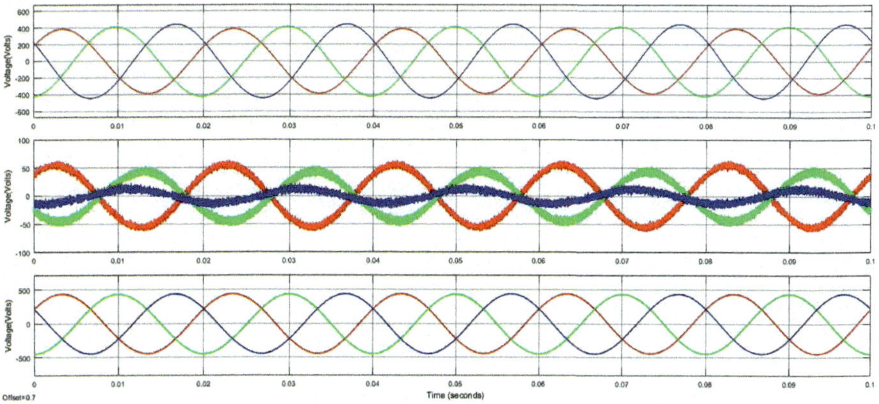

(a)

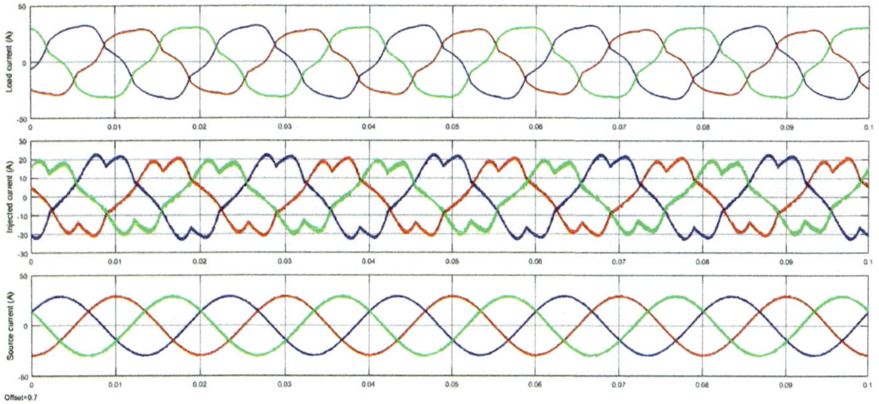

(b)

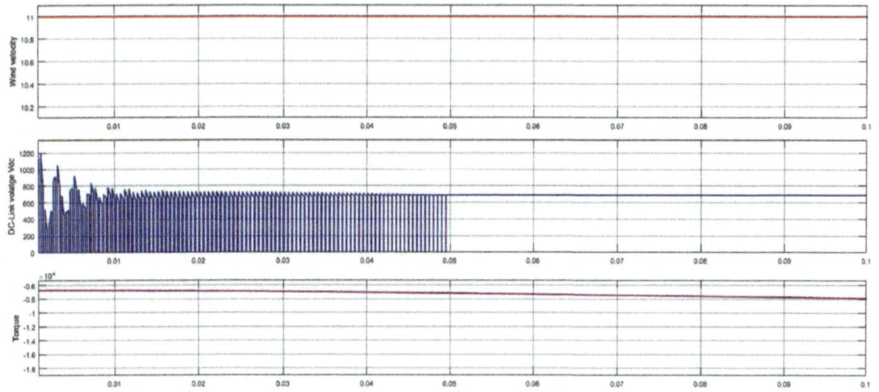

(c)

Fig. 15 Waveform of U-WBS for case 3; **a** V_S, V_{Se}, V_l; **b** i_l, i_{sh}, i_s; **c** velocity of wind, VLDC, torque

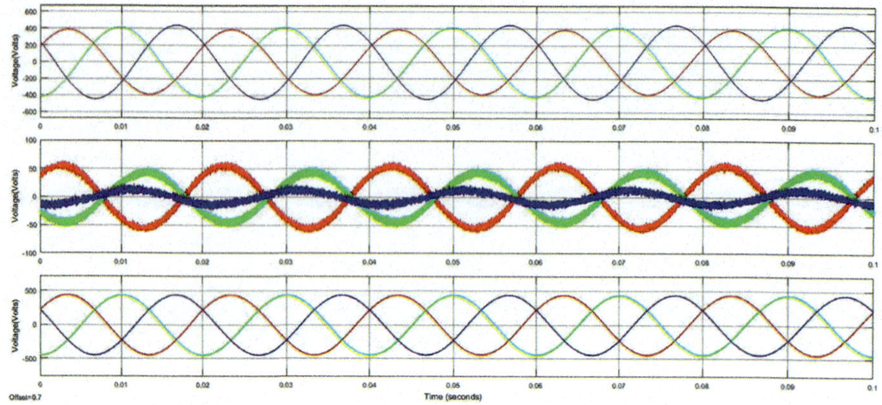

(a)

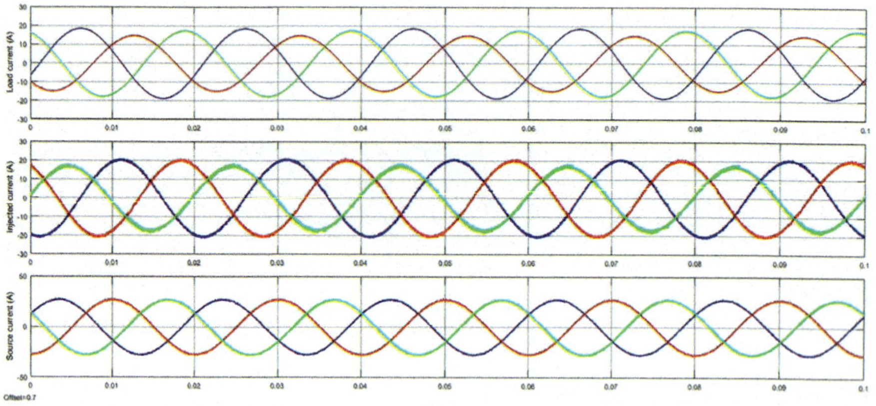

(b)

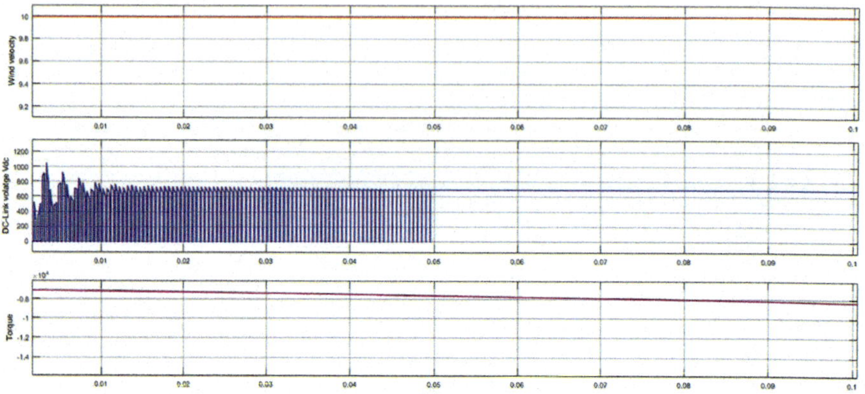

(c)

Fig. 16 Waveform of U-WBS for case 4; **a** V_S, V_{Se}, V_l; **b** i_l, i_{sh}, i_s; **c** velocity of wind, VLDC, torque

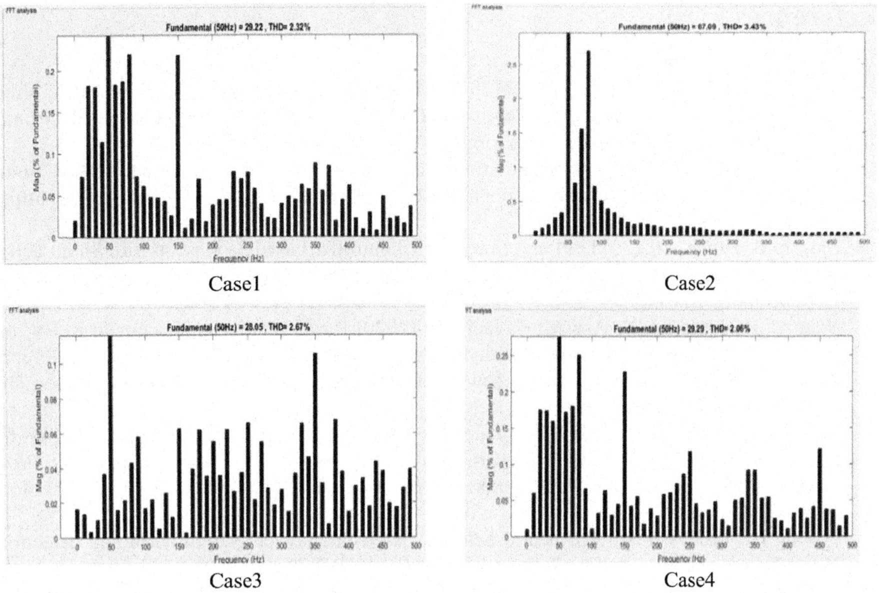

Fig. 17 **THD** spectrum

5 Conclusion

The hybrid controller comprised of FL and PI controllers with WPG and BS was designed for STF control-based U-WBS. The objective of FLPIC for the shunt controller is to quickly stabilize the DC-Link capacitance-voltage, eliminate swell, sag, and disturbances in voltage, suppress the current harmonics, and enhance the PF. By the analysis of four studies, it is noticeable that FLPIC is capable to decrease the THDs within the IEEE standards and PF to almost unity. Moreover, compared to the previous methods, it has been proved that the performance was much better than controllers like PI, SMC, and FLC. The FLPIC also sustains a steady DC-link voltage across the capacitor. The recommended configuration can be further studied with multi-level UPQC for micro-grids in the distribution network as future work.

Acknowledgements This research work was supported by "Woosong University's Academic Research Funding—2023".

References

1. Vijayakumar G, Nitin Gupta, Gupta R (2017) Mitigation of power quality problems using shunt active power filters: a comprehensive review. 12th IEEE Conference on Industrial Electronics and Applications (ICIEA). https://doi.org/10.1109/ICIEA.2017.8283004
2. Metin K, Engin O (2011) Synchronous-reference-frame-based control method for UPQC under unbalanced and distorted load conditions. IEEE Trans Industr Electron 58:3967–3975. https://doi.org/10.1109/TIE.2010.2100330
3. Suresh M, Panda AK (2011) PI and fuzzy logic controller based 3-phase 4-wire Shunt active filter for mitigation of current harmonics with Id-Iq control strategy. J Power Electr 11:914–921. https://doi.org/10.6113/JPE.2011.11.6.914
4. Soumya RD, Prakash K, Asit M, Gayadhar P (2020) Power quality enhancement in PV and battery storage based microgrid using hybrid active filter. 2020 3rd International Conference on Energy Power and Environment: Towards Clean Energy Technologies. https://doi.org/10.1109/ICEPE50861.2021.9404453
5. Suresh M, Panda AK (2012) RTDS hardware implementation and simulation of SHAF for mitigation of harmonics using p-q control strategy with PI and Fuzzy logic controllers. Front Elect Electr Eng 7: 427–437. https://link.springer.com/article/https://doi.org/10.1007/s11460-012-0198-7
6. Hsiung CL (2007) Intelligent neural network-based fast power system harmonic detection. IEEE Trans Ind Elect 54: 43–52. https://doi.org/10.1109/TIE.2006.888685
7. Dash P, Panda SK, Lee TH, Xu JX, Tray AR (1997). Fuzzy and Neural Controllers for Dynamic Systems: An Overview. https://doi.org/10.1109/PEDS.1997.627490
8. Devassy S, Singh B (2016) Design and performance analysis of three-phase solar PV integrated UPQC. ICPS- IEEE. https://doi.org/10.1109/ICPES.2016.7584022
9. Viswanth K, Santanu Kumar. D, Kumari R (2020) Development and analysis of power quality by using fuel cell based shunt active power filter. 2020 2nd International Conference on Innovative Mechanisms for Industry Applications (ICIMIA). https://doi.org/10.1109/ICIMIA48430.2020.9074886
10. Alok Kumar M, Soumya R, Prakash KR, Ranjan M, Asit M, Mishra DK (2022) PSO-GWO optimized fractional order PID based hybrid shunt active power filter for power quality improvements. IEEE Access 8:74497–74512. https://doi.org/10.1109/ACCESS.2020.2988611
11. Kumar C, Jaisiva S, Clement RA, Logeshwari V (2021) Hybrid renewable energy based smart grid system for reactive power management and voltage profile enhancement using artificial neural network. Ene Sou Part A: Rec Util Environ Effects 43:2419–2442. https://doi.org/10.1080/15567036.2021.1902430
12. Mohanraj M, Prakash R (2020) A unified power quality conditioner for power quality improvement in distributed generation network using Adaptive Distributed Power Balanced Control (ADPBC). Int J Wavelets, Multi-Resol Inf Proc 18:1–25. https://doi.org/10.1142/S0219691319410212
13. Krishna S, Debashis C, Goswami SK (2020) A modified PV-wind-PEMFCS-based hybrid UPQC system with combined DVR/STATCOM operation by harmonic compensation. Int J Model Simul 41:243–255. https://doi.org/10.1080/02286203.2020.1727134
14. Saggu TS, Singh L, Gill B, Malik OP (2018) Effectiveness of UPQC in mitigating harmonics generated by an induction furnace. Elect Power Comp Syst 46:629–636. https://doi.org/10.1080/15325008.2018.1460641
15. Dheeban S, Muthu Selvan NB (2021) ANFIS-based power quality improvement by photovoltaic integrated UPQC at distribution system. IETE J Res 67: 1–20. https://doi.org/10.1080/03772063.2021.1888325
16. Mahmoud E, Mohammad T, Reza G (2017) A new approach based on instantaneous power theory for improving the performance of UPQC under unbalanced and distortional load conditions. Automatika, J Control Meas Elect Comput Commun 56:226–237. https://doi.org/10.7305/automatika.2015.07.750

17. Mohamed A, Philippe P, Shahram K, Shahrokh S (2009) New digital reference current generation for shunt active power filter under distorted voltage conditions 79:759–765. https://doi.org/10.1016/j.epsr.2008.10.009
18. Almelian M, Mohd I, Omran M, Sheikh U (2018) Performance of unified power quality conditioner (UPQC) based on fuzzy controller for attenuating of voltage and current harmonics. IOP Conf Series Mater Sci Eng 342: 012–084. https://iopscience.iop.org/article/https://doi.org/10.1088/1757-899X/342/1/012084/pdf
19. Vinothkumar V, Kanimozhi R (2021) Power flow control and power quality analysis in power distribution system using UPQC based cascaded multi-level inverter with predictive phase dispersion modulation method. J Ambient Intell Humaniz Comput 12:6445–6463. https://doi.org/10.1007/s12652-022-03972-0
20. Samal S, Hota PK (2017) Design and analysis of solar PV-fuel cell and wind energy based microgrid system for power quality improvement. Cogent Eng 4: 1–22. https://doi.org/10.1080/23311916.2017.1402453
21. Nandhini E, Sivaprakasam A (2020) A review of various control strategies based on space vector pulse width modulation for the voltage source inverter. IETE J Res 68: 3718–3201.https://doi.org/10.1080/03772063.2020.1754935
22. Garima G, Goswami PK (2021) Self-adaptive learning based controller to mitigate PQ issues in internet of things devices. Int Trans Elect Energy Syst 31:1–20. https://doi.org/10.1002/2050-7038.12888
23. Abdelnasser A, Aya Heikal N, El-Sehiemy RA, Waleed AAS (2022) Intelligent fuzzy-based controllers for voltage stability enhancement of AC-DC micro-grid with D-STATCOM. Alexandria Eng J 61: 2260–2293. https://doi.org/10.1016/j.aej.2021.07.012
24. Sayed A, Sabha RA, Ranjan KJ (2021) Biogeography based optimization strategy for UPQC PI tuning on full order adaptive observer based control. IET Gener Transm Distrib 15:279–293. https://doi.org/10.1049/gtd2.12020
25. Renduchintala UK, Pang C, Tatikonda KM, Yang L (2021) ANFIS-fuzzy logic based UPQC in interconnected microgrid distribution systems: modeling, simulation and implementation. J Eng 21:1–29. https://doi.org/10.1049/tje2.12005
26. Devassy S, Singh B (2021) Performance analysis of solar pv array and battery integrated unified power quality conditioner for microgrid systems. IEEE Trans Ind Elect 68: 4027–4035. https://ieeexplore.ieee.org/document/9057583
27. Rajesh P, Shajin FH, Umasankar L (2021) A novel control scheme for PV/WT/FC/battery to power quality enhancement in micro grid system: a hybrid technique. Ene Sour Part A: Rec Util Environ Effects 42:1–18. https://doi.org/10.1080/15567036.2021.1943068
28. Dheeban SS, Muthu Selvan NB, Umashankar S (2021) Artificial Neural network based solar energy integrated unified power quality conditioner. Ene Sou Part A: Rec Util Envir Effects 41: 1–26. https://doi.org/10.1080/15567036.2021.1919247
29. Salkuti SR (2022) Emerging and advanced green energy technologies for sustainable and resilient future grid. Energies 15(18): 6667. https://doi.org/10.3390/en15186667
30. Fredrick N, Franklin N, Ifedayo O, Ramon Z (2021) Optimal design and performance analysis of solar PV integrated UPQC for distribution network. Eur J Elect Eng Comp Sci 5: 1–8. https://doi.org/10.24018/ejece.2021.5.5.361
31. Punna S, Mailugundla R, Salkuti SR (2022) Design, analysis and implementation of bidirectional DC–DC converters for HESS in DC microgrid applications. Smart Cities 5(2):433–454. https://doi.org/10.3390/smartcities5020024
32. Pazhanimuthu C, Ramesh S (2018) Grid integration of renewable energy sources (RES) for power quality improvement using adaptive fuzzy logic controller based series hybrid active power filter (SHAPF). J Intel Fuzzy Syst 35: 749–766. http://dx.doi.org/https://doi.org/10.3233/JIFS-171236
33. Bhukya L, Kedika NR, Salkuti SR (2022) Enhanced maximum power point techniques for solar photovoltaic system under uniform insolation and partial shading conditions: a review. Algorithms 15(10):365. https://doi.org/10.3390/a15100365

Optimal Scheduling of Micro-sources in Multi-microgrid System

C. Srinivasa Rathnam, Anil Annamraju, Battapothula Gurappa, Chandrasekhar Yammani, and Surender Reddy Salkuti

Abstract It is well-known fact that microgrid has several advantages over conventional power systems. Islanded operation is a remarkable feature of a microgrid. Considering this, in this chapter, an active islanded distribution system has been sectionalized into multiple self-sufficient microgrids. These microgrids are proposed to operate independently or in amalgamation with other microgrids. The control variables in this work are micro-sources that exist in these microgrids. The control variables are optimally selected to attain the objectives of minimization of operating cost, minimization of active power losses, and minimization of voltage deviation. The proposed work has been tested on a practical Indian 85 bus distribution system. From the results of the Jaya optimization technique and the Genetic algorithm, the results with the former are found to be promising.

Keywords Distribution network · Microgrids · Multi-microgrids · Distributed generation · Sectionalization · Optimization

---

C. S. Rathnam
Department of Electrical and Electronics Engineering, Vasavi College of Engineering, Hyderabad, Telangana, India
e-mail: crsrinu2@staff.vce.ac.in

A. Annamraju
Department of Electrical and Electronics Engineering, CBIT, Hyderabad, Telangana, India
e-mail: annamrajuanilkumar_eee@cbit.ac.in

B. Gurappa
Department of Electrical and Electronics Engineering, BVRIT, Narsapur, Telangana, India
e-mail: gurappa.b@bvrit.ac.in

C. Yammani
Department of Electrical Engineering, NIT Warangal, Warangal, Telangana, India
e-mail: chandrayammani@nitw.ac.in

S. R. Salkuti (✉)
Department of Railroad and Electrical Engineering, Woosong University, Daejeon 34606, Republic of Korea
e-mail: surender@wsu.ac.kr

S. R. Salkuti et al. (eds.), *Power Quality in Microgrids: Issues, Challenges and Mitigation Techniques*, Lecture Notes in Electrical Engineering 1039,
https://doi.org/10.1007/978-981-99-2066-2_12

Nomenclature

CEA	Central Electricity Authority
DG	Distributed Generation
EA	Evolutionary Algorithm
GA	Genetic Algorithm
GW	Gigawatts
Iter	Iteration
Itermax	Maximum number of iterations
kW	Kilo Watts
KV	Kilo Volts
MG	Microgrid
MGCC	Microgrid Central Controller
P_{size}	Population Size
P.U	Per Unit
RES	Renewable Energy Sources
SI	Swarm Intelligence

1 Introduction

It is well-known fact that human life is much reliant on electrical energy. Existing human life without electricity in the twenty-first century is impossible. The author in [1] stated that an electrical power system is a technical wonder. It is proclaimed by the National Academy of Engineering that in the twentieth century, the greatest engineering accomplishment are electricity and its accessibility in advance of computers and airplanes. From the Refs. [2, 3], the growth rate of installed capacity in India has raised from 1.36 GW in 1947 to 399.49 GW by 2022. The year-wise growth of installed capacity is presented in Fig. 1. Figure 2 characterizes India's year-wise per capita consumption of electrical energy, as per the Central Electricity Authority (CEA) report [4]. Further, Fig. 3 depicts various countries' per capita consumption of electricity in the year 2021 [5]. From these Figs. 2 and 3, one can interpret the dependency of human life on electrical energy. Thus, providing a reliable power supply for the progress of human life is essential. The per capita electrical energy consumption is a reliable indicator of the development of any country [6].

It is a well-known fact that power systems are prone to faults without prior warning. Tree contacts, animal contacts, lightning, equipment failure, wind, ice/snowfall, dig-in, vehicle accidents, vandalism, construction activities, etc. are some of the causes of faults in power systems. If a particular region of the system is affected by a fault, it may lead to overloading or isolation of other regions due to load redistribution [7]. Continuous load redistribution to other regions often leads to cascading phenomena and causes catastrophic failure, which has an enormous social and economic impact

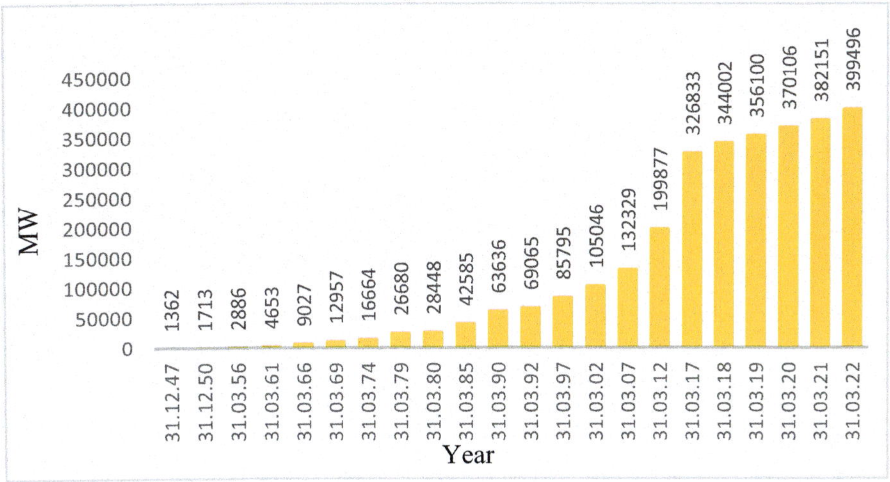

Fig. 1 All India installed capacity

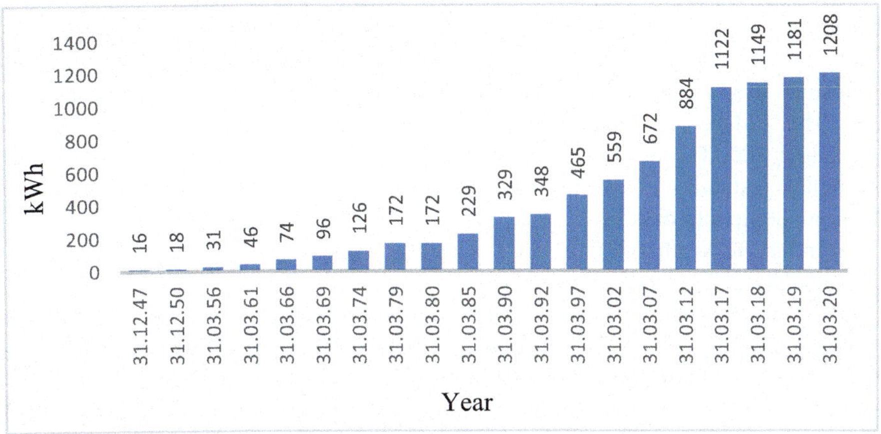

Fig. 2 Year-wise India's per capita consumption of electricity

on society due to power disruptions. As proclaimed in [8], a blackout was experienced by the Northeast United States and Canada in August 2003, which plunged thirty million people into darkness. In Sir Adam Beck Station-2 in Ontario, Canada, the mal-operation of a single faulty relay is found to be the cause for this blackout. Mal-operation of the relay in this substation leads to the opening of a transmission line. This fault, in turn, propagated to other areas, caused the overloading of other transmission lines, and in a sequel caused the automatic shutdown of power generation plants throughout the region.

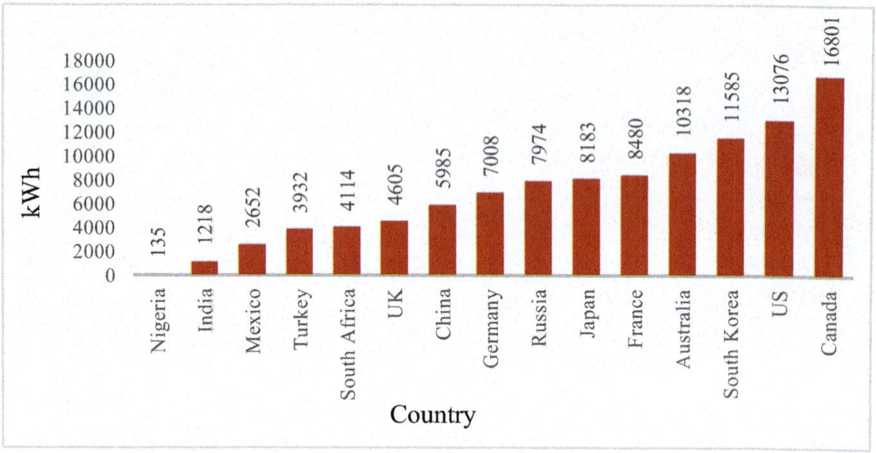

Fig. 3 Various countries' per capita consumption of electricity in 2021

Similarly, on 28th September 2003, Italy was in a blackout. The root cause for the blackout, as reported by UTEC in [9], was tripping off the first line from Switzerland—the so-called "Lukmanier" line—caused by a tree flashover. The automatic reclosure failed to switch on the breaker, even after the carbonization of the affected tree. This failure of reclosure is due to the phase angle difference over the line exceeding 300. The islanding operation of the system was found to be failed due to angular instability and voltage collapse, which caused a blackout, and pushed 56 million people into darkness.

In India, the Northern region consisting of 8 states, experienced a major blackout on 30th July 2012 affecting 400 million people. The cause was identified as circuit breakers on the 400 kV Bina-Gwalior line getting tripped. As this line fed into the Agra-Bareilly transmission section, breakers at that station tripped, and power failures cascaded through the grid. All major power stations were shut down in the affected states, causing an estimated shortage of 32 GW. Again on 31st July 2012, the system failed, due to a relay problem in a substation near the Taj Mahal. As a result, power stations across the affected parts of India again went offline. Over 600 million people, nearly half of India's population, in 22 out of 29 states in India, were without power [10].

To avoid such blackouts, the IEEE Power Engineering Society has made a report on remedial measures and restoration practices for blackout conditions [9]. The report proclaimed that grid strengthening, remote emergency control for severe contingencies, and out-of-step islanding were the remedial solutions. Considering the challenges faced by the Power systems i.e., (a) Electrification of remote locations where grid expansion is impossible and (b) Avoiding blackouts due to cascading effects on other parts, power systems Researchers have started contemplating a feasible solution. Instigation of deregulation in the electrical systems has provided a realizable solution to power systems researchers, thus evolving the Microgrid (MG) concept.

2 Problem Formulation

To supply power to the microgrid loads with diverse reliability requirements, the important factors that play a crucial role are the location of distributed generators, the topology of the power distribution network, and the magnitude of load at the buses. To downturn the power outage duration, periodic power outages, and level of energy undelivered to the microgrid [11], locating a distributed generator near the load is essential.

It is known that planning studies are offline studies aiming for providing quality power supply in a reliable manner to a consumer. While planning, it is imperative to acknowledge both economic analyses as well as technical aspects. An admitted fact that power systems undergo faults without any anterior warnings. Some of the causes for fault may be a failure of power system equipment, formation of ice/snow on transmission lines, falling of trees/animal contact on transmission lines, construction activities, lightning, vandalism, vehicle accidents, etc. It is known that, due to fault in one region, other regions may experience overloading or isolation. If overloading persists continuously, it leads to catastrophic failure of entire power systems. Any country experiences enormous social discomfort and huge economic loss due to catastrophic failure. Analysis of blackout in India's power systems on July 30th and 31st 2012 has been analyzed to billions of dollars to the industries [12]. As the microgrid (MG) system has several advantages, to avails its advantages and to mitigate the difficulties that emanate due to faults in the distribution network, an active islanded distribution network has been divided into multiple networks which are self-sufficient.

To sectionalize into self-sufficient microgrids, the contemplated criteria are the location of the distributed generator and its size, distribution network topology, and load demand. While making self-sufficient microgrids, the operational constraints are regarded [13]. It is considered that during normal operation, the active microgrids share power among themselves and enter into an islanded mode during a fault condition. The objective functions addressed in this chapter are (a) Operating cost minimization (b) Active power loss minimization (c) Voltage deviation minimization. The proposal is investigated, for optimal scheduling of DGs in sectionalized Microgrids. The potential of the proposed methodology is examined on a modified Practical Indian 85 Bus Distribution system. Jaya algorithm, a population-based meta-heuristic algorithm is attempted for optimal scheduling of controllable micro-sources/distributed generators (DGs). The validity of the obtained results has been made by comparison with that of the Genetic Algorithm (GA) [14, 15]. The below points are the contributions of this chapter:

- The active distribution system has been sectionalized into self-sufficient microgrids subjected to the distributed generators' position, the network topology, and the possible fault locations.

- Single-objective optimization problem as presented below has been attempted by the scheduling of the controllable distributed generators in an optimal manner in each microgrid.
  - operating cost minimization
  - active power loss minimization
  - voltage deviation minimization
- An algorithm-specific parameter-free population-based optimization, the Jaya algorithm, has been attempted for optimization.
- For examining the proposed strategy, the test system considered is a modified Practical Indian 85 bus distribution system. The validation of results has been made with the comparison with Genetic Algorithm results.

2.1 Sectionalization of Active Distribution System

The objective of the sectionalization is to cater to power supply more reliably to consumers either during normal operating conditions as well as during faulty conditions. Figure 4 illustrates the splitting of the distribution network into multiple MGs with sectionalizing static switches. The controllable DGs in the distribution system are scheduled optimally during normal operating conditions, to realize the aforementioned objective functions.

Self-healing mode is a remarkable feature of the microgrid system. The active distribution network transforms into the self-healing mode, whenever the network experiences an abnormality in the system. By reconfiguring the existing system by pulling out the faulty portion and interconnecting the on-outage areas, reliability increases. In general, radiality is the nature of the distribution network, the same has been maintained while reconfiguring. Each loop in the network composes of a sectionalizing switch and a tie switch, to retain the radial nature, at any time one of the switches is opened by closing the other [16].

While sectionalizing the distribution network, it is ensured the stability of the networks of the MG Central Controller (MGCC), by maintaining the self-adequacy in supply and demand balancing in all sectionalized networks. However, this chapter

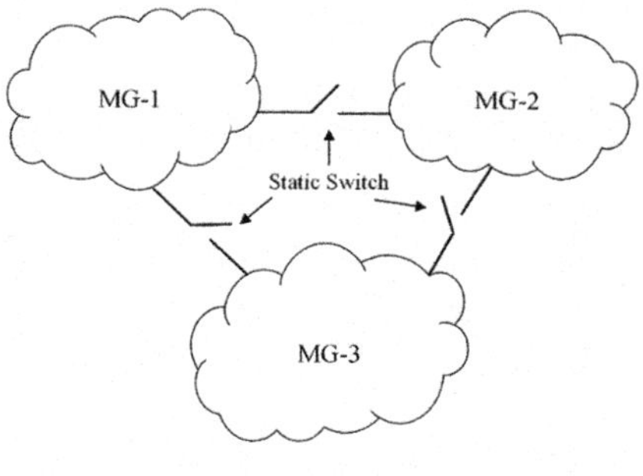

Fig. 4 Sectionalization of distribution network into multiple microgrids

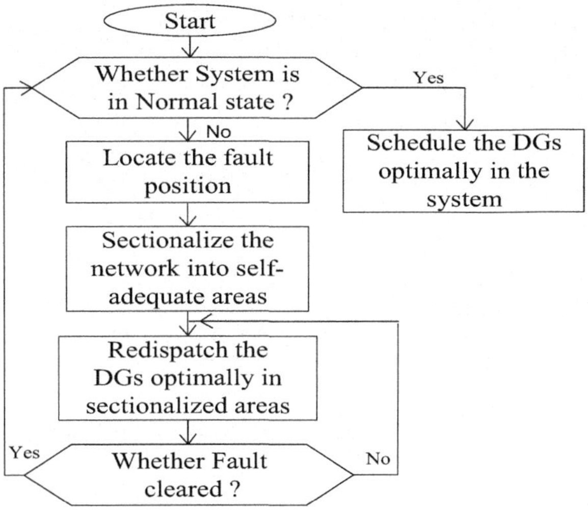

Fig. 5 Islanded active distribution system sectionalization

is not addressing the transient and dynamic stability aspects of the networks while sectionalizing [17, 18].

A simple distribution network made into three self-adequate microgrids (MGs) is depicted in Fig. 4. Whenever, the abnormality is visualized in the system, for instance in microgrid-1, the static switches of tie-lines connected between MG-1 and MG-2 and MG-1 and MG-3 isolate the MG-1 such that fault will not get propagated to other regions. A similar operation can be performed for faults in multiple microgrids. On isolating the abnormality, the network returns to the original state depending on the microgrid central controller decision. Figure 5 depicts the distribution system sectionalization flowchart.

2.2 *Objective Function*

Identifying the optimal solution from a massive set of a viable solutions is defined as the optimization problem. The motto of the optimization problem will be either to identify the minimum value or the maximum value of the function [18]. In this chapter, single-objective optimization has been executed. In a single-objective optimization problem, each of these functions (i) Operating cost, (ii) active power losses, and (iii) system voltage deviation, has been considered for optimization at a time, to get the optimal scheduled output power of DGs. The control variables for these optimization problems is the controllable DGs output. Equality constraints and inequality constraints are considered while optimally scheduling the distributed generators' output power, which is explained in Sect. 2.2.4. The formulation of the optimization problem is as follows.

2.2.1 Operating Cost Minimization

Minimization of total generation cost or total operating cost in an isolated microgrid or multi-microgrid system by optimally scheduling the controllable DGs output is considered as an objective function. The cost function of each DG is expressed as a quadratic cost function. The total generation cost is obtained by summing up the quadratic cost model of all the generators [19, 20] and is defined as per Eqs. (1) and (2).

$$F_1(X) = \sum_{q=1}^{m} \sum_{k=1}^{N_{gen}} (a_{qk} P_{qk}^2 + b_{qk} P_{qk} + c_{qk}) \frac{\$}{\text{hr}} \tag{1}$$

$$[X] = [P_{q1i} P_{q2i} P_{q3i} \ldots\ldots P_{qk-1i} P_{qki}] \tag{2}$$

where m is the total number of microgrids in the system, N_{gen} is the total number of generators in each microgrid-q. The a_{qk}, b_{qk} and c_{qk} are the fuel cost coefficients of kth generator in qth microgrid, P_{qki} is the active power generation value of kth generator in qth microgrid at ith iteration, X is a control variable relating any P_{qk}, which is defined as per Eq. (2). However, while optimizing the generation cost, equality constraints (i.e., power balance constraint), as well as inequality constraints (i.e., generators capacity constraints and bus voltage constraints), need to be satisfied.

2.2.2 Active Power Loss Minimization

This loss minimization objective function is achieved by optimally scheduling the controllable DGs such that total active power losses in the isolated microgrid or multi-microgrids are minimum and these are expressed in Eqs. (3) and (4). While scheduling to attain the objective function, the equality constraint and inequality constraints need to comply. The objective function is defined as follows:

$$F_2(X) = \text{Minimize}(P_{loss}) \tag{3}$$

$$P_{loss} = \sum_{q=1}^{m} P_{q,loss} = \sum_{q=1}^{m} \left[\sum_{k=1}^{N_{gen}} P_{qk} - P_{q\,demand} \right] \tag{4}$$

where $P_{q,loss}$ is the active power loss in qth microgrid, P_{qk} is the real power output value of kth generator in qth microgrid, $P_{q\,demand}$ is the total active power demand in qth microgrid and P_{loss} is the total active power losses of all active microgrids.

2.2.3 Voltage Deviation Minimization

Minimization of voltage deviation is attempted as a third objective function in this chapter. It is characterized as per Eq. (5). To attain the desired objective, the DGs are optimally scheduled satisfying constraints as defined in Sect. 2.2.4 below. The voltage deviation is defined as follows [21].

$$F_3(x_{i,k}) = \text{Minimize}\left[\frac{1}{n_{bus}}\sum_{q=1}^{m}\sum_{k=1}^{nbus_q}\left(|V_{qk}| - |V_{qk}^{sp}|\right)^2\right] \quad (5)$$

where V_{qk} and V_{qk}^{sp} are the absolute voltage value and the specified voltage value at kth bus of qth microgrid respectively, $nbus_q$ is the total number of buses in the qth microgrid and n_{bus} is the total number of buses in all active microgrids.

2.2.4 Constraints of the Problem

(a) ***Equality and Inequality Constraints***

The system constraints are generally defined as equality constraints and inequality constraints. The equality constraint is a power generation and demand balance whereas the inequality constraints are boundaries of active power output and voltage magnitude at the bus. These constraints are mathematically defined as follows.

(b) Power Balance Constraint

The total power generation in the microgrid(s) system must be equal to the total power demand and the total power losses in the microgrids considered for the operation, to maintain the stability of the system as per Eq. (6). It is necessary to gauge the power losses in the network, as the feeders in each radial network comprise numerous buses and loads.

$$\sum_{q=1}^{m}\left[\sum_{k=1}^{N_{gen}} P_{qk}\right] = \sum_{q=1}^{m}\left[P_{q\,demand} + P_{q\,loss}\right] \quad (6)$$

where $P_{q\,demand}$ is the total active power demand in the qth microgrid, $P_{q,loss}$ is active power losses in the qth microgrid, P_{qk} is the real power output value of kth generator, m is the total number of microgrids in the system and N_{gen} is the total number of generators in each microgrid-q.

(c) Generation Capacity Constraints

For stable operation, the real power output of each generator P_{qk} is restricted by lower and upper limits. While scheduling the DGs output, the power limits have to be enforced for the stable operation of the system [22, 23]. Thus, the active power generation boundaries of DGs are defined as per Eq. (7).

$$P_{qk}^{\min} \leq P_{qk} \leq P_{qk}^{\max} \tag{7}$$

where $P_{qk}^{\min}$ is the minimum active power generating limit, $P_{qk}^{\max}$ is the maximum active power generating limit, q is the microgrid number in the network, and k is the unit number in the microgrid q.

(d) Bus Voltage Constraints

The magnitude of the kth bus voltage in any MG system should be bounded between the minimum and maximum limits [24] so as to maintain voltage stability. Thus, the boundaries of the bus voltages magnitudes are defined in Eq. (8). The voltage limits considered in this research work are $\pm 5\%$.

$$V_k^{\min} \leq V_k \leq V_k^{\max} \tag{8}$$

3 Optimization Technique

There are certain limitations in the search mechanism of conventional algorithms, such as Dynamic Programming, Sequential Programming, Nonlinear Programming, etc. The nature of the objective function, constraint functions, and modeling of variable types, affect the search mechanism of these algorithms. The number of design variables, size of the solution space, and framework of the solution space determine the efficiency of these algorithms. Thus, over the past two decades, population-based optimization techniques become popular to surmount the difficulties associated with traditional optimization techniques.

The conventional algorithms have limitations in their search process. The conventional algorithms are computationally expensive, difficult to identify the global optimal in non-convex functions with multiple sub-optimal values and provide a sub-optimal solution for discontinuous and non-convex functions. For example, Dynamic programming takes a lot of memory to store the calculated result of every subproblem without ensuring if the stored value will be utilized or not. Many times, the output value gets stored and never gets utilized in the next subproblems while executing. The efficiency of these algorithms relies on the solution space size, the structure of the solution space, and the number of variables. Thus, population-based meta-heuristic techniques are the viable solution methods to overcome the difficulties in conventional algorithms [25, 26].

Evolutionary Algorithms (EA) and Swarm Intelligence (SI) algorithms are the types of population-based algorithms. The EA is inspired by Darwin's evolutionary theory. These algorithms comprise the following steps: selection of parents, crossover operation, and mutation operation to produce the fittest or optimal solution. On the other hand, the SI methodology to find the optimal solution is based on the collective behavior of the social animal and insects to identify the food/prey. These algorithms required common control parameters like population size and maximum generations/ iterations count and algorithm-specific parameters which are specific to the type of algorithm selected. Choosing inaccurate values of algorithm-specific parameters leads to divergence of the algorithm. Thus, in view of these facts, the Jaya algorithm, a novel meta-heuristic, algorithm-specific parameters independent algorithm has been considered in this chapter. R. Venkata Rao has proposed the Jaya Algorithm which is simple to implement and effective in identifying the global optimal solution [27, 28].

3.1 *Jaya Algorithm*

Consider an objective function '*f*' to be minimized. Let the objective function has *k* number of design variable/control variables(X). Assuming the size of the population be P_{size} (Population size). Let the design variables be confined to the lower and upper limits. Generate design variable X within their limits and evaluate the objective function value for each candidate X of the population. The evaluation of the optimal solution in this algorithm is based on the following steps discussed below [28].

3.1.1 Best Solution Candidate and the Worst Solution Candidate Identification Phase

Let $X_{i,j,\mathrm{k}}$ be the candidate solution, where i stands for iteration number, $i \in [1,$ max. no. of iterations], j points out the candidate number in the population, $j \in [1, P_{size}]$, and k refers to the design variable, $k \in [1,$ max. no. of design variables]. The best solution candidate be $(X_{i,best,\mathrm{k}})$ and the worst solution candidate is $(X_{i,worst,k})$ among the total population. $(X_{i,best,\mathrm{k}})$ and $(X_{i,worst,k})$ are the candidates in the population which provide the lowest value and highest value of f in the present iteration. On identification of $(X_{i,best,\mathrm{k}})$ and $(X_{i,worst,k})$, each candidate has to be updated as specified below.

3.1.2 Updation Phase

Equation (9) provides the mathematical formula for updating each candidate in the present iteration.

$$X^1_{i,j,k} = X_{i,j,k} + r_{i,j,1}\left(X_{i,best,\mathrm{k}} - |X_{i,j,k}|\right) - r_{i,j,2}\left(X_{i,worst,k} - |X_{i,j,k}|\right) \tag{9}$$

where

$r_{i,j,1}, r_{i,j,2}$ Represents two random nos. in iteration i, which ranges [0, 1].

$X^1_{i,j,k}$ represents updated candidate.

The other terms in Eq. (9) are the same as defined earlier. The term "$r_{i,j,1}\left(X_{i,best,k} - |X_{i,j,k}|\right)$" attributes the tendency of the candidate to shift towards the Best candidate solution. Another term "$-r_{i,j,2}\left(X_{i,worst,k} - |X_{i,j,k}|\right)$" attributes the progressing candidate movement away from the worst solution. Upon evaluation of the above, the updated candidate ($X^1_{i,j,k}$) and the original candidate ($X_{i,j,k}$) are provided as inputs to the next phase.

3.1.3 Comparison Phase

For each candidate (X) in the solution set, its fitness value is evaluated. Fitness value resembles how close the given solution is to the optimal value. A higher fitness solution is closer to the optimal solution and vice-versa. The objective functions addressed in this work are of minimization type, thus the fitness value of each candidate solution is evaluated using Eq. (10).

Upon completion of the Updation phase, compute the fitness values of $X_{i,j,k}$ and the updated candidate $X^1_{i,j,k}$. The fitness value of the candidate solution is determined as per Eq. (10).

$$fitness(X) = \frac{1}{1 + f(X)} \tag{10}$$

In this phase, the following steps are to be performed. Among original and updated candidates, if fitness($X^1_{i,j,k}$) > fitness($X_{i,j,k}$), then retain $X^1_{i,j,k}$ for the next generation, otherwise, retain $X_{i,j,k}$ for the next generation and discard the other candidate. All the candidates for the next generation will be evaluated in a similar manner.

Performing the above steps results in the completion of one iteration. The same shall be executed for all the candidates' updation. These updated candidates will be considered as input candidates for the next iteration. Until convergence criteria are satisfied, the above procedure will be repeated.

3.2 Algorithm: Implementation Steps for Proposed Methodology Using Jaya Algorithm

In this section, various steps involved in identifying the optimal solution has been presented.

1. Read input data of the multi-microgrid System: number of micro-sources in each microgrid, micro-sources location in each microgrid, initial bus voltage magnitudes, fuel cost coefficients of generators, lower and upper boundaries of active power generations of all generators, line data of the system, bus data of the system.

2. Initializing the parameters of the Jaya algorithm:
   (a) Population size(P_{size})
   (b) Generations/iteration size (iter_max).
3. Select microgrid(s) (MG-1, MG-2, and MG-3) which are active.
4. Select the scenario depending on the optimization of the objective function, i.e., Scenario-1 (Operating cost minimization), Scenario-2 (Active power losses minimization), and Scenario-3 (Voltage deviation minimization).
5. Generate population ($X_{i,j,k}$) within their limits.

$$X_{i,j,k} = \{P_{i,j,1}\, P_{i,j,2} \ldots\ldots P_{i,j,\mathrm{k}})$$

where $i \in \{1, 2, ..iter_{\max}\}$, $j \in \{1, 2, 3 \ldots P_{size}\}$ and $k \in \{1, 2, 3 \ldots N_{gen}\}$

6. Set iteration count (iter) = 1.
7. Identify the voltage at all buses, and system losses by carrying out load flows.
8. Using Eqs. (1), (2), and (4), evaluate the total generation cost, total active power losses, and voltage deviation respectively, and depending on the scenario considered, evaluate the fitness function.
9. Identify ($X_{i,best,\mathrm{k}}$) and ($X_{i,worst,k}$) in the population.
10. Using Eq. (9), update the candidate.
11. Go to the comparison phase. Evaluate the candidate for the next iteration.
12. Update iteration count i.e., iter = iter + 1.
13. Until convergence, perform Steps (7) to (12).
14. Upon convergence, terminate the program and display the optimal scheduled values of micro-sources and the objective function value for the selected scenario.

The flowchart for the Jaya algorithm is presented in Fig. 6.

4 Results and Analysis

4.1 Test System

As detailed above, in this chapter, to experiment with the proposed strategy, a practical Indian 85 bus distribution system has been examined [29, 30]. The 85-bus distribution system is presented in Fig. 7. The flow chart for solving the objective function with the Jaya algorithm is detailed in Fig. 8.

(a) Line status of modified Practical Indian 85 bus distribution system for formulation of multi-microgrid System

The modified line details of 85 bus distribution systems for the formulation of a multi-microgrid system are is presented in Table 1.

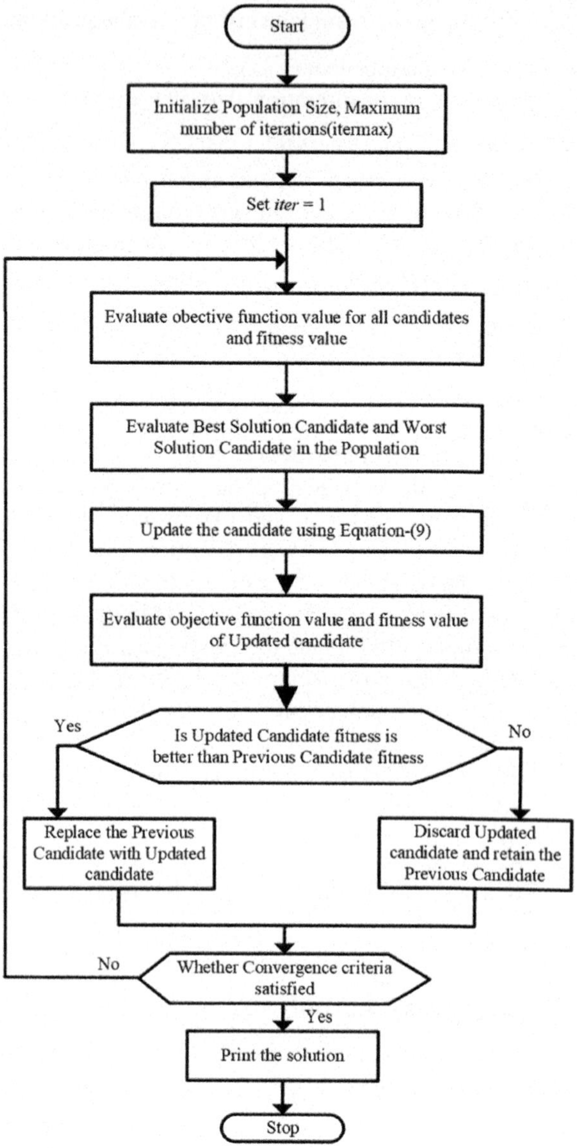

Fig. 6 Steps involved in the Jaya algorithm

(b) Assumptions made for testing the proposed algorithm

1. The DGs considered in this work are dispatchable and their locations are fixed by considering the topology of the network and the load demand in the respective microgrid. The position of DGs in the 85 bus distribution network is illustrated in Table 2.
2. Static switches are utilized for isolation and tie-line connection of the networks.

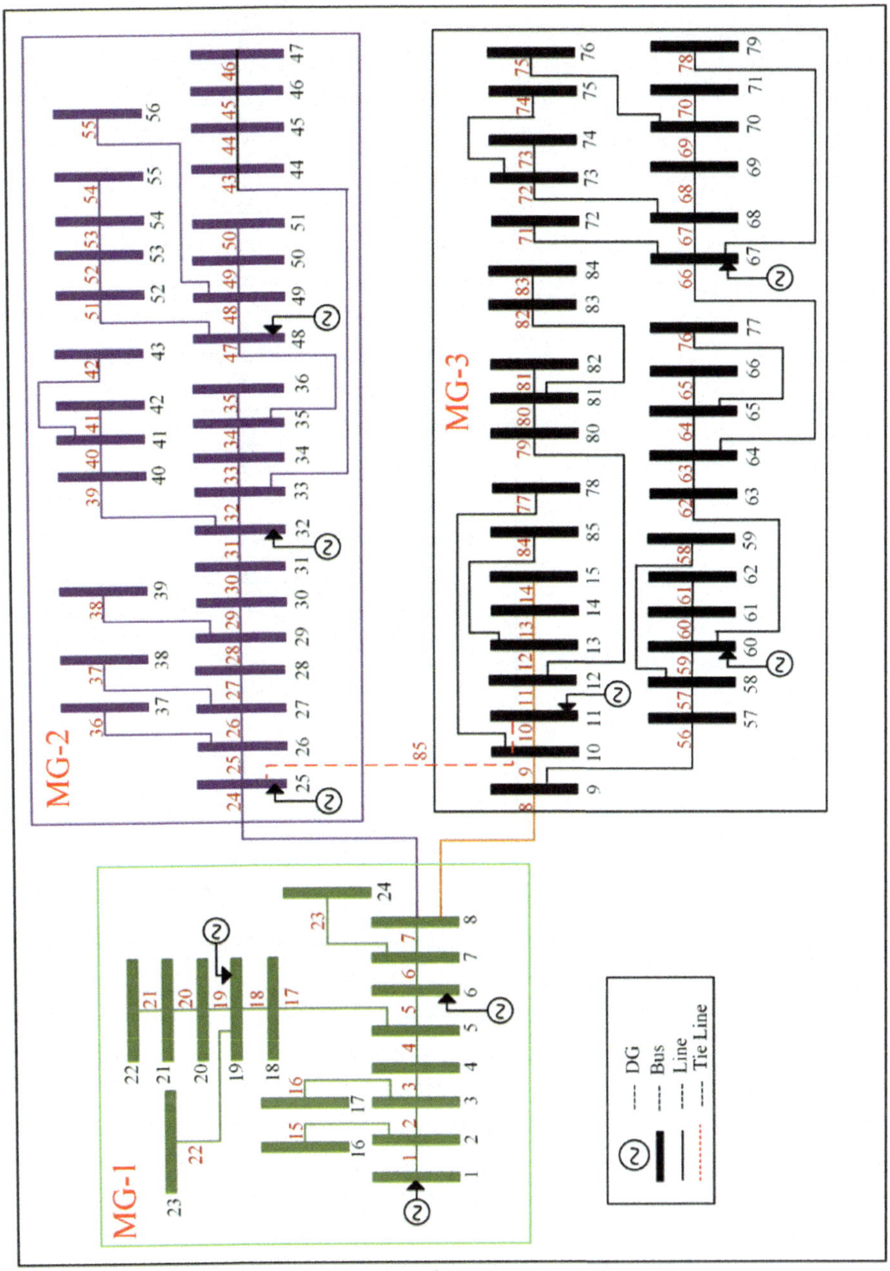

Fig. 7 Modified Indian Practical 85 bus distribution system with DGs

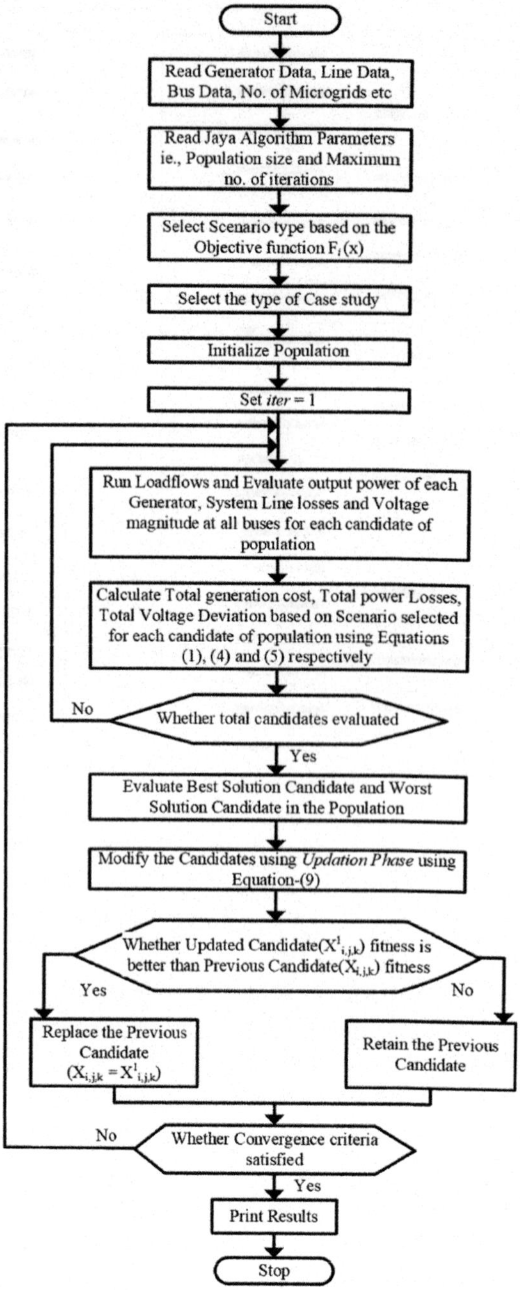

Fig. 8 Flow chart-Jaya algorithm implementation method for optimal scheduling of DGs

Table 1 85 bus distribution system line parameters

Active Microgrids	Line no.	Start Bus	End Bus	Resistance (P.U)	Reactance (P.U)
MG-1	8	8	9	—/—	—/—
	24	7	25	—/—	—/—
MG-2	24	7	25	—/—	—/—
	85	11	25	—/—	—/—
MG-3	8	8	9	—/—	—/—
	85	11	25	—/—	—/—
MG-1& MG-2	8	8	9	—/—	—/—
	85	11	25	—/—	—/—
MG-2 & MG-3	85	11	25	0.089256	0.061983
	8	8	9	—/—	—/—
	24	7	25	—/—	—/—
MG-1& MG-3	24	7	25	—/—	—/—
	85	11	25	—/—	—/—
MG-1, MG-2 & MG-3	85	11	25	—/—	—/—
Legend: Microgrid-k → MG-k, Line Open → —/—					

Table 2 DGs location

Network	Bus number at which DG located		
	MG-1	MG-2	MG-3
85 bus distribution system	1, 6, 19	25, 32, 48	11, 60, 67

Three scenarios are formulated based on objective functions. The details of the scenarios (S) are as follows:

S-1: Operating cost minimization
S-2: Active power loss minimization
S-3: Voltage deviation minimization

Further, depending on the operation of the multi-microgrid Systems, different case studies are formulated as illustrated in Table 3. The faulty microgrid(s) which are not functioning are isolated from the rest of the active system. The common control parameter of the proposed Jaya algorithm are: $P_{size} = 80$, iter_max = 200.

4.2 Scenario-1 (Operating Cost Minimization)

Operating cost minimization is regarded as the objective function in this scenario. The cost coefficients of DGs for the Indian 85 bus distribution system are taken from

Table 3 Case studies in each scenario

Case study	Operating MG(s)	Fault Type*	Faulty MG(s)
Case-1	MG-1	MMGF	MG-2 and MG-3
Case-II	MG-2	MMGF	MG-1 and MG-3
Case-III	MG-3	MMGF	MG-2 and MG-3
Case-IV	MG-1 and MG-2	SMGF	MG-3
Case-V	MG-2 and MG-3	SMGF	MG-1
Case-VI	MG-1 and MG-3	SMGF	MG-2
Case-VII	All MGs	NF	-

* SMGF—Single MG fault MMGF—Multi MG fault NF—No fault

reference [31]. The line parameters of the test system are referred to from reference [30]. The optimal scheduled power output of different DGs for attaining minimum cost considering cost minimization as the objective, for various cases as defined in Table 3, using the Jaya Algorithm and Genetic Algorithm are emphasized in Tables 4 and 5, respectively. The total generation cost obtained with optimal scheduling of DGs along with active power loss, and Voltage deviation values are presented in Tables 4 and 5. It is noticeable from the above results that the Jaya Algorithm is yielding minimum costs from Case-I to Case-VII of 23,840.38$/hr, 30,627.02$/hr, 43,449.32$/hr, 56,323.14$/hr, 75,924.69$/hr, 68,248.23$/hr and 103,099.16$/hr in comparison to GA of 23,849.92$/hr, 30,627.49$/hr, 43,450.85$/hr, 56,323.68$/hr, 75,929.71$/hr, 73,297.60$/hr, and 103,395.19$/hr, respectively. Thus, it is evident from the results that as the loading level upturns, the cost saving is significant.

Figure 9 outlined below represents the convergence characteristics of the Jaya algorithm and GA for Case-VII with Operating cost minimization as the objective function. It is perceptible from Fig. 9 that for convergence, Jaya Algorithm takes 25 iterations as against 55 iterations by the Genetic Algorithm. It is noticeable that the proposed Jaya Algorithm converges faster than Genetic Algorithm on a practical distribution system.

Figure 10 depicts case-I to case-VII voltage magnitude at different buses, by using the Jaya algorithm. It is assessed from this plot that for all cases except case VII, the voltage magnitudes are in tolerance value of $\pm 5\%$. Thus, it is evident that the individual operation of microgrids has improved the voltage profile of the system for the given locations of DGs.

4.3 Scenario-2 (Active Power Loss Minimization)

The objective function treated in this scenario is only active power loss minimization. Tables 6 and 7 encapsulate the active power losses for diverse case studies with the Jaya algorithm and GA, respectively. Tables 6 and 7 unveil the optimal outputs of each DG, total generation cost, and voltage deviation values for several case studies. It is

Table 4 Scenario-1—Jaya algorithm optimal DG values for different case studies

	Case-I	Case-II	Case-III	Case-IV	Case-V	Case-VI	Case-VII
P_{G1} (kW)	227.3001	–	–	233.2695	–	317.3417	306.4235
P_{G2} (kW)	72.9134	–	–	76.3804	–	100.0000	100.0000
P_{G3} (kW)	285.3294	–	–	291.0507	–	367.8978	358.9201
P_{G4} (kW)	–	368.1779	–	385.0844	489.1914	–	453.3466
P_{G5} (kW)	–	251.1934	–	258.6239	302.1059	–	287.2758
P_{G6} (kW)	–	300.0000	–	300.0000	300.0000	–	300.0000
P_{G7} (kW)	–	–	492.8121	–	405.8350	377.9075	367.9179
P_{G8} (kW)	–	–	189.3617	–	155.5133	143.8066	139.8869
P_{G9} (kW)	–	–	400.0000	–	400.0000	400.0000	400.0000
P_{loss} (kW)	2.5828	6.0113	9.2138	48.0889	66.3256	51.0337	144.4908
Q_{loss} (kVAR)	1.4636	2.4910	4.4277	27.2133	32.5269	31.8287	90.2311
Cost ($/hr)	**23,840.38**	**30,627.02**	**43,449.32**	**56,323.14**	**75,924.69**	**68,248.23**	**103,099.16**
VD (P.U)	7.8652E-06	1.0890E-05	2.2000E-05	5.9200E-04	3.2400E-04	6.3500E-04	1.6660E-03
P_{demand} (kW)	582.96	913.36	1072.96	1496.32	1986.32	1655.92	2569.28
Q_{demand} (kVAR)	594.74	931.81	1094.64	1526.55	2026.45	1689.37	2621.19

Table 5 Scenario-1—genetic algorithm optimal DG values for different case studies

	Case-I	Case-II	Case-III	Case-IV	Case-V	Case-VI	Case-VII
P_{G1} (kW)	218.0677	–	–	233.5952	–	338.2800	281.3239
P_{G2} (kW)	67.3828	–	–	76.5625	–	99.9756	99.9756
P_{G3} (kW)	300.0977	–	–	291.3086	–	293.6523	350.0000
P_{G4} (kW)	–	367.6758	–	384.2285	488.6719	–	430.3711
P_{G5} (kW)	–	251.7658	–	258.7891	296.7529	–	277.2217
P_{G6} (kW)	–	299.9268	–	299.9268	299.9268	–	299.9268
P_{G7} (kW)	–	–	492.0654	–	412.1094	374.8779	375.0000
P_{G8} (kW)	–	–	190.2032	–	155.3084	350.0000	200.1953
P_{G9} (kW)	–	–	399.9023	–	399.9023	249.8047	399.9023
P_{loss} (kW)	2.5882	6.0083	9.2110	48.0906	66.3517	50.6706	144.6367
Q_{loss} (kVAR)	1.4632	2.4897	4.4260	27.2144	32.5376	31.5962	90.3331
Cost ($/hr)	**23,849.92**	**30,627.49**	**43,450.85**	**56,323.68**	**75,929.71**	**73,297.60**	**103,395.19**
VD (P.U)	7.8499E-06	1.0897E-05	2.1638E-05	5.9274E-04	3.2665E-04	6.1449E-04	1.5857E-03
P_{demand} (kW)	582.96	913.36	1072.96	1496.32	1986.32	1655.92	2569.28
Q_{demand} (kVAR)	594.74	931.81	1094.64	1526.55	2026.45	1689.37	2621.19

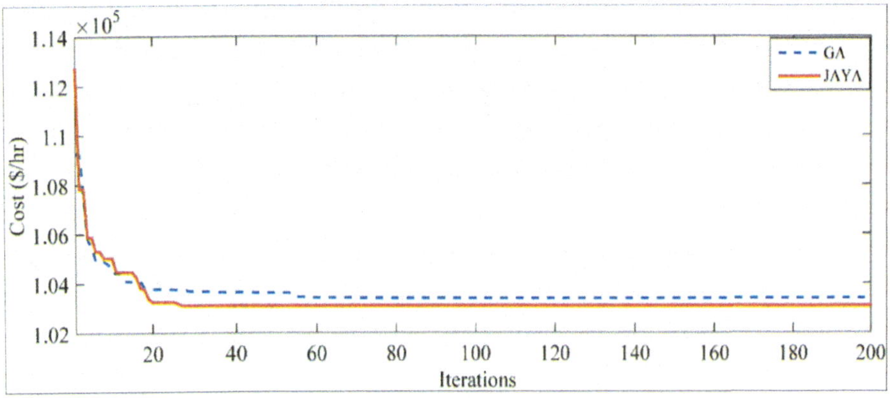

Fig. 9 Scenario-1, case-VII convergence characteristics

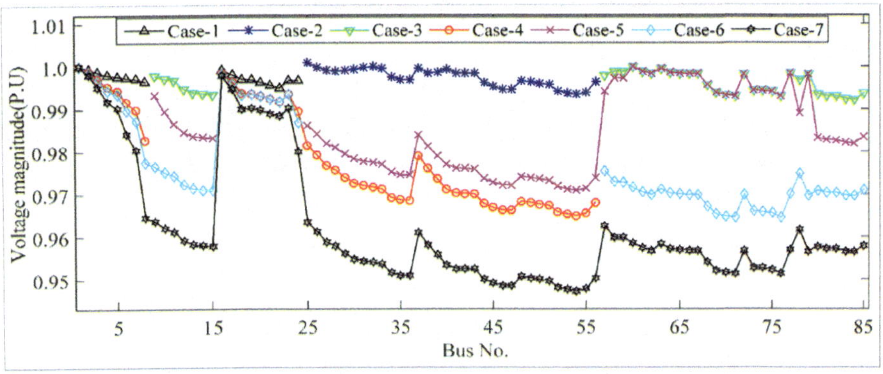

Fig. 10 Scenario-1 voltage (P.U) profile for different case studies using Jaya algorithm

apparent from the above test results, that the Jaya algorithm is scheduling the DGs optimally for attaining the desired objective function of minimizing active power losses. The active power losses obtained from Case-I to Case-VII are 2.54802 kW, 5.58956 kW, 9.060678 kW, 47.46691 kW, 65.809831 kW, 49.713619 kW and 142.374521 kW as against the Genetic Algorithm of 2.548044 kW, 5.590582 kW, 9.060680 kW, 47.467299 kW, 65.81596 kW, 49.800818 kW and 143.640221 kW, respectively.

The convergence characteristics portrayed in Fig. 11 reveals that minimum power losses are attainable with the Jaya Algorithm as compared to the Genetic Algorithm. Both algorithms converged in 33 iterations. Figure 12 exhibits the voltage profile in P.U at all buses by the Jaya Algorithm. It is notable from this figure that upon making the distribution network into several microgrid networks, the minimum P.U voltage values in these networks are superior and maintained ±5% tolerance for all the cases.

Table 6 Scenario-2—Jaya algorithm optimal DG values for different case studies

	Case-I	Case-II	Case-III	Case-IV	Case-V	Case-VI	Case-VII
P_{G1} (kW)	195.8020	–	–	152.5054	–	178.9979	155.1976
P_{G2} (kW)	100.0000	–	–	100.0000	–	100.0000	100.0000
P_{G3} (kW)	289.7060	–	–	284.1365	–	294.7551	306.9213
P_{G4} (kW)	–	231.2381	–	309.0028	244.0640	–	372.0580
P_{G5} (kW)	–	387.7115	–	398.1423	397.9214	–	397.2527
P_{G6} (kW)	–	300.0000	–	300.0000	300.0000	–	300.0000
P_{G7} (kW)	–	–	412.2778	–	448.7059	435.5996	405.3178
P_{G8} (kW)	–	–	286.6343	–	278.3296	313.5037	292.3253
P_{G9} (kW)	–	–	383.1086	–	383.1090	382.7774	382.5820
P_{loss} (kW)	**2.5480**	**5.5896**	**9.0607**	**47.4669**	**65.8098**	**49.7136**	**142.3745**
Q_{loss} (kVAR)	1.4386	2.3161	4.3252	26.9268	32.2380	31.0103	89.0266
Cost ($/hr)	23,900.20	32,455.84	44,538.12	58,033.52	79,568.28	71,145.24	106,598.08
VD (P.U)	7.4787E-06	1.300E-05	3.1218E-05	4.2800E-04	3.5700E-04	4.2200E-04	1.2680E-03
P_{demand} (kW)	582.96	913.36	1072.96	1496.32	1986.32	1655.92	2569.28
Q_{demand} (kVAR)	594.74	931.81	1094.64	1526.55	2026.45	1689.37	2621.19

Table 7 Scenario-2—genetic algorithm optimal DG values for different case studies

	Case-I	Case-II	Case-III	Case-IV	Case-V	Case-VI	Case-VII
P_{G1} (kW)	195.7864	–	–	149.9152	–	171.6144	150.6155
P_{G2} (kW)	99.9756	–	–	99.9756	–	99.9756	99.9756
P_{G3} (kW)	289.7461	–	–	287.5000	–	300.0000	305.4688
P_{G4} (kW)	–	224.8535	–	310.1074	301.1719	–	375.0000
P_{G5} (kW)	–	394.1703	–	396.3623	406.2500	–	406.2500
P_{G6} (kW)	–	299.9268	–	299.9268	299.9268	–	299.9268
P_{G7} (kW)	–	–	411.9873	–	383.3008	427.9785	406.2500
P_{G8} (kW)	–	–	286.8303	–	273.9865	406.2500	369.5313
P_{G9} (kW)	–	–	383.2031	–	387.5000	299.9023	299.9023
$\boldsymbol{P_{loss}}$ **(kW)**	**2.5480**	**5.5906**	**9.0607**	**47.4673**	**65.8160**	**49.8008**	**143.6402**
Q_{loss} (kVAR)	1.4386	2.3166	4.3252	26.9264	32.2421	31.0419	89.7885
Cost ($/hr)	23,900.17	32,632.62	44,540.53	58,004.54	78,815.15	74,903.60	109,759.28
VD (P.U)	7.4773E-06	1.2680E-05	3.1252E-05	4.9834E-04	3.5138E-04	4.3032E-04	1.2443E-03
P_{demand} (kW)	582.96	913.36	1072.96	1496.32	1986.32	1655.92	2569.28
Q_{demand} (kVAR)	594.74	931.81	1094.64	1526.55	2026.45	1689.37	2621.19

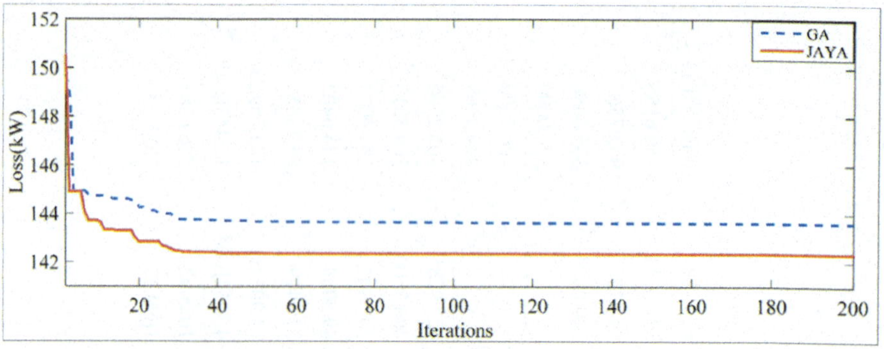

Fig. 11 Scenario-2, case-VII convergence characteristics

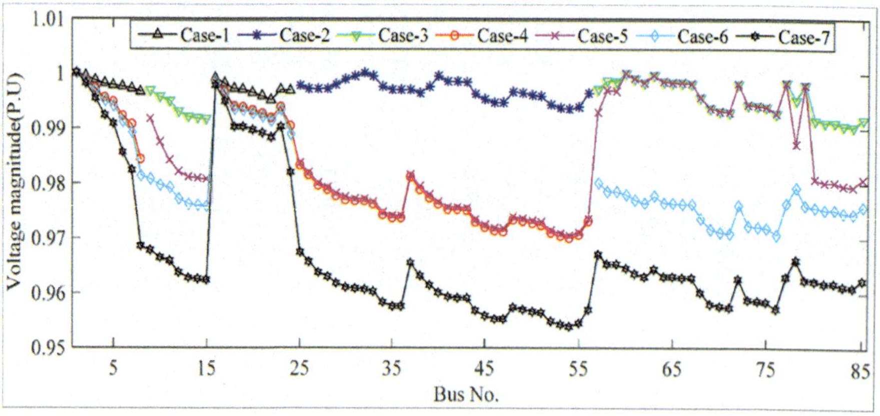

Fig. 12 Scenario-2 voltage (P.U) profile for various cases using Jaya algorithm

From the test results, it is investigated that the power losses obtained by Jaya Algorithm in Scenario-2 for various case studies are much lesser than that of Scenario-1. The reduction in power losses for Case-VII is identified as 2116.30 W. From the above test results, the proposed Jaya Algorithm's optimal allocation of micro-sources for various scenarios is promising.

4.4 Scenario-3 (Voltage Deviation Minimization)

Minimization of voltage deviation minimization is treated as the objective function in this scenario. The voltage deviation minimization values for different cases are manifested in Tables 8 and 9 for the 85-bus distribution system using the Jaya algorithm and GA, respectively.

Table 8 Scenario-3—Jaya algorithm optimal DG values for different case studies

	Case-I	Case-II	Case-III	Case-IV	Case-V	Case-VI	Case-VII
P_{G1} (kW)	85.7148	–	–	0.2728	–	0.0547	4.1948
P_{G2} (kW)	100.0000	–	–	100.0000	–	15.4684	6.8153
P_{G3} (kW)	400.0000	–	–	47.3080	–	0.3588	0.0000
P_{G4} (kW)	–	394.6044	–	600.0000	600.0000	–	275.4859
P_{G5} (kW)	–	224.9447	–	500.0000	500.0000	–	500.0000
P_{G6} (kW)	–	300.0000	–	300.0000	300.0000	–	300.0000
P_{G7} (kW)	–	–	500.0000	–	500.0000	500.0000	473.8153
P_{G8} (kW)	–	–	182.2017	–	0.7193	799.7360	761.4577
P_{G9} (kW)	–	–	400.0000	–	158.0648	400.0000	400.0000
$\boldsymbol{P_{loss}}$ **(kW)**	2.7548	6.1891	9.2417	51.2609	72.4641	59.6980	152.4891
Q_{loss} (kVAR)	1.5123	2.5646	4.4444	29.2316	35.8715	36.5468	94.8147
Cost ($/hr)	24,736.48	30,694.78	43,454.27	64,899.08	85,300.10	105,524.93	140,895.01
VD (P.U)	**3.1245E-06**	**1.0846E-05**	**2.0877E-05**	**1.5709E-04**	**7.3850E-05**	**9.0958E-05**	**5.9473E-04**
P_{demand} (kW)	582.96	913.36	1072.96	1496.32	1986.32	1655.92	2569.28
Q_{demand} (kVAR)	594.74	931.81	1094.64	1526.55	2026.45	1689.37	2621.19

Table 9 Scenario-3—genetic algorithm optimal DG values for different case studies

	Case-I	Case-II	Case-III	Case-IV	Case-V	Case-VI	Case-VII
P_{G1} (kW)	85.8365	–	–	0.0017	–	0.0456	0.0029
P_{G2} (kW)	99.9756	–	–	68.2129	–	22.2412	20.3857
P_{G3} (kW)	399.9023	–	–	100.0977	–	28.4180	0.9766
P_{G4} (kW)	–	394.6289	–	581.1035	599.5605	–	580.0781
P_{G5} (kW)	–	224.9937	–	497.9248	499.8779	–	495.3613
P_{G6} (kW)	–	299.9268	–	299.9268	299.6338	–	290.2588
P_{G7} (kW)	–	–	499.8779	–	499.8779	466.1865	290.1611
P_{G8} (kW)	–	–	182.4209	–	0.1247	799.6094	693.1641
P_{G9} (kW)	–	–	399.9023	–	160.0586	398.3398	351.4648
P_{loss} (kW)	2.7544	6.1894	9.2411	50.9473	72.8135	58.9206	152.5735
Q_{loss} (kVAR)	1.5122	2.5647	4.4441	29.0512	36.0446	36.0626	95.0140
Cost ($/hr)	24,734.95	30,695.08	43,455.47	63,940.02	85,268.54	104,679.55	135,151.63
VD (P.U)	**3.1277E-06**	**1.0851E-05**	**2.0892E-05**	**1.6503E-04**	**7.4015E-05**	**9.9281E-05**	**6.1097E-04**
P_{demand} (kW)	582.96	913.36	1072.96	1496.32	1986.32	1655.92	2569.28
Q_{demand} (kVAR)	594.74	931.81	1094.64	1526.55	2026.45	1689.37	2621.19

The optimal DGs power output in this scenario for each case study using Jaya Algorithm and GA are illustrated in Tables 8 and 9, respectively. These tables, for each case study, reveal the total system losses and the total cost of generation. From these test results, it is noticeable that the proposed algorithm is offering minimum deviation for Case-I to Case-VII, in P.U of 3.1245E-06, 1.0846E-05, 2.0877E-05, 1.5709E-04, 7.3850E-05, 9.0958E-05 and 5.9473E-04 as against Genetic Algorithm of 3.1277E-06, 1.0851E-05, 2.0892E-05, 1.6503E-04, 7.4015E-05, 9.9281E-05 and 6.1097E-04, respectively.

The convergence behavior of the proposed algorithm and GA is depicted in Fig. 13. Though both the algorithms converge for the same number of iterations, Jaya Algorithm provides a minimum voltage deviation value in P.U compared to Genetic Algorithm. From Fig. 14, it is noticed that the voltage profile is enhanced with the formulation of a multi-microgrid system compared to the operation of the entire distribution system as a single distribution network.

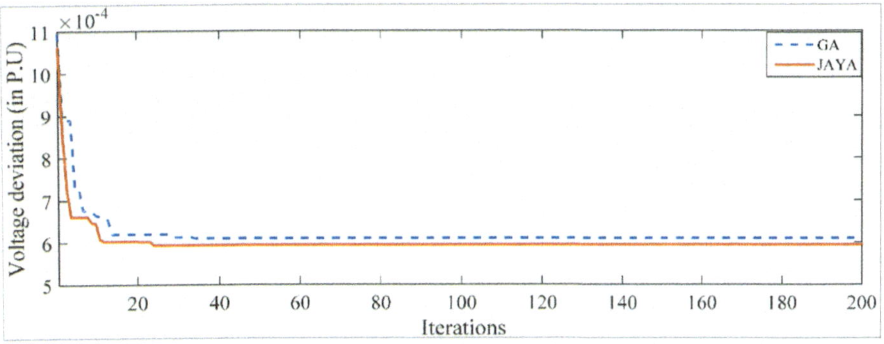

Fig. 13 Scenario-3, case-VII convergence characteristics

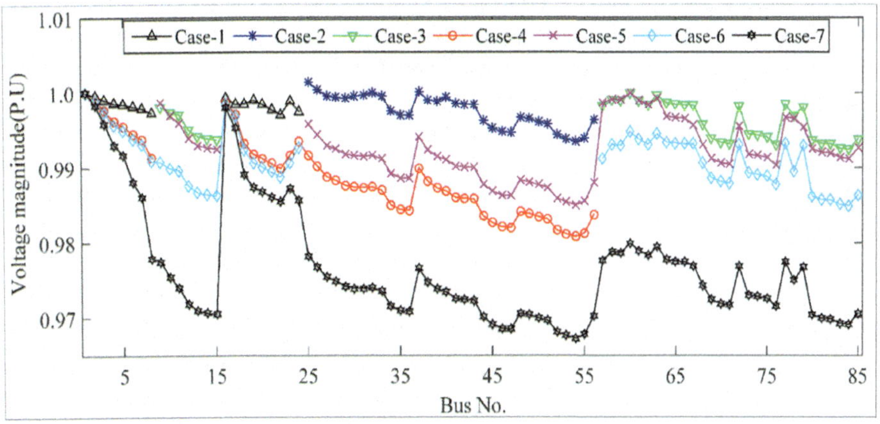

Fig. 14 Scenario-3 voltage (P.U) profile for different case studies using Jaya algorithm

On scrutiny, it is analyzed that the voltage deviation values in P.U obtained by the proposed Jaya Algorithm for Case-VII of Scenario-3 are 6.0959E-04, which is superior to that of 1.6660E-03 and 1.2680E-03 attained in Scenario-1 and Scenario-2 respectively. Thus, the optimal scheduling of DGs by the Jaya algorithm for Scenario-3 is encouraging.

The voltage profile in P.U for case-VII of scenario-1, scenario-2, scenario-3, and base case is depicted in Fig. 15. From Fig. 15, it is perspicuous that the voltage deviation in the base case is beyond the acceptable limits of ±5%, with a minimum voltage magnitude of 0.873309P.U at bus number 54. The minimum voltage magnitudes noticed in the voltage deviation minimization objective is found to be 0.96725 P.U at bus number 54, whereas, in the case of cost minimization and loss minimization objectives, the values of voltage magnitude are found to be 0.95397 P.U and 0.95142 P.U at bus number 54. These are much-improved values compared to 0.873309 P.U and this promises a healthy voltage profile at all the buses.

Based on the above studies, it may be concluded that, whenever a fault is noticed in an active islanded distribution system, by sectionalizing the network into multiple self-sufficient networks, the faulty network can be isolated such that other parts of the network can be provided with a power supply which improves the reliability of the system. The functioning of these microgrids can be the individual mode or coupled with other microgrids decided by the Microgrid Central Controller (MGCC) based on the importance of objectives.

5 Conclusions

In summary, this chapter has considered the optimal operation of a multi-microgrid system by partitioning the islanded active distribution network into numerous self-sufficient microgrids. The optimal operation of the multi-microgrid system is achieved by the optimal scheduling of controllable DGs. To attain the desired objective, these microgrids are proposed to operate alone or united with other networks. Single objective optimization has been addressed in this chapter for optimal scheduling of controllable DGs either in the individual mode or coupled with other microgrids. The objective functions attempted in this work are minimizing the total operating cost of DGs, system active power losses, and voltage deviation. A meta-heuristic algorithm, free from algorithm-specific parameters, the Jaya algorithm, is exercised to identify the optimal solution of the objective function on modified Indian 85 bus distribution systems. To validate the results obtained by the Jaya algorithm, a comparison with the well-known algorithm GA is performed. The test results reveal the supremacy of the proposed algorithm in finding the optimal solution.

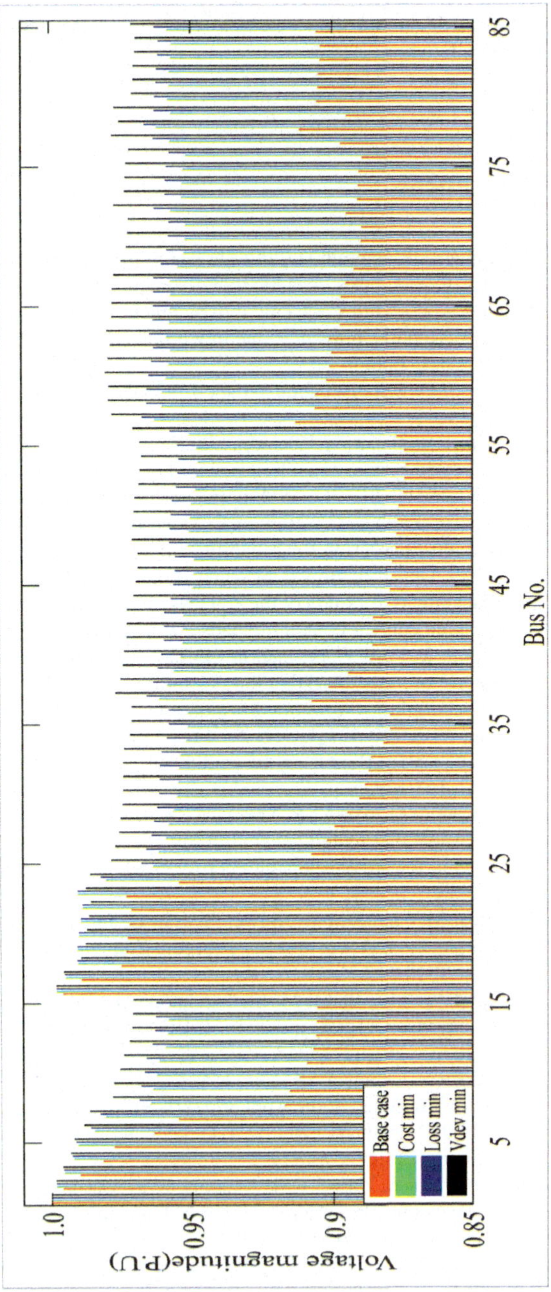

Fig. 15 Voltage magnitude in P.U for various scenarios using the Jaya algorithm

Acknowledgements This research work was supported by "Woosong University's Academic Research Funding—2023".

References

1. Kothari DP, Nagrath IJ (2008) Power systems engineering, 2nd edn. TMC
2. CEA Report. https://cea.nic.in/wp-content/uploads/pdm/2020/12/growth_2020.pdf
3. CEA Report. https://cea.nic.in/wp-content/uploads/installed/2022/03/installed_capacity.pdf
4. CEA Report. https://cea.nic.in/wp-ontent/uploads/executive/2022/01/Executive_Summary_Jan_2022-2.pdf
5. https://www.statista.com/statistics/383633/worldwide-consumption-of-electricity-by-country/
6. Srinivasarathnam C, Chandrasekhar Y, Sydulu M (2019) Load frequency control of multi-microgrid system considering renewable energy sources using grey wolf optimization. Smart Sci 7(3):198–217. https://doi.org/10.1080/23080477.2019.1630057
7. Srinivasarathnam C, Yammani C, Maheswarapu S (2019) Optimal scheduling of micro-sources in multi-microgrids for reliability improvement. In: National power electronics conference (NPEC), pp 1–6. https://doi.org/10.1109/NPEC47332.2019.9034703
8. Blackout: a case study of the 2003 North American power outage with exercises (2003). https://cip.gmu.edu/wp-content/uploads/2015/08/Blackout-Learner-Version.pdf
9. Union for the Coordination of the Transmission of Electricity (UCTE) (2003) Final report of the Investigation Committee on the 28th September 2003 Blackout in Italy, Italy. https://eepublicdownloads.entsoe.eu/clean-documents/pre2015/publications/ce/otherreports/20040427_UCTE_IC_Final_report.pdf
10. CERC (2012) Report on the grid disturbance on 30th July and grid disturbance on 31st July 2012. 167:129. https://cercind.gov.in/2012/orders/Final_Report_Grid_Disturbance.pdf
11. Makandar M, Atla CSR, Velamuri S (2016) Reliability assessment of distribution system with renewable distributed generation. In: Biennial international conference on power and energy systems: towards sustainable energy (PESTSE), pp 1–5. https://doi.org/10.1109/PESTSE.2016.7516365
12. Shuai M, Chengzhi W, Shiwen Y, Hao G, Jufang Y, Hui H (2018) Review on economic loss assessment of power outages. Procedia Comput Sci 130:1158–1163. https://doi.org/10.1016/j.procs.2018.04.151
13. Khodayar ME, Barati M, Shahidehpour M (2012) Integration of high reliability distribution system in microgrid operation. IEEE Trans Smart Grid 3(4):1997–2006. https://doi.org/10.1109/TSG.2012.2213348
14. Katoch S, Chauhan SS, Kumar V (2021) A review on genetic algorithm: past, present, and future. Multimed Tools Appl 80:8091–8126. https://doi.org/10.1007/s11042-020-10139-6
15. Srinivas M, Patnaik LM (1994) Genetic algorithms: a survey. Computer 27(6):17–26. https://doi.org/10.1109/2.294849
16. Chaouachi A, Kamel RM, Andoulsi R, Nagasaka K (2013) Multiobjective intelligent energy management for a microgrid. IEEE Trans Ind Electron 60(4):1688–1699. https://doi.org/10.1109/TIE.2012.2188873
17. Zidan A, El-Saadany EF (2012) A cooperative multiagent framework for self-healing mechanisms in distribution systems. IEEE Trans Smart Grid 3(3):1525–1539. https://doi.org/10.1109/TSG.2012.2198247
18. Tan S, Xu JX, Panda SK (2013) Optimization of distribution network incorporating distributed generators: an integrated approach. IEEE Trans Power Syst 28(3):2421–2432. https://doi.org/10.1109/TPWRS.2013.2253564

19. Li Z, Cao Y, Dai LV, Yang X, Nguyen TT (2019) Optimal power flow for transmission power networks using a novel metaheuristic algorithm. Energies 12:4310. https://doi.org/10.3390/en12224310
20. Reddy SS, Rathnam CS (2016) Optimal power flow using glowworm swarm optimization. Int J Electr Power Energy Syst 80:128–139. https://doi.org/10.1016/j.ijepes.2016.01.036
21. Carpinelli G, Mottola F, Proto D, Russo A (2017) A multi-objective approach for microgrid scheduling. IEEE Trans Smart Grid 8(5):2109–2118. https://doi.org/10.1109/TSG.2016.2516256
22. Wang D, Zhang C, Jia W, Liu Q, Cheng L, Yang H, Luo Y, Kuang N (2022) A novel interval programming method and its application in power system optimization considering uncertainties in load demands and renewable power generation. Energies 15:7565. https://doi.org/10.3390/en15207565
23. Mühlpfordt T, Faulwasser T, Hagenmeyer V (2018) A generalized framework for chance-constrained optimal power flow. Sustain Energy Grids Netw 16:231–242. https://doi.org/10.1016/j.segan.2018.08.002
24. Xie S, Wang X, Qu C, Wang X, Guo J (2013) Parking lot allocation with maximum economic benefit in a distribution network. Int Trans Electr Energy Syst 20:1–6. https://doi.org/10.1002/etep.2234
25. Yammani C, Maheswarapu S, Kumari MS (2014) Optimal placement and sizing of DER's with load variations using Bat algorithm. Arab J Sci Eng 39:4891–4899. https://doi.org/10.1007/s13369-014-1174-8
26. Salkuti SR, Kim SC (2019) Congestion management using multi-objective glowworm swarm optimization algorithm. J. Electr. Eng. Technol. 14:1565–1575. https://doi.org/10.1007/s42835-019-00206-w
27. Rao RV, Rai DP, Balic J (2017) A multi-objective algorithm for optimization of modern machining processes. Eng Appl Artif Intell 61:103–125. https://doi.org/10.1016/j.engappai.2017.03.001
28. Srinivasarathnam C, Battapothula G, Annamraju A, Yammani C (2022) Parameters estimation of solar PV using Jaya optimization technique. In: Salkuti SR, Ray P (eds) Next generation smart grids: modeling, control and optimization. Lecture notes in electrical engineering, vol 824, Springer, Singapore. https://doi.org/10.1007/978-981-16-7794-6_9
29. Das D, Kothari DP, Kalam A (1995) Simple and efficient method for load flow solution of radial distribution networks. Int J Electr Power Energy Syst 17(5):335–346. https://doi.org/10.1016/0142-0615(95)00050-0
30. Prakash DB, Lakshminarayana C (2017) Optimal siting of capacitors in radial distribution network using whale optimization algorithm. Alex Eng J 56(4):499–509. https://doi.org/10.1016/j.aej.2016.10.002
31. Srinivasarathnam C, Yammani C, Maheswarapu S (2019) Multi-objective Jaya algorithm for optimal scheduling of DGs in distribution system sectionalized into multi-microgrids. Smart Sci 7(1):59–78. https://doi.org/10.1080/23080477.2018.1540381

MPPT Algorithms for Solar PV–Drip Irrigation System

Rajagopal Veramalla, Raveena Voddamalla, Surender Reddy Salkuti, and V. Nagamalleswari

Abstract Drip irrigation for lifting irrigation water using a solar photovoltaic system based on several maximum power point tracking (MPPT) approaches is discussed in this chapter, which will improve the efficiency, stability, and accuracy of solar systems. To evaluate the system's performance, the voltage at MPP (VMPP), power at MPP (PMPP), open-circuited voltage (VOC), short-circuited current (ISC), and current at MPP (IMPP) are all measured. PV cells are designed to collect as much sunlight as possible and convert it into energy. PV systems have a nonlinear I–V curve and peak power that changes with the amount of sunlight. MPPT algorithms are used to calculate the maximum power available from a photovoltaic array. As a result, an intermediate DC–DC converter can improve efficiency by matching the PV system to demand. At their MPP, the PV system is linked to the load demand and is operational. MPPT is a critical component in increasing the use and efficiency of a PV system. The DC–DC converter output is fed into a three-phase inverter, which drives an induction motor that pumps water. The solar drip irrigation algorithms are validated using MATLAB/Simulink under various irradiance and temperature conditions.

Keywords Incremental conductance technique · P & O Algorithm · Modified perturb and observe algorithm · Current scaling algorithm · Photovoltaic system · Drip irrigation

R. Veramalla · R. Voddamalla
Department of Electrical and Electronics Engineering, Kakatiya Institute of Technology and Science, Telangana 506015, India

S. R. Salkuti (✉)
Department of Railroad and Electrical Engineering, Woosong University, Daejeon, Republic of Korea
e-mail: surender@wsu.ac.kr

V. Nagamalleswari
Department of Electrical and Electronics Engineering, Balaji Institute of Technology and Science, Warangal, Telangana, India

S. R. Salkuti et al. (eds.), *Power Quality in Microgrids: Issues, Challenges and Mitigation Techniques*, Lecture Notes in Electrical Engineering 1039,
https://doi.org/10.1007/978-981-99-2066-2_13

1 Introduction

Exploiting and using solar and wind power on a much larger scale is crucial for a long-term and reliable energy source. Solar photovoltaic is anticipated to rank among the most popular sources because of its availability, and ease of installation. Additionally, it is seen as clean power and thus allays environmental worries. Solar electricity is still more expensive than fossil fuels, though, because of the low conversion efficiency of PV modules. The maximum power point algorithm is one efficient technique to boost efficiency. There have been many MPPT techniques disclosed in the literature so far. They can be broadly divided into two categories: indirect and direct techniques. P&O and INC MPPT techniques are the two most widely used direct MPPT techniques. Fuzzy logic, and particle swarm optimization, on the other hand, are indirect techniques and are more adaptable and flexible. P&O is the simplest and shows the best convergence among the traditional MPPTs. However, the algorithm has two significant flaws. The oscillations around MPP and is susceptible to losing their tracking orientation when the irradiance rises quickly. Both issues contribute to power loss and thus reduced tracking efficiency. For improved steady-state performance and dynamics, the authors of [1] have developed a modified MPPT approach. Based on MATLAB/Simulink and dSPACE, a single-phase photovoltaic system is designed with deadbeat control to analyze its tracking response; accuracy, and impact on the system reliability as a whole. A modified P&O variable step size MPP Tracking algorithm can effectively increase the response time by a significant margin. The standard P&O approach is slower and less accurate than this algorithm [2].

For a very long time, agriculture has been and will continue to be the foundation of our nation's economy. In this regard, a fresh design idea for a smart drip irrigation system is discussed in the literature. Drip irrigation has two benefits: it increases production while using less water, and it lowers labor costs and fertilizer costs. Particularly for freshly planted sugarcane, the operation of a drip irrigation system was affordable for farmers owning a small field in arid areas or remote locations with limited water supplies and electricity. The system's cost might be reduced because it was also simple to transport and set up in other fields [3]. When insolation increases, the P&O tracking system drifts, and this drift effect is particularly severe when insolation increases quickly [4]. To tackle these issues, the work in [5] suggests a more significant modification to the traditional P&O algorithm. The modified approach keeps the very same basic structure as the conventional P&O algorithm but adds a dynamic perturbation to reduce oscillations. The effectiveness of drip irrigation and the utilization of photovoltaic panels to power drip irrigation are all demonstrated in [6]. For isolated hilly places with limited access to water and electricity, a novel concept design for smart drip systems is also put out in this context. The proposed approach is a better alternative to current systems, consequently reducing the use of both water and electricity thus saving resources. Microgrids, on the other hand, based on a central Photovoltaic system, might be a viable alternative for the electrification of isolated rural settlements. This depends on the nation's solar resources

[7]. It was shown that because the main loads on rural microgrids are residential and street lighting, they have a lower energy demand than larger standalone power systems. Likewise, these microgrids might be established as single-phase systems. Consequently, a better PV microgrid design process was established [8]. PV irrigation systems under 40 kWp are available on the market, according to an analysis of the most recent commercially available products. The European Innovation Partnerships on Water (EIP-Water) has identified two primary technological obstacles that are limiting the power of today's cutting-edge goods. One of these is the PV power's fast intermittency caused by clouds moving in and out. But at the other hand, to optimize economic gains, water usage, and PV generation must coincide. Furthermore, the hydraulic pressure, and the electric power given to the pump, must be essentially constant when it comes to drip irrigation. The design and installation of a large-power hybrid PV–diesel irrigation system are suggested in [9] as answers to these obstacles.

Over time, a modern self-adaptive INC (SAInC) MPPT technique has been suggested to overcome the obstacles of traditional INC MPPT. For maximum power harvesting from the PV array, with a new iteration of an incremental conductance algorithm that has built-in decision-making and self-adaptation capabilities. This algorithm requires relatively little data and has a low level of complexity, making it possible to implement it on low-cost microcontrollers SAInC algorithm's principle of operation is based on up to three operating points on the Power-Voltage characteristic curve [10]. An MPPT algorithm based on an optimization technique is used to maximize power output in oscillating water column wave power plants. The goal of this algorithm is to effectively maximize power output while preventing the turbine's aerodynamic stalling [11]. It is suggested that a Zero Oscillation P&O (ZOPO) MPPT approach be used to address the issue of oscillation around the MPP, which VSS PO frequently struggles with. While the duty cycle is constant at MPP, the ZOPO technique provides faster-tracking speed. These results in fewer ripples in the converter have output power and voltage, input current [12]. Adopting a drip irrigation method based on the Internet of Things (IoT) has become very popular over time. The efficiency of water use in crops is boosted as a result of harnessing solar energy [13]. Using just one tuning parameter, the modified maximum power point tracking approach is proposed in [14]. It fixes the problems with the conventional P&O MPPT technique's steady state oscillation and reduced efficacy. A brand-new grid-integrated PV system control method based on the leaky least logarithmic absolute difference (LLLAD) and a learning-based incremental conductance (LIC) MPPT algorithm effectively reduces the fixed-step-size, steady-state oscillations, and slow dynamic responses concerns that are present in the conventional INC algorithm [15]. A more advanced incremental conductance (IC) approach that is based on the mathematical residue theorem ensures low residue which enhances operation and gets rid of oscillations around MPP [16].

A solar-powered drip irrigation system with an inverter, an MPPT based on the INC algorithm, and a DC–DC boost converter used to acquire the proper PWM wave patterns are shown in analyzed 30 kW Standalone Solar Powered Irrigation System. As discussed in [17, 18], such a system is capable of employing PV panels to power the irrigation system's water pumping equipment supplying irrigation water to the solar pump. The studies suggest a voltage-oriented MPPT control with an inner current control loop for increasing tracking speed while preserving the oscillations around MPP, a straightforward modification of this algorithm that adds momentum to the hill climbing is investigated in similar works reported in the literature [19]. A derated-mode GIPVS has overcome the overvoltage problem that exists at PCC during peak electrical generation [20]. In [21], the fixed and variable step sizes are thoroughly analyzed and compared. The contemporary design of a photovoltaic pumping chain intended for drip irrigation, which is managed by an intelligent neuron-fuzzy controller with an ANFIS architecture built on a Raspberry Pi platform work, is proposed in [22]. It aims to create a new paradigm for sustainable development. Passivity-based control (PBC) is a system that combines local and global control capabilities for a solar water pumping system with battery storage and an external power source in remote areas. This system is researched in [23]. PV water pumps in the tapioca drip irrigation system; a case study is done in [24]. Both a dependable supply of clean energy and a way to reduce power and energy requirements may come from a battery-powered solar power-driven system. Such a strategy aims to reduce the discrepancy between supply and quality when demand rises [25]. Particle Swarm Optimization is investigated to identify the appropriate soil depth to lower the cost of irrigation [26, 27] in order to provide remote system monitoring. In a decoupled PV power ramp-rate calculation approach, which is discussed and addressed in [28], the impact of variable irradiance and the P&O algorithm is estimated separately.

A field study was carried out in the Punjab province of Pakistan to evaluate the socio-economic and climatic impacts of PV drip and sprinkler irrigation systems [29]. Also introduced in [30] were MPPT systems based entirely on digital CMOS circuits without the need for complicated circuits like hold units, samples, multipliers, or analog-to-digital converters. The most prevalent issue in PV systems is power loss under partial shadowing, hence a photovoltaic array reconfiguration approach based on estimating the consequences of partial shadowing on short-circuit current [31]. Additionally, global power point cannot be tracked in such situations. Therefore, a marine predator algorithm (MPA) has been described in [32] as a way to solve this issue. On a monthly and annual basis, the PVGIS simulation software utilized in [33] examines the capacity of PV power generation with ideal PV module orientation and inclination angle, as well as an analysis of the shading condition in between modules. Before adding new PV systems to the grid, it is essential to get a precise estimate of the power output of the PV system because its output is unsteady. Consequently, stacked groups utilizing three machine learning techniques, the Stack-ETR algorithm, and a generic regression neural network (GRNN) based on grey wolf optimization (GWO), predicts PV production power one day in advance with faster calculation. This idea of ML algorithms was put forth in [34–36]. Research and

Development affect how quickly new RE technologies develop [37]. The process to determine a PV system's ideal tilt angle in order to get the most output power from it was covered in [38]. A significant problem for PV module makers, installers, researchers, etc. is the aging of panels and exposure to various ambient conditions. An insight into the PV module's reliability and degradation performance is reviewed in [39]. Consumers' Motivation to switch to solar energy Tech for Domestic Usage is another obstacle to the development of renewable energy. To inspire individuals, the government should find inventive strategies to persuade prospective customers to install solar PV. One method is for government to set up lending schemes to provide financial assistance to customers at the start [40].

2 Solar PV System

2.1 Modeling of Photovoltaic Module

Solar panels are manufactured from a variety of semiconductor materials. Silicon solar cells are currently the most widely used in the production of photovoltaic modules. A solar module is made up of PV cells that are connected in series [41]. Photovoltaic module specifications include Maximum power (P_{MP} (W)), short-circuit current (I_{SC} (A)), open-circuited voltage (V_{OC} (V)), the voltage at MPP (V_{MP} (V)), series resistance (Rs), MPP current (I_{MP} (A)), number of cells connected in series (N_S), temperature coefficient of short-circuit current (KI (A/Kelvin)), temperature coefficient of open-circuit voltage (KV (V/Kelvin)) and shunt resistance (R_{sh}). The electrical circuit in Fig. 1 can be used to depict a solar cell.

The following equation expresses its current-voltage characteristic.

$$I_{pv} = I_{ph} - I_0 * \left[\exp\left(\frac{I_{pv} * R_s + V_{pv}}{N_s * \left(\frac{\eta k_b T}{q}\right)} - 1 \right) \right] - \frac{I_{pv} * R_s + V_{pv}}{R_{sh}} \tag{1}$$

where I_{pv} and V_{pv} are output current and voltage respectively, I_{ph} is Photon current. I_0 is the saturation current, q is the electron charge, k is the Boltzmann constant; T

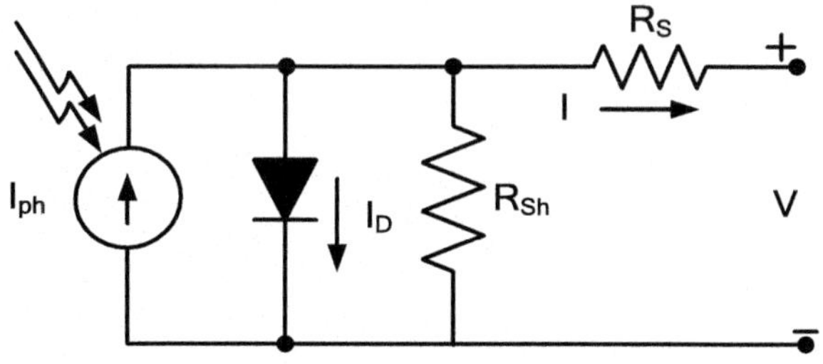

Fig. 1 Equivalent circuit of a single-diode solar cell

is the temperature. R_s is series resistance and R_{sh} shunt resistance [42]. R_s is zero and R_{sh} is infinite in an ideal case. The equations required to model the PV cell are listed below. Applying Kirchoff's current law to Fig. 1.

$$I_{shunt} = \frac{I_{pv} * R_s + V_{pv}}{R_{sh}} \tag{2}$$

I_{shunt} is the current through R_{sh} and Current through the diode I_d is given by

$$I_d = \left[\exp\left(\frac{I_{pv} * R_s + V_{pv}}{N_s * \left(\frac{\eta\, k_b\, T}{q}\right)} - 1\right)\right] \tag{3}$$

Photon current I_{ph} derived as

$$I_{ph} = I_{sc0} * \left(\frac{G}{G_0}\right) * (1 + \alpha_i * (T - T_0)) \tag{4}$$

Reverse saturation current is given by

$$I_0 = I_{0,ref} * \left(\frac{T}{T_0}\right)^3 * \exp\left[\left(\frac{q * E_g}{k_b\, \eta}\right) * \left(\frac{1}{T_0} - \frac{1}{T}\right)\right] \tag{5}$$

Figure 2 depicts the PV module realistic equivalent circuit simulated at Standard Test Conditions (STC). The Solar panel characteristics at STC are shown in Table 1.

Figure 3a, b are the I–V curves with the variation of irradiance and temperatures, and Fig. 3c, d are the PV curves with the variation of irradiance and temperatures.

Fig. 2 P–V curve at STC

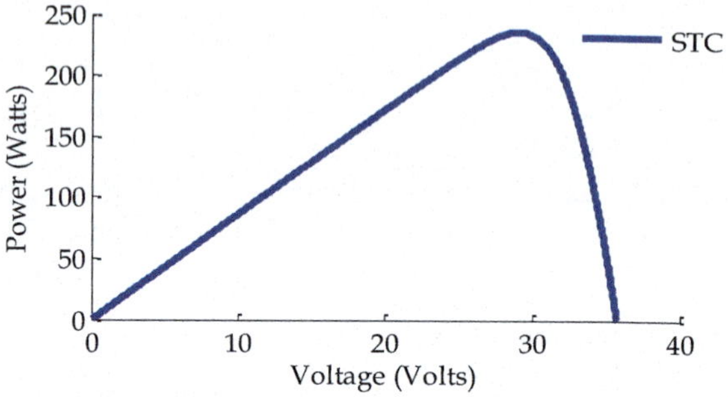

Table 1 Solar panel characteristics at STC

Peak power	235.9 W_p
Open Circuited Voltage	37.8 V
MPP Voltage	29.21 V
Short Circuited Current	8.63 A
MPP Current	8.14 A
DC-Link Capacitance	0.4 F
Solar Cells per Module (N_s)	60

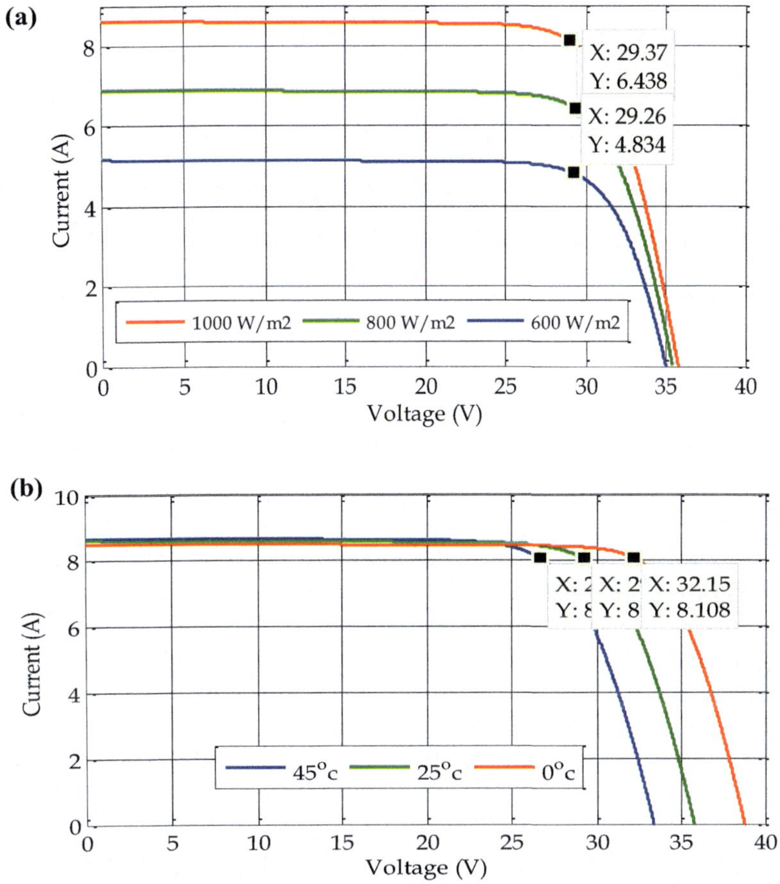

Fig. 3 **a** I–V curves at (i) 1000W/m^2 (ii) 800W/m^2 (iii) 600W/m^2 **b** I–V curves at temperatures (i) 0^0 (ii) 25^0 (iii) 45^0 **c** P–V curves at (i) 1000W/m^2 (ii) 800W/m^2 (iii) 600W/m^2 **d** P–V curves at temperatures (i) 0^0 (ii) 25^0 (iii) 45^0

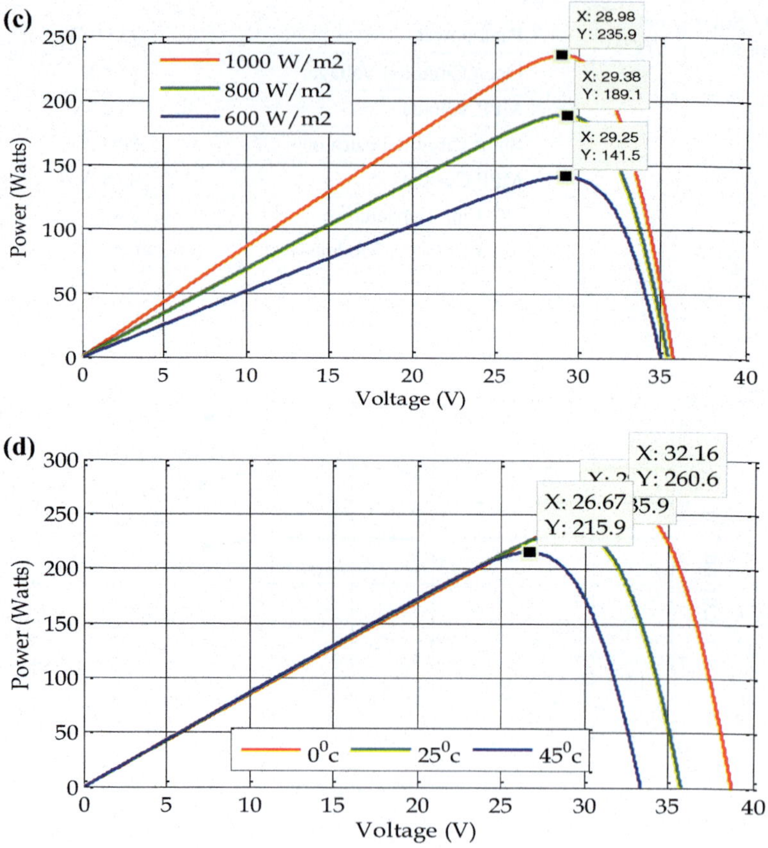

Fig. 3 (continued)

2.2 Design of Boost Converter

To regulate and enhance the PV's output voltage, DC–DC boost converters are installed between the photovoltaic module and the load. Figure 4 shows the DC–DC boost converter employed as an intermediary between the source and the load to achieve impedance matching so that MPP can be found by adjusting the duty cycle of the converter, which is controlled by the MPPT controller [43]. The converter is an interface between a PV system and a load.

The relation between the input voltage and output voltage is given by

$$M(D) = \frac{V_o}{V_s} = \frac{1}{(1 - D)} \tag{6}$$

The relevant expression when connected to load can be given as

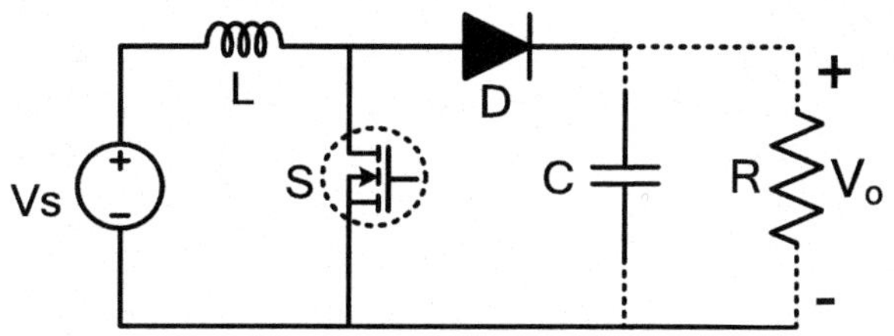

Fig. 4 DC–DC converter

Table 2 Specifications of DC–DC Converter

Component	Rating
Inductor	0.000625 H
Capacitor	0.2 F

$$R_{pv} = \frac{R_{load}}{M(D)^2} \tag{7}$$

The relationship between R_{pv} and R_{load} is given by

$$R_{pv} = R_{load}(1 - D)^2 \tag{8}$$

where 'D' is the duty cycle, as irradiance and temperature change the duty cycle is adjusted accordingly. The specifications of the boost converter are shown in Table 2.

3 MPPT Methods

3.1 P&O Algorithm

Figure 5 shows P&O Algorithm which includes fewer measurable parameters; its approach is extensively employed in MPPT [44]. It can precisely track peak power points despite variations in solar irradiance and temperature.

It operates by changing the voltage of the PV panels and then measuring the impact on the Photovoltaic panel power output. The P&O approach of MPPT is quite popular. When the difference between the previous and present power is zero, the power is said to be at its maximum. If the difference isn't really zero, it checks to see if the previous and present voltages are both zero. If it is not zero, it attempts to alter the duty cycle to extract the MPP from the model.

$$\mathrm{dP/dV} > 0 \text{ to left of MPP} \tag{9}$$

$$\mathrm{dP/dV} = 0 \text{ when MPP is achieved} \tag{10}$$

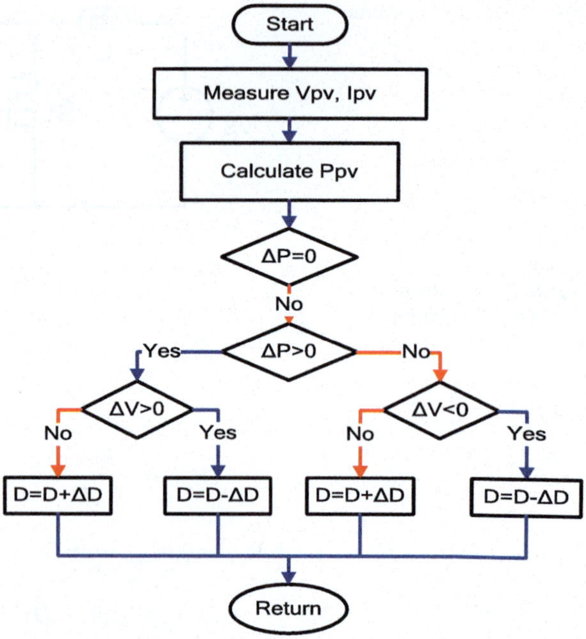

Fig. 5 Flowchart of the P&O method

Table 3 Tabulated results with P&O Algorithm

Irradiance	Voltage (V)	Current (A)	Power (W)
1000 W/m^2	29.21	8.07	235.9
800 W/m^2	29.13	6.49	189.1
600 W/m^2	29.04	4.87	141.5

$$dP/dV < 0 \text{ to right of MPP} \tag{11}$$

The performance of the P&O MPPT technique is shown in Table 3.

3.2 *Incremental Conductance MPPT Algorithm*

Figure 6 shows the INC-based MPPT technique which has the advantage of having a P–V curve slope of zero at the maximum power point.

The relationship between dI/dV & –I/V is used to compute the direction where the Maximum power point must be changed. The fact that dP/dV is positive when the MPPT is to the left and negative when it is to the right leads to the relationship as shown below.

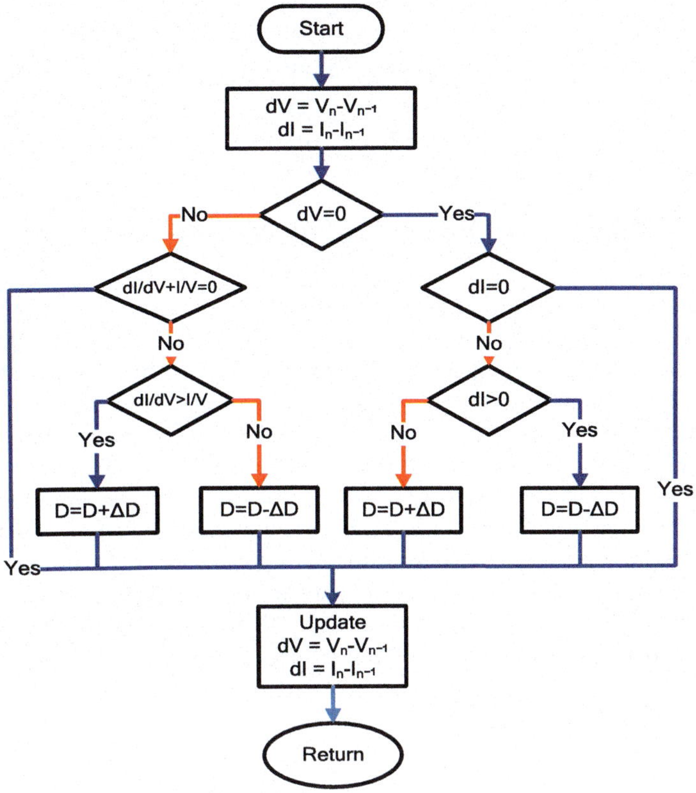

Fig. 6 Flow chart of INC-based MPPT method

$$P = V\,I \tag{12}$$

$$\delta P/\delta V = [\delta(VI)]/\delta V \tag{13}$$

$$\text{As } \delta P/\delta V = 0 \text{ at MPP} \tag{14}$$

Equation (14) becomes,

$$\delta I/\delta V = -I/V \tag{15}$$

Unlike P&O, this technique can identify when the maximum power point has reached and oscillates about the operating point. The results of the INC MPPT algorithm are tabulated in Table 4.

Table 4 Tabulated results of INC MPPT Algorithm

Irradiance	Voltage (V)	Current (A)	Power (W)
1000 W/m^2	31.38	6.98	219.1
800 W/m^2	29.14	6.48	189.1
600 W/m^2	23.07	5.11	118.1

3.3 Reference Cell—Current Scaling MPPT Algorithm

The relationship between irradiance and short-circuit current (I_{SC}) and IMPP are shown in Fig. 7. The proportionality coefficient, K, must be found for each PV array. The constant K is said to range between 0.78 and 0.92. It's difficult to calculate the short-circuited current when the PV system runs. To periodically short the PV system and record Isc, an extra switch is frequently added to the converter. A loss of power is also caused by short-circuiting the PV system. Here, $I_{MPP} \approx KI_{SC}$.

Furthermore, because the scaling relationship is an approximation, the true MPP is not obtained. To fix this problem, a reference cell with similar characteristics to that same module is used, and Isc is recorded for various irradiances. This strategy ensures that the true MPP is achieved even when there are several maxima, which increases the system's complexity. Figure 7 shows the reference cell current scaling MPPT algorithm and Table 5 shows the tabulated results of the current scaling MPPT algorithm. A PI controller is used in much of the literature that employs this MPPT approach.

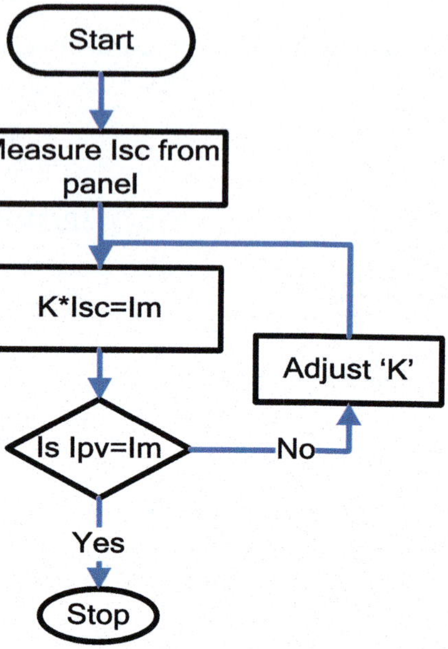

Fig. 7 Flowchart of current scaling MPPT Algorithm

Table 5 Tabulated results of current scaling MPPT Algorithm

Irradiance	Voltage (V)	Current (A)	Power (W)
1000 W/m^2	30.33	7.69	233.5
800 W/m^2	30.21	6.19	187.1
600 W/m^2	30.12	4.65	140.1

3.4 Modified (P&O) MPPT Method

The modified MPPT technique only has one tuning parameter. It overcomes the challenges of oscillation in the steady state and poorer efficiency that remains in perturb and observe technique. MPP is identified by equating the expression of dP/dV to zero.

This algorithm accomplishes the same goal as previous algorithms in a simpler manner. The duty ratio is calculated by multiplying the difference between the obtained voltages and the reference voltage by variable M.

$$D = V - C * V_{ref} \tag{16}$$

where D is the duty cycle, C is given by multiplying M by dP/dV as given below:

$$C = M * \left(\frac{dP}{dV}\right) \tag{17}$$

where M is chosen based on rate of convergence in order to achieve the best feasible result. It's just an arbitrary number between the numbers 0.3678 and 0.0000061.

$$V_{ref} = n * V_m \tag{18}$$

where 'V_m' is the voltage at MPP and 'n' is the number of cells. This strategy focuses on minimizing the voltage and current difference between the desired and current values. As a result, once the MPP is obtained, steady-state oscillation is almost neglected. Figure 8 shows the modified P&O MPPT algorithm and table 6 depicts the tabulated results of the modified (P&O) MPPT method.

4 Performance of MPPTAlgorithms

The current (PV) vs time, voltage (PV) vs time, and power output (PV) vs time curves are discussed in the following subsections. Section 4.1 discusses about mentioned three curves for perturb and observe (P&O) algorithm. Section 4.2 discusses about mentioned three curves for the incremental conductance-based MPPT algorithm. Section 4.3 discusses about mentioned three curves for reference cell-current scaling

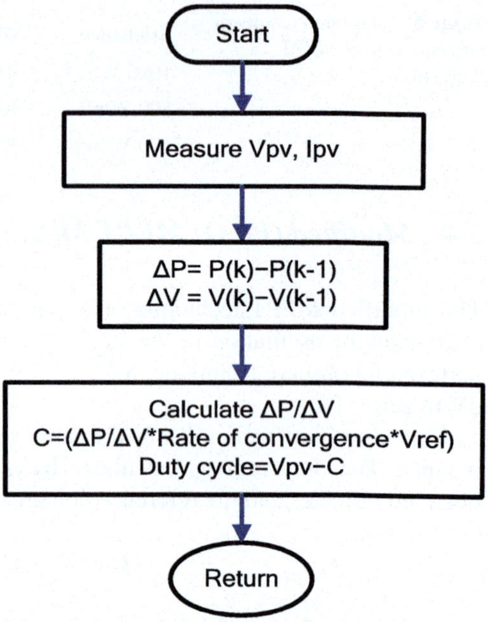

Fig. 8 Flowchart of modified (P&O) MPPT Technique

Table 6 Tabulated results of modified (P & O) MPPT Algorithm

Irradiance	Voltage (V)	Current (A)	Power (W)
1000 W/m^2	29.11	8.10	235.9
800 W/m^2	29.06	6.52	189.1
600 W/m^2	29.04	4.87	141.5

MPPT algorithm. Section 4.3 discusses about mentioned three curves for the modified perturb and observe (P&O) algorithm.

4.1 Perturb and Observe (P&O) Algorithm

Figure 9a shows the current vs time, Fig. 9b shows the voltage vs time and Fig. 9c shows the power vs time at (i) 1000W/m^2 (ii) 500W/m^2 (iii) 600W/m^2 at 25^0 for Perturb and Observe Algorithm. It has been observed that I_{PV} at maximum power point for (i) 1000W/m^2 (ii) 800W/m^2 (iii) 600W/m^2 irradiance are 8.07A, 6.49A, and 4.87A respectively. It has been learned that V_{PV} at maximum power point for (i) 1000W/m^2 (ii) 800W/m^2 (iii) 600W/m^2 insolation are 29.2 V, 29.13 V, and 29.04 V respectively and it has been viewed that P_{PV} at maximum power point for (i) 1000W/m^2 (ii) 800W/m^2 (iii) 600W/m^2 insolation are 235.9W, 189.1W and 141.5W, respectively.

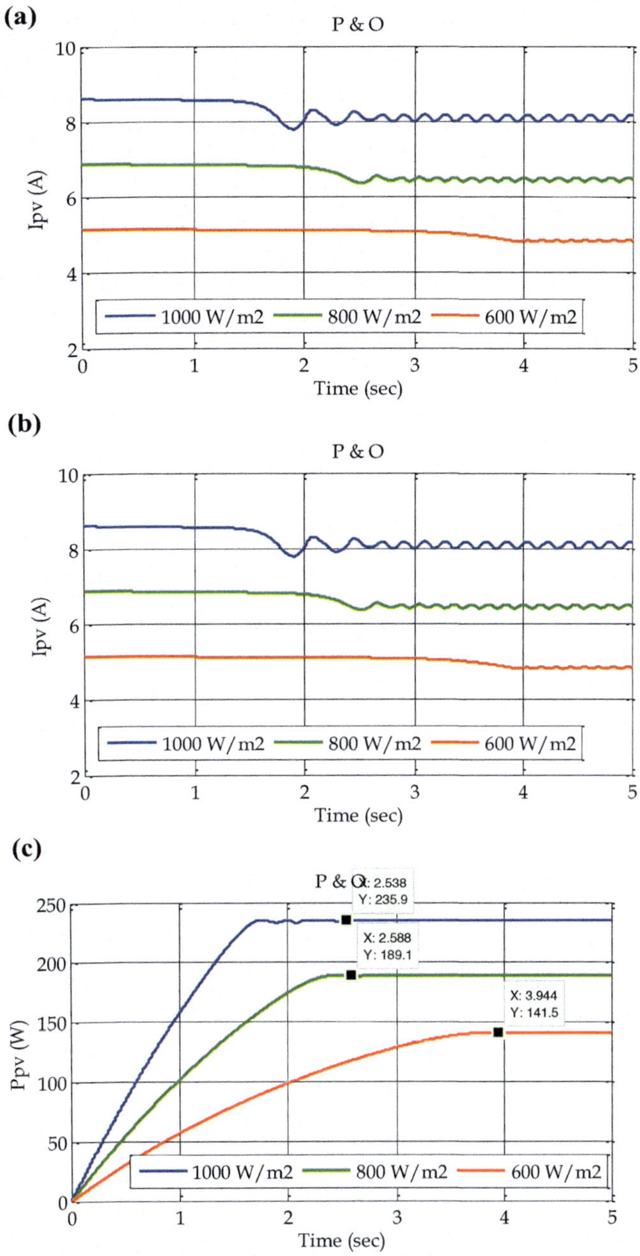

Fig. 9 Performance of P&O algorithm: **a** Current vs time curve at (i) 1000W/m^2 (ii) 800W/m^2 (iii) 600W/m^2 **b** Voltage vs time curve at (i) 1000W/m^2 (ii) 800W/m^2 (iii) 600W/m^2 **c** Power vs time curve at (i) 1000W/m^2 (ii) 800W/m^2 (iii) 600W/m^2

4.2 Incremental Conductance Based MPPT Algorithm

Figure 10a shows the current vs time curve, Fig. 10b shows the voltage vs time curve, and Fig. 10c shows the power vs time curve at (i) 1000W/m^2 (ii) 500W/m^2 (iii) 600W/m2 at 25^0 for Incremental Conductance Based MPPT Algorithm. It has been observed that I_{PV} at maximum power point for (i) 1000W/m^2 (ii) 800W/m^2 (iii) 600W/m^2 irradiance are 6.98A, 6.48A, and 5.11A, respectively. It has been learned that V_{PV} at maximum power point for (i) 1000W/m^2 (ii) 800W/m^2 (iii) 600W/m^2 insolation are 31.38 V, 29.14 V, and 23.07 V respectively and it has been viewed that P_{PV} at maximum power point for (i) 1000W/m^2 (ii) 800W/m^2 (iii) 600W/m^2 insolation are 219.1W, 189.1W and 118.1W, respectively.

4.3 Reference Cell—Current Scaling MPPT Algorithm

Figure 11a shows the current vs time curve, Fig. 11b shows the voltage vs time curve, and Fig. 11c shows the power vs time curve at (i) 1000W/m^2 (ii) 500W/m^2 (iii) 600W/m^2 at 25^0 for Reference Cell—Current Scaling MPPT Algorithm. It has been observed that I_{PV} at maximum power point for (i) 1000W/m^2 (ii) 800W/m^2 (iii) 600W/m^2 irradiance are 7.69A, 6.19A, and 4.65A respectively. It has been learned that V_{PV} at maximum power point for (i) 1000W/m^2 (ii) 800W/m^2 (iii) 600W/m^2 insolation are 30.33 V, 30.21 V, and 30.12 V respectively and it has been viewed that P_{PV} at maximum power point for (i) 1000W/m^2 (ii) 800W/m^2 (iii) 600W/m^2 insolation are 233.5W, 187.1W and 140.1W, respectively.

4.4 Modified P&O MPPT Algorithm

Figure 12a shows the current vs time curve, Fig. 12b shows the voltage vs time curve, and Fig. 12c shows the power vs time curve at (i) 1000W/m^2 (ii) 500W/m^2 (iii) 600W/m^2 at 25^0 for modified (P&O) Algorithm. It has been observed that I_{PV} at maximum power point for (i) 1000W/m^2 (ii) 800W/m^2 (iii) 600W/m^2 irradiance are 8.10A, 6.52A, and 4.87A, respectively. It has been learned that V_{PV} at maximum power point for (i) 1000W/m^2 (ii) 800W/m^2 (iii) 600W/m^2 insolation are 29.11 V, 29.06 V, and 29.04 V respectively and it has been viewed that P_{PV} at maximum power point for (i) 1000W/m^2 (ii) 800W/m^2 (iii) 600W/m^2 insolation are 235.9W, 189.1W and 141.5W, respectively. Table 7 shows the comparison of the results for various algorithms.

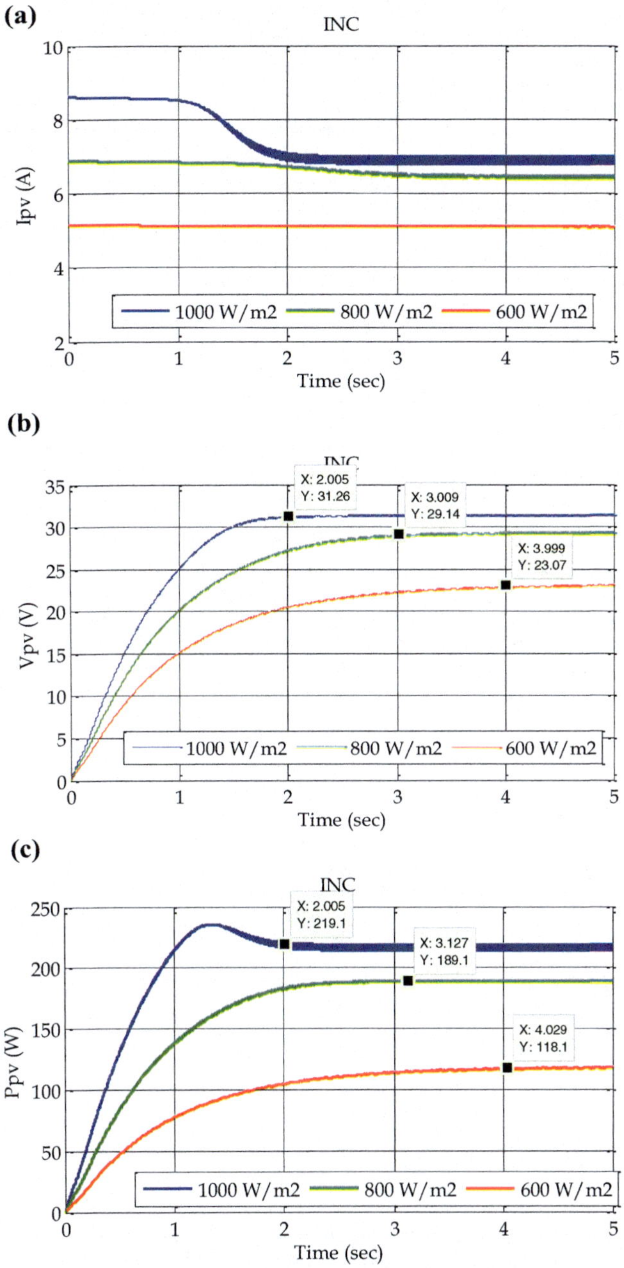

Fig. 10 Performance of Incremental Conductance based MPPT algorithm: **a** Current vs time curve at (i) 1000W/m^2 (ii) 800W/m^2 (iii) 600W/m^2 **b** Voltage vs time curve at (i) 1000W/m^2 (ii) 800W/m^2 (iii) 600W/m^2 **c** Power vs time curve at (i) 1000W/m^2 (ii) 800W/m^2 (iii) 600W/m^2

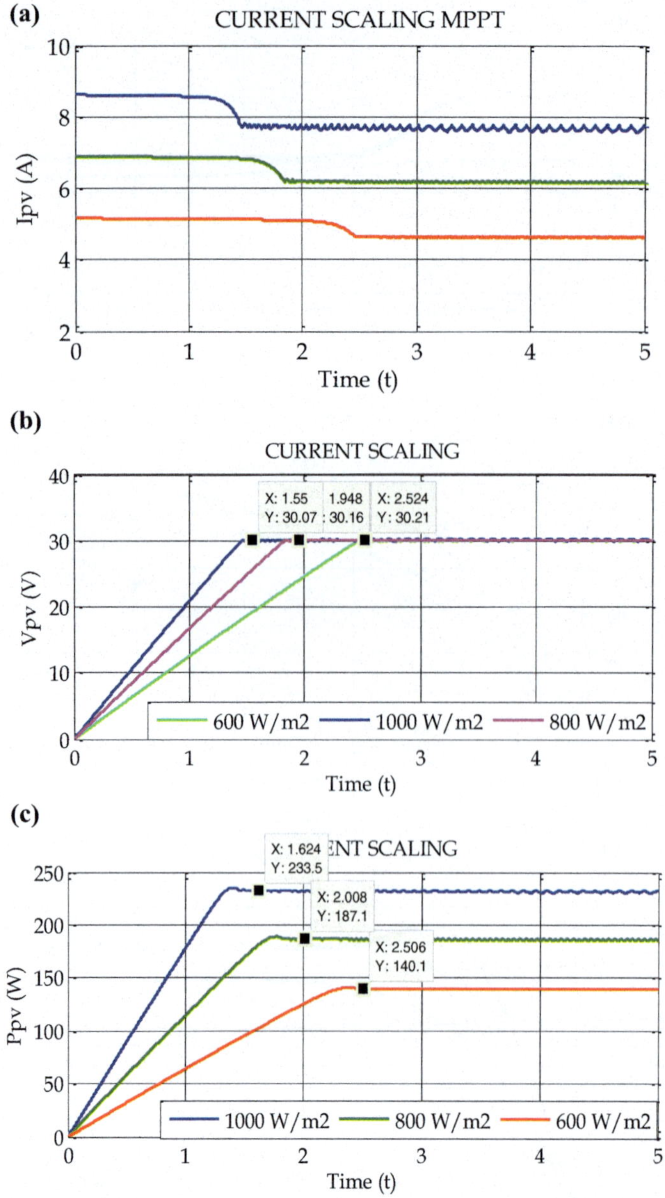

Fig. 11 Performance of reference cell—current scaling MPPT algorithm: **a** Current vs time curve at (i) 1000W/m^2 (ii) 800W/m^2 (iii) 600W/m^2 **b** Voltage vs time curve at (i) 1000W/m^2 (ii) 800W/m^2 (iii) 600W/m^2 **c** Power vs time curve at (i) 1000W/m^2 (ii) 800W/m^2 (iii) 600W/m^2

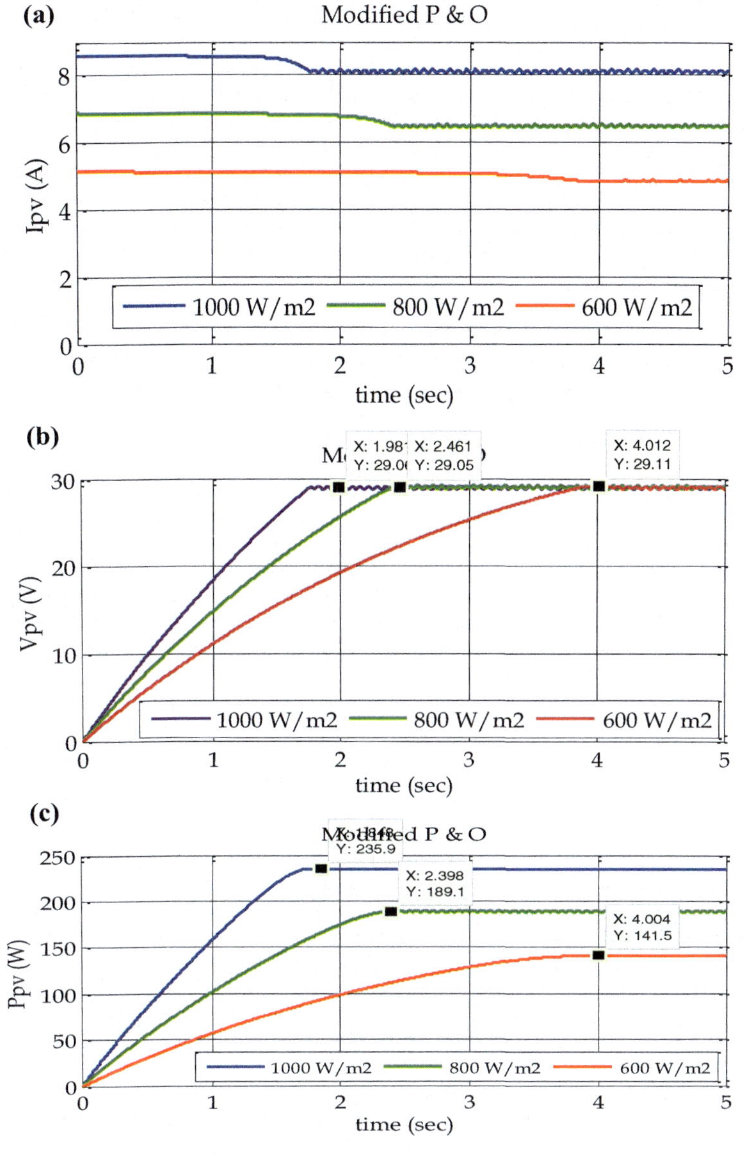

Fig. 12 Performance of modified P&O MPPT algorithm: **a** Current vs time at (i) 1000W/m^2 (ii) 800W/m^2 (iii) 600W/m^2 **b** Voltage vs time curve at (i) 1000W/m^2 (ii) 800W/m^2 (iii) 600W/m^2 **c** Power vs time curve at (i) 1000W/m^2 (ii) 800W/m^2 (iii) 600W/m^2

Table 7 Comparison of results of various MPPT Algorithms

MPPT	Tuning parameters	Type	Steady-state oscillations	Tracking speed	Global MPP tracking	P_{MPP} at STC
Perturb and observe	Voltage and Current	Direct	Yes	Slow	No	235.9
Incremental conductance	Voltage and Current	Direct	Occasionally	Medium	No	219.1
Current scaling	Current	Indirect	Occasionally	Fast	No	233.5
Modified P&O	Voltage	Indirect	Rarely	Fast	No	235.9

5 Drip Irrigation

Solar photovoltaic (PV) cells flip daylight into energy and are engineered to gather as plenty electricity as possible. Solar energy, whether in the form of thermal or electric energy, is useable energy generated by the sun. A solar module consists of a layer of silicon solar cells, silicon is a conductive nonmetal that absorbs and converts light from the sun into electrical energy. When sunlight hits a PV panel, it causes electrons in the semiconductor to move, resulting in a current flow. This is known as the "photovoltaic effect", and it governs the operation of solar panels. The generated electricity is unregulated and may not extract the maximum PV power, so a converter (DC–DC) with an MPPT algorithm is applied to extract maximum power. This output is stored in a small battery to operate a dc motor and pump the water for drip irrigation. Figure 13 shows the schematic of a solar PV system for drip irrigation.

Figure 14 shows the output voltage, state of charge (SOC), speed of the motor, and armature current of the solar PV panel after passing through the DC–DC converter which operates with the MPPT algorithm.

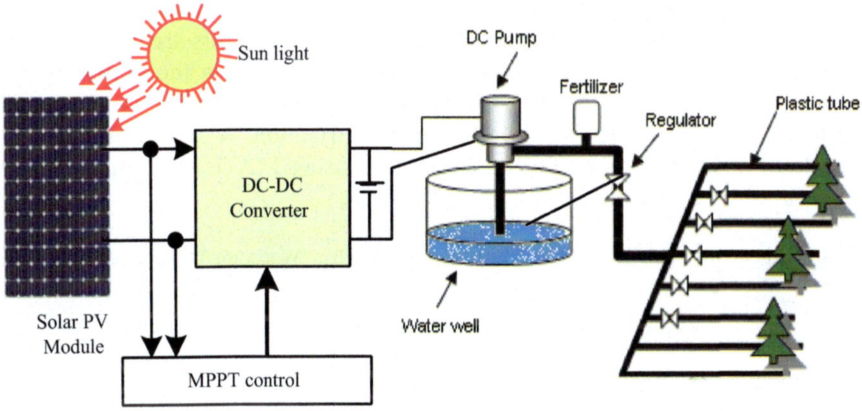

Fig. 13 Schematic of solar PV system for drip irrigation

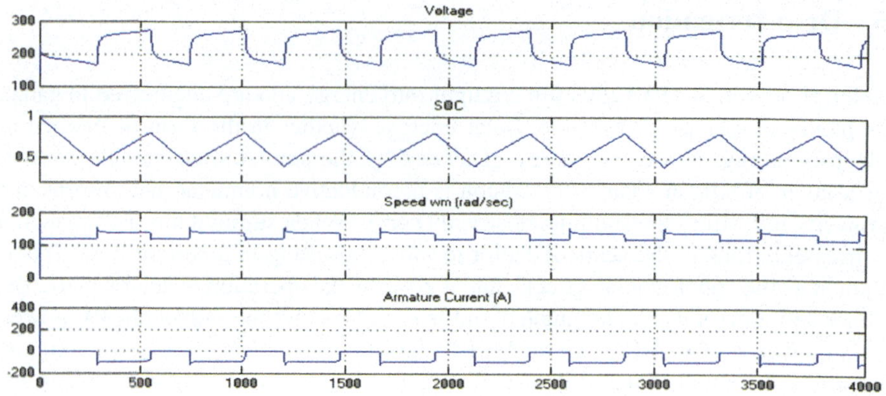

Fig. 14 Performance of drip irrigation with the solar PV system

6 Conclusion

Solar PV system converts solar energy to electrical energy, the study attempts four MPPT algorithms for drip irrigation-fed solar PV system. The P&O MPPT algorithm, incremental conductance-based MPPT algorithm, reference cell-current scaling MPPT algorithm, and modified P&O MPPT algorithm are applied to boost converter to obtain the maximum possible power and the results are tabulated. The above-said algorithms were analyzed and compared based on the output results. The output of the boost converter shall be charging a small battery; this battery in turn feeds the dc motor to pump water for drip irrigation. A single solar module is modeled and MPPT algorithms are validated using MATLAB/Simulink under various irradiance conditions.

Acknowledgements This research work was supported by "Woosong University's Academic Research Funding—2023."

References

1. Yang Y, Blaabjerg F (2012) A modified P&O MPPT algorithm for single phase pv systems based on deadbeat control. In: IET Conference Publications (592): 0–4. https://doi.org/10.1049/cp.2012.0197
2. Abdourraziq MA, Ouassaid M, Maaroufi M, Abdourraziq S (2013) Modified P&O MPPT technique for photovoltaic systems. In: Proceedings 2013 International Conference on Renewable Energy Research and Applications ICRERA 728–733. https://doi.org/10.1109/ICRERA.2013.6749849
3. Tangwongkit B, Tangwongkit R, Chontanaswat P (2014) Drip irrigation powered by solar cell for dry rainfed and no electricity area. In: Proceedings 2014 International Conference and Utility Exhibition. Green Energy Sustain. Dev. 19–21

4. Killi M, Samanta S (2015) Modified perturb and observe MPPT algorithm for drift avoidance in photovoltaic systems. IEEE Trans Ind Electron 62(9):5549–5559. https://doi.org/10.1109/TIE.2015.2407854
5. Ahmed J, Salam Z (2016) A modified P and O maximum power point tracking method with reduced steady-state oscillation and improved tracking efficiency. IEEE Trans. Sustain. Energy 7(4):1506–1515. https://doi.org/10.1109/TSTE.2016.2568043
6. Ahmed DI, Fernando X (2017) Solar-powered drip-irrigation for rural Bangladesh. IEEE Canada Int. Humanit. Technol. Conf. 193–196. https://doi.org/10.1109/IHTC.2017.8058186
7. Singh D, Thakur A (2018) Designing of smart drip irrigation system for remote hilly areas. In: 5th International Conference Parallel Distributed Grid Computing 90–94. https://doi.org/10.1109/PDGC.2018.8745934
8. Bhagavathy SM, Pillai G (2018) PV microgrid design for rural electrification. Designs 2(3):1–22. https://doi.org/10.3390/designs2030033
9. Almeida RH (2018) Development and test of solutions to enlarge the power of PV irrigation and application to a 140 kW PV-diesel representative case. Energies 11(12). https://doi.org/10.3390/en11123538
10. Kumar N, Hussain I, Singh B, Panigrahi BK (2018) Self-Adaptive incremental conductance algorithm for swift and ripple-free maximum power harvesting from PV array. IEEE Trans. Ind. Informatics 14(5):2031–2041. https://doi.org/10.1109/TII.2017.2765083
11. Mishra S, Purwar S, Kishor N (2018) Maximizing output power in Oscillating water column wave power plants: an optimization based MPPT algorithm. Technologies 6(1):15. https://doi.org/10.3390/technologies6010015
12. Krishnan R, Mohammed S, Manafudeen S (2019) Comparison of PO MPPT Based Solar PV System with Interleaved Boost Converter. In: 2nd International Conference Intell Computing Instrumentation Control Technologies. 1370–1376. https://doi.org/10.1109/ICICICT46008.2019.8993209
13. Kholifah R, AlbarSarosa KI, Fitriana R, Rochmawati I, Sarosa M (2019) Drip Irrigation System Based on Internet of Things (IoT) using solar panel energy. In: proceedings 2019 24th International Conference Informatics Computing. https://doi.org/10.1109/ICIC47613.2019.8985886
14. Kumari N, Kumar SS, Laxmi V (2019) Modified MPPT topology: A comparative analysis. IEEE 16th India Counc. Int. Conf. INDICON 2019—Symp. Proc. 7–10. https://doi.org/10.1109/INDICON47234.2019.9029042
15. Kumar N, Singh B, Panigrahi BK, Xu L (2019) Leaky-least-logarithmic-absolute-difference-InC MPPT technique for grid-integrated. IEEE Trans Ind Electron 66(11):9003–9012. https://doi.org/10.1109/TIE.2018.2890497
16. Alsumiri M (2019) Residual incremental conductance based nonparametric mppt control for solar photovoltaic energy conversion system. IEEE Access 7: 87901–87906. https://doi.org/10.1109/ACCESS.2019.2925687
17. Wanquan S, Niringiyimana E, Simiyu P, Ndayishimiye V, Dushimimana G (2020) Modelling a 30kw standalone solar powered irrigation system. IEEE PES/IAS Power Africa 1–5. https://doi.org/10.1109/PowerAfrica49420.2020.9219921
18. Rezkallah M (2021) Hardware implementation of composite control strategy for Wind-PV-Battery hybrid off-grid power generation system. Clean Technol 3(4):821–843. https://doi.org/10.3390/cleantechnol3040048
19. Raiker GA, Loganathan U, Reddy SB (2021) Current control of boost converter for PV interface with momentum-based perturb and observe MPPT. IEEE Trans Ind Appl 57(4):4071–4079. https://doi.org/10.1109/TIA.2021.3081519
20. Kumar V, Singh M (2021) Derated mode of power generation in PV system using modified perturb and observe MPPT algorithm. J Mod Power Syst Clean Energy 9(5): 1183–1192. https://doi.org/10.35833/MPCE.2019.000258
21. Gupta K (2021) Effect of various incremental conductance MPPT methods on the charging of battery load feed by solar panel. IEEE Access 9:90977–90988. https://doi.org/10.1109/ACCESS.2021.3091502

22. Bellahirich S, Mezghani D, Mami A (2021) Design and implementation of an intelligent anfis controller on a raspberry pi nano-computer for photovoltaic pumping intended for drip irrigation. Energies 14(17). https://doi.org/10.3390/en14175217
23. Shchur I, Lis M, Biletskyi Y (2021) Passivity-based control of water pumping system using bldc motor drive fed by solar pv array with battery storage system. Energies 14(23):1–25. https://doi.org/10.3390/en14238184
24. Roekrai NC, Khan-Ngern W (2021) Proposed model of STEM associated with social issues: case study on tapioca drip irrigation system. In: 6th Int. STEM Educ. Conf. iSTEM-Ed. 10–13. https://doi.org/10.1109/iSTEM-Ed52129.2021.9625089
25. Mujoo S (2021) Smart irrigation system using IoT based control valve. In: Asian Conf. Innov. Technol. ASIANCON 1–5. https://doi.org/10.1109/ASIANCON51346.2021.9544988
26. Lunaria MAR, Vallestero GC, Castro RCC, Cayanan RAF (2021) Solar powered automated drip irrigation system using particle swarm optimization, In: 2021 IEEE 13th Int. Conf. Humanoid, Nanotechnology, Inf. Technol. Commun. Control. Environ. Manag. HNICEM 2021, 2021, https://doi.org/10.1109/HNICEM54116.2021.9732025
27. Chen M, Jin Z, Fu T, Luo L, Li Y (2021) Water-saving drip irrigation system for photovoltaic power generation based on Internet of Things. In: IEEE Int. Conf. Comput. Sci. Electron. Inf. Eng. Intell. Control Technol. 233–236. https://doi.org/10.1109/CEI52496.2021.9574523
28. Riquelme-Dominguez JM, Gonzalez-Longatt FM, Martinez S (2022) Decoupled Photovoltaic Power Ramp-rate Calculation Method for Perturb and Observe Algorithms. J Mod Power Syst Clean Energy 10(4): 932–940. https://doi.org/10.35833/MPCE.2021.000603
29. Raza F (2022) The socio-economic impact of using photovoltaic (PV) energy for high-efficiency irrigation systems: a case study. Energies 15(3):1–21. https://doi.org/10.3390/en15031198
30. Mandourarakis I, Gogolou V, Koutroulis E, Siskos S (2022) Integrated maximum power point tracking system for photovoltaic energy harvesting applications. IEEE Trans Power Electron 37(8):9865–9875. https://doi.org/10.1109/TPEL.2022.3156400
31. Nguyen-Duc T, Le-Viet T, Nguyen-Dang D, Dao-Quang T, Bui-Quang M (2022) Photovoltaic array reconfiguration under partial shading conditions based on short-circuit current estimated by convolutional neural network. Energies 15(17): 6341. https://doi.org/10.3390/en15176341
32. Vankadara SK, Chatterjee S, Balachandran PK, Mihet-Popa L (2022) Marine predator algorithm (MPA)-Based MPPT technique for solar PV systems under partial shading conditions. Energies 15(17):6172. https://doi.org/10.3390/en15176172
33. Miran S, Tamoor M, Kiren T, Raza F, Hussain MI, Kim JT (2022) Optimization of standalone photovoltaic drip irrigation system: a simulation study. Sustainability 14(14):8515. https://doi.org/10.3390/su14148515
34. Abdellatif A, Mubarak H, Ahmad S, Ahmed T, Shafiullah GM, Hammoudeh A, Abdellatef H, Rahman MM, Gheni HM (2022) Forecasting photovoltaic power generation with a stacking ensemble model. Sustainability 14(17):11083. https://doi.org/10.3390/su141711083
35. Tu CS, Tsai WC, Hong CM, Lin WM (2022) Short-Term solar power forecasting via general regression neural network with grey wolf optimization. Energies 15(18):6624. https://doi.org/10.3390/en15186624
36. Hajiabadi M, Farhadi M, Babaiyan V, Estebsari A (2020) Deep learning with loss ensembles for solar power prediction in smart cities. Smart Cities 3(3):842–852. https://doi.org/10.3390/smartcities3030043
37. Crijns-Graus W, Wild P, Amineh MP, Hu J, Yue H (2022) International comparison of research and investments in new renewable electricity technologies: a focus on the European union and China. Energies 15(17):6383. https://doi.org/10.3390/en15176383
38. Hua LG, Memon QA, Shaikh MF, Shaikh SA, Rahimoon RA, Shah SHH, Qadir A (2022) Comparative analysis of power output, fill factor, and efficiency at fixed and variable tilt angles for polycrystalline and monocrystalline photovoltaic panels—the case of sukkur IBA university. Energies 15(11):3917. https://doi.org/10.3390/en15113917
39. Alimi OA, Meyer EL, Olayiwola OI (2022) Solar photovoltaic modules' performance reliability and degradation analysis—a review. Energies 15(16):5964. https://doi.org/10.3390/en15165964

40. Bouaguel W, Alsulimani T (2022) Understanding the factors influencing consumers' intention toward shifting to solar energy technology for residential use in Saudi Arabia using the technology acceptance. Sustainability 14:11356. https://doi.org/10.3390/su141811356
41. Islavatu S, Kumar P, Kumar A, Salkuti SR (2022) Power quality analysis by H-Bridge DSTATCOM control by Icosθ and ESRF SOGI-FLL methods for different industrial loads. Smart Cities 5(4):1590–1610. https://doi.org/10.3390/smartcities5040081
42. Bhukya L, Kedika NR, Salkuti SR (2022) Enhanced maximum power point techniques for solar photovoltaic system under uniform insolation and partial shading conditions: a review. Algorithms 15(10):365. https://doi.org/10.3390/a15100365
43. Rout KK, Mishra DP, Mishra S, Patra S, Salkuti SR (2023) Perturb and observe maximum power point tracking approach for microgrid linked photovoltaic system. Indones J Electr Eng Comput Sci 29(2): 635–643. http://doi.org/https://doi.org/10.11591/ijeecs.v29.i2.pp635-643
44. Rout KK, Mishra DP, Mishra S, Rishitha GT, Salkuti SR (2023) Application of fuzzy logic technique to track maximum power point in photovoltaic systems. Indones J Electr Eng Comput Sci 29(1): 101–109. http://doi.org/https://doi.org/10.11591/ijeecs.v29.i1.pp101-109

Analysis of Various Speed Control Methods for PMSM Drive-Based Electric Vehicles

A. Viswa Teja, W. Razia Sultana, and Surender Reddy Salkuti

Abstract This study discusses various permanent magnet synchronous motor speed control strategies. As opposed to other motors used in drives, permanent magnet synchronous motor is typically employed synchronous type of electric motors with feedback encoders and industries in the view of the fact that the application requires high-quality speed regulators featuring apex precision, starting torque coefficient, lower ripple torque, in addition to a large power density and high reliability in the design and deployment. There are numerous methods for controlling speed, and they vary depending on the permanent magnet synchronous motor controller utilised and the software/hardware used to implement them. This work emphasises the investigation of various control strategies with regard to speed control as well as execution.

Keywords Sensorless control · Field-oriented control · Artificial neural network controller · Fuzzy logic control · Permanent magnet synchronous motor

Nomenclature

ACIM	AC Induction Motor
PI	Proportional Integral
PID	Proportional Integral Derivative
FL	Fuzzy Logic

A. Viswa Teja · W. Razia Sultana
School of Electrical Engineering, VIT University, Vellore, Tamil Nadu, India
e-mail: viswateja.anjuru2020@vitstudent.ac.in

W. Razia Sultana
e-mail: wraziasultana@vit.ac.in

S. R. Salkuti (✉)
Department of Railroad and Electrical Engineering, Woosong University, Daejeon 34606, Republic of Korea
e-mail: surender@wsu.ac.kr

S. R. Salkuti et al. (eds.), *Power Quality in Microgrids: Issues, Challenges and Mitigation Techniques*, Lecture Notes in Electrical Engineering 1039,
https://doi.org/10.1007/978-981-99-2066-2_14

IPM	Interior Permanent Magnet
FOC	Field Oriented Control
MMF	Magneto Motive Force
BLDC	Brushless DC motor
MARS	Multivariate Adaptive Regression Splines
EMF	Electro-Motive Force
EKF	Extended Kalman Filter
SNR	Signal-to-Noise Ratio
ANN	Artificial Neural Network
FLC	Fuzzy Logic Controller
FPGA	Field Programmable Gate Array
PMSM	Permanent Magnet Synchronous Motor
FW	Field Weakening

1 Introduction

PMSMs are peak-performance electromechanical devices, due to their great performance capacity, really supplant to conventional DC servomotors and fractional horsepower induction machines. In many applications, the performance of PMSM that calls for the quick transient reactions has to be increased. Fast acceleration and deceleration, full torque control at zero speed, and trouble-free spinning over the whole, particularly high motor control features include the speed limit of the motor [1]. Numerous speed control methods have been developed in an effort to find PMSM drive control options with high speed as well as precise torque response features in order to maximise the speed-control performance of the PMSM system with various disturbances and uncertainties [2].

Instead of utilising the rotor's windings to create a magnetic field, a PMSM uses permanent magnets. A brushless DC motor and an AC induction motor hybrid can be used to describe the PMSM (BLDC) [3]. The rotor designs of PMSM are akin to BLDC motors with PMs in their rotors. However, their stator design is similar to that of its ACIM [4], where the windings are made in a fashion that creates a sinusoidal flux density in the machine's airgap. They, therefore, function best when powered by sinusoidal waves [5]. There are two different types of PMSM based on how PMs are mounted. One is an inside permanent magnet, while the other is a surface-mounted PMSM. The most prevalent kind in PMSM is Interior Permanent Magnet (IPM) [6]. High-efficiency and high-performance motor drives are frequently employed with PMSM.

PMSM picked them for variable speed applications because of their outstanding performance/cost ratio. Recent studies have shown that the PMSM is mostly used in peak torque per current ratio and high-end applications, such as industrial machines as well as robots [7], which require speed control systems with high precision, outstanding reliability, flexibility, as well as efficiency in the creation and deployment processes. In several PMSM-based drives, the traditional PI and PID controllers were employed as speed regulators [8]. A realistic PMSM is a dynamic system; therefore, there are numerous disturbances and uncertainties as well as parameter variations. These disturbances can be caused by unmodelled dynamics, parameter variations, friction force, and load disturbances, among other things [9].

Therefore, in order to achieve appropriate adjustment performances under a variety of operating settings, a proper control approach is required. However, because of the complexity of the PMSM servo system, it is exceedingly challenging to discern system features and active control parameters in actual problem-solving time. To improve the system's control performances for such a range of disturbances and uncertainties, several nonlinear control techniques of PMSM have been developed. These techniques involve adaptive control [10], I/O linearisation control, robust control [11], back-stepping control, NN control, FLC and finite-time control sliding mode control [12].

Each of these control strategies has been created to enhance PMSM performance under various load conditions and speed ranges. This chapter presents the FOC, sensorless regulation, NN controller and FLC speed regulator for the Drive system. These controllers were selected to have a rapid settling time, little overshoot, and 0% steady-state error. Comparatively to the current control methods, these methods may ensure closed-loop stability as well as reliable control of the PMSM drive systems. To validate the system outcomes, a MATLAB environment can be used. The dynamic performance of the PMSM drive system was investigated under varying loads and parameter settings.

2 Modelling of Permanent Magnet Synchronous Motor

A number of control schemes have been created due to the need for drives with inconsistent speed in both low as well as high-powered applications in order to fulfil the requirements of each application and improve the working strength of PMSM at varying loads, speed circumstances. To employ the various PMSM speed control methods, the authentic non-linear numerical method must be given a linear form [13]. The computational formulation of the PMSM drive may be represented by such equations in a d-q reference frame. At every instant t, the spinning stator MMF forms an angle with the rotor d-axis, and the revolving rotor d-axis makes an angle with the constant stator phase axis [14]. The PMSM modelling on the rotor frame of reference was created using the following assumptions:

- The saturation is ignored.
- Sinusoidal distribution of windings is used.
- Sinusoidal EMF is produced.
- Hysteresis losses and eddy currents are insignificant.

There aren't any field current fluctuations in PMSM. The rotor (dq) frame both mechanical as well as electrical equations for the PMSM are,

$$V_d^s = i_d^s R_d + d'\left(\varphi_d^s\right) - \omega_r \varphi_q^s \tag{1}$$

$$V_q^s = i_q^s R_q + d'\left(\varphi_d^s\right) + \omega_r \varphi_d^s \tag{2}$$

$$\varphi_d^s = L_d^s i_d^s + \varphi_{fr} \tag{3}$$

$$\varphi_q^s = L_q^s i_q^s \tag{4}$$

$$\varphi_{fr} = L_{md} + i_{fr} \tag{5}$$

where V_d^s, V_q^s are stator voltages of dq axis, φ_d^s, φ_q^s are flux linkages, φ_{fr} is the field flux, i_d^s, i_q^s are stator currents of dq axis, L_d^s, L_q^s are stator inductances of dq axis, L_{md} rotor inductance of d-axis, i_{fr} is the current through the field.

In matrix form,

$$\begin{bmatrix} V_d^s \\ V_q^s \end{bmatrix} = \begin{bmatrix} R_d + d' L_d^s & -\omega_r L_q^s \\ \omega_r L_d^s & R_q + d' L_q^s \end{bmatrix} \begin{bmatrix} i_d^s \\ i_q^s \end{bmatrix} + \begin{bmatrix} 0 \\ \omega_r \varphi_{fr} \end{bmatrix} \tag{6}$$

$$T_e = \frac{3p}{4}\left(\varphi_d^s i_d^s - \varphi_q^s i_q^s\right) \tag{7}$$

The PMSM's torque equation is similar to that of a standard DC motor. Consequently, it may make machine control very simple and effective. In the rotor-based dq coordinates system, *Id* and *iq* components are separated from the motor currents. The condition when the fluxes of the rotor and stator are transverse corresponds to the value of $i_d = 0$, which yields the maximum torque [15]. The drive then functions similarly to an armature current regulated DC motor.

3 Strategies and Methods for Control

The different control approaches aim to achieve high performance and high efficiency control by controlling rotor speed, torque, and position. This analysis has highlighted a wide range of speed control methods. One of the fundamental control methods used to regulate the speed of PM synchronous motors is the vector control approach. FOC is another name for the vector control approach. The approach specifically suggests measuring motor currents and converting them into a coordinate system that rotates with the machine's rotor. Another technique for speed control is sensorless control [16]. A rotor position detector upon the shaft can sometimes reduce the resilience and dependability of the overall system, even though it is typically required to effectively manage the performance of the PMSM. As a result, the objective is not to directly detect the rotor position using this mechanical sensor, but rather to approximate it indirectly. One of the techniques used to get around the base speed restriction is called the field or flux weakening control approach. MPC is also one of the most beneficial state-of-the-art control approaches in firm applications. PMSM also employs artificial intelligence-based techniques for speed control, such as FLC, NN, and Adaptive Neuro-fuzzy inference hybrid systems. Other regulating methods, such as MRAS, or hybrid methods that combine FL with sensor-free conversely fuzzy-neuro methods may also be used [17].

Based on the existing approaches and methodologies for hardware design, the speed modulation schemes can be broadly categorised as smart control techniques (such as FLC, NN), sensorless speed control scheme, hybrid approaches, and other techniques/methods. In this study, a review of several control techniques is highlighted with regard to speed control and the use of a speed controller.

3.1 Field-Oriented Control

By converting the stator phase currents from a fixed reference frame to the torque and flux generating current components in a rotating reference frame, much like a dc machine, FOC is a technique used to separately decouple control of flux as well as torque. It enhances PMSM's steady-state and transient responses [18].

$$T_e = \frac{3p}{4}\left(\frac{1}{2}(L_d^s - L_q^s)i_s^2 \sin 2\delta + L_{md} i_{fr} \sin\delta\right) \tag{8}$$

When choosing the $\delta = \frac{1}{2}$ for field-oriented control, we obtain the torque expression as,

$$T_e = \frac{3p}{4}\left(L_{md} i_{fr} i_s\right) \tag{9}$$

Stator current has two components,

$$i_s^* = i_T^* i_F^* \tag{10}$$

where i_T^* is the torque component and i_F^* is the flux component.

The PMSM drive with vector-controlled speed is seen in Fig. 1. Actual speed is contrasted with the reference speed. The magnitude of the torque, which is adjusted according to speed, will be provided by the PI controller. Up to the base speed, the torque won't change. The torque will then be reduced for the region where the field is deteriorating. This is multiplied by the FW performance parameter to provide the testimonial torque Te*. The torque component of the current IT is obtained by dividing this Te* by a constant. Similar to how the flux component is obtained, the magnetising flux (*m*) is created by multiplying the function generator's output by the constant *Kf*. This magnetic flux is utilised to produce an IF (Flux Component of Current) in a function block. As a result, the production of reference stator current is transformed from the IT and IF. Using only a magnitude and angle resolver, IT and IF are transformed into polar units known as amplitude IS and phase shift (delta prime), which is produced by adding rotor angle *r* to it. To create the stator current reference, Ias*, Ibs*, and Ics*, the currents Is* and *s* are fed into the stator current generator's input. These currents must be in the machine's position [19]. Therefore, the current controller is used to run the inverter. The inaccuracy is determined by comparing the reference current with the actual currents received by the hall sensor in the motor. This error is sent to the hysteresis controller, which then controls the inverter. For closed-loop control, the synchronous machine just needs position feedback. Therefore, the position will be measured by an encoder that is installed on the rotor shaft and utilised to drive the inverter. In order to get the real rotor position in electrical angle *r*, the position detector (rotor position encoder) multiplies the mechanical position rm by the pole pair. In order to determine the mechanical speed rm, this location is maintained throughout another computing block (speed calculator). This is used as feedback to compare the speed to the standard. The speed controller in this case is a PI controller [20]. The block diagram has the ability to regulate speed both above and below the standard speed. The magnetising flux will be persistent below the base speed. *M* stands for the flux that magnetises. This is the outcome of the stator and the field. FOC has the following benefits:

- Fast transient responses.
- Excellent steady-state and dynamic performance.
- During start-up, it's having low currents and high torque.
- Efficiency will be apex.
- Broad speed range through FW.
- Control over torque and flux is independent, comparable to a DC motor.

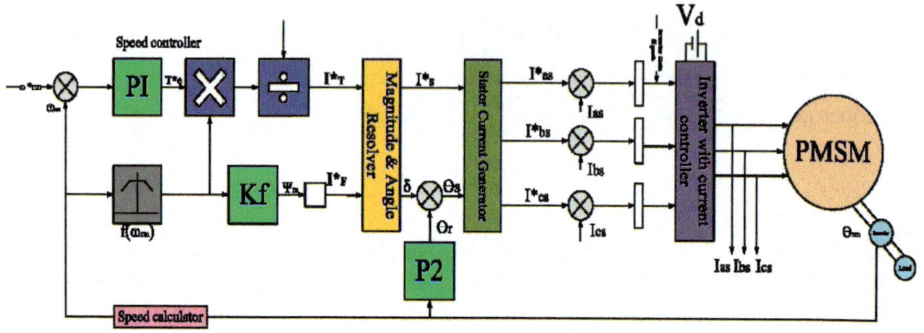

Fig. 1 Vector speed control of PMSM

3.2 Sensorless Control

When used without damper winding, the process of driving a PMSM requires position sensors in the rotor shaft. The creation of position measurement tools is required due to the requirement of knowing the rotor position. It is essential to estimate the rotor position precisely since if done incorrectly, the motor's starting torque may be lowered and it may briefly revolve in the wrong direction. The following tools are available for measuring position: potentiometers, resolvers, optical encoders, and linear variable differential transformers. The most widely used of these to measure the position in motors are encoders and revolvers. Numerous methods for sensorless PMSM drive operation have been reported because sensors have a number of drawbacks, including decreased reliability, high cost, size, and weight, as well as increased drive system difficulty.

The existence of the detector on shaft may significantly lessen the drive's overall ruggedness in many industrial installations. Others may see a large increase in the expense of the drive. In recent years, there has been a major advancement in the techniques used to locate the rotor in electric machines using just voltage and current measurements. Typically, these techniques are referred to as sensor-less, encoder-less, or self-sensing. To execute an efficient MMF control, which is controlled by stator current control, the PMSM torque control has to be aware of the rotor position [21]. Additionally, the speed signal is necessary for speed control. Induction motor, BLDC, and PMSM motor drives without position sensors or speed sensors have attracted a lot of study attention recently.

The back EMF, the position of the inductances acting on them, flux linkage detection, etc. may all be used to detect the position of the rotor in a PMSM. Model-based adaptive observers are a widely utilised approach in the many sensorless control strategies used for PMSMs. The observer-based estimate has upper hand over the alternatives since it disregards dedicated conditions and the steadiness of the assessment can be verified using formal control theory. Schematic diagram of a sensorless control design is presented in Fig. 2.

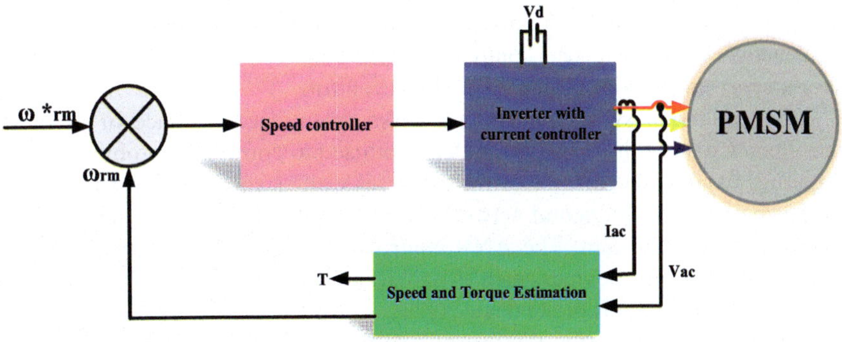

Fig. 2 PMSM drive sensor-less operation

The EKF is a useful and computationally efficient choice for the online estimation of the rotor speed as well as position of a PMSM. The EKF is a fantastic predictor in the least-square interpretation for predicting the states of dynamic quasi-system applications. The theoretical underpinnings and digital implementation of EKF have undergone extensive study.

For low-speed EKF sensor-less control of PMSM drives, a novel technique known as adjustable DC bus voltage was introduced in 2009. Assumed here is that there is no correlation between measurement noise and disturbance noise. Inaccuracies in measurement and modelling are accounted for by the noise sources. A mathematical model is used to anticipate the states at the time of computations in the first stage, and a feedback correction system is used to continually correct the predicted states in the second stage. By adding a term to the anticipated states that are acquired in the first step, this technique employs real measured states [22].

The weighted difference of the output signals from measurement and calculation is contained in the extra term. EKF provides the best output value at the following input instant based on the divergence achieved from the predicted value. The PM flux linkage error has a significant impact on the EKF estimation. The poor performance at low speed (5 Hz), however, is at least one significant limitation of the EKF application to sensorless drives that has not yet been resolved. In addition to the estimating approach, the system observability has a significant impact on the accuracy of system error estimations. Voltage and current are the EKF filter's measurements.

Current data moving at low speed is unaffected. The harmonic component has an apex ratio in the voltage data, which is one issue. Due to information on system errors that are affected by rapidly changing random errors, this estimation changes significantly. Because of this, a voltage with a greater signal-to-noise (SNR) ratio would function better at low speeds. Increasing the signal-to-noise ratio may be achieved easily by lowering the DC bus's voltage level.

4 Neural Network Control

One of the most difficult tasks in the field of control systems now involves using neural networks to the speed control of PMSM motors. Figure 3 shows PMSM motor's generalised block diagram for speed control.

In the traditional system, neural networks are utilised in place of PI or PID controllers. Creating the error signal that is utilised to correct the problem involves comparing the intended set values for torque as well as speed with the true values. This has been used with neural networks to predict the flux, motor's torque or flux angle at any given instant. The ANN controller produced the control signals for the firing of electronic devices, such as a SCR, power transistors, current controlled device, or any other appropriate devices [23].

Here, we briefly describe the use of two different artificial intelligence-based estimators, such as ANN and fuzzy neural networks. For the estimate of the rotor location as well as rotor angle, a supervised multi-layer feed-forward ANN may be

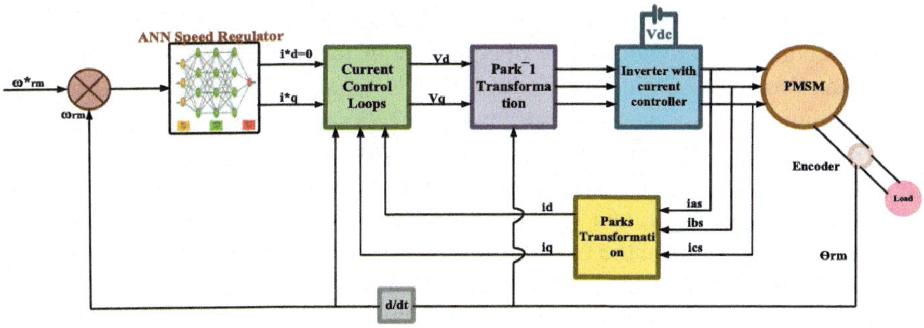

Fig. 3 ANN-based speed control of PMSM

trained via back-propagation training. The square of the error allying the expected and true ANN output is minimised by utilising the backpropagation technique [24]. Then, real-time applications may take advantage of the trained ANN. An ANN of this type has hidden layers in addition to input and output layers. Although it should be noted as a general rule that the total number of hidden layers is often 1 or 2 in electrical engineering applications, the hidden layers to be employed are not known in advance; this must be decided by trial and error method. Furthermore, it is again determined by trial and error how many hidden nodes are present in the hidden layers as this information is not known beforehand.

The kind of PMSM (SMPM or IPMSM) determines the number of input nodes. It is feasible to build a neural network of this type that also uses the machine's stator currents as inputs, but each stator current must have two inputs, one for the present and one for the past. Such a method has the benefit of not requiring a mathematical model of the machine, in contrast to other traditional procedures. There are no recommendations for the selection of the number of hidden layers as well as nodes in the ANN-based method, making it challenging to connect the network's structure to the physical process [25]. By utilising a fuzzy-neural estimator, it is feasible to get around some of the drawbacks of the ANN-based technique. Fuzzy neural systems combine the benefits of neural networks and fuzzy logic. The fundamental benefit of a fuzzy-neural network is that the number of layers and nodes is known [20].

5 Fuzzy Logic Controller

Compared to traditional PI or PID controllers, FLC are more resistant to changes in system plant parameters and are better at noise rejection. The fuzzy adaptive strategies, which represent the knowledge and experience of experts, are more similar to them. In high-performance drive applications, fuzzy controllers are particularly competitive as contemporary smooth control techniques get more complicated [26]. As a result, adaptive fuzzy controllers often have a higher system performance/ complexity ratio [22].

Since it doesn't need to understand the mathematical model of the plant and has advantages that include steady operation both for nonlinear and linear plants, a technique for representing information that is suitable for context-sensitive and challenging-to-describe ideas. Fuzzy logic is a collection of well-defined rules that may be used with the membership function to solve a particular electrical task for an electrical issue [27]. There are several controllers used to control the speed of PMSM. Traditional P, PI, and PID controllers are unable to handle changes in the system's parameters since the level of fine-tuning required is minimal. Additionally, these controllers' performance is impacted by changes in physical characteristics like as saturation, noise and temperature, etc.

For PMSM drive applications, many control systems employ adaptive controllers, which can only monitor linear systems. Therefore, a FL-based controller may be utilised to manage complicated non-linear features and obtain effective, accurate, and quick solutions. The block diagram of the fuzzy control system is shown in Fig. 4 and includes defuzzification, rule database, fuzzification, and rule inference, respectively [28]. The motor speed is controlled by FLC with a straightforward topology.

According to Fig. 3, the fuzzy PI controller was put into practice. There are seven triangle memberships that stand in for the inputs Error (e), Change of Error (e). By translating the actual range of error (e) as well as change of error (De) to FLC inputs, membership functions are shown in Fig. 6, which demonstrates the normalisation.

Real-time operations are necessary for a number of fuzzy control applications, including those involving physical systems. Large quantities of fuzzy logic may be implemented on a single integrated circuit using programmable logic devices like FPGA with higher density. A direct adaptive controller directly includes linguistic fuzzy control principles into the FLS it employs as a controller [29]. As illustrated

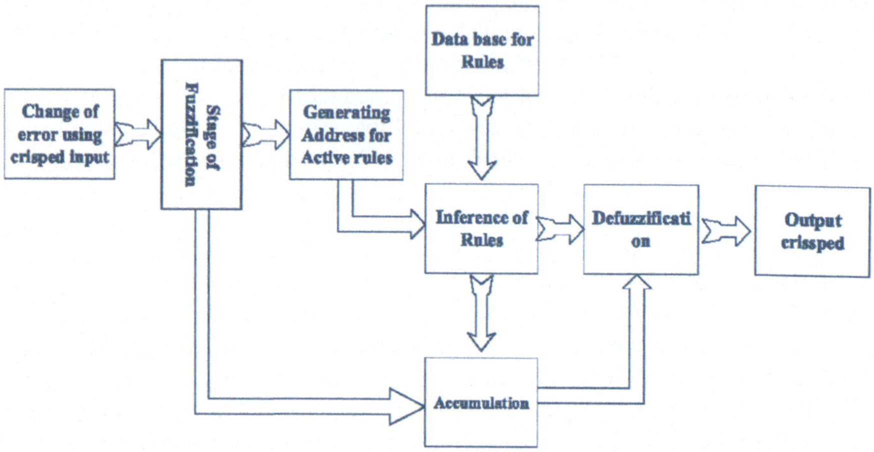

Fig. 4 Architecture of FLC

in Fig. 5, the direct adaptive fuzzy control method we employed in this section comprises of an FLC, a reference model, and an adjusting mechanism.

It is the controller's primary component, and it is divided into the following sections. Fuzzification, Centre of Gravity Defuzzification, Active Rules Address Generator, and Rule Inference. Equations (9) and (10) describe the tracking error (e) and change of error (e), respectively. Figure 6 depicts the normalisation of these membership functions.

$$e(k) = \sigma_{\text{ref}}\sigma_{\text{act}} \tag{11}$$

$$\Delta e(k) = e(k) - e(k-1) \tag{12}$$

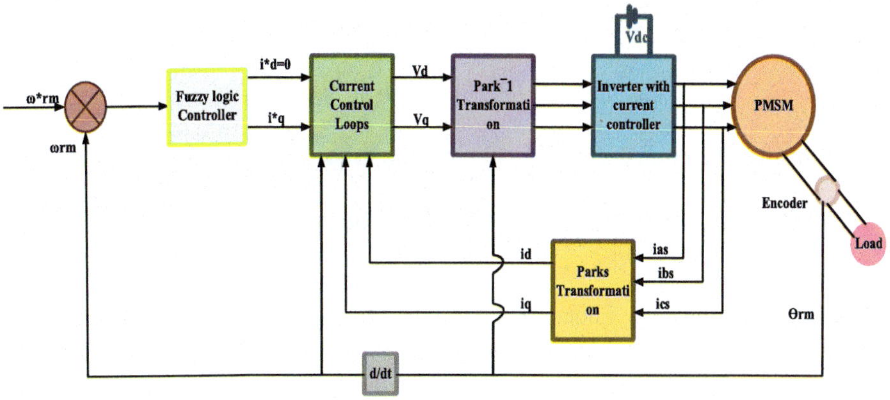

Fig. 5 AFLC block diagram

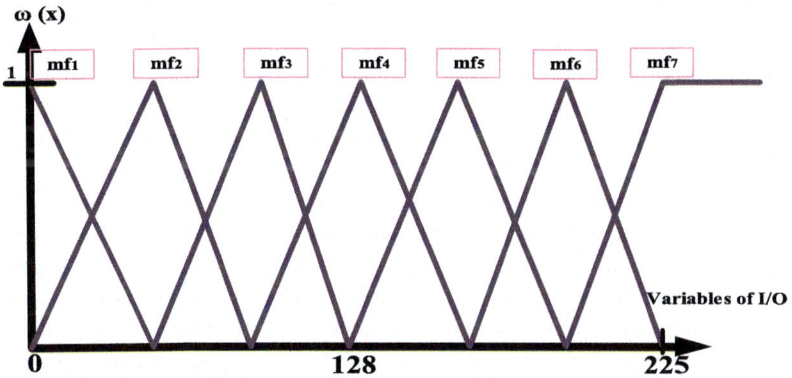

Fig. 6 Membership function for input and output

To build an initial control solution, some information about appropriate plant control actions is used rather than knowledge about the plant (PMSM motor) [30]. The control knowledge is expected to be stated as a collection of "m" fuzzy IF–THEN rules with the following structure:

$$\text{IF } e \text{ is } A_i \text{ and } \Delta e \text{ is } B_i \text{ THEN } \cup_f \text{ is } Z_i \tag{13}$$

$U_f = Z_i$ (13) with $I = 1, 2, 3$, etc. if $e = A_i$ and $e = B_i$. There are seven fuzzy sets and 49 fuzzy control regulations for the two inputs as well as one output fuzzy system U_f for each linguist values A_i, B_i, and Z_i. At any given moment, a maximum of four regulations will be in effect.

$$\mu_z(\cup_f) = \max_{(j=1,i+1)}^{\min\left(\min\left(\mu_{A_j}^{(e)},\mu_{B_j}^{(e)}\right)\right)\mu_{Z_j}^{(\cup_f)}} \tag{14}$$

$$\mu_z(\cup_i) = \left(\mu_{A_j}^{(e)}, \mu_{B_j}^{(\Delta e)}\right) \tag{15}$$

Defuzzifier then transforms Eq. (11) into the following expression:

$$\cup_f = \frac{\sum_{i=1}^{n} \mu(\cup_i) C_i}{\sum_{i=1}^{n} \mu(\cup_i)} \tag{16}$$

where C_i represents the values of the output MF centres and (U_i) represents the values of the output membership function (MF). The PMSM motor then receives this result.

6 Conclusion

This chapter reports research on several permanent magnet synchronous motor speed-control strategies. High-performance applications typically use permanent magnet synchronous motor in servo drives and industries, thus these applications call for speed regulator with high accuracy, apex performance, flexibility, and efficiency. The many controller types that are available, like Fuzzy logic, neural networks, as well as conventional proportional integral derivative control algorithms, are all examined in the research. The majority of the research under review lacked a thorough modelling of the control system, nevertheless. Hardware implementation, on the other side, differs from the design and simulation of the controller. This has an influence on the accuracy, potency, and efficacy of the permanent magnet synchronous motor's control strategy. As a result, the objective for the future is to work towards filling the aperture in the literature by evolving a completely new system that will do away with the stumbling of the existing ones.

Acknowledgements This research work was supported by "Woosong University's Academic Research Funding—2023".

References

1. Abdellah B, Abdeldjebar H, Medjdoub K (2022) An application for nonlinear control by input-output linearization technique for pm synchronous motor drive for electric vehicles. Int J Power Electron Drive Syst 13(4):1984–1992. https://doi.org/10.11591/ijpeds.v13.i4.pp1984-1992
2. Mwasilu F (2020) Direct predictive speed control of salient PMSM drives in constant torque and constant power regimes for electric vehicles applications. Tanzan J Eng Technol 39(2):127–143. https://doi.org/10.52339/tjet.v39i2.700
3. Madanzadeh S, Abedini A, Radan A, Ro JS (2020) Application of quadratic linearization state feedback control with hysteresis reference reformer to improve the dynamic response of interior permanent magnet synchronous motors. ISA Trans 99:167–190. https://doi.org/10.1016/j.isatra.2019.08.067
4. Vuddanti S, Shastri S, Salkuti SR (2022) Design of permanent magnet brushless DC motor for electric vehicle traction application. In: Kumar S, Singh B, Singh AK (eds) Recent advances in power electronics and drives. Lecture notes in electrical engineering, vol 852. Springer, Singapore. https://doi.org/10.1007/978-981-16-9239-0_24
5. Shanthi R, Kalyani S, Devie PM (2021) Design and performance analysis of adaptive neuro-fuzzy controller for speed control of permanent magnet synchronous motor drive. Soft Comput 25(2):1519–1533. https://doi.org/10.1007/s00500-020-05236-5
6. Yang Z, Shang F, Brown IP, Krishnamurthy M (2015) Comparative study of interior permanent magnet, induction, and switched reluctance motor drives for EV and HEV applications. IEEE Trans Transp Electrif 1(3):245–254. https://doi.org/10.1109/TTE.2015.2470092
7. Justo JJ, Mwasilu F, Kim EK, Kim J, Choi HH, Jung JW (2017) Fuzzy model predictive direct torque control of IPMSMs for electric vehicle applications. IEEE ASME Trans Mechatron 22(4):1542–1553. https://doi.org/10.1109/TMECH.2017.2665670
8. Salkuti SR (2022) Emerging and advanced green energy technologies for sustainable and resilient future grid. Energies 15(18):6667. https://doi.org/10.3390/en15186667
9. Sandulescu P, Meinguet F, Kestelyn X, Semail E, Bruyere A (2014) Control strategies for open-end winding drives operating in the flux-weakening region. IEEE Trans Power Electron 29(9):4829–4842. https://doi.org/10.1109/TPEL.2013.2283107
10. Wang K, Wu W (2022) Simplified current control strategy for permanent-magnet synchronous motors with rotor flux distortion. IEEE Trans Ind Electron 69(4):3407–3417. https://doi.org/10.1109/TIE.2021.3075861
11. Dong H, Xiao M, Li Z, Zhang S, Chen J (2021) FPGA-based high-performance double-loop PDF control strategy for PMSM. IEEE Access 9:59822–59834. https://doi.org/10.1109/ACCESS.2021.3071130
12. Han Z, Liu J (2021) Comparative analysis of vibration and noise in IPMSM considering the effect of MTPA control algorithms for electric vehicles. IEEE Trans Power Electron 36(6):6850–6862. https://doi.org/10.1109/TPEL.2020.3036402
13. Liu S, Liu C, Huang Y, Xiao Y (2022) Direct modulation pattern control for dual three-phase PMSM drive system. IEEE Trans Ind Electron 69(1):110–120. https://doi.org/10.1109/TIE.2021.3053880
14. Chaoui H, Khayamy M, Okoye O, Gualous H (2019) Simplified speed control of permanent magnet synchronous motors using genetic algorithms. IEEE Trans Power Electron 34(4):3563–3574. https://doi.org/10.1109/TPEL.2018.2851923
15. Bilewski M, Fratta A, Vagati A, Villata F, Giordano L (1993) Control of high-performance interior permanent magnet synchronous drives. IEEE Trans Ind Appl 29(2):328–337. https://doi.org/10.1109/28.216540

16. Koganti S, Koganti KJ, Salkuti SR (2022) Design of multi-objective-based artificial intelligence controller for wind/battery-connected shunt active power filter. Algorithms 15(8):256. https://doi.org/10.3390/a15080256
17. Chai S, Wang L, Rogers E (2013) Model predictive control of a permanent magnet synchronous motor with experimental validation. Control Eng Pract 21(11):1584–1593. https://doi.org/10.1016/j.conengprac.2013.07.008
18. Wu BF, Lin CH (2019) Ad neural predictive control for permanent magnet synchronous motor systems with long delay time. IEEE Access 7:108061–108069. https://doi.org/10.1109/ACCESS.2019.2932746
19. Meher H, Panda AK, Ramesh T (2013) Performance enhancement of the vector control based permanent magnet synchronous motor drive using hybrid PI-Fuzzy logic controller. In: Students conference on engineering and systems (SCES), pp 1–6. https://doi.org/10.1109/SCES.2013.6547542
20. Nguyen NK, Wira P, Flieller D, Ould Abdeslam D, Merckle J (2011) Harmonics identification with artificial neural networks: application to active power filtering. Int J Emerg Electr Power Syst 12(5). https://doi.org/10.2202/1553-779X.2783
21. Pereyra V, Scherer G, Wong F (2006) Variable projections neural network training. Math Comput Simul 73(1–4):231–243. https://doi.org/10.1016/j.matcom.2006.06.017
22. Patan K, Patan M (2020) Robustness of neural-network-based nonlinear iterative learning control. In: 16th IEEE International conference on control, automation, robotics and vision, vol 2, pp 578–583. https://doi.org/10.1109/ICARCV50220.2020.9305505
23. Fasil M, Antaloae C, Mijatovic N, Jensen BB, Holboll J (2016) Improved dq-axes model of PMSM considering airgap flux harmonics and saturation. IEEE Trans Appl Supercond 26(4). https://doi.org/10.1109/TASC.2016.2524021
24. Rocha-Osorio CM, Solís-Chaves JS, Casella IRS, Capovilla CE, Azcue Puma JL, Sguarezi Filho AJ (2017) GPRS/EGPRS standards applied to DTC of a DFIG using fuzzy—PI controllers. Int J Electr Power Energy Syst 93:365–373. https://doi.org/10.1016/j.ijepes.2017.05.033
25. Tarczewski T, Grzesiak LM (2016) Constrained state feedback speed control of PMSM based on model predictive approach. IEEE Trans Ind Electron 63(6):3867–3875. https://doi.org/10.1109/TIE.2015.2497302
26. Formentini A, Trentin A, Marchesoni M, Zanchetta P, Wheeler P (2015) Speed finite control set model predictive control of a PMSM fed by matrix converter. IEEE Trans Ind Electron 62(11):6786–6796. https://doi.org/10.1109/TIE.2015.2442526
27. Sutikno T, Idris NRN, Jidin A (2014) A review of direct torque control of induction motors for sustainable reliability and energy efficient drives. Renew Sustain Energy Rev 32:548–558. https://doi.org/10.1016/j.rser.2014.01.040
28. Wang WC, Liu TH, Syaifudin Y (2016) Model predictive controller for a micro-PMSM-based five-finger control system. IEEE Trans Ind Electron 63(6):3666–3676. https://doi.org/10.1109/TIE.2016.2543179
29. Xie G, Lu K, Dwivedi SK, Riber RJ, Wu W (2015) Permanent magnet flux online estimation based on zero-voltage vector injection method. IEEE Trans Power Electron 30(12):6506–6509. https://doi.org/10.1109/TPEL.2015.2439718
30. Korkmaz F, Cakır MF, Topaloglu I, Gurbuz R (2013) Artificial neural network based DTC driver for PMSM. Int J Instrum Control Syst 3(1):1–7. https://doi.org/10.5121/ijics.2013.3101

Short-Term Load Forecasting Using Jaya Algorithm

Papia Ray and Surender Reddy Salkuti

Abstract Load forecasting plays an essential role in the power system energy management system. Important factors that are crucial for short-term load forecasting (STLF) are unit commitment and economic allotment of the generation preservation schedule. At present, many techniques have been used for load forecasting, but Artificial Intelligence-based techniques such as Fuzzy Logic (FL) and Artificial Neural Networks (ANNs) gave better efficiency as compared to conventional techniques such as Regression and Time Series approaches. This chapter solves the STLF problem using FL, ANNs, and Jaya optimization algorithm methods and compares their errors to find out which one is more accurate for load forecasting. MATLAB, Simulink, the Fuzzy Logic Toolbox, and the ANN Toolbox are the programs utilized for this work.

Keywords Load forecasting · Fuzzy logic · Artificial neural network · Evolutionary algorithm · Jaya algorithm

Nomenclature

$f(x)$	Activation function
N	Time period
P_a	Actual load
P_f	Forecasted load
μ_A	Membership function of a fuzzy set A

P. Ray
Department of Electrical Engineering, Veer Surendra Sai University of Technology, Burla, Odisha, India
e-mail: papiaray_ee@vssut.ac.in

S. R. Salkuti (✉)
Department of Railroad and Electrical Engineering, Woosong University, Daejeon, Republic of Korea
e-mail: surender@wsu.ac.kr

S. R. Salkuti et al. (eds.), *Power Quality in Microgrids: Issues, Challenges and Mitigation Techniques*, Lecture Notes in Electrical Engineering 1039,
https://doi.org/10.1007/978-981-99-2066-2_15

$F(x)$	Objective function
W_i	Weights
X	Input term
y_k	Output of *k*th layer
Θ_i	Bias of neuron *i*

1 Introduction

Transmitting consistent power to the consumers is the pre-eminent duty of the electrical company. In electrical distribution systems, the load consumption of the consumers varies due to human activities on a day-to-day basis. Generally, the load consumption is more during the daytime and late afternoons due to high industrial loads, and less during the late evenings and early mornings as human activities are less during that time. This prognosis of active loads in the future at different load buses in the power system is called load forecasting (LF) [1, 2]. The LF is an important part of planning the power system for all developing countries. This is required to utilize the power more efficiently, minimize generation cost, spin reserve capacity, and elevate the reliability of the power supply.

LF is almost identical to weather forecasting in which experts tell us about future weather conditions. Similarly, in LF, the load conditions can be estimated in the future, and power system planning can be done. The energy requisite of customers varies every minute as human activities vary throughout the day. This prioritizes the need for the estimation of future load demand to fulfill desired future energy requirements. Thus, LF has a major role in the planning, control, and operation of an electrical power system. LF represents the evaluation of the value of variables at any time in the future. The input variables include the historical load data which comprises time and the load consumed [3, 4]. They are determined from the typical load curve for a day. Also, weather factors like temperature, humidity, etc., affecting the load are also taken into consideration. The output is the forecasted load and by comparing both the actual and forecasted loads, one can know the error and find ways to minimize it so that energy loss can be minimized. There are 4 sorts of LF [5]:

- Very short-term LF (VSTLF): With this technique, the load is predicted for a few minutes to several hours.
- Short-term LF (STLF): The load is anticipated using this way for a period of a few hours to a few days.
- Medium-term LF (MTLF): The load is anticipated using this manner for a period of a few weeks to a few months.
- Long-term LF (LTLF): With this strategy, the load is anticipated for a time frame ranging from one year to more than one year.

The primary field of study in STLF is the hourly combined system load. Furthermore, to forecast the hourly load, STLF also involves forecasting hourly or half-hourly value and daily or weekly energy of a power system. STLF is an inherent component of the energy scheduling area. Demand scheduling of generation, transmission, and distribution helps in determining the future power sector [6]. It is required to study STLF for the future power market operation because improper load forecasting (LF) will overburden the power utilities with enormous financial costs. STLF plays an important role in deciding the work plan of power plants as well as selecting the finest production group. Balancing of an electrical network is done a day in advance based on the forecasted values provided by the demand side. Due to the behavior of various load types, the power system characteristics are highly nonlinear and quite dissimilar, which in turn creates an error in the forecast resulting in an unbalance in the system. This results in an increase in the overall cost of the power network. Forecasting of electrical loads in power systems has been executed by using different techniques [7].

1.1 Related Work

Substantial literature is available on STLF using various conventional methods as well as recently developed computing techniques. Multiple Linear Regression (MLR) is generally used as a conventional method for LF which incorporates weighted least squares estimation techniques. In this technique, the statistical linkage between load and weather parameters like temperature, humidity, etc., can be evaluated. Several regression models are proposed in reference [8]. These models use the determinants such as holidays, stochastic determinants such as average loads, and exogenous determinants such as weather. Most frequently used time series (TS) models such as Autoregressive Moving Average (ARMA), Autoregressive Integrated Moving Average (ARIMA), Autoregressive Moving Average with exogenous variables (ARMAX), and autoregressive integrated moving average with exogenous variables (ARIMAX) has been proposed in [9].

Fuzzy Logic (FL) system can identify and approximate any unknown dynamic system on the compact set to arbitrary accuracy [10–12]. The FL-based forecaster works in two stages—training and online forecasting. The historical load data are used to train the inputs and outputs in the training stage using a fuzzy rule base [13, 14]. After performing the training procedure, it will be linked with the FL controller to predict the forecasted load. Reference [15] introduced the Mamdani implication in FL approach for STLF. The rule base of the FL approach is devised based on time duration, temperature, and identical previous day load. They used MATLAB and the Simulink platform for the implementation of this method. Reference [16] proposed a forecasting technique that is focused on decreasing errors between actual and forecasted values. An FL-based STLF method that reduced forecasted error and time taken for processing has been proposed in [17]. This process involved the use of Gaussian membership function, fuzzy-based rules, and FL operation. Reference

[18] proposed an STLF method that incorporates an operational mechanism under the changed power sector framework in India. It also exhibits the influence of load and temperature changes on generation, transmission, and distribution. The LF has been implemented using the triangular membership function. An STLF approach for the holidays and various models using the FL method without any data on weather [19].

An advanced meta-heuristic optimization algorithm called Jaya which has outstanding characteristics such as easier application without any requirement of control techniques has been proposed in [20]. This method is found to be efficient in solving constrained and unconstrained optimization tasks. Jaya algorithm (JA) is widely used in many domains like mechanical engineering, ANN training [21], structural optimization [22], and power systems problems. The artificial Neural Network (ANN) model learns to carry out a required task instantly from examples using special training algorithms. A neural network (NN) approach that involves the amalgamation of linear, as well as nonlinear terms to map past load and temperature inputs to the load forecast output is proposed in [23]. Reference [24] developed NNs exhibiting nonlinear curve fitting which is nonlinear circuits. A river flow application that uses NN as a forecasting tool has been proposed in [25]. It has used a multi-layer network which is trained using the backpropagation algorithm (BPA).

A Back Propagation Neural Network (BPNN) learning technique which is implemented by interchanging the interconnection between the processors is proposed in [26]. This technique has extraordinary mapping (forecasting) abilities due to which it approximates any continuous nonlinear function. NNs extensively used for solving the STLF, fault diagnosis/fault location, economic load dispatch, and security assessment problems of power systems [27]. A recurrent high-order NN (RHONN) has been proposed in [28] which adapts quickly to changing load patterns. A multi-layered feed-forward NN for STLF is proposed in [29]. In LF, the accuracy primarily relies on training and the time duration of the forecast. The feed-forward NN has presented good results for STLF and they are reported in [30]. A supervised NN-based model for STLF in the Nigerian power system has been proposed in [31].

It is evident from the aforementioned literature review that LF is crucial to a power system's ability to operate economically. Time Series and Multiple Linear Regressions (MLR) are the standard techniques for STLF (TS). The MLR method's primary drawback is that it is extremely sensitive to changes in temperature. A minute change in temperature causes a significant change in the load estimate; hence it requires a very precise temperature forecast. The TS method's main drawback is that it requires a lot of time and effort to use. This promotes the employment of several advanced computing approaches for the STLF problem that are more effective and accurate than the aforementioned standard techniques, such as FL, ANN, and JA.

1.2 Contributions to the Work

In this research work, some conventional techniques for solving the STLF problem such as Multiple Linear Regression (MLR) and Time Series (TS) are described. They are found to have some disadvantages like sensitivity to temperature fluctuations, time-consuming, and high error rate. Hence, advanced computing methods which are more accurate than the conventional ones are used in this work. Some of the advanced computational intelligence methods used in this chapter are the Fuzzy Logic (FL), Jaya algorithm (JA), and Artificial Neural Network (ANN) methods. These three methods have been described and implemented for solving the STLF problem. Out of these three techniques, JA is found to be the most effective and accurate method for STLF as the percentage error of JA is found to be the lowest among the three proposed methods. To meet the main objective of solving the STLF problem, the following sub-objectives are to be achieved:

- Historical load data containing time period and load data of a day are collected.
- Pre-processing of the input data is done and forecasted loads are obtained by the various proposed methods for STLF.
- To make a comparison of the errors of the computational techniques and find out the most efficient technique among them for STLF.
- The above objectives are achieved by studying the three proposed methods for STLF which are FL, ANN, and JA, and they are implemented in MATLAB and Simulink platforms.

The rest of this chapter is arranged as follows. Section 2 discusses the FL, JA, and ANN and carries out STLF using the proposed methodologies. Simulation results and discussions are presented in Sect. 3. Section 4 is the conclusion of the chapter. It contains the summary and the contribution of the chapter along with the future scope of work.

2 STLF Using Fuzzy Logic

For STLF, a mathematical model is applied in FL. FL is a generalized method that uses Boolean logic used in designing digital circuits which take either "0" or "1" values. It uses weather information (temperature and humidity), time in hours, and load as input information and forecasted load as output information for LF. Figure 1 describes the framework of FL [32]. The main purpose of this chapter is to carry out STLF by taking time and load as input variables and output as forecasted load as output variables. The categorization of load data is done by fetching fuzzy set techniques.

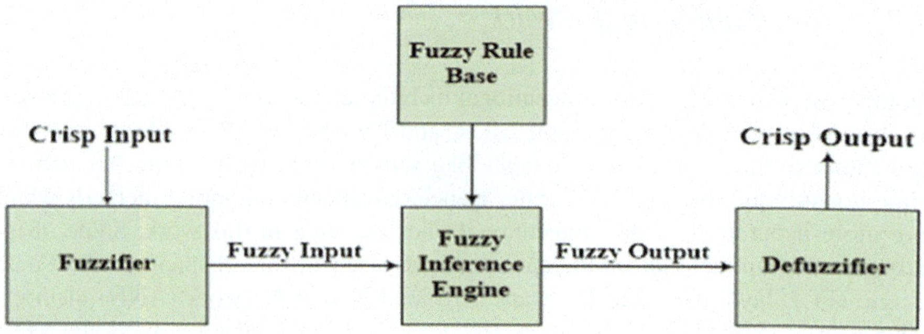

Fig. 1 Fuzzy logic structure

2.1 Block Diagram and Flowchart of Fuzzy Logic

Figure 2 shows the block diagram of the FL technique. The time and actual load at that particular time are given as input to the FL model. The fuzzy interface system is the heart of the system [33]. The fuzzy rules determine the precision of the forecasted load. The defuzzification block shown in Fig. 2 transforms fuzzified output to fresh output and it is displayed in a graph, this graph is known as the load curve.

A fuzzy set is characterized by its membership functions which give them flexibility in modeling frequently used linguistic variables. The characteristic function of the set generally has values between zero and one which expresses the degree of membership of an element in the set [34]. If X is a collection of objects denoted by x, then a fuzzy set A in X is defined as,

$$A = \{(x, \mu_A(x))\,|x \in X\} \tag{1}$$

where μ_A is the membership function (MF) for set A. The MF maps each element of X to a membership grade between 0 and 1. The fuzzy rule base converts fuzzy input to fuzzy output. This method of converting the real value to membership value with the help of MFs is known as fuzzification. These MF values are subsequently utilized in

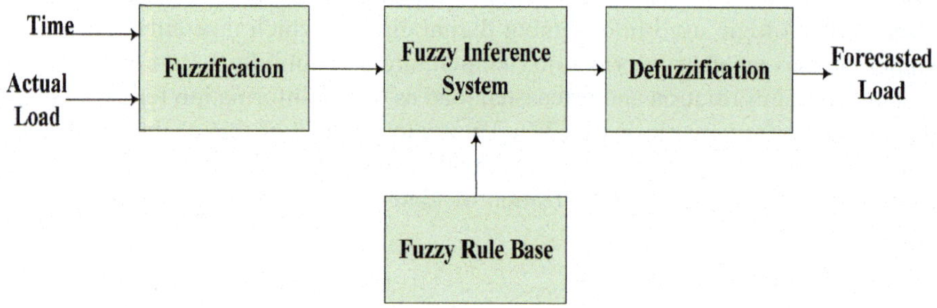

Fig. 2 Block diagram of fuzzy logic

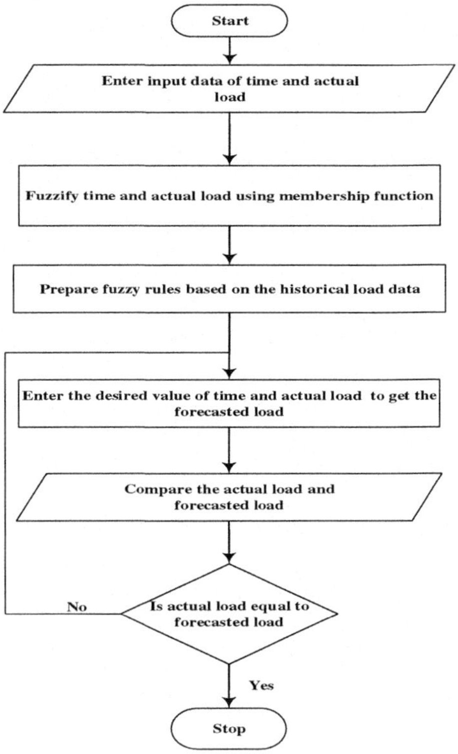

Fig. 3 STLF using fuzzy logic (FL)

the process of a fuzzy inference system, which in turn after being achieved produces results in membership units [35]. The process of defuzzification is generally required to obtain the actual output from fuzzy value for ultimate decision-making whenever the fuzzy process output is needed to be a scalar quantity. The most widely used defuzzification technique is the centroid method. The forecasted load is the result, and it is then compared to both the actual and predicted load. Figure 3 depicts the flowchart of FL which is used for STLF.

2.2 Detailed Modeling of STLF for Fuzzy Logic

The steps involved in solving the STLF using fuzzy logic are described next.

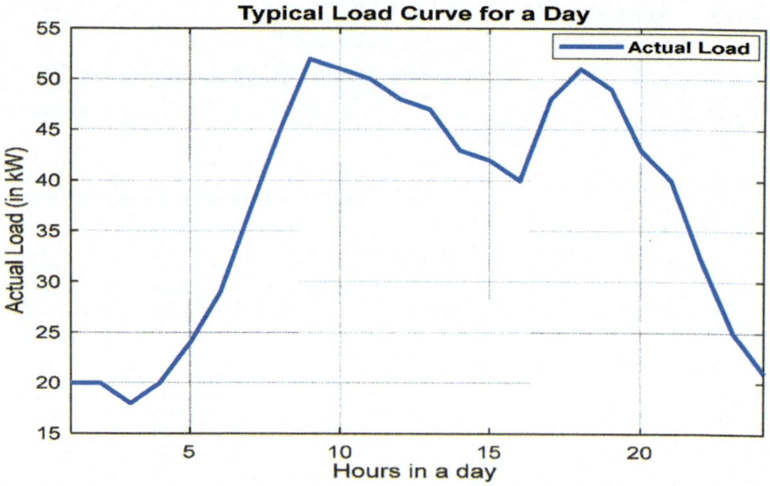

Fig. 4 Typical load curve for a day

2.2.1 Method of Data Collection

The historical load data of a particular day in a week of a college in 2019 is collected and used to carry out the STLF using the proposed model [36]. Figure 4 shows the actual load (in kW) versus the time (in hours) of the day.

2.2.2 Fuzzification of Input and Output Data

Fuzzification represents the procedure of conversion of actual values to membership values related to FL. The nonlinear input and output may be obtained but the MF is always considered to be linear for simplicity. Rectangular MF is utilized in this chapter as both input and output values. Time and load are the two inputs used in the current work for STLF [37]. The time is divided into six fuzzily defined sets, as shown in Figs. 5 and 6: midnight, morning, noon, afternoon, evening, and night. The four fuzzy sets of load are very low, low, average, and high. Figure 7 illustrates how the output or anticipated load is split into four fuzzy sets once the fuzzy rule base has been applied: extremely low, low, average, and high.

2.2.3 Assigning of Membership Functions (MFs)

The membership function (MF) represents a curve that establishes the mapping of points in input space to their respective membership values between zero and one. In this chapter, the time and load parameters are considered fuzzy sets. Various MFs used in FL are generally triangular and trapezoidal. In the present work, triangular

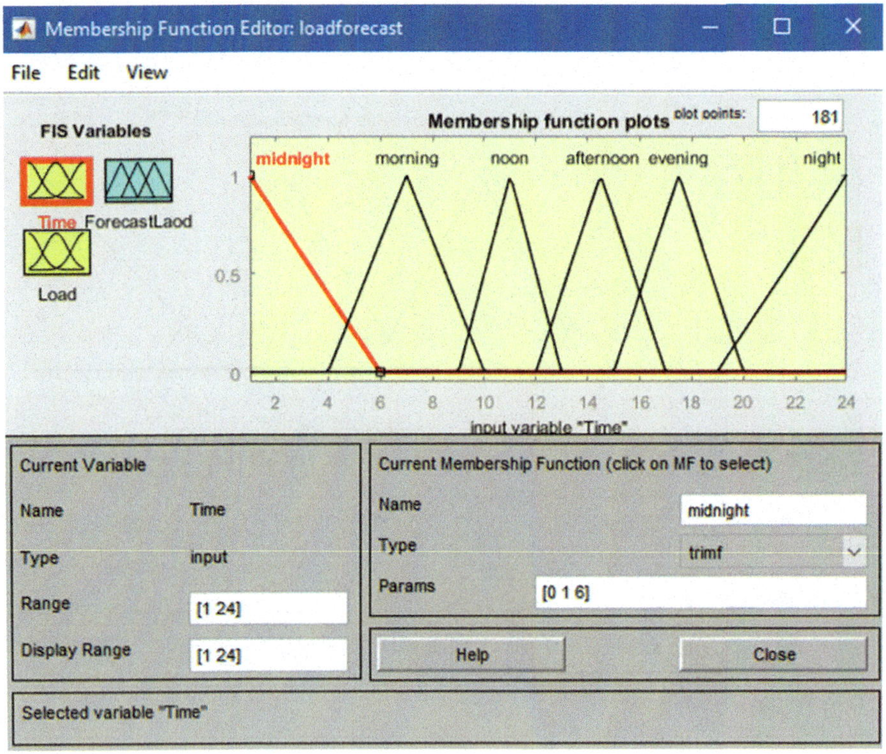

Fig. 5 MF of time

MF is used. Time is classified in the ranges of midnight (0–6), morning (4–10), noon (9–13), afternoon (12–17), evening (15–20), and night (19–24). The actual load is classified as very low (20–25), low (23–35), average (33–47), and high (43–53). The implementation of the above membership functions is shown in Figs. 3 and 4, respectively. Figure 5 shows the forecasted load (output) classified into four fuzzy sets after the implementation of the fuzzy rule base. They are very low (3.8–24), low (22–34), average (32–44), and high (42–52), and they are shown in Figs. 5, 6, and 7.

2.2.4 Fuzzy Rule Base

This is the heart of the FL system. In this base, the If–Then rule is applied to obtain forecasted load as output. Figure 8 depicts the fuzzy rule base and Fig. 9 shows the three-dimensional surface view of MFs for different inputs and outputs. The rules in the fuzzy rule base are described below:

- If it is midnight and load is extremely low (forecasted load is very low).
- If it is morning and load is extremely low (forecasted load is very low).

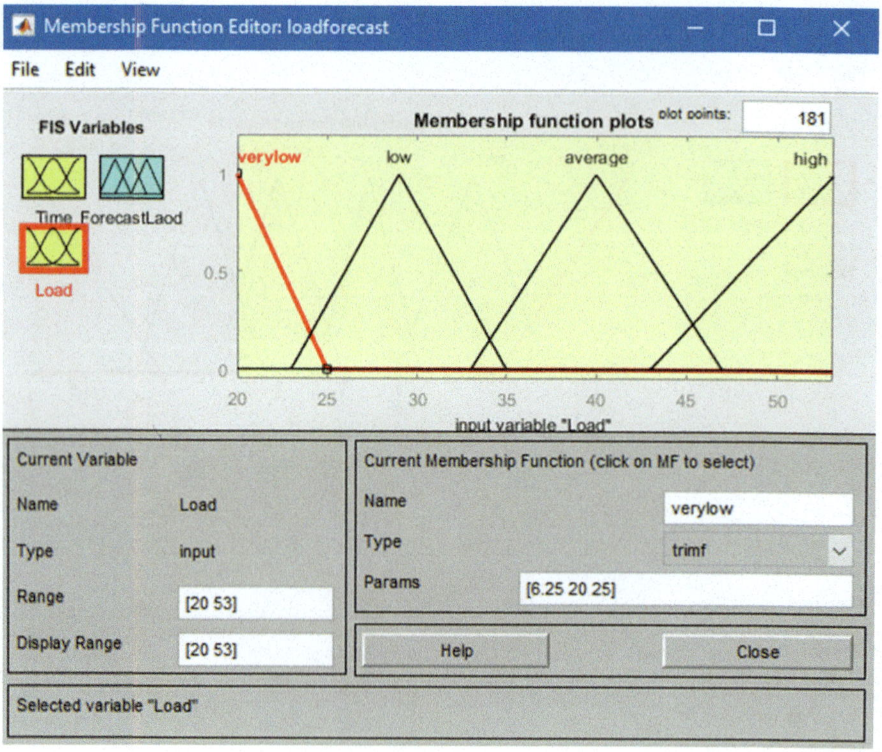

Fig. 6 MF of load

- If it's early and load is low (forecasted load is low).
- If the time is evening and load is typical (forecasted load is average).

2.2.5 Performance Criterion for Error Analysis

Two benchmarks (i.e., absolute percentage error (APE) and mean absolute percentage error (MAPE)) are needed to be outlined for the evaluation of the performance of the proposed LF method. APE stands for the difference between a day's actual and predicted load, and it is formulated as,

$$\text{APE} = \left(\frac{P_a - P_f}{P_f}\right) \times 100 \tag{2}$$

The average of all APE's for 24 h is called MAPE, and it is formulated as,

$$\text{MAPE} = \frac{1}{N}\sum_{i=1}^{N}\left(\frac{P_a(i)}{P_f(i)}\right) \tag{3}$$

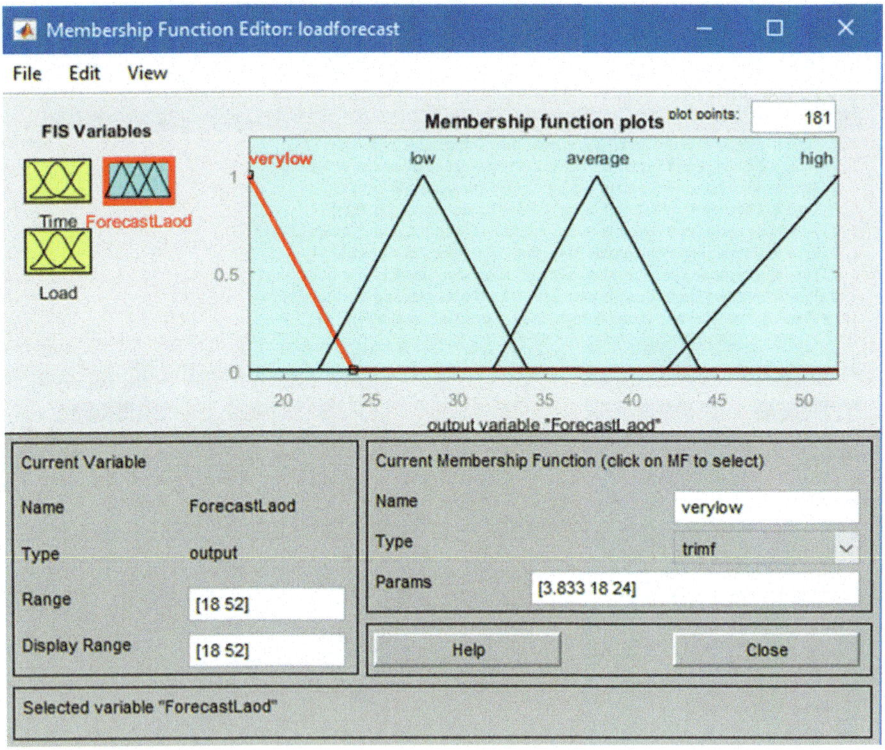

Fig. 7 MF of forecasted load

where P_f represents the forecasted load, P_a represents the actual load, and N represents the total time interval or the forecast period. These performance criteria determine whether the proposed LF method is accurate or not. Our main aim in STLF is to keep these errors as low as possible as it is one of the key factors when considering energy management. If the APE, as well as MAPE in the day, can be accurately found then we can devise new methods of reducing it. It eventually leads to a better understanding of power usage and potentially improves bill management for the consumer.

3 Jaya Algorithm (JA)

The JA represents a metaheuristic technique for solving optimization tasks. JA comprises a single-phase adherence the application is much simpler. The methodology of the algorithm as described is somewhat similar to the teaching learning-based optimization (TLBO) algorithm. TLBO algorithm has been introduced in [28] which does not need any specific parameters. The only dissimilarity between TLBO

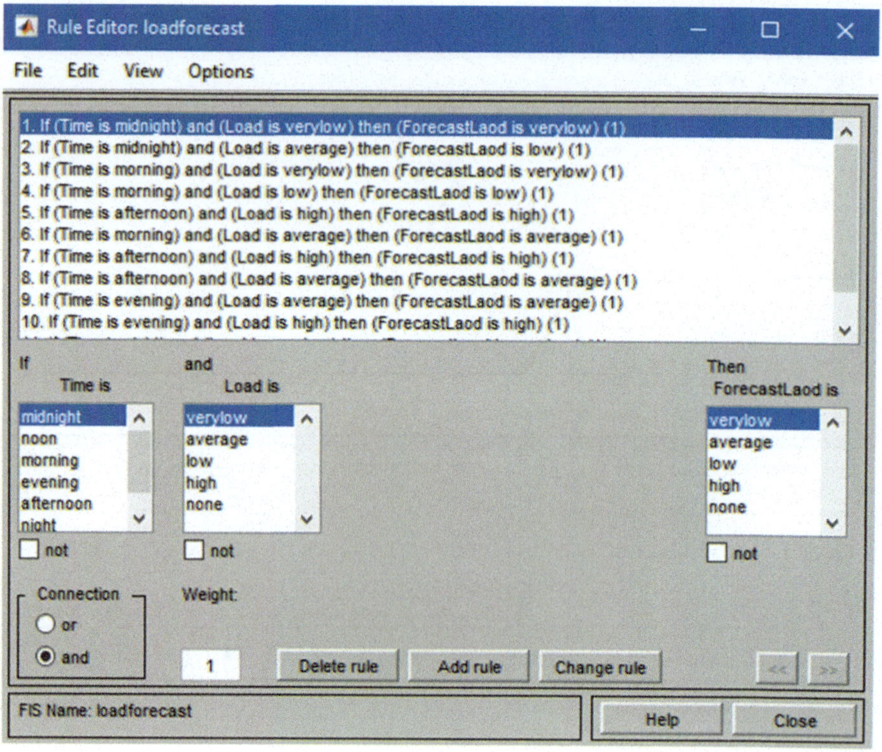

Fig. 8 Fuzzy rule base

and JA is that the former has two phases (that is teacher and learner phase), whereas the latter has only one phase, that is the teacher phase, making it simpler to implement. This algorithm takes into account the fact that candidates of the population strive toward the best solution for the population and prevent them from attaining the worst one. Hence, this algorithm is designated as Jaya, which is a Sanskrit word signifying victory.

The JA is a powerful global optimization algorithm that is used to solve constrained and unconstrained problems. This algorithm needs just a few control parameters like the maximum number of iterations, population size, and the number of design variables No specific control parameters are essential for tuning it before real-time experiments. The working principle of the JA is summarized in a simplified manner as follows:

- Step 1: Initialize the load size, number of design variables, and termination condition or number of iterations.
- Step 2: Input the previous day's loads in the week.
- Step 3: Initialize the best solution and worst candidate solution.
- Step 4: Until the termination condition is satisfied repeat steps 4 to step 5.

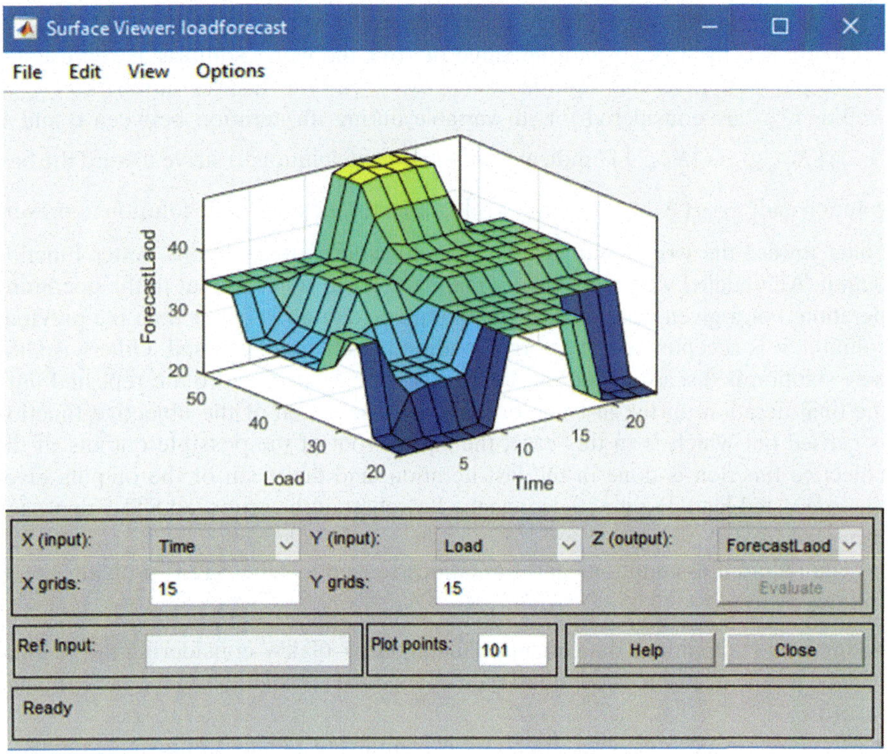

Fig. 9 Three-dimensional surface view

- Step 5: Modify the solution based on the best and worst solutions.

$$X'_{j,k,i} = X_{j,k,i} + r_{1,j,i}\left(X_{j,\text{best},i} - \left|X_{j,k,i}\right|\right) - r_{2,j,i}\left(X_{j,\text{worst},i} - \left|X_{j,k,i}\right|\right) \quad (4)$$

- Step 6: If $X'_{j,k,i} > X_{j,k,i}$, then update the previous solution else no update in the previous solution.
- Step 7: Reverse action of the objective function of the best possible outputs after the last iteration is done.
- Step 8: Display the forecasted load.

Minimization and maximization of an objective function represented by $F(x)$ is the first and foremost objective of this algorithm. In the present work, the proposed algorithm is demonstrated by the Sphere function. First, we assume 'm' number of design variables (i.e., $j = 1, 2, \ldots, m$) and 'n' number of load size, ($k = 1, 2, \ldots, n$) for i iterations. Then, we have to input the earlier day loads in a week according to load size. Then, we initialize the best and worst values of $f(x)$ in the entire population. $X_{j,k,i}$ represents the value of the jth variable for the kth candidate during ith iteration, which is thereafter modified according to Eq. (4) where the value of j for the best solution is represented by $X_{j,\text{best},i}$ and the value of j for the worst solution is represented by $X_{j,\text{worst},i}$. $X'_{j,k,i}$ is the modified data of $X_{j,k,i}$. Two random numbers $r_{1,j,i}$ as well as $r_{2,j,i}$ are considered for jth variable during ith iteration between 0 and 1. "$r_{1,j,i}\left(X'_{j,\text{best},i} - |X_{j,k,i}|\right)$" indicates inclination of solution to strive toward the best solution and "$r_{2,j,i}\left(X'_{j,\text{worst},i} - |X_{j,k,i}|\right)$" indicates inclination of solution to prevent going toward the worst solution. $X'_{j,k,i}$ is considered if it obtains better function output. All updated values are finally maintained and sent as input to the upcoming iteration. For a given candidate, if the modified solution is better than the previous solution, it is accepted and the corresponding population is revised. Otherwise, the new solution is discarded and the old one is kept. Steps 5 and 6 are repeated until the final iteration. In the last iteration, the reverse action of the objective function is carried out which is in this case, the square root of the possible outputs of the objective function is done in the last iteration, and the mean of the outputs gives the forecasted load. Figure 10 depicts the flowchart of the proposed STLF approach using the JA.

The required assumptions and limitations to exercise the STLF problem are as follows:

- Input and output load data are normalized on 10–60 kW considering the original load profile to ensure accommodation of any load for calculating the performance of LF.
- The time period considered for STLF is confined to 24 h to diminish the requisite amount of load data.
- The weather has a negligible impact on the output and hence is discarded.
- Input and output load data are normalized on 10–60 kW considering the original load profile to ensure accommodation of any load for calculating the performance of LF.
- The time period considered for STLF is confined to 24 h to diminish the requisite amount of load data.
- The weather has a negligible impact on the output and hence is discarded.

4 Artificial Neural Network (ANN)

ANN are advanced models, either software or hardware, which take inspiration from the human biological systems by the structure and behaviour of the neurons. The resemblance to the neurons, however, ceases after this point. Feedforward neural networks (FFNNs) are mainly utilized to solve nonlinear regression problems, considering learning from the data fed to the model.

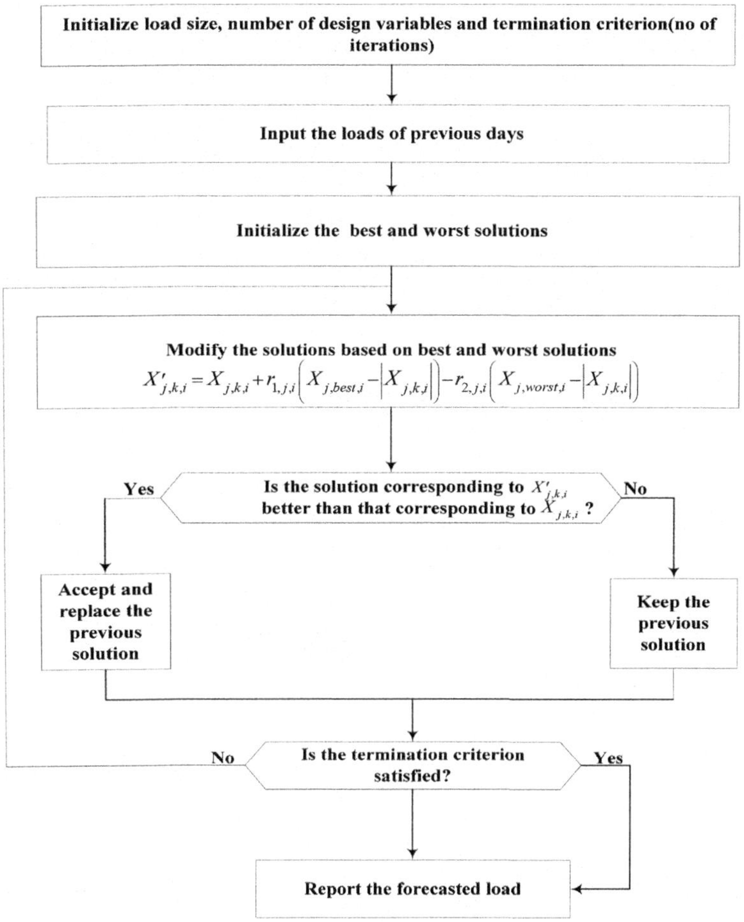

Fig. 10 Flow chart of Jaya algorithm for STLF

4.1 Proposed Method of Implementing ANN

The technique implemented for the present work is ANN. It comprises a vast number of processing units connected in parallel and feeds forwarding in various layers. These inputs and biases combine and are fed to the nonlinear activation function $f(x)$ and produce neurons as the output. The training of the model requires an increasing and differentiable activation function. Figure 11 depicts the architecture of the proposed multi-layer perceptron (MLP) method, in which the ANN model comprises an output layer and a hidden layer. This model has more than one output in connection to the network architecture.

A training algorithm helps determine the weights and biases of the proposed ANN model to reduce loss function. The weights and biases are corrected to minimize

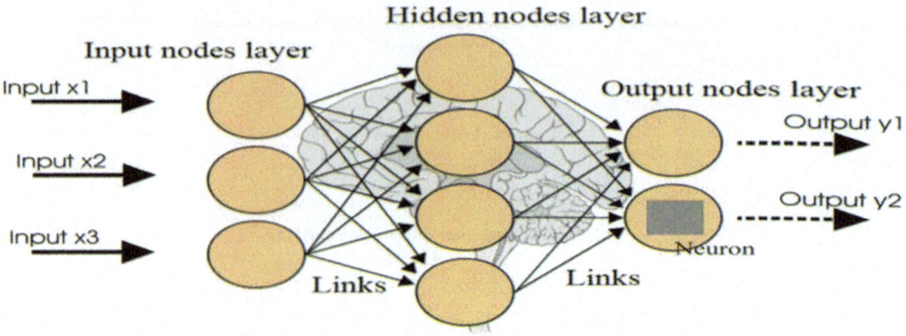

Fig. 11 The architecture of MLP

errors between the values of output and the target. After network training is done, it is fed with the inputs and estimates the error after comparing the forecasted to the actual load. Forecast error can be represented using root mean square error (RMSE) as given in Eq. (5), where y_t represents actual load, and $\widehat{y_t}$ represents forecasted load at time t.

$$\text{RMSE} = \sqrt{1/N \times \sum_{t=1}^{N} (y_t + \hat{y}_t)} \tag{5}$$

However, the most important parameter used for error calculation is MAPE given by,

$$\text{MAPE} = \frac{1}{N} \times \sum_{t=1}^{N} \left(\frac{|y_t + \hat{y}_t|}{y_t} \times 100 \right) \tag{6}$$

The present work aims to verify the proposed model by comparing it with the previous models for STLF. It requires two variables, namely actual load data and time (for 1 day). The actual data learns NN to provide it with input signals and desirable outputs. The built-in Matlab® Levenberg Marquardt (LM) optimization training function is required for the backpropagation training of the ANN model. Updating of internal weight and bias values is done to get a low error output by using the dataset of training predictor and target. The dataset of target contains actual load values for the given dataset of the predictor.

5 Results and Discussion

As mentioned earlier, in this chapter, STLF is achieved by fuzzy logic (FL), Artificial Neural Networks (ANNs), and Jaya Algorithm (JA). The results obtained with these approaches are described next.

5.1 Results of STLF Using Fuzzy Logic (FL)

FL simulation for STLF is demonstrated in Fig. 12. MATLAB/SIMULINK is the program utilized for the simulation work. The workspace file is used to retrieve the actual load and input data, which are then entered into the simulation diagram shown in Fig. 12. The FL block receives the input data. The fuzzy inference system's ".fis" The anticipated output is provided by the fuzzy rules developed in the FL block. The output of the entire fuzzy inference system is aggregated to create the membership function used by the inference engine. Figure 13 depicts the rule viewer for a sample set of output forecast data.

Figure 14 depicts the comparison between the actual and forecasted load for a day after implementing the FL method. The *x*-axis represents time in hours, and the *y*-axis depicts load in kW. Actual load data, predicted load data, and % error for a 24-h period are shown in Table 1 using the FL technique. From obtained results from FL, it is concluded that MAPE is 3.75%.

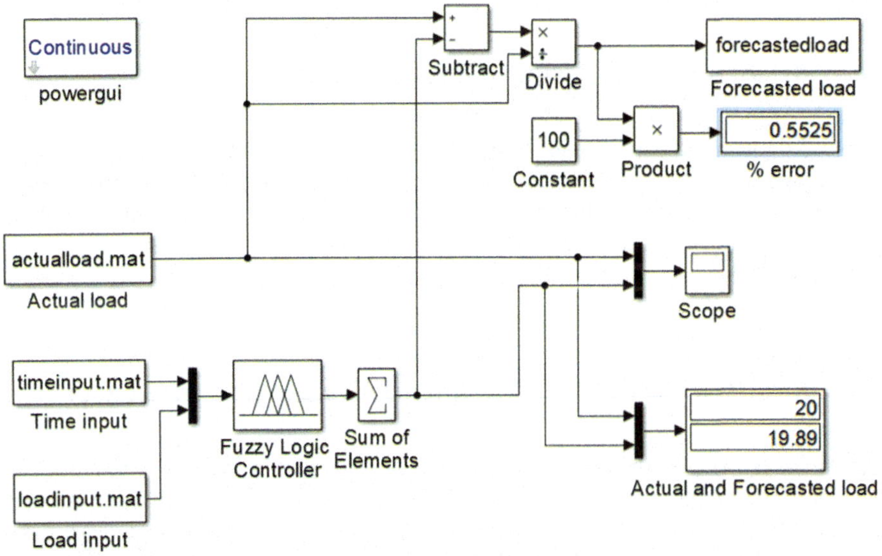

Fig. 12 Fuzzy logic simulation of STLF

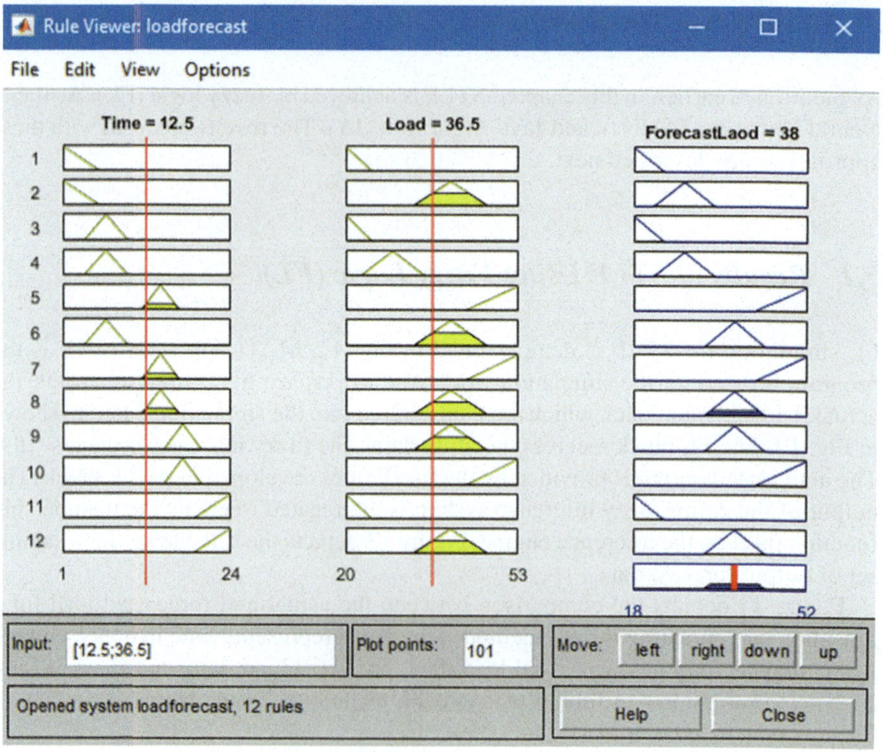

Fig. 13 One sample of data with a rule viewer

5.2 *Results of STLF Using Jaya Algorithm (JA)*

Table 2 depicts the actual load, forecasted load, and percentage error for 24 h for the proposed JA. From Table 2, it is concluded that the MAPE found from results obtained for JA is 2.42%. Figure 15 depicts a comparison between the actual and forecasted load in a day by JA. The X-axis depicts the time period in hours, and Y-axis represents the load in kW.

5.3 *Results of STLF Using Artificial Neural Network (ANN)*

Figure 16 is the plot fit for output and targets versus input using ANN. Errors are shown in the graph, which shows the difference between actual and predicted data. Figure 17 is the error histogram that shows the errors in various instances, eventually determining the MAPE of the ANN model implementation. The MAPE can be calculated from the error histogram (Fig. 17) as 3.04192% (i.e., 73.00062/24). Figure 18 demonstrates the plot of forecasted and actual load for the ANN method.

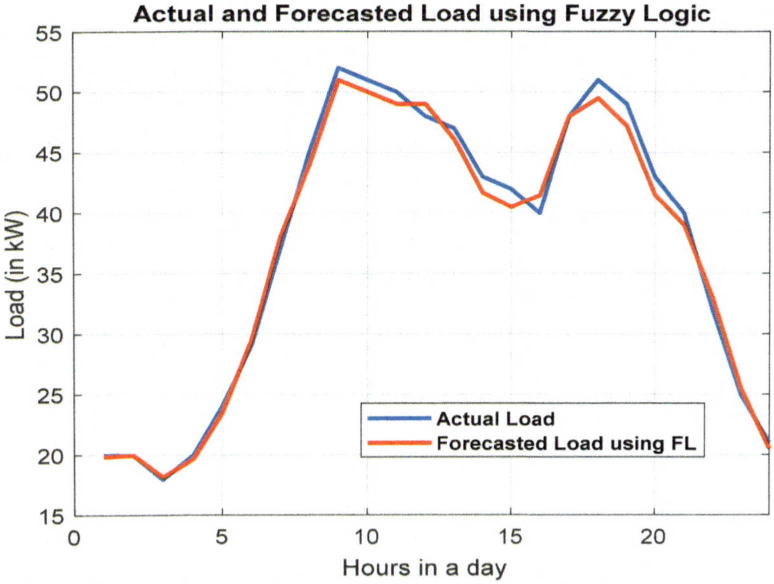

Fig. 14 Comparison of actual and forecasted load of a day for Fuzzy Logic

Table 3 depicts the actual load, forecasted load, and percentage error for 24 h for the proposed ANN model. It can be noticed from Table 3 that MAPE is 3.04%.

5.4 Comparison of MAPE of Three Proposed Computing Methods

Figure 19 shows the comparison of the APE of the three proposed methods for STLF, which are FL, JA, and ANN. The APE of the three proposed methods is 3.47, 2.42, and 3.04%, respectively. Table 4 depicts the comparison of MAPE between the FL Method, JA, and ANN.

Table 4 depicts that the MAPE of JA is less as compared to that of FL and ANN. So JA is preferred over FL and ANN for STLF as it is more accurate and efficient than other computing methods implemented in the present work.

Table 1 Forecasted load and error for 24 h using fuzzy logic

Time (in h)	Load (in kW)	Actual Load (in kW)	Forecasted Load (in kW)	APE (in %)
1	20	20	19.9	0.5
2	20	20	20	0
3	20	18	18.2	1.11
4	26	20	19.7	1.5
5	31	24	23.5	2.08
6	34	29	29.5	1.72
7	42	37	38	2.70
8	53	45	44	2.22
9	49	52	51	1.92
10	48	51	50	1.96
11	46	50	49	2
12	46	48	49	2.08
13	44	47	46.1	1.91
14	43	43	41.7	3.02
15	42	42	40.5	3.57
16	45	40	41.5	3.75
17	47	48	48	0
18	53	51	49.5	2.94
19	46	49	47.2	3.67
20	39	43	41.5	3.48
21	37	40	39	2.5
22	32	32	33	3.12
23	24	25	25.5	2
24	22	21	20.5	2.38
			MAPE	3.75

Table 2 Input and output values of the Jaya algorithm (JA) for the STLF problem

Time (in h)	Best load (in kW)	Worst load (in kW)	Possible loads (in kW)	Actual load (in kW)	Forecast load (in kW)	APE (in %)
1	20	15	16; 17; 19; 20	20	20.3	1.48
2	20	15	16; 17; 18; 20	20	20.02	0.10
3	20	15	16; 18; 18; 20	20	19.36	3.31
4	20	20	24; 25; 26; 28	26	26.85	3.17
5	32	28	30; 31; 35; 33	32	32.77	2.35
6	34	28	33; 34; 35; 34	35	36.19	3.29
7	43	35	38; 39; 42; 43	43	42.28	1.70
8	53	43	43; 49; 52; 53	53	53.84	1.56
9	52	45	52; 53; 48; 45	52	50.52	2.93
10	50	45	48; 49; 50; 55	50	52.22	4.25
11	46	45	46; 47; 48; 49	46	46.55	1.18
12	48	45	48; 50; 55; 54	48	49.37	2.77
13	45	45	50; 51; 52; 54	45	45	0.00
14	43	43	42; 45; 44; 50	43	45	4.44
15	44	43	42; 44; 43; 46	44	44.22	0.50
16	42	42	44; 42; 46; 48	42	42.97	2.26
17	41	40	40; 41; 48; 55	41	42	2.38
18	52	40	50; 51; 52; 54	55	55.48	0.87
19	46	40	48; 44; 43; 49	46	48.33	4.82
20	39	35	35; 36; 38; 45	39	40	2.50
21	37	35	35; 36; 37; 40	37	37.55	1.46
22	24	20	28; 25; 30; 35	25	25.93	3.59
23	25	20	21; 20; 26; 25	25	26.29	4.91
24	22	20	21; 22; 22; 24	22	22.53	2.35
					MAPE	2.42

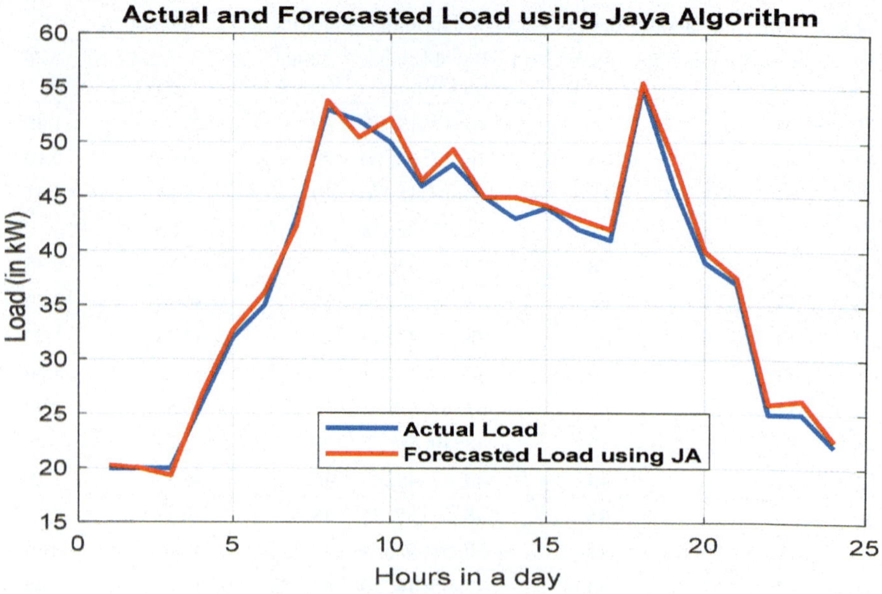

Fig. 15 Comparison between actual and forecasted load for 24 h for the Jaya algorithm

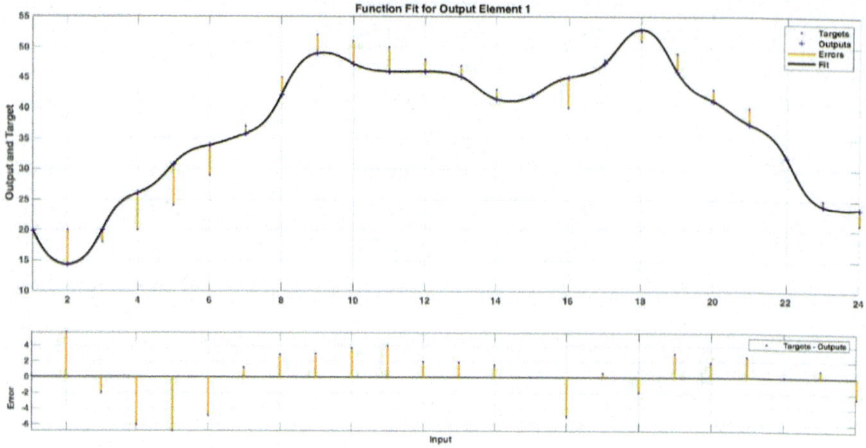

Fig. 16 Plot fit for output and targets

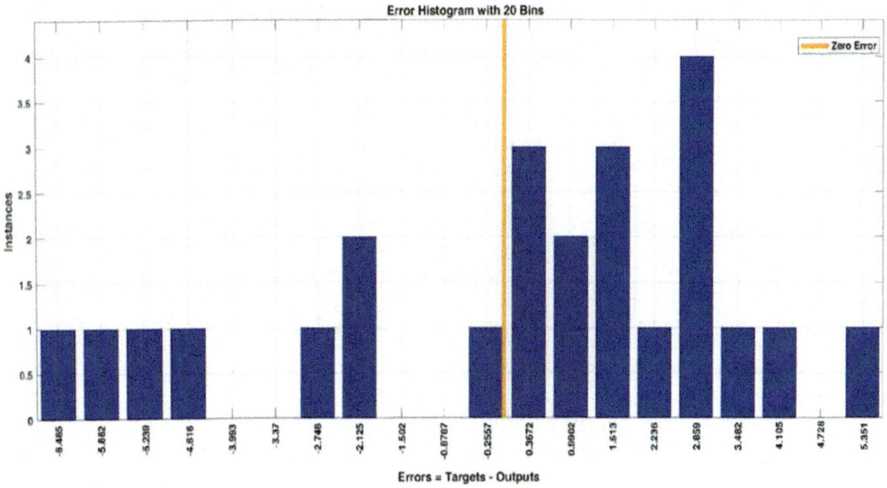

Fig. 17 Error histogram

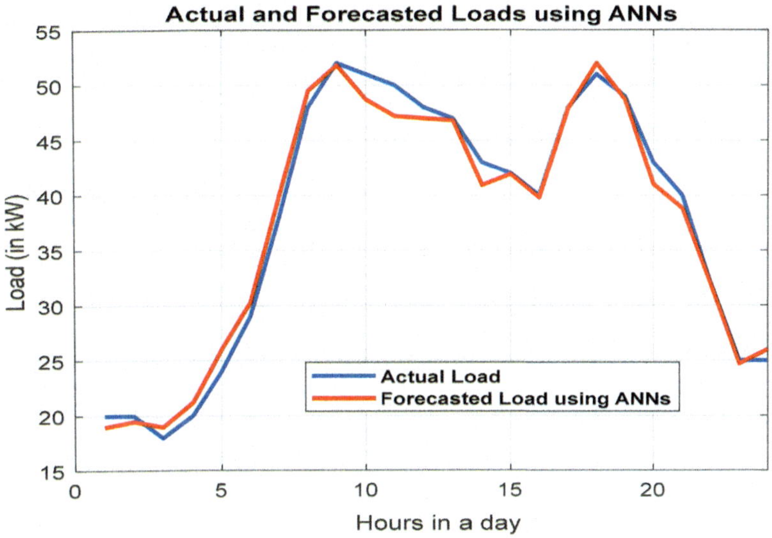

Fig. 18 Comparison of ANN actual and forecasted load model

Table 3 Forecasted load and error for 24 h of ANN model

Time (in h)	Actual Load (in kW)	Forecasted Load (in kW)	APE (in %)
1	20	19	5
2	20	19.5	3
3	18	19	5
4	20	21.2	6
5	24	26	8
6	29	30.3	4
7	38	40	5
8	48	49.5	3
9	52	51.8	0
10	51	48.7	5
11	50	47.2	6
12	48	47	2
13	47	46.8	0
14	43	40.9	5
15	42	41.9	0
16	40	39.8	1
17	48	47.9	0
18	51	52	2
19	49	48.7	1
20	43	41	5
21	40	38.8	3
22	32	31.9	0
23	25	24.7	1
24	25	26	4
		MAPE	3.04

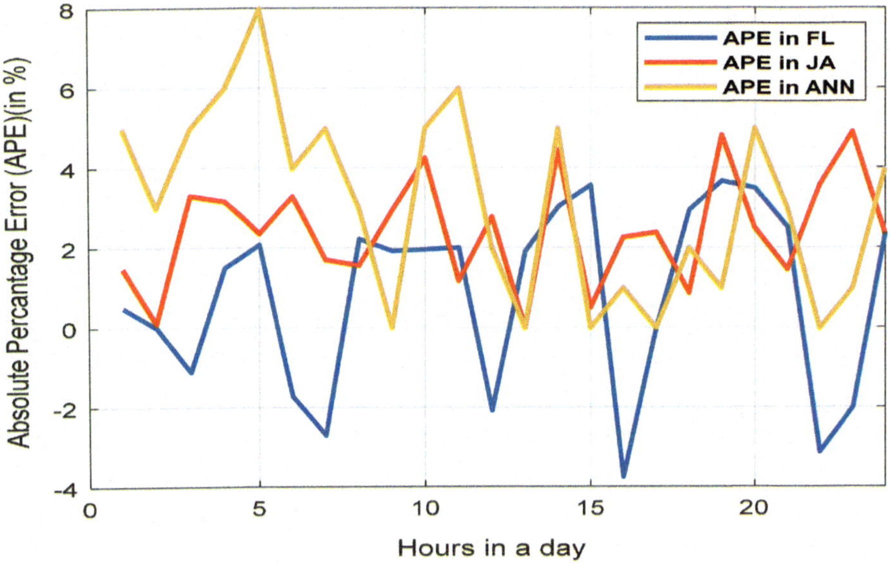

Fig. 19 Comparison of APEs of the three proposed methods

Table 4 Comparison of the MAPE of the three proposed computing methods

Methods	MAPE (in %)
Fuzzy logic	3.75
Artificial neural network	3.04
Jaya optimization algorithm	**2.42**

6 Conclusion

In this chapter, we studied and implemented STLF using three computing methods: FL, ANN, and JA. After comparing the three methods above, JA is the most preferred method among the three computing methods for STLF in terms of accuracy and efficiency. STLF forms the foundation of energy management and decisions in markets. So, it is necessary to carry out LF efficiently. In this chapter, STLF using FL, ANN, and JA is discussed. Their performances are evaluated in MATLAB and Simulink platforms. FL, ANN, and JA MAPE are 3.75, 3.04, and 2.42%, respectively. From Table 4, it can be observed that the MAPE of JA is less than that of the MAPE of FL and ANN for STLF. This work can be extended by designing the STLF using Evolutionary algorithms like PSO, FF, genetic algorithm (GA), etc., and comparing them with JA to determine which one is more efficient and accurate. Also, designing the hybrid model for STLF in which two computing methods would be combined to perform load forecasting is the scope for future work.

Acknowledgments This research work was supported by "Woosong University's Academic Research Funding—2023".

References

1. Jahan IS, Snasel V, Misak S (2020) Intelligent systems for power load forecasting: a study review. Energies 13(22):1–12. https://doi.org/10.3390/en13226105
2. Hernandez L, Baladron C, Aguiar J. M, Carro B, Sanchez-Esguevillas AJ, Lloret J, Massana J (2014) A survey on electric power demand forecasting: future trends in smart grids, microgrids and smart buildings. IEEE Commun Surv Tutor 16(3):1460–1495. https://doi.org/10.1109/SURV.2014.032014.00094
3. Chen Y, Zhang D (2021) Theory-guided deep-learning for electrical load forecasting (TgDLF) via ensemble long short-term memory. Adv Appl Energ 1:1–15. https://doi.org/10.1016/j.adapen.2020.100004
4. Singh P, Dwivedi P, Kant V (2019) A hybrid method based on neural network and improved environmental adaptation method using controlled Gaussian mutation with real parameter for short-term load forecasting. Energy 174:460–477. https://doi.org/10.1016/j.energy.2019.02.141
5. Ali D, Yohanna M, Puwu M, Garkida B (2016) Long-term load forecast modelling using a fuzzy logic approach. Pac Sci Rev A Natl Sci Eng 18(2):123–127. https://doi.org/10.1016/j.energy.2019.02.141
6. Tayab UB, Zia A, Yang F, Lu J, Kashif M (2020) Short-term load forecasting for microgrid energy management system using hybrid HHO-FNN model with best-basis stationary wavelet packet transform. Energy 203:117857. https://doi.org/10.1016/j.energy.2020.117857
7. Chafi ZS, Afrakhte H (2021) Short-term load forecasting using neural network and particle swarm optimization (PSO) algorithm. Math Probl Eng 2021:1–10. https://doi.org/10.1155/2021/5598267
8. Engle RF, Mustafa C, Rice J (1992) Modelling peak electricity demand. J Forecast 11(3):241–251. https://doi.org/10.1002/for.3980110306
9. Janacek G, Swift L (1993) Time series: forecasting, simulation, applications (mathematics & its applications). Ellis Horwood, New York, pp 1–331. ISBN: 0131035819, 9780131035812
10. Zou Y, Feng W, Zhang J, Li J (2022) Forecasting of short-term load using the MFF-SAM-GCN model. Energies 15(9):3140. https://doi.org/10.3390/en15093140
11. Mamlook R, Badran O, Abdulhadi E (2009) A fuzzy inference model for short-term load forecasting. Energy Policy 37(4):1239–1248. https://doi.org/10.1016/j.enpol.2008.10.051
12. Lee WJ, Hong J (2015) A hybrid dynamic and fuzzy time series model for mid-term power load forecasting. Int J Electr Power Energy Syst 64:1057–1062. https://doi.org/10.1016/j.ijepes.2014.08.006
13. Aggarwal S, Kumar M, Saini L, Kumar A (2011) Short-term load forecasting in deregulated electricity markets using fuzzy approach. J Eng Technol 1(1):1–7. https://doi.org/10.4103/0976-8580.74559
14. Kosko B (2006) Neural networks and fuzzy systems. Prentice Hall of India Private Limited, New Delhi. ISBN: 978-1-4615-6253-5.
15. Manoj PP, Shah AP (2014) Fuzzy logic methodology for short-term load forecasting. Int J Res Eng Technol 3(4):2321–7308. eISSN: 2319-1163
16. Kaur J, Brar YS (2014) Short-term load forecasting using fuzzy logic of 220 kV transmission line. Int J Eng Res Technol 3(9):336–343. ISSN: 2278-0181
17. Ahmadi S, Bevrani H, Jannaty H (2012) A fuzzy inference model for short-term load forecasting. In: proceedings of IEEE second Iranian conference on renewable energy and distributed generation, vol 39-44, Tehran, Iran. https://doi.org/10.1109/ICREDG.2012.6190465

18. Holmukhe RM, Dhumale S, Chaudhari PS, Kulkarni PP (2010) Short-term Load forecasting with fuzzy logic systems for power system planning, a review. In: International conference on modeling, optimization and computing, vol 1298, no 1, pp 445–458. https://doi.org/10.1063/1.3516348
19. Wang S, Zhang Z (2021) Short-term multiple load forecasting model of regional integrated energy system based on QWGRU-MTL. Energies 14(20):6555. https://doi.org/10.3390/en14206555
20. Çevik HH, Çunkas M (2016) A fuzzy logic based short-term load forecast for the holidays. Int J Mach Learn Comput 6(1):57–61. https://doi.org/10.18178/ijmlc.2016.6.1.572
21. Rao RV (2016) Jaya: a simple and new optimization algorithm for solving constrained and unconstrained optimization problems. Int J Ind Eng Comput 7:19–34. https://doi.org/10.5267/j.ijiec.2015.8.004
22. Suraj SRK, Ghosh S (2016) Jaya based ANFIS for monitoring of two class motor imagery task. IEEE Access 4:9273–9282. https://doi.org/10.1109/ACCESS.2016.2637401
23. Rao RV, Saroj A (2017) A self-adaptive multi-population based Jaya algorithm for engineering optimization. Swarm Evol Comput 37:1–26. https://doi.org/10.1016/j.swevo.2017.04.008
24. Esclapez AC, García ML, Valero Verdú S, Senabre Blanes C (2022) Reduction of computational burden and accuracy maximization in short-term load forecasting. Energies 15:3670. https://doi.org/10.3390/en15103670
25. Salkuti SR (2022) Emerging and advanced green energy technologies for sustainable and resilient future grid. Energies 15(18):6667. https://doi.org/10.3390/en15186667
26. Reddy SS (2018) Bat algorithm-based back propagation approach for short-term load forecasting considering weather factors. Electr Eng 100:1297–1303. https://doi.org/10.1007/s00202-017-0587-2
27. Feinberg EA, Genethlion D (2005) Load forecasting. In: Applied mathematics for power systems: optimization, control and computational intelligence. Springer Science and Business Media, New York
28. Bento PMR, Pombo JAN, Calado MRA, Mariano SJPS (2021) Stacking ensemble methodology using deep learning and ARIMA models for short-term load forecasting. Energies 14:7378. https://doi.org/10.3390/en14217378
29. Huang S, Zhang J, He Y, Fu X, Fan L, Yao G, Wen Y (2022) Short-term load forecasting based on the CEEMDAN-sample entropy-BPNN-transformer. Energies 15(10):3659. https://doi.org/10.3390/en15103659
30. Salkuti SR (2018) Short-term electrical load forecasting using radial basis function neural networks considering weather factors. Electr Eng 100:1985–1995. https://doi.org/10.1007/s00202-018-0678-8
31. Reddy SS, Momoh JA (2014) Short term electrical load forecasting using back propagation neural networks. In: North American power symposium (NAPS), Pullman, WA, USA, pp 1–6. https://doi.org/10.1109/NAPS.2014.6965453
32. Hu B, Xu J, Xing Z, Zhang P, Cui J, Liu J (2022) Short-term combined forecasting method of park load based on CEEMD-MLR-LSSVR-SBO. Energies 15(8):2767. https://doi.org/10.3390/en15082767
33. Zhang J, Liu C, Ge L (2022) Short-term load forecasting model of electric vehicle charging load based on MCCNN-TCN. Energies 15:2633. https://doi.org/10.3390/en15072633
34. Zhou M, Hu T, Bian K, Lai W, Hu F, Hamrani O, Zhu Z (2021) Short-term electric load forecasting based on variational mode decomposition and grey wolf optimization. Energies 14(16):4890. https://doi.org/10.3390/en14164890
35. Reddy SS, Jung CM, Seog KJ (2016) Day-ahead electricity price forecasting using back propagation neural networks and weighted least square technique. Front Energy 10:105–113. https://doi.org/10.1007/s11708-016-0393-y

36. Panda SK, Ray P, Salkuti SR (2022) A review on short-term load forecasting using different techniques. In: Gupta OH, Sood VK, Malik OP (eds) Recent advances in power systems. Lecture notes in electrical engineering, vol 812. Springer, Singapore. https://doi.org/10.1007/978-981-16-6970-5_33
37. Li Y, Guo P, Li X (2016) Short-term load forecasting based on the analysis of user electricity behavior. Algorithms 9(4):80. https://doi.org/10.3390/a9040080

Optimal Power Flow by Different Modern Optimization Techniques

Bibhu Prasad Nanda, Debani Prasad Mishra, and Surender Reddy Salkuti

Abstract The power flow solution can be optimized in different ways, providing the safest point of operation based on specific objective functions while meeting the system's operational restrictions. Many objective functions in power structures can be improved, including the total amount of power production, involved flexible AC transmission device maintenance cost, power flow capability of the system, the residue effect of the generating station, and many more. During the optimization phase, variables like true power (P), the voltage at generating station bus (V), and tap settings of the transformer, which are controllable, can be pinched. Several processes are considered earlier for the solution of the power flow issue. Classification and innovation of some recent or new techniques have been developed for finding the solution to those optimal power flow (OPF) problems. Several classical and modern optimization and meta-heuristic strategies can be used to optimize and prevent entrapment locally. In this section, a detailed recent optimization technique overview has been made, among which nature, evolutionary, human, and physics-encouraged methods with artificial neural network (ANN) strategies are discussed.

B. P. Nanda · D. P. Mishra
Department of Electrical and Electronics Engineering, IIIT Bhubaneswar, Bhubaneswar, Odisha, India
e-mail: debani@iiit-bh.ac.in

S. R. Salkuti (✉)
Department of Railroad and Electrical Engineering, Woosong University, Daejeon, Republic of Korea
e-mail: surender@wsu.ac.kr

S. R. Salkuti et al. (eds.), *Power Quality in Microgrids: Issues, Challenges and Mitigation Techniques*, Lecture Notes in Electrical Engineering 1039,
https://doi.org/10.1007/978-981-99-2066-2_16

Keywords Optimal power flow · Power flow · Optimization techniques · Economic dispatch · Artificial neural network

Nomenclature

FAPSO	Fuzzy Adaptive Particle Swarm Optimization
GSA	Gravitational Search Algorithm
CNN	Convolutional Neural Network
ALO	Ant-Lion optimization
ED	Economic Dispatch
f, x, c, m	Objective function, state variable, control variable vector, and power flow equation
s, n	Parameter vector and operation and the physical constraint
λ	Lagrange multiplication factor for power management
μ	Lagrange multiplication factor for operational and physical constraints
g	Power Flow Equation
KKT	Karush–Kuhn–Tucker
£	Lagrange function
EICC	Equal Incremental Cost Approach
VD	Voltage Drop

1 Introduction

Energy grids are diversifying because of cross-country grid connections, the deregulation of the entire energy sector, and rising energy costs. As a result, utilities are looking for ways to make better use of their existing transmission networks. Cramming or burden on transmission lines may occur because of an interaction difference between energy production and transmission supplying firms, as well as unanticipated necessities such as development outages, a rapid increase in associated load, and occasionally, kit failure. With increased competition in power networks, controlling transmission congestion has become a major issue. As a result of network consolidation and the expansion of peripheral facilities, a competitive spirit emerges. Power flow optimization via generation redistribution is a sensible attempt to make better use of the current network. To relieve network line congestion, rescheduling generators and drop-off loads were implemented as management acts. This point of view employs a local optimization strategy [1].

Additionally, the sensitivity of section congestion caused by current injection on any bus is frequently considered which is correlating with previous frameworks [2]. Similarly, the Locational Marginal Pricing (LMP) policy is applied to an optimal point, and range reorganized power generation for power groups in the energy sector

[3]. LMP is applied for regional obstruction organization, highlighting natural and feasible congestion transfer variables [4]. Due to the high level of competition in the power markets, necessity-side regulation is preferred [5, 6] over supply-side management to reduce transmission bottlenecks [7, 8]. As a result, additional energy suppliers are being introduced into the existing vibrant energy market to improve the strength and effectiveness of the power grids [9, 10]. Several scientific studies show that DGs in distribution grids can improve voltage and reduce genuine power losses, resulting in higher grid performance. Soft computing solutions with single and multi-goal functions are implemented simultaneously to accomplish goals such as improved power streams, loss reduction, voltage increase, and running costs [11]. A PSO-based hybrid technique was employed to lower the system load to accomplish the optimal standing and sizing of DGs for the network to be inspected [12]. Reference [13] includes a network reconfiguration process as well as DG integration. Simultaneously, a GA is employed to evaluate the optimal capability for communication obstruction organization by making voltage modifications and contractions of actual power indulgence into account [14]. The authors of [15] proposed a hybrid method that deals with traffic congestion by merging methodologies such as the GSA and FAPSO. In [16], it is suggested that LMP-dependent DGs work together to manage transmission bottlenecks. Encouraging customers to interact constructively with decentralized energy infrastructure is a significant challenge in the development of efficient electricity market systems, which can be accomplished through the implementation of location-based marginal valuing in retail energy situations [17, 18]. OPF is a challenge in which specific factors such as active power generation costs, including losses, can be designed to alleviate the best possible solution. Numerous effective OPF methods have been developed, such as B. the well-known method with reduced gradients [19], evolutionary scheduling [20], Newton procedure [21], and others. The requirement impacts are evaluated when a system turns out to be jammed and LMP boosts [22]. LMP, also called interface difference, is a principle for determining the overload of the associated lines. The more significant the LMP difference, the more congested the link [23]. The LMP methodology can reduce network administration costs while increasing network overhead [24]. The authors of [25] suggest that a DG is used to relate the difference between production and necessity in the distributed LMP discrepancy. Fostering functional consumer participation in distributed energy resources (DERs), which can be accomplished through the deployment of LMPs in comprehensive power markets, is a key task in the development of intelligent power flow organizations [26, 27]. A reorganized generation helps to cover the increasing demand. Placement and size of the DG unit limit network expansion [28]. The scientists in [29] used the ALO approach to place wind turbines as DG in the power grid. To address the DG assignment challenge, the researchers of [30]-[32] used heuristic techniques, while those of [33, 34] used meta-heuristic procedures. Simultaneously, managing DG allocation and network reconfiguration [35] proved more beneficial than investigating them separately. While creation-side managing of transmission grid congestion is an established method, according to literature research, demand-side management is developing into an essential tool for efficient and economical management of transmission grid congestion. As a result, various

optimization techniques are used. This chapter proposes a multi-objective policy to optimize the abilities of decentralized generators (DG) as well as to redesign conventional generators for efficient congestion organization and to improve the power grid operation in terms of power failure, power stream and electrical energy profile. The LMP and TCC approach determines the best GD placement locations.

Distribution networks (DN) are constantly linked to DG, which uses renewable energy sources and is involved with sustainability and the environment. Wind, PV, biomass, and mini-hydro are common renewable energy sources used to generate power as DG sources. Because of its close vicinity to loads, DG inclusion not only favors the environment but also enhances electricity delivery dependability and decreases overall power damage. Despite these benefits, incorporating a huge number of DG resources into distribution systems will present new challenges to network security mechanisms. As a result, we now have an efficient protection scheme that is regarded as an essential element of the grid. A decent safety scheme inhibits power outages and safeguards expensive equipment, resulting in less money being lost. Protection relay organization and seclusion of the defective zone are also required to secure the energy structure with sufficient limits and allowable time delays. As a result, we now have an efficient protection scheme that is considered an essential element of the power grid. A good protection scheme avoids power outages and safeguards costly apparatus, resulting in less money being lost.

In an incorporated network, the directional over-current relay has been selected as the primary protection relay to detect faults in both directions. For power generation, DG is an imperative viewpoint. Numerous reasonable power resources get on it exclusive of interruption to the circulation network or at the client's counter facet [36]. It might also be dispersed production, reorganized energy, and on-site innovation can be identified as distributed energy [37, 38]. The power resource that can be anticipated in a distributed generation is a crucial task in the distribution system in the case of power pressure [39, 40]. In literature, several scientific methodologies like novel heuristic approach [41], new power stability index approach [42] and novel adaptive shuffled frogs leaping algorithm [43] have been proposed.

Because of its relevance in power system functioning, much research has been done on OPF during the last half-century. OPF is considered a vital element by which various technology come up from the operation and electrical network design point of view [44]. The primary objective is to find the certain and fundamental operating point known as the control variables for a few numbers of specific functions by maintaining the system constraint condition. Variables such as FACTS device cost, line loss, component stability cost, electricity generating cost, and so on are considered during optimization. Aside from that, system-controlled components that can be extracted include FACTS device sizing, voltages of various buses, power-produced values, and transformer tap setting [45]. To overcome the problem related to power flow, many deterministic and non-deterministic heuristic optimization processes are implemented [46]. A combination of sensitivity-based analysis and gradient-based methods is used. Due to the nonlinear characteristics of optimal power flow, sometimes traditional methods may need to be revised with accuracy and certainty for which the developed technologies or meta-heuristic optimization techniques are extensively used.

2 OPF: A General Overview

The OPF was proposed for the first time by Carpentier. Following that, various conventional techniques were used to solve the problems associated with optimal power flow. The Newton method, Linear Programming, Nonlinear Programming, Quadratic Programming, and the interior point are all popular techniques. They do, however, have several drawbacks, most of which are related to stabilization, such as a high degree of nonlinearity and multi-modality, and thus may become trapped in local minima [47]. In the present scenario, some intelligent optimization techniques like evolutionary-encouraged algorithm [48, 49], human-encouraged algorithm [50], nature-encouraged algorithm, and ANN methods [51] are applied to nonlinear systems, and corresponding models are developed to allow the system to optimally go over the power transmission, thereby refining the management of the system can be done appropriately. The traditional and contemporary optimization schemes for optimal power flow problems are depicted in Fig. 1. This chapter provides a wide-ranging overview of all strategies as well as a qualitative presentation of each. The latest techniques for resolving such power flow issues in the distribution system are discussed here. Many more discussions are provided in various sections where parameters of concern include power flow optimization issues, objective function, and system constraint. Similarly, conventional OPF approaches can be used to generate ideas or provide a brief overview. The other section demonstrates the recent widespread use of optimization models to solve power flow problems optimally.

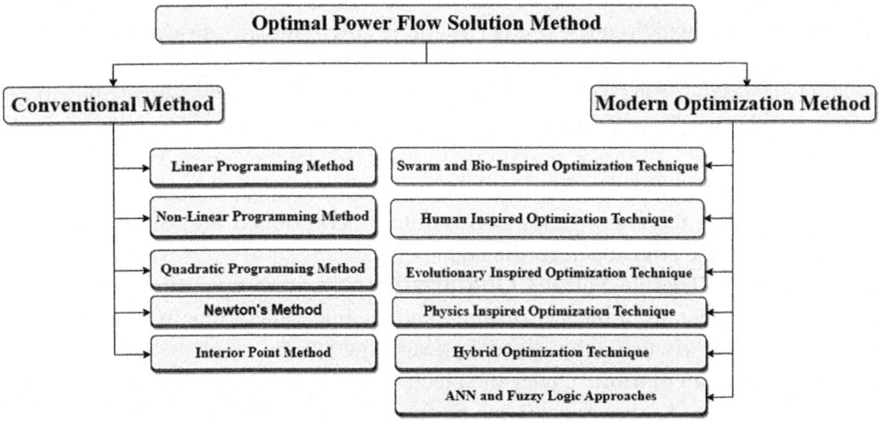

Fig. 1 Methods involved in OPF problem solution

The OPF was initially envisioned as a modified version of the standard economic dispatch (ED), in which the ED with the concerns of power flow is addressed concurrently. This traditional model of OPF can be presented as a challenge with nonlinearity whose purpose is to discover related linked variables like state and control variables. When designing an OPF problem, variables, restrictions, and the objective function must all be determined.

2.1 Variables

Active and reactive power, generator voltage magnitudes, transformer tap, and phase shifters are all examples of control variables (u). State variables include bus voltage magnitudes and relative bus voltage angle deviations from the reference angle (x).

2.2 Constraints

In this problem, there are both equality and inequality restrictions. Inequality constraints are associated with upper limitations and the lower one of the equipment connected in the system, like transmission lines, generators, and transformers, whereas equality constraints are associated with power flow balance.

2.3 Objective Function

Some illustrations of objective functions are the minimization of fuel cost, active and reactive loss, etc., which can be used in a generic OPF issue. Because of its versatility, the OPF may be used to unravel various problems. The following sections provide a detailed discussion of the issue, which is resolved by various OPF control algorithms.

Economic Dispatch: To reduce operating costs, including the cost of the security of the network.

Preventive Dispatch: To keep the voltages and current flow in the branch in the acceptable range, the contingency constraint is incorporated into the problem and the control is to be adjusted accordingly.

Active and Reactive Voltage Optimization: The best operating point for the generator voltage, taps of the transformer, and the injection of reactive power can be determined regularly using the optimal power flow.

Spot Price Calculation: Under the electrical market regime, during the regular interval, the per-node energy price can be found with the help of the OPF.

3 Generic OPF Problem Formulation

The essential requirement of resolving the OPF is the optimization of the objective function by using the best correction of the control variables of the power system by keeping both equality and inequality constraints into account. The representation of optimization issues in mathematical format may be written as follows.

$$minimize\ f(x, c) \quad (1)$$

Subjected to

$$m(x, c, s) = 0 \Leftrightarrow \lambda \quad (2)$$

$$n(x, c, s) \leq 0 \Leftrightarrow \mu \quad (3)$$

The function of Lagrangian can be written as follows.

$$\pounds = \mathrm{f(x,\ c,\ s)} + \lambda \mathrm{t\ m\ (x,\ c,\ s)} + \mu \mathrm{t\ n\ (x,\ c,\ s)} \quad (4)$$

The first-order Karush–Kuhn–Tucker (KKT) optimality condition will take the different differential equations for optimization. They are:

$$\frac{\partial f}{\partial x} + \left[\frac{\partial m}{\partial x}\right]^t \lambda + \left[\frac{\partial n}{\partial x}\right]^t \mu = 0 \quad (5)$$

$$\frac{\partial f}{\partial c} + \left[\frac{\partial m}{\partial c}\right]^t \lambda + \left[\frac{\partial n}{\partial c}\right]^t \mu = 0 \quad (6)$$

$$m(x,\ c,\ s) = 0 \quad (7)$$

$$n(x,\ c,\ s) \leq 0 \quad (8)$$

$$\mu\_i\ h\_i\ (x,\ c,\ s) = 0 \quad (9)$$

$$\mu\_i \geq 0 \quad (10)$$

Equations (5) and (6) indicate that the relation between the lagrangian gradient with state variable (x) and control variable (c) is zero with the optimal condition. Equations (7) and (8) represent the restriction attached to the problem, and Eqs. (9) and (10) are the equations that complement the system equation for the handling of active and non-active limitations.

4 The Solution to the Problem of OPF

The first algorithm for estimating least fuel cost was developed in the 1930s, and since then, numerous methods for ensuring optimal power system function have been developed. The Equal Incremental Cost Approach (EICC) is regarded as the prototype of the OPF. The first OPF techniques, developed in the 1960s, have many nonlinearities in which an iterative procedure is maintained, and nonlinear constraints are included for the formulation of the problem. From that period, so many techniques have evolved, which can be classified as follows.

- Heuristic approaches
- Primal approaches
- Penalty and barrier techniques
- Linear programming-based systems

The most generally used approaches are described next. The simple approach tries to discover a solution in the smallest possible area. The method's fundamental flaw is the requirement for obtaining a first viable location. When equality constraints are the key factor, it is crucial and critical for maintaining a viable region with a typical iterative process. Dommel and Tinney proposed a technique based on the gradient method in 1968. From then, a drastic movement was found in research, which finds a viable power flow with no inequality constraints as an initial solution. Then, these constraints are applied to the OPF problem utilizing penalty functions. The gradient approach to the OPF problem can be represented numerically as,

$$\text{Minimize } f\ (\text{x, c}) \tag{11}$$

$$\text{Subjected To: m(x, c, s)} = 0 \Leftrightarrow \lambda \tag{12}$$

Lagrangian is

$$\pounds = \text{f (x, c, s)} + \lambda' \text{ m (x, c, s)} \tag{13}$$

and the order Karush–Kuhn–Tucker (KKT) conditions are

$$\nabla \pounds_x = \frac{\partial f}{\partial X} + \left[\frac{\partial m}{\partial X}\right]^t \lambda = 0 \tag{14}$$

$$\nabla \pounds_c = \frac{\partial f}{\partial c} + \left[\frac{\partial m}{\partial c}\right]^t \lambda = 0 \tag{15}$$

$$\text{m(x, c, s)} = 0 \tag{16}$$

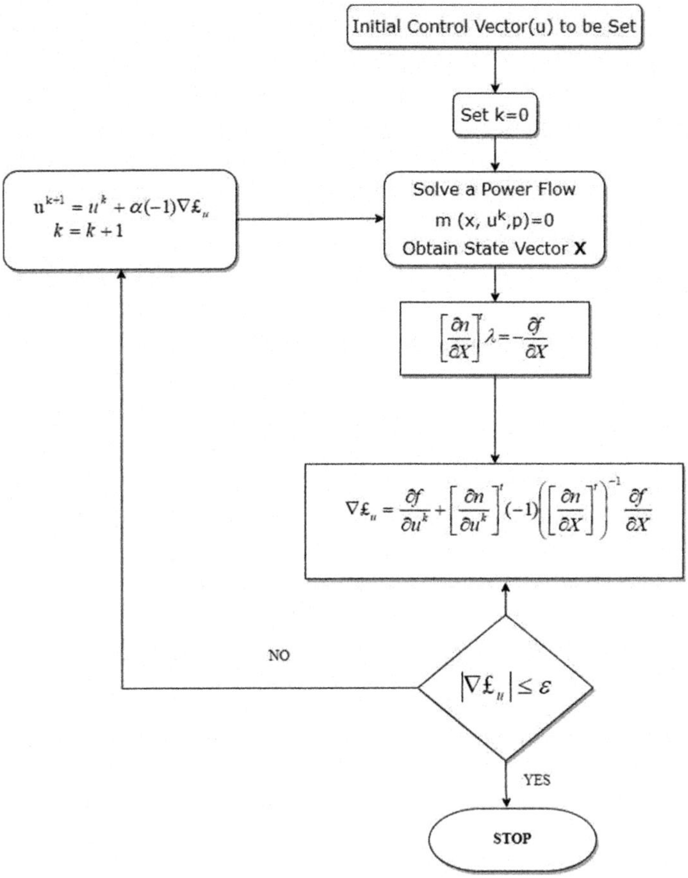

Fig. 2 Primal-based OPF algorithm

The primal-based optimization process is presented in Fig. 2.

5 Optimal Power Flow Recent Optimization Methods

Because the OPF problem is multi-modal, nonlinear, or non-convex, standard methods are only sometimes proper and cannot guarantee a coherent approach. Many empirical optimization techniques to deal with the OPF problem have been proposed to overcome the constraints of standard methods. The following is a summary of the advantages of modern optimization techniques:

- These approaches apply to small and large-scale systems combing.
- They are highly reliable in terms of obtaining optimal solutions.

- They are rarely locked in local minimum solutions.
- When compared to traditional procedures, they arrive at the optimal solution quickly.

This section will discuss how to employ current (non-deterministic) heuristic optimization methodologies to obtain the optimum power stream. The empirical optimization techniques used for the OPF difficulty can be categorized as follows based on their sources of inspiration.

5.1 Bio-Inspired and Swarm Optimization Techniques

Nature-encouraged and bio-encouraged optimization algorithms are obtained in connection with the motions and probing behavior of swarms of creatures for sustenance springs. Table 1 presents a summary of the nature-encouraged techniques used to solve power system issues coherently related to the optimality of the transmission line.

5.2 Techniques for Human-Inspired Optimization

Several optimization strategies, particularly in thinking and decision-making, are designed to emulate human behavior. Table 2 focuses on the review of human-encouraged algorithms for different systems for the solution of OPF problems.

5.3 Optimization Techniques Inspired by Physics

Algorithms based on physics rules or natural occurrences in space are known as physics-encouraged algorithms. Table 3 lists the applied physics-encouraged optimization approaches for OPF.

5.4 Evolutionary-Inspired Optimization Techniques

Natural selection mechanics and living organism genetics are used to develop evolutionary optimization techniques, some of which are listed in Table 4. These are some of the mechanisms that are used to solve OPF problems.

Table 1 Nature-encouraged algorithms for OPF problem

S.No	Involvement of algorithm with objective Function	IEEE bus system involved	Scope/Gap
1	Improved artificial bee colony (ABC) processes may increase fuel costs, valve effects and fuel costs with P_{loss} [53]	30,118	Constant data set
2	Use the Moth Swarm algorithm to calculate fuel cost using valve effect and emissions, L-index, P_{loss}, piecewise cost, and VD [54]	30,57,118	Dynamic data needed
3	Adaptive split pollination algorithm focuses on fuel cost, P_{loss} and VD [55]	30,57	Data sets are not flexible
4	Fuel costs and emissions are calculated using the glow-worm-swarm optimization algorithm [56]	30,75	Static data gives less accuracy
5	The Best-guided artificial bee colony algorithm can reduce fuel costs [57]	30,57	Dynamic data needed
6	An opposing krill swarm algorithm was used to improve fuel cost, valve effect emission fuel cost, and P_{loss} [58]	30,57	Limited to the lower bus system
7	The chaotic krill swarm algorithm increases fuel and fuel costs with valve effects, P_{loss}, and VD [59]	26,57	Lower bus application with static data
8	The Improved group search optimization technique considers both the fuel cost and the valve effect [60]	26,30,118	The higher bus can be taken but is applicable only for static data sets
9	The Chaotic Artificial Bee Colony's algorithm improves fuel cost and transient stability [61]	30,39 (New England)	Limited to the lower bus system
10	Adaptive clone selection is used to improve fuel cost, P_{loss}, L-Index, and VD [62]	30	Limited to the lower bus system
11	Multihive's bee hunting algorithms increase fuel costs by leveraging valve effects, emissions and pros [63]	30	Applicable only for the single bus system
12	The improved flower pollination algorithm maximizes P_{loss}, VD, and fuel costs [64]	30	Limited to the lower bus system
13	The moth flame optimization algorithm can be used to enhance fuel cost, L-index, Ploss, and VD [65]	30	Limited to the lower bus system

(continued)

Table 1 (continued)

S.No	Involvement of algorithm with objective Function	IEEE bus system involved	Scope/Gap
14	Gray Wolf's optimization algorithm reduces fuel costs by utilizing the valve effect, P_{loss}, and Q_{loss} [66]	30, 118	Limited to the lower bus system

Table 2 Human-encouraged algorithms for OPF problem

S. No	Involvement of algorithm with the objective function	IEEE bus system involved	Scope/Gap
1	Methods based on biogeography enhance fuel costs and emissions, as well as the L-Index, P_{loss}, and VD [67]	30,57	Static data is needed
2	In the presence of prohibited zones, the symbiotic organism's search algorithm enhances the cost of fuel and the valve effect [68]	30	Limited to the lower bus system

Table 3 Physics-encouraged algorithms for OPF problem

S. No	Involvement of algorithm with the objective function	IEEE bus system involved	Scope/Gap
1	An improved impactor optimization algorithm uses valve effect, emissions, L-index, P_{loss}, unit price, and VD to improve fuel and fuel costs [69]	30,57,118	Dynamic data needed with a high no of iteration
2	Opposition-based gravitational search algorithms improve fuel cost, valve effect fuel cost, emissions, L-index, P_{loss}, piecewise cost, VD, and more [70]	30	Limited to the lower bus system
3	The improved mechanism method improves fuel cost, L-index, P_{loss}, Q_{loss}, unit price, and VD [71]	30,57	Limited to the lower bus system

5.5 *Hybrid Optimization Techniques*

To take advantage of several strategies and achieve superior outcomes over a single strategy, many numbers of hybrid optimization techniques have been presented. Table 5 lists the hybrid optimization strategies that have been used for OPF.

Table 4 Evolutionary encouraged algorithms for OPF problem

S. No	Involvement of algorithm with the objective function	IEEE bus system involved	Scope/Gap
1	Streamline fuel costs and emissions, L-Index, and P_{loss} with an improved self-adaptive delta advance calculation method [72]	30,57	Speedy processing time with a static data set
2	Improved evolutionary algorithm techniques reduce drastically the emission and fuel cost [73]	30,57	Processing time is limited to the lower-order bus system
3	Backtracking search optimization uses fuel cost, fuel cost with valve effect emissions, L-index, piecewise cost, and VD [74]	30,57,118	Operation time requirement is more
4	The decomposition based on the modified evolutionary algorithm enhances emissions, L-index, VD and Ploss and fuel costs [75]	30	Reduced structure with a static data set
5	The differential search algorithm is used to enhance fuel costs, emissions, L-Index, Ploss and VD [76]	9,30,57	Characteristics are different with different data sets

Table 5 Hybrid encouraged algorithms for OPF problem

S. No	Involvement of algorithm with objective Function	IEEE bus system involved	Scope/Gap
1	The cost and severity index is powered by a fuzzy logic system with a harmonic search algorithm [77]	30,57,118	Dynamically responded. The response cycle can be improved
2	PSO with aging leader and challenger algorithms to improve fuel cost, valve effect fuel cost, P_{loss} and VD [78]	30,118	
3	Implement artificial bee colony using quantum theory method to support fuel cost and valve effect fuel cost [79]	30,118	Better response with a low number of iterations
4	A hybrid model of genetic algorithm and PSO can keep fuel cost and valve effect fuel cost well [80]	30	Better response but limited to lower order system

5.6 *Artificial Neural Networks (ANN) and Fuzzy Logic Approach*

ANN methods can simulate the function of biologic brain networks, while the fuzzy set theory tool can be used to represent inexact relationships. Some of the optimization strategies based on ANN and Fuzzy logic are presented in Table 6.

Table 6 ANNs and Fuzzy approach encouraged algorithms for OPF problem

S.No	Involvement of algorithm with the objective function	IEEE bus system involved	Scope/Gap
1	The fuzzy logic model enhances emissions and fuel costs [81]	14	Traditional model used for a lower-order bus system
2	Fuzzy linear programming maximizes generation reserve and reduces fuel costs [82]	5,14	The model can be developed for dynamic data

6 Comparison Between the Recent Approaches

Different algorithms, such as the artificial bee colony, moth swarm, and glow worm swarm optimization, are used in the bio-encouraged optimization techniques utilized by taking different systems as per IEEE standard, and in the end, it is observed that the fuel cost, valve effect, emission, P_{loss}, and VD are optimized up to certain level. In the case of human and physics-encouraged techniques, different algorithms are implemented with slight development in the processing time with different static kinds of bus systems. But in evolutionary and ANN-encouraged models, the data set and number of iteration is depending on the user and that model is dynamic to a large data set generated depending upon different load flow conditions and so the power flow issue can be resolved optimally by using the advancement in the said encouraged algorithms. Also, the time needed for a higher order system is minimized slightly in these models and the advanced CNN model can be implemented or the implementation of Distributed Generation algorithm can be done to minimize the power loss, cost, emission, and other parameters.

7 Conclusion

The OPF problem was the subject of this survey. However, some of the basic points are there which is always considered. OPF problem formulation, which includes the power system's common goal functions, control variables, and operational restrictions. The traditional ways that has been used to tackle the OPF problem, as well as the benefits and drawbacks of these strategies. The most modern optimization techniques used in the OPF are classified depending on evolutionary and human-encouraged strategies. Also, natural, bio-encouraged, and physics-encouraged strategies are considered for classification purposes. Recent optimization techniques, on the other hand, outperform traditional procedures in some ways and can be used in both small and large-scale systems. High dependability in achieving the best possible results.

Acknowledgements This research work was supported by "Woosong University's Academic Research Funding—2023".

References

1. Zaeim-Kohan F, Razmi H, Doagou-Mojarrad H (2018) Multi-objective transmission congestion management considering demand response programs and generation rescheduling. Appl Soft Comput 70:169–181. https://doi.org/10.1016/j.asoc.2018.05.028
2. Nesamalar JJD, Venkatesh P, Raja SC (2016) Energy management by generator rescheduling in congestive deregulated power system. Appl Energy 171:357–371. https://doi.org/10.1016/j.apenergy.2016.03.029
3. Siddiqui AS, Saxena A (2021) Optimal intelligent strategic LMP solution and effect of DG in deregulated system for congestion management. Int Trans Electr Energy Syst 31(11):1–51. https://doi.org/10.1002/2050-7038.13040
4. Zhao J, Wang Y, Song G, Li P, Wang C, Wu J (2019) Congestion management method of low voltage active distribution networks based on distribution locational marginal price. IEEE Access 7:32240–32255. https://doi.org/10.1109/ACCESS.2019.2903210
5. Patnam BSK, Pindoriya NM (2021) DLMP calculation and congestion minimization with EV aggregator loading in a distribution network using bilevel program. IEEE Syst J 15(2):1835–1846. https://doi.org/10.1109/JPROC.2022.3177230
6. Zhou S, Han Y, Yang P, Mahmoud K, Lehtonen M, Darwish MMF, Zalhaf AS (2022) An optimal network constraint-based joint expansion planning model for modern distribution networks with multi-types intermittent RERs. Renew Energy 194:137–151. https://doi.org/10.1016/j.renene.2022.05.068
7. Abubakr H, Vasquez JC, Mahmoud K, Darwish MMF, Guerrero JM (2022) Comprehensive review on renewable energy sources in Egypt—current status, grid codes and future vision. IEEE Access 10:4081–4410. https://doi.org/10.1109/ACCESS.2022.3140385
8. Al-Gabalawy M, Mahmoud K, Darwish MMF, Dawson JA, Lehtonen M, Hosny NS (2021) Reliable and robust observer for simultaneously estimating state-of-charge and state-of-health of LiFePO4 batteries. Appl Sci 11(8):3609. https://doi.org/10.3390/app11083609
9. Xu J, Xie B, Liao S, Yuan Z, Ke D, Sun Y, Li X, Peng X (2021) Load shedding and restoration for intentional island with renewable distributed generation. J Mod Power Syst Clean Energy 9(3): 612–624. http://dx.doi.org/https://doi.org/10.35833/MPCE.2019.000062
10. Park B, Zhou Z, Botterud A, Thimmapuram P (2020) Probabilistic zonal reserve requirements for improved energy management and deliverability with wind power uncertainty. IEEE Trans Power Syst 35(6):4324–4334. https://doi.org/10.1109/TPWRS.2020.2992763
11. Esmailia M, Firozjaee EC, Shayanfar HA (2014) Optimal placement of distributed generations considering voltage stability and power losses with observing voltage-related constraints. Appl Energy 113:1252–1260. https://doi.org/10.1016/j.apenergy.2013.09.004
12. Gampa SR, Das D (2015) Optimum placement and sizing of DGs considering average hourly variations of load. Int J Elect Power Energy Syst 66:25–40. https://doi.org/10.1016/J.IJEPES.2014.10.047
13. Aman MM, Jasmon GB, Bakar AHA, Mokhlis H (2014) A new approach for optimum simultaneous multi-DG distributed generation units placement and sizing based on maximization of system loadability using HPSO (hybrid particle swarm optimization) algorithm. Energy 66:202–215. https://doi.org/10.1016/J.IJEPES.2014.10.047
14. Rosseti GJS, Oliveira, Oliveira LW, Silva IC, Peres W (2013) Optimal allocation of distributed generation with reconfiguration in electric distribution systems. Elect Power Syst Res 103: 178–183. https://doi.org/10.1016%2Fj.epsr.2013.05.017

15. Peesapati R, Yadav A, Yadav VK, Kumar N (2019) GSA–FAPSO-based generators active power rescheduling for transmission congestion management. IEEE Syst J 13(3):3266–3273. https://doi.org/10.1109/JSYST.2018.2869672
16. Salkuti SR (2022) Emerging and advanced green energy technologies for sustainable and resilient future grid. Energies 15(18):6667. https://doi.org/10.3390/en15186667
17. Shojaei F, Rastegar M, Dabbaghjamanesh M (2021) Simultaneous placement of tie-lines and distributed generations to optimize distribution system post-outage operations and minimize energy losses. CSEE J Power Energy Syst 7(2): 318–328. https://doi.org/10.17775/CSEEJPES.2019.03220
18. Gu C, Yan X, Yan Z, Li F (2017) Dynamic pricing for responsive demand to increase distribution network efficiency. Appl Energy 205:236–243. https://doi.org/10.1016/j.apenergy.2017.07.102
19. Afkousi-Paqaleh M, Fard AAT, Rashidinejad M (2010) Distributed generation placement for congestion management considering economic and financial issues. Electr Eng 92(6):193–201. https://doi.org/10.1007/s00202-010-0175-1
20. Mohanty AK, Suresh Babu P, Salkuti SR (2022) Optimal allocation of fast charging station for integrated electric-transportation system using multi-objective approach. Sustainability 14(22):14731. https://doi.org/10.3390/su142214731
21. Salkuti SR (2021) Optimal operation of smart distribution networks using gravitational search algorithm. Inter J Adv Comput Sci Appl 12(6): 531–538. https://doi.org/10.14569/IJACSA.2021.0120661
22. Gautam D, Mithulananthan N (2007) Locating distributed generator in the LMP-based electricity market for social welfare maximization. Electr Power Compon Syst 35(5):489–503. https://doi.org/10.1080/15325000601078146
23. Liyanapathirane U, Khorasany M, Razzaghi R (2021) Optimization of economic efficiency in distribution grids using distribution locational marginal pricing. IEEE Access 9:60123–60135. https://doi.org/10.1109/ACCESS.2021.3073641
24. Tushar W, Zhang JA, Smith DB, Poor HV, Thiebaux S (2014) Prioritizing consumers in smart grid: a game theoretic approach. IEEE Trans Smart Grid 5(3): 1429–1438. https://doi.org/10.48550/arXiv.1312.0659
25. Nematshahi S, Mashhadi HR (2019) Application of distribution locational marginal price in optimal simultaneous distributed generation placement and sizing in electricity distribution networks. Inter Trans Elect Energy Syst 29(5):e2837. https://doi.org/10.1002/2050-7038.2837
26. Abubakr H (2022) Adaptive LFC incorporating modified virtual rotor to regulate frequency and tie-line power flow in multi-area microgrids. IEEE Access 10:33248–33268. https://doi.org/10.1109/ACCESS.2022.3161505
27. M. N. Ali, M. Soliman, K. Mahmoud, J. M. Guerrero, M. Lehtonen, and M. M. F. Darwish (2021) Resilient design of robust multi-objectives PID controllers for automatic voltage regulators: D-decomposition approach. IEEE Access 9: 106589–106605. https://doi.org/10.1109/ACCESS.2021.3100415
28. F. U. Din, A. Ahmad, H. Ullah, A. Khan, T. Umer, and S. Wan (2019) Efficient sizing and placement of distributed generators in cyber-physical power systems. J Syst Archit 97: 197–207. https://doi.org/10.1016/j.sysarc.2018.12.004
29. Ali ES, Elazim SMA, Abdelaziz AY (2017) Ant lion optimizational algorithm for optimal location and sizing of renewable distributed generations. Renew Energy 101:1311–1324. https://doi.org/10.1016/j.renene.2016.09.023
30. Prabha DR, Jayabarathi T (2016) Optimal placement and sizing of multiple distributed generating units in distribution networks by invasive weed optimization algorithm. Ain Shams Eng J 7(2):683–694. https://doi.org/10.1016/j.asej.2015.05.014
31. Barin A, Canha L, Abaide A, Machado R (2012) Methodology for placement of dispersed generation systems by analyzing its impacts in distribution networks. IEEE Latin Amer Trans 10(2):1544–1549. https://doi.org/10.1109/ACCESS.2022.3187723
32. El-Khattam W, Bhattacharya K, Hegazy Y, Salama MMA (2004) Optimal investment planning for distributed generation in a competitive electricity market. IEEE Trans Power Syst 19(3):1674–1684. https://doi.org/10.1109/TPWRS.2004.831699

33. H. Falaghi and M. Haghifam (2007) ACO based algorithm for distributed generation sources allocation and sizing in distribution systems. Proceedings on IEEE Lausanne Power Tech, pp 555–560. https://www.academia.edu/26718152
34. Golshan MEH, Arefifar SA (2007) Optimal allocation of distributed generation and reactive sources considering tap positions of voltage regulators as control variables. European Trans Elect Power 17(3):219–239. https://doi.org/10.1002/etep.130
35. Rao RS, Ravindra K, Satish K, Narasimham SVL (2013) Power loss minimization in distribution system using network reconfiguration in the presence of distributed generation. IEEE Trans Power Syst 28(1):317–325. https://doi.org/10.1109/TPWRS.2012.2197227
36. McIlwaine N, Foley AM, Morrow DJ, Al Kez D, Zhang C, Lu X, Best RJ (2021) A state-of-the-art techno-economic review of distributed and embedded energy storage for energy systems. Energy 229: 120461. https://doi.org/10.1016/j.energy.2021.120461
37. Shutari H, Saad N, Nor NBM, Tajuddin MFN, Alqushaibi A, Magzoub MA (2021) Towards enhancing the performance of grid-tied vswt via adopting sine cosine algorithm-based optimal control scheme. IEEE Access 9: 139 074–139 088. https://doi.org/10.1109/ACCESS.2022.3208583
38. Sanjay R, Jayabarathi T, Raghunathan T, Ramesh V, Mithulananthan N (2017) Optimal allocation of distributed generation using hybrid grey wolf optimizer. IEEE Access 5: 14 807–14 818. https://doi.org/10.1109/ACCESS.2017.2726586
39. Iweh CD, Gyamfi S, Tanyi E, Effah-Donyina E (2021) Distributed generation and renewable energy integration into the grid: prerequisites, push factors, practical options, issues and merits. Energies 14(17):5375. https://doi.org/10.3390/en14175375
40. Natarajan M, Balamurugan R, Lakshminarasimman L (2015) Optimal placement and sizing of DGs in the distribution system for loss minimization and voltage stability improvement using cabc. Int J Elect Eng Inf 7(4): 679. https://doi.org/10.15676/ijeei.2015.7.4.11
41. Bayat A, Bagheri A (2019) Optimal active and reactive power allocation in distribution networks using a novel heuristic approach. Appl Energy 233:71–85. https://doi.org/10.1016/j.apenergy.2018.10.030
42. Aman M, Jasmon G, Mokhlis H, Bakar A (2012) Optimal placement and sizing of a DG based on a new power stability index and line losses. Int J Electr Power Energy Syst 43(1):1296–1304. https://doi.org/10.1016/j.ijepes.2012.05.053
43. Onlam A, Yodphet D, Chatthaworn R, Surawanitkun C, Siritaratiwat A, Khunkitti P (2019) Power loss minimization and voltage stability improvement in electrical distribution system via network reconfiguration and distributed generation placement using novel adaptive shuffled frogs leaping algorithm. Energies 12(3):553. https://doi.org/10.3390/en12030553
44. Carpentier J (1979) Optimal power flows. Int J Electr Power Energy Syst 1(1):3–15. https://doi.org/10.1016/0142-0615(79)90026-7
45. Al Harari ME, Ramababu Adapa, Momoh JA (1993) A review of selected optimal power flow literature to 1993. II. Newton, linear programming and interior point methods. IEEE Transaction on Power System, pp 105–111. https://doi.org/10.1109/59.744495
46. Momoh JA, Adapa R, El-Hawary ME (1999) A review of selected optimal power flow literature to 1993, I- Nonlinear and quadratic programming approaches. IEEE Trans Power Syst 14:96–104. https://doi.org/10.1109/59.744492
47. Maria GA, Findlay JA (1987) A Newton optimal power flow program for Ontario Hydro EMS. IEEE Trans Power Syst 2:576–582. https://doi.org/10.1109/TPWRS.1987.4335171
48. Stott B, Hobson E (1978) Power system security control calculations using linear programming, Part I. IEEE Trans Power Appar Syst 5:1713–1720. https://doi.org/10.1109/TPAS.1978.354664
49. Shen CM, Laughton MA (1970) Power-system load scheduling with security constraints using dual linear programming. Proc Inst Elect Eng 117:2117–2127. https://doi.org/10.1049/piee.1970.0382
50. Zhu J (2015) Optimization of power system operation. John Wiley & Sons, Hoboken, p 47
51. Wells D (1968) Method for economic secure loading of a power system. Proceedings of Institution of Electrical Engineers, pp. 1190–1194. https://doi.org/10.1049/piee.1968.0210

52. Ross Baldick (2009) Applied optimization: formulation and algorithms for engineering systems. 1st Edition, Cambridge University Press. http://users.ece.utexas.edu/~baldick/ntroductory_Chapters
53. Wenlei Bai, Ibrahim Eke, Kwang Y. Lee (2017) An improved artificial bee colony optimization algorithm based on orthogonal learning for optimal power flow problem. Control Eng Practice, pp 163–172. https://doi.org/10.1016/j.conengprac.2017.02.010
54. Al-Attar Ali Mohamed, Yahia S Mohamed, Ahmed AM El-Gaafary, Ashraf M Hemeid (2017) Optimal power flow using moth swarm algorithm. Elect Power Syst Res 190–206. https://doi.org/10.1016/j.epsr.2016.09.025
55. Belgacom Mahdad K Srairi (2016) Security constrained optimal power flow solution using new adaptive partitioning flower pollination algorithm. Applic Soft Comput, 501–522. https://doi.org/10.1016/j.asoc.2016.05.027
56. Salkuti Surender Reddy, Ch Srinivasa Rathnam (2016) Optimal power flow using glowworm swarm optimization. Intern J Electr Power Energy Syst, 128–139. https://doi.org/10.1016/j.ijepes.2016.01.036
57. Jadhav HT, Bamane PD (2016) Temperature-dependent optimal power flow using g-best guided artificial bee colony algorithm. Int J Elect Power Energy Syst, 77–90. https://doi.org/10.1016/j.ijepes.2015.11.026
58. Aparajita Mukherjee, V Mukherjee (2016) Solution of optimal power flow with FACTS devices using a novel oppositional krill herd algorithm. Int J Elect Power Energy Syst, 700–714. https://doi.org/10.1016/j.ijepes.2015.12.001
59. Aparajita Mukherjee V Mukherjee (2015) Solution of optimal power flow using chaotic krill herd algorithm. Chaos, Solitons Fractals, 78: 10-21. https://doi.org/10.1016/j.chaos.2015.06.020
60. Yi, Tan, Li C, Cao Y, Lee KY, Li L, Zhou STL (2015) Improved group search optimization method for optimal power flow problem considering valve-point loading effects. Neuro Comput 148:229–239. https://doi.org/10.1016/j.neucom.2013.09.065
61. Ayan K, Baraklı UKB (2015) Chaotic artificial bee colony algorithm based solution of security and transient stability constrained optimal power flow. Intern J Elect Power Energy Syst 64:136–147. https://doi.org/10.1016/j.ijepes.2014.07.018
62. Srinivasa Rao B, Vaisakh K (2014) Multi-objective adaptive clonal selection algorithm for solving optimal power flow considering multi-type FACTS devices and load uncertainty. Appl Soft Comput 13:286–297. https://doi.org/10.1016/j.asoc.2014.06.043
63. Chen H, Zhu MLBY (2014) Multi-hive bee foraging algorithm for multi-objective optimal power flow considering the cost, loss, and emission. Intern J Electr Power Energy Syst 60:203–220. https://doi.org/10.1016/j.ijepes.2014.02.017
64. Barocio E, Regalado J, Uribe EC, Pavel Zuniga ero J. Ramirez Torres (2016) Modified bio-inspired optimisation algorithm with a centroid decision-making approach for solving a multi-objective optimal power flow problem. IET Gener Transm Distrib 11. http://dx.doi.org/https://doi.org/10.1049/iet-gtd.2016.1135
65. Salkuti SR (2020) Multi-objective based economic environmental dispatch with stochastic solar-wind-thermal power system. Intern J Electr Comput Eng 10(5): 4543–4551. http://doi.org/https://doi.org/10.11591/ijece.v10i5.pp4543-4551
66. El-Fergany AA, Hasanien HM (2015) Single and multi-objective optimal power flow using grey wolf optimizer and differential evolution algorithms. Electr Power Compon Syst 43:1548–1559. https://doi.org/10.1080/15325008.2015.1041625
67. Kumar AR, Premalatha L (2015) Optimal power flow for a deregulated power system using adaptive real coded biogeography-based optimization. Intern J Electr Power Energy Syst 73:393–399. https://doi.org/10.1016/j.ijepes.2015.05.011
68. Duman S (2017) Symbiotic organisms search algorithm for optimal power flow problem based on valve-point effect and prohibited zones. Neural Comput Applic 28: 3571–3585. https://doi.org/10.1007/s00521-016-2265-0
69. Bouchekara HREH, Chaib AE, Abido MA, El-Sehiemy RA (2016) Optimal power flow using an improved colliding bodies optimization algorithm. Applic Soft Comput 42:119–131. https://doi.org/10.1016/j.asoc.2016.01.041

70. Bhowmik AR, Chakraborty AK (2015) Solution of optimal power flow using non-dominated sorting multi-objective opposition based gravitational search algorithm. Intern J Electr Power Energy Syst 64:1237–1250. https://doi.org/10.1016/j.ijepes.2014.09.015
71. Wang XY, Yuan Y, Huang Y, Zhang X (2015) A new quantum-inspired chaotic artificial bee colony algorithm for optimal power flow problem. Energy Conv Manag 100:1–9. https://doi.org/10.1016/j.enconman.2015.04.051
72. Harish Pulluri R, Naresh VS (2017) Anenhanced self-adaptive differential evolution based solution methodology for multiobjective optimal power flow. Applic Soft Comput 54:229–245. https://doi.org/10.1016/j.asoc.2017.01.030
73. Yuan X, Zhang B, Wang P, Liang J, Yuan Y, Huang Y, Lei X (2017) Multi-objective optimal power flow based on improved strength Pareto evolutionary algorithm. Energy 122:70–82. https://doi.org/10.1016/j.energy.2017.01.071
74. Chaib AE, Bouchekaraa HREH, Mehasni R, Abido MA (2016) Optimal power flow with emission and non-smooth cost functions using back tracking search optimization algorithm. Intern J Electr Power Energy Syst 81:64–77. https://doi.org/10.1016/j.ijepes.2016.02.004
75. Zhang J, Tang Q, Li P, Deng D, Chen Y (2016) A modified MOEA/D approach to the solution of multi-objective optimal power flow problem. Applic Soft Comput 47:494–514. https://doi.org/10.1016/j.asoc.2016.06.022
76. Abaci K, Yamacli V (2016) Differential search algorithm for solving multi-objective optimal power flow problem. Intern J Electr Power Energy Syst 79:1–10. https://doi.org/10.1016/j.ijepes.2015.12.021
77. Singh RP, Mukherjee V, Ghoshal SP (2016) Particle swarm optimization with an ageing leader and challengers algorithm for the solution of optimal power flow problem. Applic Soft Comput 40:161–177. https://doi.org/10.1016/j.asoc.2015.11.027
78. Pandiarajan K, Babulal CK (2015) Fuzzy harmony search algorithm based optimal power flow for power system security enhancement. Inter J Electr Power Energy Syst 78:72–79. https://doi.org/10.1016/j.ijepes.2015.11.053
79. Akay B, Karaboga D (2012) A modified artificial bee colony algorithm for real-parameter optimization. Sci Direct Inf Sci 192:120–142. https://doi.org/10.1016/j.ins.2010.07.015
80. Gacem A, Benattous D (2017) Hybrid genetic algorithm and particle swarm for optimal power flow with non-smooth fuel cost functions. Inter J Assur Eng Manag 8:146–153. https://doi.org/10.1007/s13198-014-0312-8
81. Ramesh VC, Li X (1997) Optimal power flow with fuzzy emissions constraints. Electric Mach Power Syst 25:897–906. https://doi.org/10.1080/07313569708955784
82. Reddy SS, Panigrahi BK, Bijwe PR, Abhyankar AR (2010) Comparison and application of swarm intelligent techniques to optimal power flow. Joint International Conference on Power Electronics, Drives and Energy Systems & Power India, New Delhi, India 1–6. https://doi.org/10.1109/PEDES.2010.5712552

Review on Microgrids: Types, Challenges, Opportunities, Uncertainties, and Their Modeling

Kunal Shankar, Surender Reddy Salkuti, and Seong-Cheol Kim

Abstract The world needs an easily attainable, dependable, and cheaper electricity network in the coming days and this can be realized with the help of microgrids. These microgrids are rising rapidly and the factors backing up this rise includes several mechanical enhancements, sinking price, and environment-friendly solutions driven by renewable sources of energy in place of conventional sources like petroleum and fossil fuels. With factors like various geographic locations, availability of resources, consumer demand, and existing trends of transmission and distribution, microgrids are able to perform in isolated mode as well as when connected to the grid. In this chapter various topics like the definition of microgrids, their classification based on sites, the challenges faced in their establishment and functioning and the opportunities that the future of the microgrid withholds, the uncertainties in the operation of microgrids, and modeling of various components to deal with those uncertainties.

Keywords Microgrids · Renewable energy · AC–DC Microgrids · Modeling · Uncertainties · Opportunities

Nomenclature

DERs	Distribution energy resources
MG	Microgrid
PEI	Power electronic interface
CHP	Combined heat and power

K. Shankar
Department of Electrical Engineering, Birla Institute of Technology Mesra, Patna, Bihar, India

S. R. Salkuti (✉) · S.-C. Kim
Department of Railroad and Electrical Engineering, Woosong University, Daejeon 34606, Republic of Korea
e-mail: surender@wsu.ac.kr

S.-C. Kim
e-mail: kmin@wsu.ac.kr

S. R. Salkuti et al. (eds.), *Power Quality in Microgrids: Issues, Challenges and Mitigation Techniques*, Lecture Notes in Electrical Engineering 1039,
https://doi.org/10.1007/978-981-99-2066-2_17

DR	Demand response
DMS	Distribution management system
PV	Photovoltaic
LVDC	Low-Voltage Direct Current
SDN	Software-Defined Networking
CFLs	Compact fluorescent tubes
FC	Fuel cell
DEG	Diesel engine generator
ESD	Energy storage devices
BES	Battery energy storage
FES	Flywheel energy storage
EV	Electric vehicles
PEC	Power electronic converters

1 Introduction

Conserving energy and ensuring its compatibility with the environment has become one of the necessary conditions in today's world [1]. The microgrid concept is based on single electric power systems that have the involvement of DERs, i.e., distributed energy resources such as photovoltaic cells, wind energy turbines, fuel cells, and gas turbines. These resources associated can be either of conventional or renewable nature. Storage systems like capacitor banks, chargeable battery arrangements, and flyways also play an important role in completing their load requisition by the use of the diverse feature and the dimensions at the culmination manipulators of electrical energy. If we use individual DERs for micro-generation there might be issues like a rise in local voltage, the capability of exceeding the limit of various lines thermally, high capital cost, and these problems could be dealt with more easily with the help of a microgrid. In a microgrid, the power quality and output of energy are maintained very well because of DERs equipped with proper power electronic interface (PEI) and control. The main advantage of this system is that it functions as a single controlled entity along with the contribution from the power system that helps them to function as a single load. The benefits that a user gets at their end include improved power quality, feeder loss is reduced, smooth flow of power, and the electricity demand being met easily and locally. The microgrid system provides benefits to users that include selling surplus energy, buying it from other users, or even producing it [2]. The flexibility of microgrids provides a suitable portfolio for demand regulation among the producers and end users [3]. Microgrids are capable of selling as well as buying power from upstream [4]. By adding technology to the microgrid, a connection gets established between the energy management system and microgrid as well as with the consumers [5]. Microgrids with the aid of demand-side management systems utilize renewable energy resources to the full extent [6].

Issues caused by the microgrid system are power quality issues. They use renewable sources for operation and these sources are generally dependent on nature. Thus, the availability might vary from place to place but we definitely cannot overlook the advantage of renewable energy systems like low cost, clean, and environmentally friendly energy. Hence, for this, further research is needed to sort out this problem of reliability. Another important point for the smooth functioning of a microgrid system is its storage system. Most microgrids have systems like battery storage, flywheels, etc., for storage. It was noticed that some microgrid systems are made up of more than two to three storage systems, while some might not have it. Thus, it was concluded that a system should have at least a DG source in case no storage systems are present. It was also discovered that most of the test beds used by these microgrid systems are either AC or DC in nature. AC had the upper hand as the grid and even the loads are mostly AC in nature, it is easy to integrate with the grid. However, AC systems might face issues with power quality. DC is used in a very limited number due to the unavailability of enough DC loads. However, the advantage in DC is that there are fewer cases of power quality issues and thus less control and component are required.

This chapter aims at discussing all the kinds of challenges be they political, economic, social, or environmental that acts as a hindrance to the smooth functioning of these systems and the opportunities that can be attained in the future with the help of some small regulations in the arrangements.

2 What Is a Microgrid?

It is an aggregate of loads, distributed generator (DG) sources, and storage devices and utilizes DG resources and utilizes few kilowatts or megawatts that can be integrated into the microgrid at various levels of loads like local, feeder, and substation. Considering the perspective of the power system, as compared to the traditional energy systems that followed the path of electricity getting produced at the power plants and then going to the transmission systems and ultimately getting distributed the system of a microgrid is relatively new [7]. As envisaged, a large number of interconnected microgrids, having Renewable energy resources and storage devices, could make up the future infrastructure [8]. In a broader sense, the microgrid can be termed as distributed and small electrical power generators collectively working in a system [9]. It generally has the capability to operate in two modes, i.e., islanded and grid-connected modes. If the microgrids are being fed by the utility grid then the mode is said to be grid-connected and if it gets separated from the main grid and starts acting on its own, while the fault occurs then the mode is called islanded mode. In islanded mode distributed generators like diesel generators, micro-turbines and these help in feeding the local requirement and these help in providing reliable and regular connection to the customers. Microgrid is a promising development that has been proven very reliable after the connection is removed from the main grid [10]. In recent times, in order to aid the supply of electricity to the areas which are rural by nature as the traditional way of providing electricity to them would cost a lot and make it difficult for them, microgrids can be a solution to this problem [11]. And, this is the reason why these microgrids are mostly preferred to be established in rural areas where the power supply, if provided by conventional means, would cost a lot [12]. The microgrid system generally has the following characteristics:

- These subsystems are capable of operating independently at various topographical locations.
- In these systems, we are capable of adding or removing the subsystems at any instant in time.
- The emergency requirement of microgrids is to cater to all the load requirements at any instant of time [13].

Microgrids make use of distributed energy resources like electricity generators, energy storage, load control, and electric interfaces between generators and distribution grids. It is made up of power electronic interfaces like inverters, load controllers, energy storage, and made up of generators [14]. The major benefit of the microgrid is it combines the advantage of highly efficient heat and power (CHP) systems and non-conventional or renewable low-carbon generation technologies. The distributed generator used in a system is chosen on the basis of the climate and the geographical locations of a system. Other factors affecting the sustainability of a microgrid are strategy, energy scenario, and policy of the country and it usually varies from region to region [15].

The most essential importance of microgrids is that it provides local energy security in remote communities. In this way, more secure energy is assured and makes them less dependent on importing energy from the main grid. This also enables no additional effort for frequency regulation which makes it more in demand. Distributed generation plays an essential role in today's environment as it helps in sorting two most difficult factors which include carbon emission concern and increasing energy demand which can be solved by using DERs thus this leads to the development and establishment of small-scale grid networks and their technologies [16]. With the decrease in the resources of fossil fuels and due to the rising economic and environmental issues, researchers pay more emphasis on using renewable energy resources. The type of renewable energy used by the engineers is mostly found at the local level at distribution power systems and that helps in generating power from them. These energy resources are also termed distributed energy resources [17]. The areas where existing power grids are difficult to reach can also be provided with electricity by the use of locally present renewable energy sources and this is getting a lot of attention in today's world. However, there has been continuous development in the field of renewable energy resources is it optimizing the use of these energy resources or developing the power conversion system [18].

For a long time only the conventional structure of the power systems which are generally dependent on centralized generation resources that are connected to the consumers via long transmission lines has been defined. The losses in such systems generally include poor visibility, huge power losses, low penetration of distributed energy resources (DERs), and electrochemical devices attached to slow the response time [19]. These challenges concerning the economic aspect impact caused on the environment and the advancements in technology have led to the demand for using the distributed generation instead of conventional centralized generation. Since power operation companies pay importance to such factors as the requirement of the load, the satisfaction of the customer in the product, and the effect of the product on the environment and that leads to the preferring of distributed energy resources as it helps in reducing the pressure from the main grid [20].

3 Classification of Microgrids

3.1 Remote Microgrids

Remote microgrids are generally grids that operate without being connected to a nation's electric power system. There are certain areas where the main power plant supply may not be present and these are the areas where remote microgrids can play an essential part. It generally works on extra low voltage and operates at a very cheap price. These also can be set up in areas where infrastructure cost is high or even if there are other factors that may prevent the connection from the main grid. The main advantage of these grids is effectiveness and their reliability in cases of blackout or in cases of connection issues with the main grid. These have been proven as the best solution for providing electricity to the villages where connection to the main grid can't be made. Other ways of increasing the efficiency of these systems are using Low-Voltage Direct Current (LVDC) which can increase the safety of the systems or smart metering technology that can be useful in providing better control and management of the microgrid systems [21, 22].

3.2 Campus Microgrids

The sun as the primary source can be utilized in the universities by tapping the sunlight and utilizing it in the microgrids thus giving rise to campus microgrids. Because of the increasing concerns about global warming and the prevailing economic recession, a microgrid can be established inside the campus. Here, the need for microgrids can be met easily inside the campus as there are enough resources to provide to the system. Thus network reliability, reduced cost, and a pollution-free environment—all these three factors can be achieved inside the campus with the help of this microgrid. Microgrids are the initial building blocks of the smart grids and since we know that all the universities are good contributors to electrical consumption, which can be efficiently controlled by the use of an energy management system [23, 24].

3.3 Military Microgrids

The forward bases that operate in the military are located in deserted environments and lack connection to the heavy power systems that also require the need of microgrids so that at least they can act independently. However, these FOBs should avoid steady wireless communications for security. The power is distributed using field distribution boxes that are also extended to provide network communications across the base. Other issues associated are that there are substantial monetary, safety, and logistic costs. So not only microgrids were needed but also, to cut the cost and reduce pollution, these were integrated with renewable energy resources. The military microgrids generally consist of low power, i.e., 10 s of kWs of powered generators which have relatively equal capacity. These systems are mostly available in sizes of 30 kW and 60 kW. The nature of loads in these types of microgrids is inductive in nature and that ultimately provides pressure on the grid system. Microgrids that are meant for serving in the military also face several unique challenges that are not discussed here in this chapter in detail [25, 26].

3.4 Residential Microgrids

A residential microgrid is a combination of cooling, heating, and power (CCHP) system which also includes the power generation of renewable energy and is most prevalent in residents or communities in such cases better consumption of these renewable energy resources can be facilitated by microgrid energy management and that also helps in reducing microgrid operation and also this management allows us to optimize other energy resources that might be available. The optimization of these renewable sources is maintained at the local levels with the help of these microgrids. Also, other services like ancillary service are provided by these systems that help in making the voltage and frequency stable in case of islanding conditions. They also act as micro-networks and work as serve the infrastructure for smart grids which is generally harder to implement in larger grids and thus play a very essential role in the deployment of smart grids [27].

3.5 Commercial Microgrids

There are buildings that have mixed residential as well as commercial units that show relevant power peaks and thus microgrids that are designed for such kind of buildings are commercial microgrids. That is proof that big businesses are also utilizing microgrids for energy generation mainly to protect them from sudden outages and also reduce or minimize their costs. They adopt distributed energy resources and local consumptions near the user. They are not similar to other microgrids as they mostly aim at attending to the demand of businesses and corporations and do not cater to the needs of a particular population or group of people in a society. This also intends in generating revenue by cutting the cost of company energy expenditure and preventing revenue losses incurred in a microgrid [28].

4 Microgrid Architectures

Microgrids on the basis of market segments can be divided into different categories like remote microgrids, campus microgrids, military-based microgrids, residential and commercial microgrids, and when divided on the basis of system topology they can be classified into AC microgrids, DC microgrids or a combination of AC and DC microgrid. AC microgrid interconnects various distributed generators which are AC in nature like wind turbines and DC distributed generators like PV, and fuel cells using the inverter. In the case of a DC microgrid rectifier is used to convert the output of a distributed generator which is AC by nature to DC. Their architecture is mainly determined by the nature of loads, the to-be-implemented or existing generators, the existing communications, the place where the energy storage devices are added and their needed requirement in case of power and energy, and the barriers that would come up in building new electrical lines [29, 30].

4.1 AC Microgrids

AC microgrids are the most used and most conventional type of microgrids. The AC microgrid comprises various renewable-based energy sources such as wind turbine generators (WTGs), photovoltaic (PV) cells, microturbines (MTs), aqua electrolyzer based fuel cells (FC), and diesel engine generators (DEGs) along with various Energy storage devices (ESD) like battery energy storage (BES), flywheel energy storage (FES) and electric vehicles (EV) [31]. Power electronic converters (PECs) are used here to integrate different DERs such as wind-turbine generators, microturbines, fuel cells, and turbines. Minimum modifications or changes are required to establish these because already there exist AC conventional power networks. Generally, they are connected to medium voltage and low voltage networks which reduces transmission losses by enhancing power flow in distribution networks. They come with disadvantages like system stability, synchronization, and power quality which of course can be resolved by applying various techniques in control.

For most of the cases related to the implementation of microgrids generally, AC microgrids are preferred. The existing AC infrastructure that includes distribution, protection, and transformers makes it easier for AC microgrids to get implemented and thus the people are sure of what they are constructing as these microgrids already have shown reliable and assured results in the past. Consortium for electric reliability technology solutions which is also known as CERTS was the body that developed the first microgrid in 1998, which was basically made by bringing many micro-generators together and they beheld the capability of isolating themselves from the main grid without any effect on the loads working on it. Taking this model of CERTS as a reference, architecture has been proposed as shown in Fig. 1 which is made up of 3 AC feeders where two of them have critical loads like distributed generation (DG) and energy storage systems (ESSs) and the other one focuses on the non-critical loads. The best feature of this microgrid is that it was made to be accustomed to any kind of operating conditions and holds the capability of generating and providing the demands in such conditions which is mainly done by changing the topology with the help of circuit breakers. The static switch is been provided to manage the connection of microgrids to distribution grids. This also holds the capability of disconnecting itself and functions in islanded conditions when poor-quality of electricity distribution is provided.

The result of which is the good and reliable quality of electricity provided to the load. In order to remove the connection from the non-critical loads the static switch and the circuit breaker of the bus which is third in position are disconnected in case of grid faults and protect from further damage. The series-connected converter is generally avoided in AC microgrids, instead, there is a direct flow of power from and toward the grid which makes these grids even more reliable. The feeders are

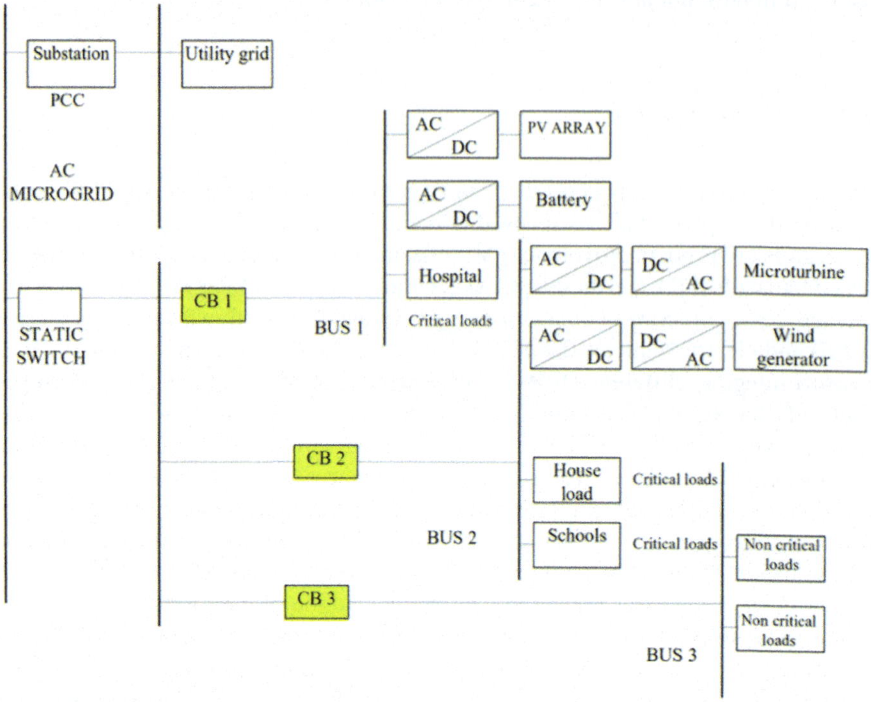

Fig. 1 Proposed model for AC microgrid

also fed with the same frequency and voltage so that the ESS, DG, and loads are grid-compliant. There are certain drawbacks of these microgrids that include that in order to provide good quality AC currents that do not have harmonics, DERs have to be synchronized with the AC utility grid and this can be possible with only a high amount of complex power electronics interfaces which will cause the decline of reliability and efficiency of these microgrids and also the conversion steps in case of AC is much higher than that of DC.

4.2 DC Microgrids

Various developments in power electronic devices in DC microgrids are putting these microgrids to various applications as well. And also, DC based DERs and various ESS create more opportunities for using DC microgrids. It highly reduces the number of stages that are needed in the conversion and there is the absence of active current circulations which are the major benefits of using DC microgrids. DC microgrids and dc distribution is becoming a system that has become very common in recent days [32].

DC is preferred in cases at places where the Energy Storage Devices (ESS) which has terminal-point connection points are mainly supercapacitors, batteries, and fuel cells which are completely DC in nature as well the power electronic interfaces on the Distributed Energy Resources (DERs) which have intermediate DC link. We also cannot overlook the fact that most of the consumers have loads that are DC supplied. It is preferred to integrate this device into the DC microgrids with the help of DC-DC converters as this increases the efficiency as the conversion steps are reduced and also a better power quality is achieved. We have discussed DC microgrid architecture in Fig. 2 where the power electronic converter of this architecture is the AC–DC interface which is also known as the interlinking converter as the point of common coupling connects the DC microgrid to the AC utility grid post voltage shift provided by the transformer. This power-electronics converter has to be bidirectional so that power could flow in both directions. These power electronic devices generally play the role of linking the DG and ESS to a DC bus with a pre-given voltage. This given voltage of the DC bus is regulated by the IC which provides high quality to the microgrid irrespective of the quality of the main grid. The main benefits of these microgrids that are they are cheaper than the AC microgrids have a simpler construction and the efficiency as we saw earlier, get increased [33]. The losses in the distribution losses are much less compared to AC microgrids as there is no reactive current component. Due to the voltage regulations done by the IC and the energy stored in the DC capacitors, DC microgrids make their way through the faults and are not affected by the decline in voltage or even the blackouts. However, they are incompatible compared to AC microgrids to the actual power systems and there is a need of building DC distribution lines. Since DC microgrids don't have a very assured and proven past, generally there are uncertainties regarding immaturity in standards and guidelines [34]. Another disadvantage is voltage is not standardized because the

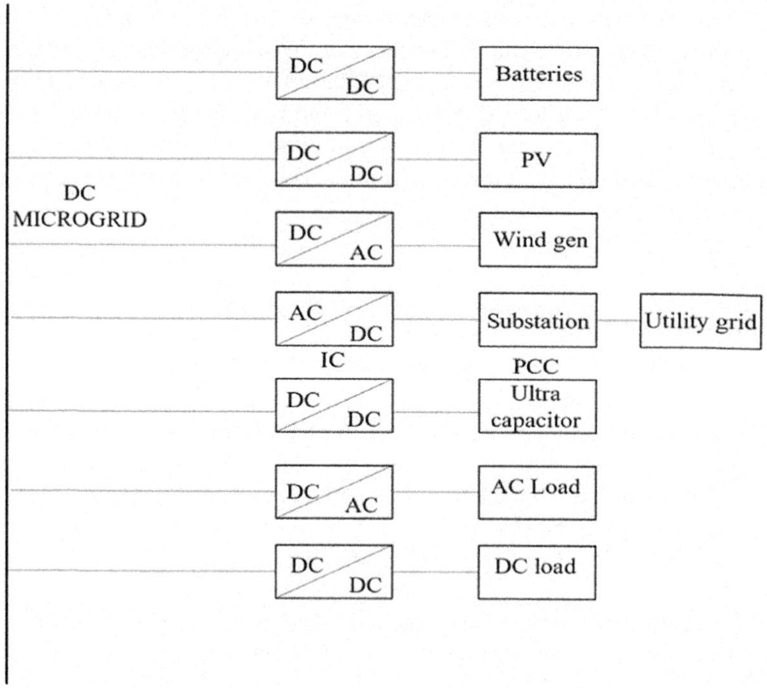

Fig. 2 Proposed model for DC microgrid

AC loads cannot be connected to the system of microgrids which generally leads to the need for additional converters.

4.3 *AC–DC Microgrids*

Hybrid microgrids combine the benefits of AC and DC microgrids that come with benefits like increased reliability, economical aspects, and efficiency without a doubt. Direct integration is possible by the usage of these hybrid microgrids of AC and DC based DERs and ESSs and loads with the existing distribution system. In domestic use, there is the vast application of DC appliances and they would require conversion of AC to DC [35]. Power losses are reduced in hybrid microgrids by reducing the number of converters.

Nowadays hybrid microgrids are getting attention because they combine the benefits of AC and DC microgrids together which ultimately helps in the direct and advantageous integration of both AC and DC based DERs and loads. The architecture of a hybrid microgrid has been discussed in Fig. 3 where an AC–DC converter has been used to link the AC and DC grids. This converter is without a doubt bidirectional in nature. DC loads are given connection to DC buses and AC loads are connected to the AC buses with the aid of a power converter so that changes can be made according to the required voltage. The connection of ESS and the DG can be done on either AC or DC buses in the best way possible to reduce the conversion steps and as we discussed the advantages get added, so AC buses enable the use of existing equipment and DC bus helps in reducing the number of simple converters [36]. More robust loads are connected in the AC feeder and alongside the connection of reasonable loads to the DC feeder. The electronic surfaces got reduced while combining these two architectures and so did the energy losses. The quality of AC is maintained in the utility grid by controlling the harmonic injections by including DC devices. With the help of these hybrid grids, a communication network and a majorly needed AC–DC converter can be introduced, which makes the cost higher but returns the investment faster compared to the AC microgrids. These grids are also not devoid of disadvantages like protection and management issues [30, 37].

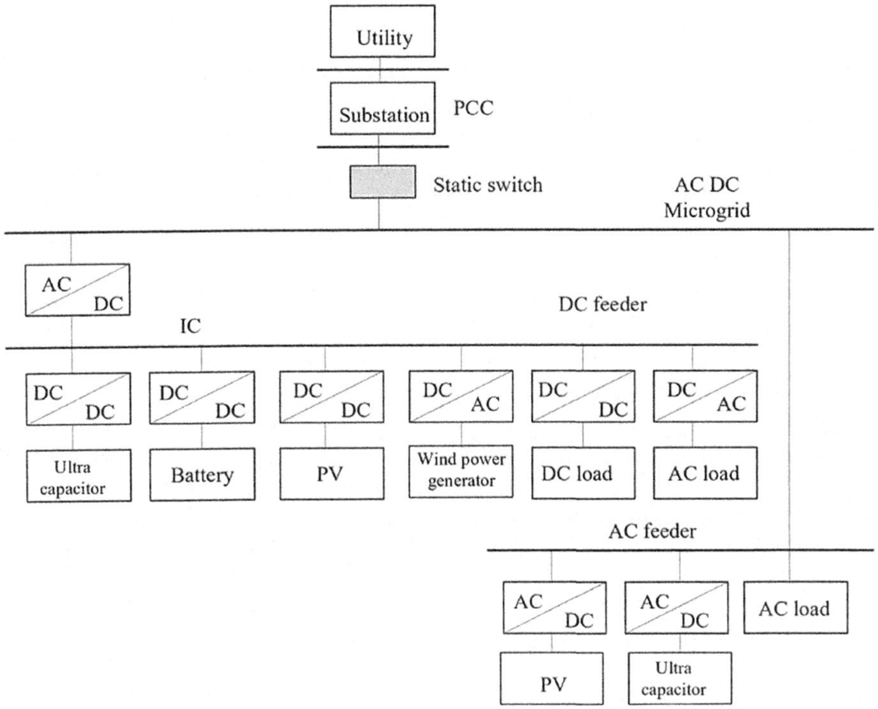

Fig. 3 Proposed model for AC–DC microgrid

5 Challenges in Microgrids

Microgrids are gaining popularity day by day and a great number of them are going to be installed in the future for consumers. But before that issues related to technical faults, control, and protection has to be addressed properly. Technical faults occur when the balance between active and reactive power is not maintained and the unbalance is the result of inequality between power generated and the power demand resulting in deviation of the systems. Also, factors like the usage of renewable energies don't always guarantee constant intensity throughout, like solar energy intensity is not the same everywhere, and also solar energy is not available at night [38]. The same goes for other renewable energy sources like wind, water, etc. It is mainly because fluctuations in weather or the flow of the river give rise to fluctuating voltage supply which is not profitable for end users thus making them less reliable. In offshore wind power plants, problems like non-linear power circuitry and the nature of wind can bring power quality (PQ) issues in the system and the harmonics produced to affect the stability of the system. So in a nutshell there is a high usage of renewable resources meant for the generation and ultimately increasing its contribution to electricity networks on a global scale.

However, these networks are prone to changes in climatic and weather conditions which is serving as a challenge in terms of reliability. Protection issues might arise in islanded conditions where the fault current in inverter-based microgrids might not have sufficient magnitudes to make use of traditional over-current [39]. The protection systems of microgrids also generally face problems like false separation and as a result, failure in the network or microgrid cannot predict if there is a fault in the network or there is a fault in network distribution. The parts of the DC Microgrids might be damaged by high fault current and this may lead to loss of voltage and current control and this can increase the cost as a result of a high power rating [40]. Factors like the supply and storage of energy that is supplied in form of batteries must be planned in accordance with load demand on the microgrid. In addition, the flow in both directions, tripping of the system in a certain manner, a different mode of operations causing changes in the level of short circuit, and very limited contribution of fault current in the microgrids is also a serious issue in these systems. Other issues also arise in cyber security, designing appropriate grounding systems, and fast fault detection [41]. These faults causing factors can be broadly classified into social, economic, technical, and environmental causes and are discussed briefly below.

5.1 Social Causes

The areas where microgrids are established are mostly rural and there is a lack of education that prevails a lot among the people. They don't really know the benefits of using microgrids and tend to stick to the traditional way of using inefficient applications. They are not educated on energy conservation and a lot of energy is wasted as a result. They are not aware of how these systems can be maintained as they are interpreted mostly as no maintenance system by the majority of people and as a result, most of the systems get abandoned in no time. The same goes for the community may be mostly because there is a communication gap during the initial stages of system planning [42]. The community also can lead to problems like land conflicts as there is a lack of mutual agreement between the several parties who are to be involved in the project such as the developers, government, or investors.

Another major problem is that the installation of these microgrid systems is done by inexperienced installers or practitioners as they think that this is a simple plug-and-play game that leads to the malfunctioning of the systems. They don't pay attention to essential factors like energy consumption patterns, variable power output, and the possibility of users' load growth. Last but not least vandalism practiced by the people of the society and theft of the components of the microgrid systems lead to the failure of smooth functioning of microgrids and thus taking away the opportunity of having a rich energized environment from the poor people of rural societies.

5.2 *Technical Causes*

The architecture of renewable energy microgrids is completely different from the setup of conventional energy resources. When the designing of such systems is handed over to technicians who are not aware as to what are the sources of the renewable energy available, in what quantity is the renewable energy available, and is not aware of the climatic conditions of that particular area, can cause the formation of poorly designed microgrids that may not function properly as they have not been made in accordance to the weather conditions it will be prone to. Another factor is poor and improper maintenance of these systems as these microgrids are generally established in areas which usually are dusty and rural and this ultimately causes the microgrids to cover in dust and malfunction. Even the use of low-quality sub-standard materials affects the functioning of microgrids as the initial cost of setting up a microgrid is usually high but in order to reduce that people are hampered by the quality of these systems which don't function properly in long run. Although at some places corrective precautions are carried out for maintenance lack of monitoring systems ultimately causes the system's failure as corrective measures are not enough because these systems need to be monitored 24*7 even when people are not present around but since they are unfortunately not present at most of the places, that also is one of the reasons for making microgrid functioning a failure.

5.3 *Economic Causes*

In most countries, in order to set up a microgrid system help from the government is required in some form of donation. But still, some government bodies are unsure because of the associated uncertainties attached to these systems which stop them from investing and providing aid. No one wants to bear the expenses that come along with the systems and thus making people reluctant and unsure. As mentioned earlier since the users as well as the community is not aware of the benefits and thus are also not concerned with the maintenance of these and this leads ultimately to poor revenue generation. Lastly, not only the initial establishing cost of these systems are high but even the parts associated are highly expensive so if they stop functioning, it usually takes a huge sum to mend them and that generally leads to the systems remaining unattended for a long time and slowly every part of the system keep on malfunctioning and thus making the system useless.

5.4 Environmental Causes

It is very vital to consider environmental factors while installing the microgrids. In most areas, people are unaware of the available energy resources that are available in the site of construction and if they are, they mostly know about solar energy. This leads to inefficient energy generation in those areas as they only remain dependent on solar energy and don't explore other renewable energies that can be equally beneficial. Another factor contributing to the environmental causes is the cradle-to-grave cycle of the microgrid systems, i.e., when these are manufactured as well as de-commissioned there is a primary energy demand and carbon emissions which naturally has an impact on nature. Similarly when these are disposed harmful materials like Compact fluorescent tubes (CFLs) and battery cells are simply thrown or disposed as normal materials by the people instead of taking the steps of proper disposal and this also is a factor contributing to the harm [43].

6 Opportunities in Microgrids

Opportunities for microgrids are increasing dramatically with time as features including integration of renewable and environment-friendly energy have greatly attracted customers' attraction towards them. People are likely to avoid the hustle of building larger and centralized power plants which cause pollution. The opportunities of the microgrid are numerous in the future if the defects are addressed properly. Some of the opportunities of the microgrid include.

6.1 Safe Islanding Systems

One of the most attractive features of the microgrids is the safe islanding feature which allows them to operate independently from the main grids. But there are associated issues related to feeder fault concerns in the case of synchronous generators which however can be avoided by the usage of inverters that allow the systems to operate even when a larger grid goes down.

6.2 Load Fluctuation Absorbent

On using dedicated sources of energy storage and output sufficiently in microgrid systems, the system becomes prone to fluctuations in the load. It means the system becomes capable 13 of handling load fluctuations be it of any size and this also helps the generators to function in a more optimally loaded state.

6.3 Harmonics Free System

There are certain nonlinear loads when connected to microgrid systems that can cause power quality problems that may arise due to the presence and propagation of harmonics in the systems which however may be solved by the usage of inverters in the system.

6.4 Software-Defined Systems

Microgrid has been defined as a group of renewable energy resources and the loads existing as a system which can be controlled independently that can serve a local area and the usage of renewable energy has highly increased its value in the market. This along with the coupling of smart grid infrastructure makes it superior. The progress in technologies like Software Defined Networking (SDN) has made microgrid architecture to be more distributed and intelligent.

6.5 Cost-Efficient DERs

The feeder-based microgrids can be deployed in the distribution system with the help of technology as well as political drivers [44]. It takes into account existing architectures and what can be or has been integrated into them. DER Technology is improving on a daily basis making the performance and production better while reducing the cost.

6.6 Stable Systems

The various needs of the customers and various type of distributed generators used in a microgrid generally leads to the demand for varying active and reactive power that may affect the stability of systems. However, integration of Energy Storage Systems (ESS) into DG can help in stabilizing the system in terms of both voltage and frequency.

6.7 Healthcare Benefits

These increase the power quality in urban health subunits where major operations or surgeries get paused due to power outages [45]. Not only that the usage of renewable energies reduces carbon dioxide emissions and thus helps in addressing environmental concerns.

6.8 Artificial Intelligence Efficient

They will be capable of predicting energy usage on the demand side by the usage of several factors that involves weather, activity in commerce and industry, the density of human presence, patterns in the traffic of vehicles, and energy in the historical usage information as well as Energy production prediction for essential weather signals, historical generation information and installed generation capacity.

6.9 Controlled Systems

Usage of AC–DC microgrids where the generation units and the load units are connected on different sides [46], i.e., PV units and electric vehicles (EVs) are connected to the DC side and wind turbine and refrigerator is connected to the AC side, and divides the grid into two sub-grids and that can be controlled very easily in a separate manner.

6.10 Nature-Friendly Systems

The advantages of using microgrid make a nature-friendly impact as it uses renewable energy as a resource. However, this efficiency can be increased by the addition of Plug-in electric vehicles, which decreases greenhouse emissions as compared to other hybrid Electric vehicles (EVs) and also decreases carbon emissions considerably [47].

7 Uncertainties in Microgrid and Modeling

Smart grid technologies come with certain technical and economic parameters and therefore it is essential to discuss the various uncertainties that can occur while the application of these microgrids. This section briefly discusses the common techniques to model those uncertainties in order to increase the efficiency of the system. The various sources of uncertainties mainly include Renewable energy generation from uncertain sources like wind solar and hydro energy, the load, generation units, distribution lines, feeders, Market price, and many other factors that contribute to the uncertainty in the system, The major factor of the uncertainties are the renewable energy resources as they are non-dispatchable in nature and as discussed earlier are intermittent and uncertain in nature. The various factors are discussed below along with the methods to resolve such uncertainties.

7.1 Uncertainty on the Side of Source

The volatility of the renewable energy output and the errors in predicting the outcome of the renewable energy resources is the main reason for the confusion on the side of the source. The output of renewable energy resources is influenced by the surrounding environmental conditions and the wind turbines also get affected by the randomness in the weather. Similarly, in the case of photovoltaic arrays conditions like different solar irradiances, temperatures, air pressure, and cloud cover also affect the efficiency of these systems. It is considered that the light intensity obeys Beta distribution and the distribution of Weibull is also obeyed by the speed of the wind. The stable operation and the required control of these distribution systems are affected by the randomness and volatility of these renewable energy resources.

The consumption of renewable energy resources can be increased and the economic operation's safety can be ensured by active control of the distribution system with the help of coordinated dispatch of "Source-grid-load-storage". In the intraday optimization scheduling process [48], previous-day data of the output of the renewable energy resources are taken as a standard. There is always a difference between the real value and the estimated value of output in renewable energy. The renewable energy outputs' next-day forecast is around 25–40% of the installed capacity. High precision is one of the methods to decrease the bad impact of renewable energy resources incorporation in the grids. The previous researches also prove that the forecasting methods can be split into physical methods and statistical methods. The time series method which comes under the statistical forecasting methods and other methods which are included in this forecasting method is the artificial neural network and support vector machine which helps in reducing the uncertainties.

7.2 Uncertainty on the Grid Side

One of the types of uncertainties in the grid is the equipment operating status. The equipment operating status and the uncertainties related to it mainly include two types which are the equipment variable operating conditions and equipment failure. The equipment shows variable condition characteristics which prove that the energy consumption of equipment is varying due to factors like power, load rate, environmental factors, and many others. In research, a RESs optimal scheduling model has been proposed which could operate in varying conditions it also increased the model's accuracy, and the error that occurred in predicting the cost that occurred in the scheduling scheme was also reduced to a great extent. An economic optimization model had been proposed in the past equipment characteristics of several different conditions were considered for small generator sets, CHP, and energy production equipment. There is a coupling relation that is linear in nature between the input and the output of the device can be preserved by the usage of linearization which takes place piecewise or even linearization meant for polynomial fitting. Although in the recent condition, the integrated energy resources take less consideration of the various variable conditions. The main single factor which affects the equipment has been taken into consideration and thus more conditions should be taken into consideration in the future to refine the efficiency of the model.

7.3 Uncertainty in the System

There are certainties in the case of energy demand and supply in each subsystem but along with this come uncertainty of natural gas and varying outdoor temperatures and the parameters of the heating pipeline. The operating status of the system affects the parameters of the pipelines and they are not constant values. There may be a deviation of the true value of the parameter for the pipeline from the original parameters but since there is an absence of monitoring devices, such changes are very hard to observe. The building parameters and the outdoor temperature affect the requirements needed for heating and cooling and comfort in the energy of the consumers. Therefore the thermal inertia of the building and the prediction error should be given importance and taken into consideration while the uncertainty in cold and the demand of the heat is taken into consideration [49]. The actual working of the Integrated Energy System is affected by uncertain factors in each energy system through the coupling of energy flows multiple in nature and multiple energy scales. It is also a deep integration application of physical systems in the cyber.

7.4 Uncertainty on the Energy Storage Side

The main three types of energy storage systems are divided into fixed energy storage, storage of mobile energy, and storage of energy virtually. Most of the research that has taken place till now uses energy storage which acts as a resource that helps in the scheduling of IES optimally. The distribution of load in the space and the time dimensions can be changed by the setup of various energy storage equipment and this will cause to reduce the difference in the distribution of electricity, heat, and load. But the factor of uncertainty that exists in energy storage is not considered very commonly. Good mobility and easy installation are the characteristics of mobile energy storage and can be set up in different time periods and spaces. The electric vehicles have an actual charging load which represents the mobile energy storage that varies randomly due to the various factors contributed by vehicle operation, traffic, environment, and other factors which cause strong uncertainties. The Monte Carlo simulation, fuzzy methods, and combination of these methods mentioned are used for modeling uncertainty analysis of electric vehicles. Virtual energy storage helps in achieving load peak reduction and the controllable resources are integrated which helps in valley fitting on the consumer side [50]. Nowadays there is more active participation of the users in supervising the activities in the market but when this happened that the consumers take part in the demand response, they will be faced with the factors like Market prices, policy incentives, and other factors. This even increases the uncertainty more and makes it difficult for us to predict.

7.5 Trading Methods

The transactions of the market include bilateral transactions, e-centralized interactions, and peer-to-peer interactions. In bilateral transactions, parties on both sides sign a contract with the aid of negotiation, and a particular amount for a particular subject matter is sold as well as bought with respect to a condition that has been pre-agreed and takes place in the time mentioned in that agreement. The market rules and transaction cost decides the transaction scale. Factors like the load by the IES, the equipment failure, and the transmission capacity causes an error between the volumes of the transaction that does not go in accordance with the terms mentioned in the contract and the volume of the real demand. There comes the need to establish centralized transactions where the quotation rules help the participants of the market where quotations are presented to the organizers of the market. The growth in the methods meant for trading and the entities in trading increase the uncertainties in IES [48, 49].

8 Modeling of Various Components in Microgrids

Microgrid is often referred to as a building block of the smart grids in the coming future and it is termed as the network of power generating units, storage devices, and loads. Various methods for controlling and modeling these components of microgrids which enable them to get incorporated into the existing grid have been discussed here.

8.1 Modeling of Controllable Loads

Electric cars Controllable loads of which electric cars are an example, always are in the power supply systems, and that too mostly in low voltage levels in a very small quantity. These electric cars have an exceptionally great future in both medium as well as low-voltage distribution systems. This modeling is taken into consideration along with the feature of its bidirectional operation. The maximum loading capacity of the car as well as the current state of charge which is also referred to as SOC is transferred automatically to the system with the arrival of an electric vehicle at a charging station. There is a user interface with the help of which the car's user can communicate its departure time. The charging characteristics are taken out for the car which gets loaded during the time of the service on the basis of information collected on the system.

According to the different states of the system, it is preferable to load different charge characteristics to the car during lunch the load should be minimized. The specific power requirement is calculated after calculating the charge curve for different conditions. The power consumed by the car is controlled by the coefficient of attenuation that is calculated for each car individually. This can exceed the charging power at the maximum point, so energy is returned from the car to the host system. On the basis of the departure time and the loading capacity of the cars, the characteristics curve can be generated and it would help us to predict which time would be the required time for the car to switch from "Maximum" to the newest and the battery gets SOC less than 100% at that mentioned departure time. In the case of minimum SOC, i.e., SOC less than 100%, an additional parameter can be set up for the leaving time by the user, and that leads to growth in the maximum time for the charging of the power by increasing its degree of freedom. However, there might be unexpected trips or emergency conditions and that is why a minimum SOC should always be obtained. This mentioned requirement should be provided for all the loads because without this the car at the departure time will not be completely charged.

8.2 Modeling of Storage

For storage modeling, it is required to tune the generation and the consumption to each other on a permanent basis in the case of microgrids. If the feature-dependent sources are the support system of the production either partially or completely, the balancing of these sources with the help of storage systems can be possible. Various technologies have already been discussed in the past. The needed and appropriate storage technology can be predicted for a microgrid on the basis of the type of work that the microgrid has to perform. Batteries which are an example of electrochemical storage can perfectly work as stationary storage in the microgrids because of the factors which support it like their maturity on the technical side, market availability, and dependency. The system requires adequate power to be obtained from the storage to its limit technically. As the batteries are connected to the microgrid with the aid of inverters, it can be predicted that there is no delay between the demand of load and the storage in-feed. The four key parameters are taken into consideration while modeling is the Stage of Charge (SOC), actual energy content, and reactive and active power. The storage works as an element reactive in nature for the system and provides energy to the system and also uses up the energy to its maximum load. The inverter has its own technical limitations which cause the limit between the transport and the uptake to be restricted. Physical effects like reductions in the voltage can be avoided for the calculation of energy balance. In the testing of the needed size of the storage, voltage-related provisions are included.

8.3 Modeling of Feature-Dependent Generation

Feature-dependent generations like photovoltaic (PV) systems, mean by their variability strong challenges to the secure network operation are modeled. In order to maintain the condition of safe and sound operation of the grid, renewable energy resources can definitely be disconnected but they should be used as the last option on the basis of their widely accepted grid codes. In order to show the strong variability of feeds provided by the photovoltaic systems, it is crucial to keep the calculation size for less than one second. Losing information regarding the system dynamics is not acceptable at longer simulations and that can lead to unacceptable errors in the system. If the available storage is fully charged and if the supply exceeds the load then it becomes necessary for the PV infeed to be cut down by the control system in the microgrid. And in cases of supply-dependent feed, the actions that are required must be taken. The model below is the model for implementation 19 of feature-dependent generation in Fig. 4 which shows the block diagram of the simulation for implementation of the PV system and the model is basically formed of 4 sections. The irradiance values are entered into the simulation and using the factor K of the gain block, the PV system dimensions was set. The maximum power feed, real-time power requirement, and the real-time power requirement of the load were added together in

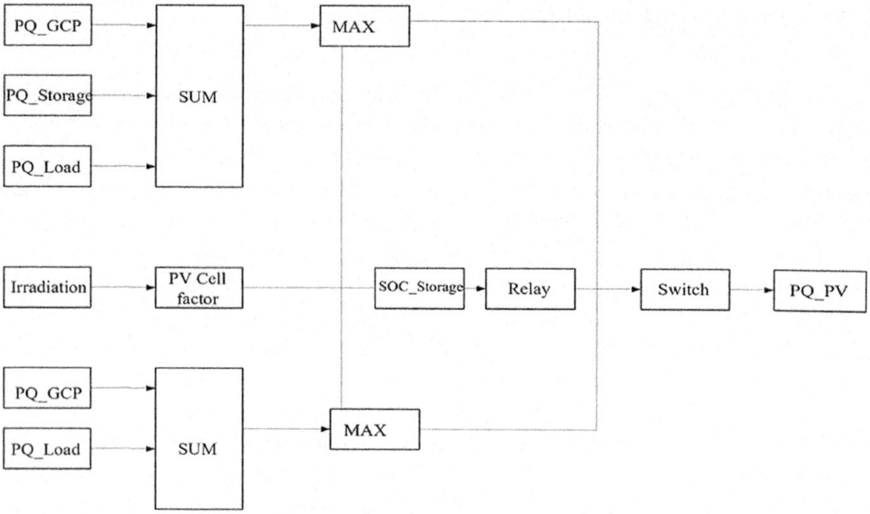

Fig. 4 Block diagram of the feature-dependent generation

order to balance the power in the microgrid. This signifies the upper limit of power consumed and that is why the block of min–max is used to ensure that power does not go beyond the predefined limit. The PV irradiation can be calculated by comparing input and system set values. When the existing system storage can't absorb any more energy limit on PV must be adjusted. Relay block helps in comparing the SOC of storage to a definite value. And as soon as it is achieved the switch block changes its path A second threshold value is formed in the relay after it is switched on again. On reaching the limit of the SOC the power can no longer be provided to the grid and giving real-time power requirement of the load and real-time power demand from the grid. This demand when added up is equal to the highest power which is fed into the system. This simulation helps to compare the possible and the actual PV output. A separate measuring device is attached in order to measure the damage caused due to the depression in the plot of PV and that also plays a role in recording the history of the irradiation. This kind of measuring system is formed of one or more than one cell meant for measurement which are mostly established near the PV systems.

8.4 Modeling of Controllable Generation

Controllable feeds also known as feature-independent feeds are responsible for supplying the needed power to the system for the wanted time or the reaction to the need for support arriving from the control systems. Variants in controllable fields represent the current controlled CHPs. Once the control system issues a start command for the synchronization of the micro CHP at least once in the standby state. After this synchronization is done, the mini CHP can be adjusted to the desired feed. The device decides the constant of the time for a state change. If heat storage is available in the system the heat production can be passed on to this system. The unit meant for the storage of heat acts as a limit to the mini CHP but when the storage for the heat gets completely filled up it is necessary for the mini CHP to shut down. The current controlled micro- as well as the mini CHP is considered as a simple infeed that can be used to provide the power at any required time in case of a simulation. It is very necessary to differentiate between active and reactive power as they both can be supplied independently and both have their individual highest and lowest points. Capacitance or inductance can be served as a reactive power. As long as no measurements are available reaction times can be guessed with 5 s in maximum when the values for the active power which are already set are changed. In the case of network synchronization a time constant of less than 30 s can be assumed. In case of unavailability of the manufacturer's data, a ratio of 1:2 is assumed and this can help us in getting a rough approximation of whether an available memory has the preferable size that is required in the system. When the storage in the system is filled completely it becomes difficult for the mini CHP to get operated and when the heating network of the district is connected to the micro CHP production of heat doesn't have to be taken into consideration. The model for CHP model has been designed with the block diagram for the simulation as shown in Fig. 5.

The values of the power in real time of the active component were summed up to find the active and the reactive model. The mini CHP has storage available for real power only. The saturation block limits the value of Active power for minimum and maximum points. Since the reactive power has different independent limits than the active power another saturation block is required for that purpose. The OR operator

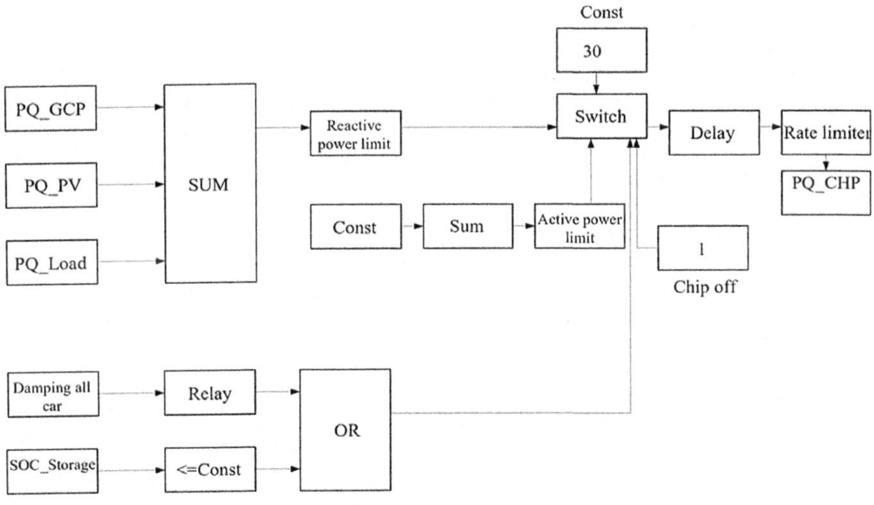

Fig. 5 Block diagram of mini CHP

decides under which condition would be mini CHP be required [50]. The rate of change and the reaction time are managed by a rate limiter and delay blocks.

9 Conclusions

In this chapter, efforts have been made to make the viewers understand the definition and the importance of microgrids and how they can impact the future of power generation and consumption. The classification of these systems on the basis of various factors, for example, one being energy availability in the area of the setup and several others. Also, the challenges that can be faced while the operation of microgrids has been discussed in detail from every aspect be it social, political, economic, and many more, and the opportunities that microgrid can guarantee in the future if certain precautions are taken into consideration and the uncertainties that come with various parts of the microgrids and the modifications and modeling done to overcome such difficulties have been discussed in brief.

Acknowledgements This research work was supported by "Woosong University's Academic Research Funding – 2023."

References

1. Jha SK, Kumar D (2020) Assessment of battery energy storage system with hybrid renewable energy sources to voltage control of islanded microgrid considering demand-side management capability. Iran J Sci Technol Trans Electr Eng 44(2):861–877.https://doi.org/10.1007/s40998-019-00273-9
2. Myung S, Lee JH (2020) Ethereum smart contract-based automated power trading algorithm in a microgrid environment. J Supercomput 76(7):4904–4914. https://doi.org/10.1007/s11227-018-2697-7
3. Cansiz A, Faydaci F, Qureshi MT, Usta O, McGuiness DT (2018) Integration of a smes–battery-based hybrid energy storage system into microgrids. J Supercond Novel Magn 31(5):1449–1457. https://doi.org/10.1007/s10948-017-4338-4
4. Deng X, Jiang W (2019) Evaluating green supply chain management practices under fuzzy environment: a novel method based on d number theory. Int J Fuzzy Syst 21(5):1389–1402. https://doi.org/10.1007/s40815-019-00639-5
5. Mohammadjafari M, Ebrahimi R, Darabad VP (2020) Optimal energy management of a micro-grid incorporating a novel efficient demand response and battery storage system. J Electr Eng Technol 15(2):571–590. https://doi.org/10.1007/s42835-020-00345-5
6. Ndwali PK, Njiri JG, Wanjiru EM (2020) Optimal operation control of microgrid connected photovoltaic-diesel generator backup system under time of use tariff. J Control Automat Electr Syst 31(4):1001–1014. https://doi.org/10.1007/s40313-020-00585-w
7. Khosravi S, Beheshti MTH, Rastegar H (2020) Robust control of islanded microgrid frequency using fractional-order pid. Iran J Sci Technol Trans Electr Eng 44(3):1207–1220. https://doi.org/10.1007/s40998-019-00303-6

8. Vikash G, Ghosh A (2022) Parallel inverters control in standalone microgrid using different droop control methodologies and virtual oscillator control. J Inst Eng (India): Series B 103(1):163–171. https://doi.org/10.1007/s40031-021-00613-6
9. Shah P, Mehta B (2020) Microgrid optimal scheduling with renewable energy sources considering islanding constraints. Iran J Sci Technol Trans Electr Eng 44(2):805–819. https://doi.org/10.1007/s40998-019-00254-y
10. Sarkar SK, Badal FR, Das SK (2018) A comparative study of high performance robust pid controller for grid voltage control of islanded microgrid. Int J Dyn Control 6(3):1207–1217. https://doi.org/10.3390/en14051401
11. Khadanga RK, Padhy S, Panda S, Kumar A (2018) Design and analysis of tilt integral derivative controller for frequency control in an islanded microgrid: a novel hybrid dragonfly and pattern search algorithm approach. Arab J Sci Eng 43(6):3103–3114. https://doi.org/10.1007/s13369-018-3151-0
12. Maulana R, Helms-Lorenz M, Irnidayanti Y, van de Grift W (2016) Autonomous motivation in the Indonesian classroom: relationship with teacher support through the lens of self-determination theory. Asia Pac Educ Res 25(3):441–451. https://doi.org/10.1007/s40299-016-0282-5
13. Hemamalini S (2016) Review on microgrid and its protection strategies. Int J Renew Energy Res (IJRER) 6(4):1574–1587. https://doi.org/10.20508/ijrer.v6i4.4353.g6911
14. Banerji A, Sen D, Bera AK, Ray D, Paul D, Bhakat A, Biswas SK (2013) Microgrid: a review. In: IEEE global humanitarian technology conference: South Asia satellite (GHTC-SAS), pp 27–35. https://doi.org/10.1109/GHTC-SAS.2013.6629883
15. Mariam L, Basu M, Conlon MF (2013) A review of existing microgrid architectures. J Eng 2013. https://doi.org/10.1155/2013/937614
16. Zhou Y, Ngai-Man Ho C (2016) A review on microgrid architectures and control methods. In: 2016 IEEE 8th international power electronics and motion control conference (IPEMC-ECCE Asia), pp 3149–3156. https://doi.org/10.1109/IPEMC.2016.7512799
17. Borazjani P, Wahab NIA, Hizam HB, Soh ABC (2014) A review on microgrid control techniques. In: 2014 IEEE innovative smart grid technologies Asia (ISGT ASIA), pp 749–753. https://doi.org/10.1109/ISGT-Asia.2014.6873886
18. Salkuti SR (2022) Emerging and advanced green energy technologies for sustainable and resilient future grid. Energies 15(18):6667.https://doi.org/10.3390/en15186667
19. Beheshtaein S, Cuzner R, Savaghebi M, Guerrero JM (2019) Review on microgrids protection. IET Gener Transm Distrib 13(6):743–759. https://doi.org/10.1049/iet-gtd.2018.5212
20. Mahmoud MS, Rahman MSU, AL-Sunni FM (2015) Review of microgrid architectures–a system of systems perspective. IET Renew Power Gener 9(8):1064–1078. https://doi.org/10.1049/iet-rpg.2014.0171
21. Decuir J, Michael P (2017) draft IEEE standard for dc microgrids for rural and remote electricity access applications. In: IEEE conference on technologies for sustainability (SusTech), pp 1–5. https://doi.org/10.1109/SusTech.2017.8333510
22. Brijesh P, Jiju K, Dhanesh PR, Joseph A (2019) Microgrid for sustainable development of remote villages. In: IEEE region 10 conference (TENCON), pp 2433–2438. https://doi.org/10.1109/TENCON.2019.8929455
23. Talei H, Zizi B, Abid MR, Essaaidi M, Benhaddou D, Khalil N (2015) Smart campus microgrid: Advantages and the main architectural components. In: 3rd international renewable and sustainable energy conference (IRSEC), pp 1–7. https://doi.org/10.1109/IRSEC.2015.7455093
24. Salkuti SR (2020) Optimal day-ahead renewable power generation scheduling of hybrid electrical power system. In: Ray P, Biswal M (eds) Microgrid: operation, control, monitoring and protection. Lecture notes in electrical engineering, vol 625. Springer, Singapore. https://doi.org/10.1007/978-981-15-1781-5_2
25. Podlesak T, Vitale J, Wilson B, Bohn F, Gonzalez M, Bosse R, Siegfried S, Lynch J, Barnhill W (2019) Auto-tuning for military microgrids. In: IEEE energy conversion congress and exposition (ECCE), pp 6270–6277. https://doi.org/10.1109/ECCE.2019.8912776

26. Shabshab SC, Lindahl PA, Nowocin JK, Donnal J, Blum D, Norford L, Leeb SB (2019) Demand smoothing in military microgrids through coordinated direct load control. IEEE Trans Smart Grid 11(3):1917–1927. https://doi.org/10.1109/TSG.2019.2945278
27. Laour M, Akel F, Bendib D, Chikh M (2018) Residential microgrid load management and optimal control in grid connected and islanded mode. In: 6th international renewable and sustainable energy conference (IRSEC), pp 1–4. https://doi.org/10.1109/IRSEC.2018.8702847
28. Martirano L, Habib E, Parise G, Greco G, Cianfrini M, Parise L, Massarella F, Frattura PDL (2017) Demand side management in mixed residential/commercial buildings with pv on site generation. In: IEEE/IAS 53rd industrial and commercial power systems technical conference (I&CPS), pp 1–7. https://doi.org/10.1109/ICPS.2017.7945093
29. Jadav KA, Karkar HM, Trivedi IN (2017) A review of microgrid architectures and control strategy. J Inst Eng (India): Series B 98(6):591–598. https://doi.org/10.1007/s40031-017-0287-3
30. Abdelgawad H, Sood VK (2019) A comprehensive review on microgrid architectures for distributed generation. IEEE Electr Power Energy Conf (EPEC) 1–8. https://doi.org/10.1109/EPEC47565.2019.9074800
31. Sahu PC, Prusty RC, Panda S (2021) Improved-gwo designed fo based type-ii fuzzy controller for frequency awareness of an acmicrogrid under plug in electric vehicle. J Ambient Intell Humaniz Comput 12(2):1879–1896. https://doi.org/10.1007/s12652-020-02260-z
32. Garcés A, Montoya OD (2019) A potential function for the power flow in dc microgrids: an analysis of the uniqueness and existence of the solution and convergence of the algorithms. J Control Autom Electr Syst 30(5):794–801. https://doi.org/10.1007/s40313-019-00489-4
33. Lede AMR, Molina MG, Martinez M, Mercado PE (2017) Microgrid architectures for distributed generation: a brief review. In: IEEE PES innovative smart grid technologies conference-Latin America (ISGT Latin America), pp 1–6. https://doi.org/10.1109/ISGT-LA.2017.8126746
34. Patel H, Chowdhury S (2015) Review of technical and economic challenges for implementing rural microgrids in South Africa. IEEE Eindhoven PowerTech 1–6. https://doi.org/10.1109/PTC.2015.7232341
35. Saldarriaga-Zuluaga SD, Lopez-Lezama JM, MuñozGaleano N (2020) Protection coordination in microgrids: current weaknesses, available solutions and future challenges. IEEE Latin Am Trans 18(10):1715–1723. https://doi.org/10.1109/tla.2020.9387642
36. Ali SW, Sadiq M, Terriche Y, Naqvi SAR, Mutarraf MU, Hassan MA, Yang G, Su C, Guerrero JM (2021) Offshore wind farm-grid integration: a review on infrastructure, challenges, and grid solutions. IEEE Access. https://doi.org/10.1109/ACCESS.2021.3098705
37. Srivastava A, Mohanty R, Ghazvini MAF, Steen D, Carlson O (2021) A review on challenges and solutions in microgrid protection. IEEE Madrid PowerTech 1–6. https://doi.org/10.1109/PowerTech46648.2021.9495090
38. Almutairy (2016) A review of coordination strategies and techniques for overcoming challenges to microgrid protection. In: 2016 Saudi Arabia smart grid (SASG), pp 1–4. https://doi.org/10.1109/SASG.2016.7849681
39. Altaf MW, Arif MT, Islam SN, Haque ME (2022) Microgrid protection challenges and mitigation approaches-a comprehensive review. IEEE Access. https://doi.org/10.1109/ACCESS.2022.3165011
40. Rath SS, Panda G, Ray PK, Mohanty A (2021) A comprehensive review on microgrid protection: issues and challenges. In: 3rd international conference on energy, power and environment: towards clean energy technologies, pp 1–6. https://doi.org/10.1109/ICEPE50861.2021.9404520
41. Swain A, Salkuti SR, Swain K (2021) An optimized and decentralized energy provision system for smart cities. Energies 14(5):1451. https://doi.org/10.3390/en14051451
42. Salkuti SR (2019) Multi-objective based optimal scheduling of microgrid considering uncertainties. In: International conference on cutting-edge technologies in engineering (ICon-CuTE), Uttar Pradesh, India, pp 50–56. https://doi.org/10.1109/ICon-CuTE47290.2019.8991509

43. Akinyele D, Belikov LY (2018) Challenges of microgrids in remote communities: a steep model application. Energies 11(2):432. https://doi.org/10.3390/en11020432
44. Onai K, Ojo O (2017) Vehicle-to-grid technology assisted microgrid in Ghana: opportunities and challenges. IEEE PES PowerAfrica 341–346. https://doi.org/10.1109/PowerAfrica.2017.7991248
45. Asmus P (2010) Microgrids, virtual power plants and our distributed energy future. Electr J 23(10):72–82. https://doi.org/10.1016/j.tej.2010.11.001
46. Shabshab SC, Nowocin JK, Lindahl P, Leeb SB (2018) Microgrid modeling and fuel savings opportunities through direct load control. In: 44th annual conference of the IEEE industrial electronics society, pp 231–236. https://doi.org/10.1109/IECON.2018.8591850
47. Wang J, Costa LM, Cisse BM (2016) From distribution feeder to microgrid: an insight on opportunities and challenges. In: IEEE international conference on power system technology (POWERCON), pp 1–6. https://doi.org/10.1109/POWERCON.2016.7753897
48. Verma PP, Srinivasan D, Swarup KS, Mehta R (2018) A review of uncertainty handling techniques in smart grid. Int J Uncertainty. Fuzziness Know-Based Syst 26(03):345–378. https://doi.org/10.1142/S0218488518500186
49. Fan H, Wang C, Liu L, Li X (2022) Review of uncertainty modeling for optimal operation of integrated energy system. Front Energy Res 9:641337. https://doi.org/10.3389/fenrg.2021.641337
50. Porsinger T, Janik P, Leonowicz Z, Gono R (2017) Modelling and optimization in microgrids. Energies 10(4):523. https://doi.org/10.3390/en10040523

Load Frequency Control in Two-Area Interconnected Systems Using DE-PID and PSO-PID

Solomon Feleke, Raavi Satish, Surender Reddy Salkuti, and Almoataz Y. Abdelaziz

Abstract Interconnected network in the power system is used to share loads if any generator is overloaded. Load frequency control (LFC) is a function of automatic generation control (AGC) and the frequency is maintained in the interconnected system using this. The interconnected system has an advantage over a single-area system because, during the peak load of the system, the excess load can be easily exchanged and shared with the other interconnected system. However, if this interconnected system is not properly supported by the controller, then it fails to succeed. In this chapter, the proportional, integral and derivative (PID) parameters in two-area TPS are optimized by particle swarm optimization (PSO) and differential evolution (DE) optimization methods using a single-objective optimization approach. During optimization 0.01 p.u. load disturbance is considered in area-1. The output response parameters such as rise time, undershoot, settling time and overshoots have been compared for 1% step load disturbance in area-1. Test results show that the DE-PID controller method gives better performance when compared with the PSO-PID controller and without a controller.

S. Feleke
Department of Electrical and Computer Engineering, Debre Berhan University, Debre Berhan, Ethiopia
e-mail: sole12@dbu.edu.et

R. Satish
Department of Electrical and Electronics Engineering, Anil Neerukonda Institute of Technology and Sciences (A), Visakhapatnam, India
e-mail: satish.eee@anits.edu.in

S. R. Salkuti (✉)
Department of Railroad and Electrical Engineering, Woosong University, Daejeon, Republic of Korea
e-mail: surender@wsu.ac.kr

A. Y. Abdelaziz
Faculty of Engineering and Technology, Future University in Egypt, Cairo, Egypt
e-mail: almoataz.abdelaziz@fue.edu.eg

S. R. Salkuti et al. (eds.), *Power Quality in Microgrids: Issues, Challenges and Mitigation Techniques*, Lecture Notes in Electrical Engineering 1039,
https://doi.org/10.1007/978-981-99-2066-2_18

Keywords Area control error · Load frequency control · Proportional–integral–derivative · Particle swarm optimization · Differential evolution · Two-area model

Nomenclature

LFC	Load frequency control
PSO	Particle swarm optimization
ACE	Area control error
DE	Differential evolution
ITAE	Integral of time-weighted absolute error
EP	Evolutionary programming
AGC	Automatic generation control
PID	Proportional–integral derivative
MBA	Mine blast algorithm
WOA	Whale optimization algorithm
DFIG	Doubly fed induction generator
IHA	Improved harmony algorithm
TPS	Thermal power system
RT	Rise time
ST	Settling time
US	Undershoot
OS	Overshoot

1 Introduction

The need for power systems is growing quickly in the modern world and, as a result, its complexity is also increasing daily. Therefore, continuous monitoring of the system operation is required. The generation and load balance is maintained by AGC, which is an important technique in power system control. The system will break down if the variation between demand and generation is uncontrolled. The authors of [1] present the basic concepts of automatic generation control (AGC). Load frequency control (LFC) is the basic function of AGC and is briefly described with the compressive coverage of tie-line power and frequency control problems in a wide range of system operating conditions. Authors of [2] presented multi-area thermal power system (TPS) dynamic control of frequency using optimal control theory with a brief mathematical model strategy. The mathematical modeling of area control error (ACE) is presented in [3]. It is the ACE, which must be minimized to settle the change of tie-line power and frequency in normal operating conditions [4]. Authors of [5, 6] present the advantages and controlling performance of the recently emerged artificial intelligent techniques such as GA, DE, and PSO in single- and multi-objective approaches in separate involvement and in a hybrid way to satisfy better control as compared to the traditional. In [7, 8], the PSO technique is presented to utilize in the linear time-invariant and nonlinear functions. Authors of [9] present the control of the multi-unit non-reheat thermal plant using single and hybrid optimization techniques by optimizing PID controller. Authors of [10] discuss the different types of tuning performance criteria used as objective functions and found the settling time of integral of time-weighted absolute error (ITAE) was obtained lower as compared with others.

PID controller has a great role in the closed-loop system of industrial automation controllers due to its three modes of controller operations such as proportional, integral, and derivative [11]. Many researchers use different optimization techniques for tuning the controller of PID. In [12], the design of a power system stabilizer in the presence of PID and bat algorithm performance is good in the enhancement of signal stability. The rise time (RT), settling time (ST), and convergence characteristics have been compared between PSO- and EPSO-tuned PID controller. In [13], the tuning of fuzzy-PID in a multi-area AGC system using the grasshopper optimization algorithm was well performed with good dynamic response and with fast system operation. Authors of [14] reviewed the different types of PID tuning methods such as conventional methods, GA, DE, evolutionary programming (EP), and swarm intelligence methods. Authors of [15–17] present different optimization techniques such as GA, PSO, DE, EP, ABC, and ACO for tuning the PID controller. Authors of [18] present the design of fuzzy-PID controller using the DE algorithm for reheat type three-area TPS. In [19], the control of two-area systems was designed using fuzzy logic and GA algorithm. The application of sliding mode controller of the second order which can support rejection chattering phenomena and good effect on the power system network was studied in [20]. Authors of [21] studied the control role of LFC in the grid when renewable energies such as solar thermal, photovoltaic, and wind were integrated into the system. The powerfulness of control ability of fuzzy-PID in the presence of nonlinearities such as GRC and governor dead zone of LFC system was studied in [22]. In [23] proportional–integral (PI) controller was optimized using the PSO-whale optimization algorithm (WOA) in an interconnected system and the result was compared with each WOA and PSO separately. The control participation role of the doubly fed induction generator (DFIG) in an interconnected system along with PI controller tuning in the system was studied in [24]. In [25], the cuckoo search method was involved to tune the PI controller and the LFC system using the time domain objective function and robust results were obtained. In [26], the distributed economic model predictive control (EMPC) is proposed in LFC of interconnected power systems to economic dispatch and thus achieving an optimal result in the real-time operation. High dynamic performance in LFC of interconnected multi-area systems was achieved using active disturbance rejection control (ADRC) which is studied by authors in [27]. In [28], the control of the grid frequency in the presence of wind, HVDC in which with a coordination control strategy was studied in the LFC of the power system and improved grid frequency stability. In

[29], the frequency control of wind farms was studied using the multi-objective PSO method to tune PID as a controller. The stability analysis study of the ALFC system in the renewable rich system along with small-signal analysis was studied and the superior performance of ALFC was observed in [30].

Even though a lot of research work has been available on AGC, there is still a limitation in identifying better optimization methods by measuring performances. Considering this gap, this chapter is aimed to identify the better optimization technique by comparing PSO and DE in tuning PID for LFC controller. The objective function chosen in this chapter is ITAE. The remaining parts of the chapter are arranged as: In Sect. 2, the system modeling is described. Section 3 presents the problem formulation and solving methods. Results and discussion are described in Sect. 4. Conclusions are outlined in Sect. 5.

2 Modeling of System

The power system is interconnected due to various reasons, one of the reasons is to compensate for the power reduced by one of the loaded generators, i.e., during normal operation, two or more interconnected generators maintain the system frequency in both dynamic and static conditions, if the amount of load increases, the control system allows to speed up the system generator by allowing more water or steam in hydro and thermal generators respectively. In this regard, LFC's main goal is maintaining tie-line power and frequency. In a single-area system, if a mismatch between load and generation happens, it will be solved by the kinetic energy that is extracted from the system. On the other hand, an interconnected two-area TPS has the advantage of maintaining the power system in each area by sharing the excess power from one area to the loaded area of the other by changing the information through tie lines. Therefore, the LFC, in controlling the power system, has the advantage of maintaining the system tie-line power and frequency to its predefined value.

2.1 Primary Loop of LFC

The main control components in the LFC of the primary loop, such as the speed sensor, governor, control valve, and reference speed, are working together. If the generator unit is loaded due to an increase in customer demand, the speed sensor senses the turbine speed and compares it with the reference speed, and the governor allows the valve to enter more steam or more water in the case of a steam turbine or hydro turbine, respectively, and vice versa.

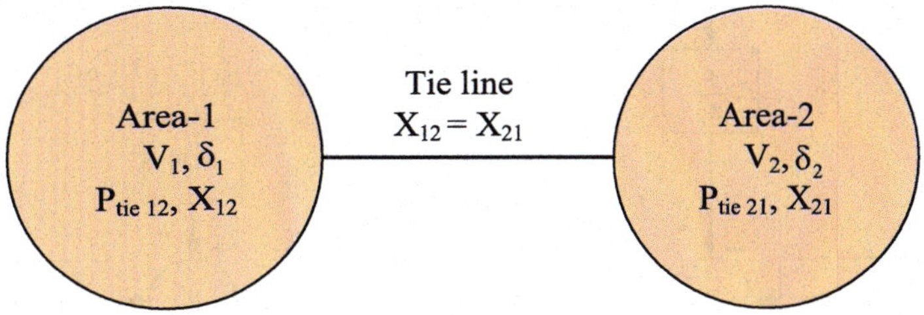

Fig. 1 General representation of the two-area TPS model

2.2 Secondary Loop of LFC

In the secondary LFC control loop, ACE has a main role as its output signal is sent to the speed changer through the integrator, the frequency which is sensed by the frequency sensor compared with the frequency of tie-line power in the signal mixer in which the reference speed is given to the governor by the speed changer, using this information signal, the steady-state change in frequency is reduced to zero with the integral controller. Figure 1 represents the general representation of the two-area TPS model. The tie-line power exchange depends on the voltages V_1, V_2; the reactance X_{12}; and the sine of the angle difference between the two voltages and this is mathematically expressed in Eq. (1).

$$P_{\text{tie}12} = \frac{|v_1|\,|v_2|}{X_{12}} \sin(\delta_1 - \delta_2) \tag{1}$$

where δ_1 and δ_2 are the angles for V_1 and V_2, respectively.

2.3 System Representation

Figure 2 is the two-area TPS with a controller which is used to study in this chapter. Equations (2) and (3) represent the ACE in area-1 and ACE in area-2 [9]. They depend on the frequency bias, tie-line power, and change in frequency.

$$ACE_1 = B_1 \Delta f_1 + \Delta P_{12} \tag{2}$$

$$ACE_2 = B_2 \Delta f_2 + \Delta P_{21} \tag{3}$$

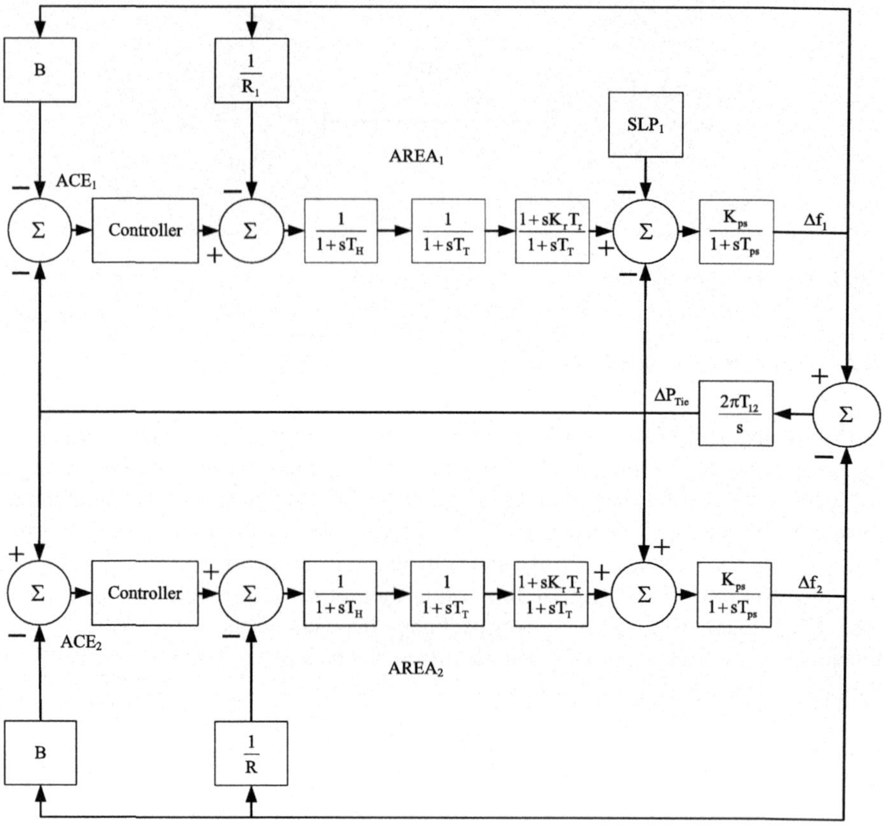

Fig. 2 Two-area TPS model

3 Problem Formulation and Solving Methods

Controlling the total ACE, which includes ACE_1 and ACE_2 will help to stabilize the operation of the system by maintaining the interchange of power and frequency. The objective of this study is to reduce the *i*th area control error (ACE_i) which is mathematically given in Eq. (4).

$$ITAE = \int_0^T t(|ACE_1| + |ACE_2|) \tag{4}$$

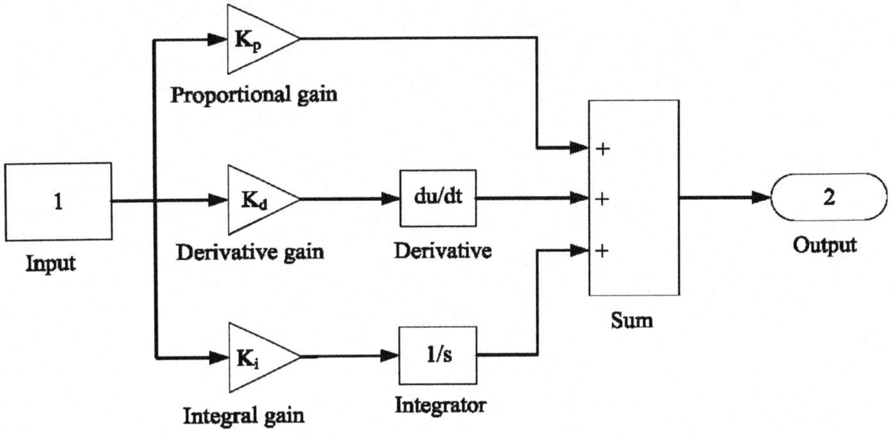

Fig. 3 Block diagram for PID controller

Table 1 Characteristics of PID controller parameters

Components	RT	OS	ST	Steady-state error
K_p	Shrink	Grow	Small change	Shrink
K_i	Shrink	Grow	Grow	Eliminate
K_d	Small change	Shrink	Shrink	Small change

3.1 *PID Controller*

PID controller is used to improve dynamic response and involves different industrial control activities and its block diagram is shown in Fig. 3, in which differential gain results in greater system stability, integral gain results in the elimination of steady-state errors, and proportional gain reduces the rise time. The general characteristics of PID controller parameters are summarized as shown in Table 1.

3.2 *Tuning Methods*

3.2.1 Particle Swarm Optimization (PSO)

PSO is a heuristic optimization technique based on the swarm theory of birds and the social behavior of schooling fish [8]. This method is a simple one to search for an optimal solution in the solution space. In this method, the main parameters of PSO are position vector $X_i(t)$, velocity vector $V_i(t)$, personal best (*pbest*), and global best (*gbest*). The optimization is required to obtain the optimal which is based on updating the iteration of the current values by using the previous values. The updating

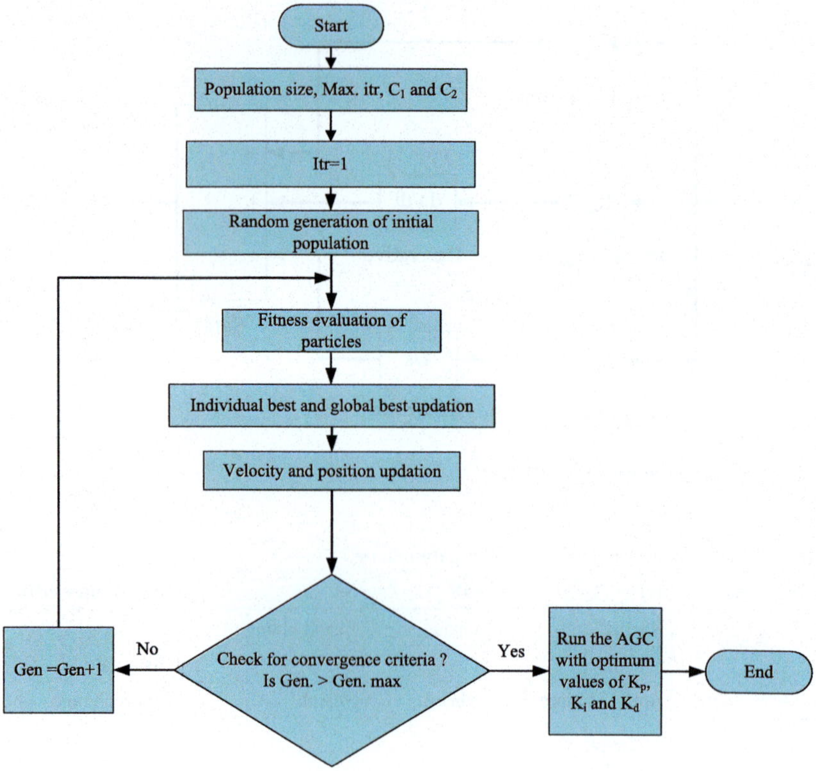

Fig. 4 PSO flowchart

procedure for the velocity v_{ik} and the position x_{ik} for kth dimension of ith particles is given by Eqs. (5) and (6), respectively.

$$V_{ik}^{(t+1)} = W \cdot V_{ik} + C_1 \cdot rand_{1,ik} \cdot \left(pbest_{ik}(t)\right) + C_2 \cdot rand_{2,ik}\left(gbest_k(t) - X_{ik}(t)\right) \tag{5}$$

$$X_{ik}(t+1) = X_{ik}(t) + V_{ik}(t+1) \tag{6}$$

In this work, during the initialization of the PSO algorithm, 50 population size and iterations of 100 are considered. Figure 4 presents the flowchart of PSO.

3.2.2 Differential Evolution (DE)

DE is a stochastic population-based optimization technique and it is first developed in [31]. The main features of DE that make it attractive in optimization are its speediness, robustness, efficiency, and real coding for solving problems with simplicity. It has four steps, briefly explained in the following sections. In this work, 50 population size and 100 generations are considered in DE coding [32]. Figure 5 presents the flowchart of DE.

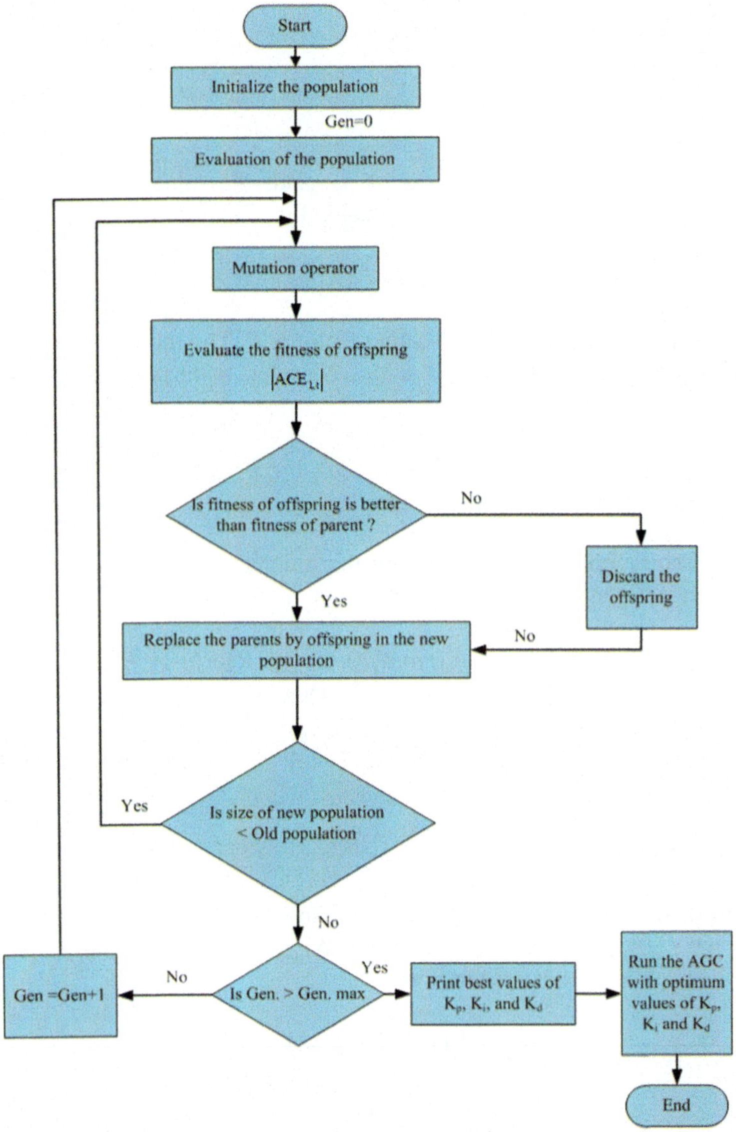

Fig. 5 Flowchart of DE

Initialization

In initialization, the two important parameters are the lower bound X_j^L and the upper bound X_j^U. From the interval $\left[X_j^L, X_j^U\right]$ random and uniform values are selected as the initial parameter.

Mutation Operation

Each target vector symbolized as $X_{i,G}$ and the mutant vector $V_{i,G} = \left\{V_{1i,G}, V_{2i,G}, \ldots V_{D1,G}\right\}$ is computed by using Eq. (7).

$$V_{i,G} = X_{r1,G} + F\left(X_{r2,G} - X_{r3,G}\right) \tag{7}$$

where G is the generation and the scaling factor is F. The indices r_1, r_2, and r_3 are randomly generated different integer values, NP refers to the population number, and D is the control variable.

Crossover Operation

In this crossover operation, by utilizing the mutant vector and target vector, the trailing vector is generated using Eq. (8).

$$U_{j,i,G} = \begin{cases} V_{j,i,G} \text{ if} \left(rand_j[0, 1] \leq CR\right)\text{or}(j = j_{rand}) \\ X_{j,i,G} \quad \text{Otherwise} \end{cases} \tag{8}$$

where $j = 1, 2, \ldots, D$.

Selection Operation

In DE, the selection operation is the last stage in which the comparison between the trial vector $f\left(U_{i,G}\right)$ and the target vector $f\left(X_{i,G}\right)$ is made to identify the best one for involvement in the next generation.

$$X_{i,G+1} = \begin{cases} U_{i,G} \text{ if } f\left(U_{i,G}\right) \leq f\left(X_{i,G}\right) \\ X_{i,G} \quad \text{Otherwise} \end{cases} \tag{9}$$

where $i \in [1, N_P]$.

4 Test Results and Discussions

Initially, two equal areas reheat the TPS model which is considered in this study as depicted in Fig. 2. Two PID controllers are incorporated in two areas and DE and PSO techniques are used to tune the PID controllers in both areas. In the case of DE, the parameter values in the programming are crossover rate (CR) which is 0.98 and the number of particles involved is 6. In the case of PSO, the regulation factor (R) is 30, constant (c_3) is 1, the number of particles is 6, and cognitive (c_1) and social (c_2) learning parameters are 2. In both DE and PSO techniques, the number of population and iterations are taken as 50 and 100, respectively. The optimized gain parameters of PID presented in Table 2 are obtained by simulating the power system model with optimization techniques using MATLAB/Simulink interfacing software 30 times by considering one percent step load change in area-1 only. Table 3 presents time response values such as ST, RT, US, and OS for two-area systems without a controller. Table 4 presents the time response values with PSO-PID, and Table 5 presents the time response values with DE-PID.

Table 2 PID gain values

Optimization method	Area	K_p	K_i	K_d
PSO-PID	A1	0.9042	0.4703	0.1203
	A2	0.9590	0.6020	0.1529
DE-PID	A1	0.9999	0.9999	0.1192
	A2	0.9999	0.0001	0.9999

Table 3 Time response output values without controller

System state variables	RT	ST	US	OS
Δ in f_1	8.5592e−05	16.5433	−0.0047	0.001
Δ in f_2	2.9392e−05	17.3441	−0.0017	6.8610e−04
Δ in P_{tie12}	0.0091	25.7324	−0.0050	2.2109e−04
Δ in ACE_1	5.3693e−05	21.2969	−4.0847e−04	0.0124
Δ in ACE_2	0.0092	20.1005	−0.0021	0.0046

Table 4 Time response values for PSO-PID

System state variables	RT	ST	US	OS
Δ in f_1	0.0792	27.9243	−0.0216	−0.0031
Δ in f_2	0.3168	28.7290	−0.0205	−0.0025
Δ in P_{tie12}	0.2799	33.6246	−0.0055	0.0131
Δ in ACE_1	0.1454	29.7489	0.0038	0.0131
Δ in ACE_2	6.5427e−05	33.2074	−0.0025	0.0054

Table 5 Time response values for DE-PID

System state variables	RT	ST	US	OS
Δ in f_1	2.2694e−05	11.1892	−0.0047	0.001
Δ in f_2	4.5043e−05	13.1448	1.8869e−07	−0.000627
Δ in P_{tie12}	2.0898e−04	18.2450	−0.0046	2.1868e−04
Δ in ACE_1	3.5724e−05	14.9611	−0.000452	−4.6207e−08
Δ in ACE_2	6.0889e−04	17.2586	−9.4966e−04	0.0025

Table 6 Cost function values comparison

Analyzed methods	ITAE objective function values
Cost function values without controller	6.5992
Cost function values of PSO-PID	1.6008
Cost function values of DE-PID	0.2424

Table 6 presents the objective function values in PSO-PID, DE-PID, and without a controller. Figure 6a–e shows the time responses with PSO-PID, DE-PID, and without a controller. Figure 6f shows convergence characteristics in PSO-PID and DE-PID. The US, OS, and ST responses with PSO-PID, DE-PID, and without controller are shown in Fig. 7a–c, respectively. The test findings show that DE produces superior outcomes when compared to PSO.

5 Conclusions

In this chapter, the simulation for tuning of PID controller is developed and carried out on two equal area reheat thermal power system (TPS) models using PSO and DE methods. In comparison with the traditional methods, these two techniques are fast in convergence. The dynamic parameters such as overshoot (OS), undershoot (US), settling time (ST), and rise time (RT) are compared for DE-PID, PSO-PID, and without the controller. The results of the test studies prove that the DE-PID is giving better performance compared to PSO-PID. Therefore, the proposed DE-PID optimization method yields better performance in load frequency control in optimizing the PID controller for suppressing the oscillations as compared to PSO-PID and without a controller.

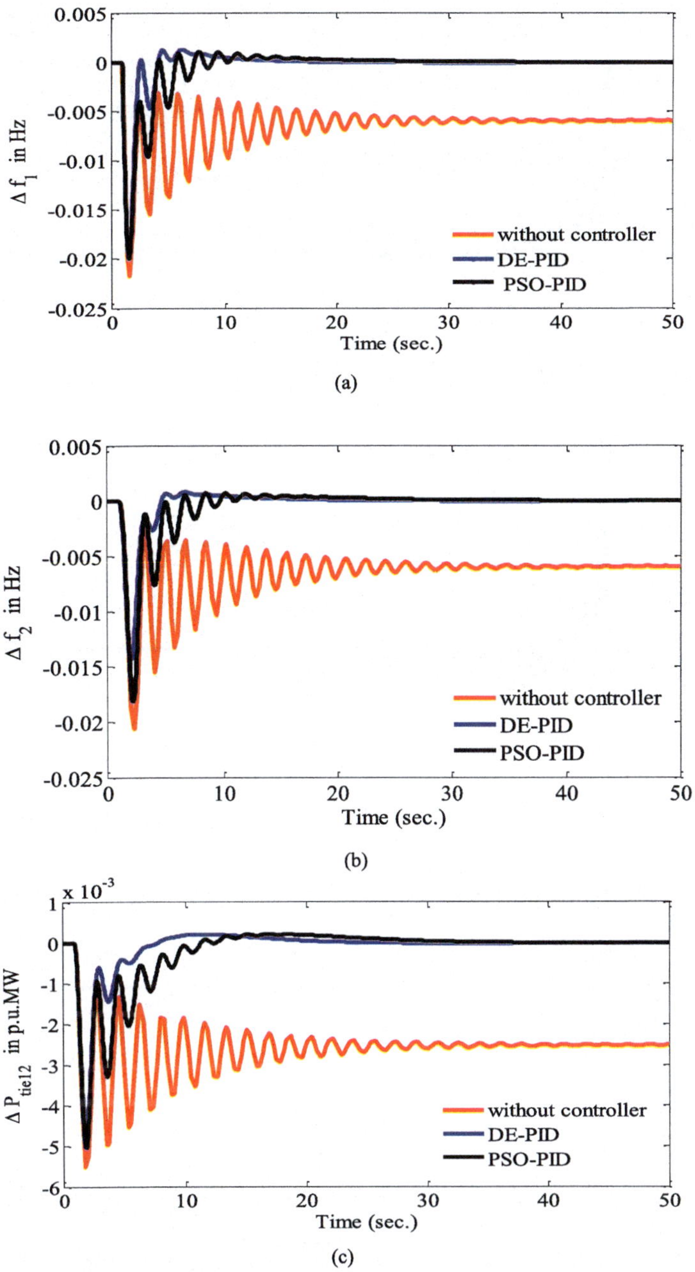

Fig. 6 Output responses **a** Δf_1 versus time **b** Δf_2 versus time **c** P_{tie-12} versus time **d** ΔACE_1 versus time **e** ΔACE_2 versus time **f** convergence characteristics

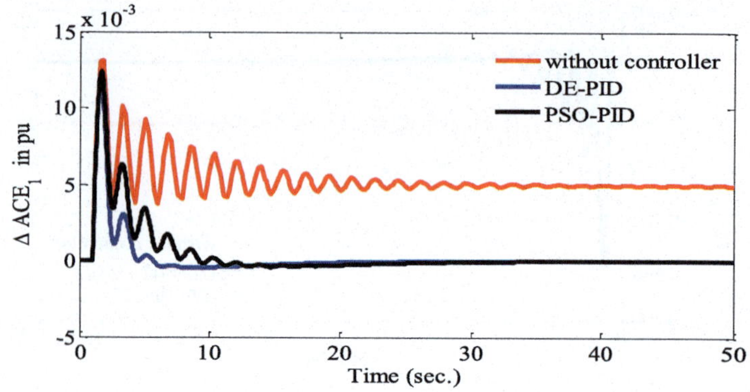

(d)

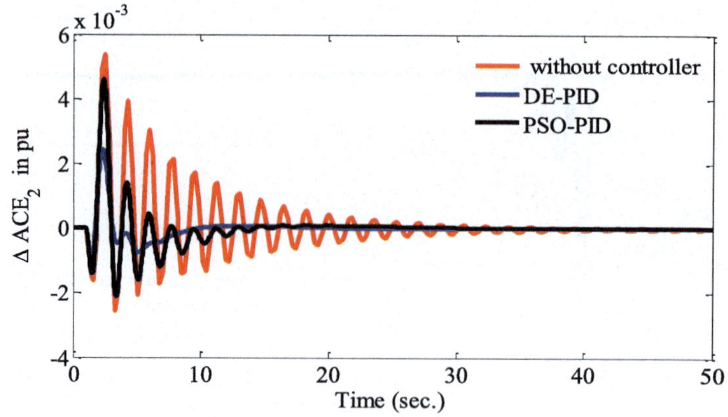

(e)

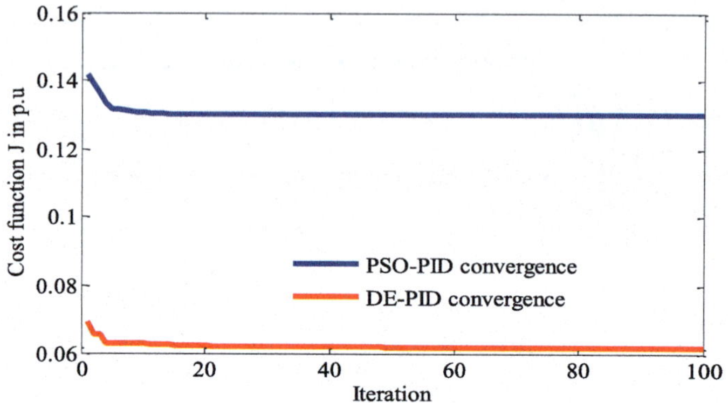

(f)

Fig. 6 (continued)

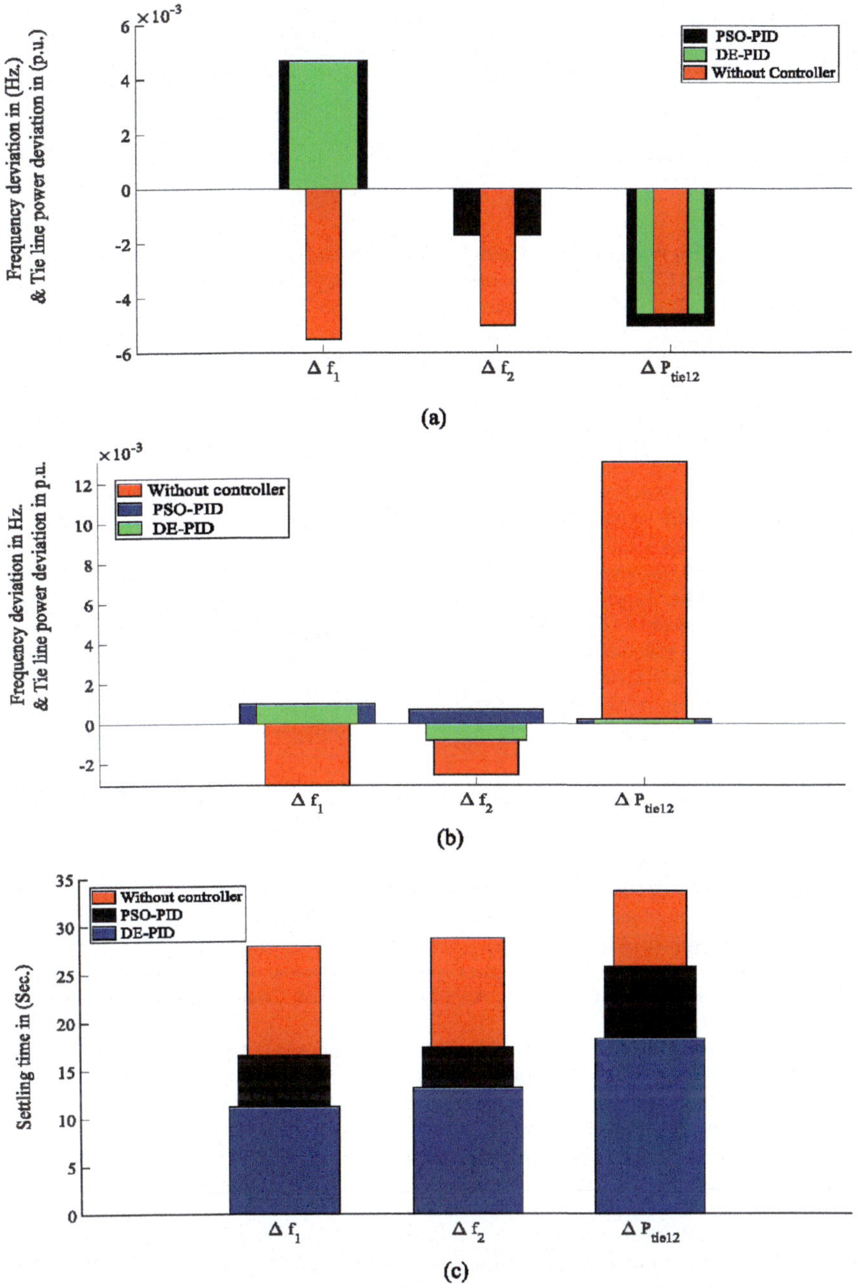

Fig. 7 Performance comparisons for DE-PID, PSO-PID, and without controller **a** Undershoot, **b** Overshoot **c** Settling time for without controller, PSO-PID, and DE-PID

Acknowledgements This research work was supported by "Woosong University's Academic Research Funding—2023".

References

1. Bevrani H (2016) Robust power system frequency control, 2nd edn. Springer, New York
2. Elgerd OI, Fosha C (1970) Optimum megawatt frequency control of multi-area electric energy systems. IEEE Trans Power Apparatus System 89:556–563. https://doi.org/10.1109/TPAS.1970.292602
3. Mohanty B (2014) Differential evolution algorithm based automatic generation control for interconnected power systems with non-linearity. Alex Eng J 53:537–552. https://doi.org/10.1016/j.aej.2014.06.006
4. Bevrani H, Hiyama T (2011) Intelligent automatic generation control, 1st edn. CRC Press, Boca Raton, FL
5. Kouba Y (2015) Load frequency control in multi-area power system based on fuzzy logic-PID controller. In: IEEE international conference on smart energy grid engineering. https://doi.org/10.1109/SEGE.2015.7324614
6. Solihin M (2011) Tuning of PID controller using particle swarm optimization. Int J Adv Sci Eng Inf Technol 1. https://doi.org/10.18517/ijaseit.1.4.93
7. Eberhart R, Kennedy J (1995) A new optimizer using particle swarm theory. In: Proceedings of sixth international symposium, micro machine and human science, Nagoya, Japan. https://doi.org/10.1109/MHS.1995.494215
8. Hasan NI, Hussein AA (2014) Decentralized automatic generation control of interconnected power systems incorporating asynchronous tie-lines. Springer Plus 3. https://doi.org/10.1186/2193-1801-3-744
9. Kumar S, Pati BS, Panda S (2014) Hybrid differential evolution particle swarm optimization optimized fuzzy proportional–integral derivative controller for automatic generation control of interconnected power system. IET Gener Transm Distrib 8:1789–1800. https://doi.org/10.1049/iet-gtd.2014.0097
10. Sahu RK, Panda S, Padhan S (2014) Optimal gravitational search algorithm for automatic generation control of interconnected power systems. Ain Shams Eng J 5:721–733. https://doi.org/10.1016/j.asej.2014.02.004
11. Sambariya DK, Prasad R (2016) Design of optimal proportional integral derivative based power system stabilizer using bat algorithm. Appl Comput Int Soft Comput. https://doi.org/10.1155/2016/8546108
12. Illias HA, Zahari AFM, Mokhlis H (2016) Optimization of PID controller for load frequency control in two-area power system using evolutionary particle swarm optimization. J Electr Syst 12:315–324
13. Kumar R, Sharma VK (2021) Automatic generation controller for multi area multisource regulated power system using grasshopper optimization algorithm with fuzzy predictive PID controller. Int J Numer Model Electron Netw Devices Fields 34. https://doi.org/10.1002/jnm.2802
14. Kasilingam G, Pasupuleti J (2015) Coordination of PSS and PID controller for power system stability enhancement–overview. Indian J Sci Technol 8:142–151. https://doi.org/10.17485/ijst/2015/v8i2/58441
15. Nagaraj B, Muruganath N (2010) A comparative study of PID controller tuning using GA, EP, PSO and ACO. In: International conference on communication control and computing technologies. https://doi.org/10.1109/ICCCCT.2010.5670571
16. Herrero JM, Blasco X, Martínez M, Sanchis J (2008). Multi-objective tuning of robust PID controllers using evolutionary algorithms. In: Applications of evolutionary computing.

EvoWorkshops, Lecture notes in computer science, Springer, Berlin, Heidelberg. https://doi.org/10.1007/978-3-540-78761-7_57
17. Theja BS, Rajasekhar A, Kothari DP, Das S (2013) Design of PID controller based power system stabilizer using Modified Philip-Heffron's model: an artificial bee colony approach. In: IEEE symposium on swarm intelligence, pp 228–234. https://doi.org/10.1109/SIS.2013.6615183
18. Feleke S, Satish R, Tatek W, Abdelaziz AY, El-Shahat A (2022) DE-algorithm-optimized fuzzy-PID controller for AGC of integrated multi area power system with HVDC link. Energies 15. https://doi.org/10.3390/en15176174
19. Cam E, Gorel G, Mamur H (2017) Use of the genetic algorithm-based fuzzy logic controller for load-frequency control in a two area interconnected power system. Appl Sci 7:308. https://doi.org/10.3390/app7030308
20. Huynh VV, Tran PT, Minh BLN, Tran AT, Tuan DH, Nguyen TM, Vu PT (2020) New second-order sliding mode control design for load frequency control of a power system. Energies 13:6509. https://doi.org/10.3390/en13246509
21. Sobhy MA, Abdelaziz AY, Hasanien HM, Ezzat M (2021) Marine predators algorithm for load frequency control of modern interconnected power systems including renewable energy sources and energy storage units. Ain Shams Eng J 12:3843–3857. https://doi.org/10.1016/j.asej.2021.04.031
22. Fathy A, Kassem AM, Abdelaziz AY (2020) Optimal design of fuzzy-PID controller for deregulated LFC of multi-area power system via mine blast algorithm. Neural Comput Appl 32:4531–4551. https://doi.org/10.1007/s00521-018-3720-x
23. Abo-elyousr F, Abdelaziz AY (2019) A novel modified robust load frequency control for mass-less inertia photovoltaics penetrations via hybrid PSO-WOA approach. Electric Power Compon Syst J 47:1744–1759. https://doi.org/10.1080/15325008.2020.1731867
24. Abo-Elyousr FK, Youssef AM, Abdelaziz AY (2018) Multi area hydrothermal interconnected load frequency control with double fed induction generator based wind turbine via improved harmony algorithm. Electric Power Compon Syst J 46:615–628. https://doi.org/10.1080/15325008.2018.1462867
25. Salkuti SR (2022) Emerging and advanced green energy technologies for sustainable and resilient future grid. Energies 15(18):6667. https://doi.org/10.3390/en15186667
26. Jia Y, Dong ZY, Sun C, Meng K (2019) Cooperation-based distributed economic MPC for economic load dispatch and load frequency control of interconnected power systems. IEEE Trans Power Syst 34:3964–3966. https://doi.org/10.1109/TPWRS.2019.2917632
27. Liu F, Li Y, Cao Y, She J, Wu M (2016) A two-layer active disturbance rejection controller design for load frequency control of interconnected power system. IEEE Trans Power Syst 31:3320–3321. https://doi.org/10.1109/TPWRS.2015.2480005
28. Fan L, Miao Z, Osborn D (2009) Wind farms with HVDC delivery in load frequency control. IEEE Trans Power Syst 24:1894–1895. https://doi.org/10.1109/TPWRS.2009.2030256
29. Gholamrezaie V, Dozein MG, Monsef H, Wu B (2018) An optimal frequency control method through a dynamic load frequency control (LFC) model incorporating wind farm. IEEE Syst J 12:392–401. https://doi.org/10.1109/JSYST.2016.2563979
30. Syed MH, Guillo-Sansano E, Mehrizi-Sani A, Burt GM (2020) Load frequency control in variable inertia systems. IEEE Trans Power Syst 35:4904–4907. https://doi.org/10.1109/TPWRS.2020.3014778
31. Storn R, Price K (1997) Differential evolution—a simple and efficient heuristic for global optimization over continuous spaces. J Global Optim 11:341–359. https://doi.org/10.1023/A:1008202821328
32. Das AK, Majumdar R, Panigrahi BK, Reddy SS (2011) Optimal power flow for Indian 75 bus system using differential evolution. In: SEMCCO'11: proceedings of the second international conference on swarm, evolutionary, and memetic computing, pp 110–118. https://doi.org/10.1007/978-3-642-27172-4_14

A Comprehensive Analysis of the Application of Swarm Intelligence Techniques to the Economic Load Dispatch Problem

Arun Kumar Sahoo, Bibhu Prasad Nanda, Debani Prasad Mishra, and Surender Reddy Salkuti

Abstract The power flow solution can be optimized in different ways, providing the safest point of operation based on specific objective functions while also meeting the system's operational restrictions. In power systems, many objective functions can be optimized, which include the complete cost of power generation, involved flexible AC transmission devices maintenance cost, power flow capability of the system, residue effect of the generating station, etc. During the optimization phase, involved variables like true power (P), the voltage at generating station bus (V), tap settings of the transformer, etc. which are controllable can be pinched. Several processes are considered earlier for the solution of the power flow issue. Classification and innovation of some recent or new techniques have been developed for finding the solution to those OPF problems. Several classical and modern optimization and meta-heuristic strategies can be used to optimize and prevent entrapment locally. In this section, a detailed recent optimization technique overview has been made among which nature-, evolutionary-, human-, and physics-inspired methods with ANN strategies are discussed.

Keywords Optimal power flow · Power flow · Optimization techniques · Economic dispatch · Artificial neural network

---

A. K. Sahoo · B. P. Nanda · D. P. Mishra
Department of Electrical and Electronics Engineering, IIIT Bhubaneswar, Bhubaneswar, Odisha, India
e-mail: debani@iiit-bh.ac.in

S. R. Salkuti (✉)
Department of Railroad and Electrical Engineering, Woosong University, Daejeon, Republic of Korea
e-mail: surender@wsu.ac.kr

S. R. Salkuti et al. (eds.), *Power Quality in Microgrids: Issues, Challenges and Mitigation Techniques*, Lecture Notes in Electrical Engineering 1039,
https://doi.org/10.1007/978-981-99-2066-2_19

Nomenclature

OPF	Optimal Power Flow
ELD	Economic Load Dispatch
ANN	Artificial Neural Network
ED	Economic Dispatch
ACO	Ant Colony Optimization
ACSA	Artificial Ant Colony Search Algorithms

1 Introduction

The generation of electrical power to achieve the load demand is highly reliant on fossil fuel-based power plants. Limited resources of fossil fuel encourage optimum and reliable power generation. The minimum cost is achieved by the most efficient distribution of electricity among the devoted generating units. The optimum power arrangement among the generating units to meet the necessities of the load demand and transmission losses with the consideration of all the practical and operative restrictions is known as Economic Load Dispatch [1]. The practical effect of the steam valve, prohibited operating zones, ramp rate limit, and spinning reserve constraints make the economic dispatch a complex optimization problem. With the adoption of all practical limitations and the size of the power system, the characteristics curve of large generators and the ELD problem performs as a non-smooth and non-convex optimization problem. The input–output characteristics of the generator are having several local minima for the effect of multiple fuels and valve point loading. The ELD problem has been solved using a variety of optimization strategies. The optimization methods are classified into three major categories: (a) conventional techniques, (b) non-conventional techniques, and c) hybrid approaches with the combination of two or more techniques [2]. A deterministic approach was applied by a conventional method, which comprises Dynamic Programming [3], Lagrange Multiplier [4], Linear Programming [5], Quadratic Programming [6], Gradient Method [7], and Lambda Iteration Technique [8]. The methods like linear programming, lambda iteration method, Lagrange multiplier method, and gradient method remained not able to find the optimal cost for non-convex and non-smooth characteristics, as all the techniques are calculus based and work effectively for smooth and convex functions. However, Dynamic Programming had no limitations for solving the non-smooth cost curve but undergoes problems for the high dimension of larger units.

These conventional approaches were unsuccessful in solving the complex problem with the practical and operational constraints as they are very subtle for an early guess. Thus, they trap in their respective local optima, which causes poor convergence property. Recently many evolutionary techniques were employed in ELD problem with the inspiration of nature. These methods are quite helpful for cracking the complex Economic Dispatch problem considering all the practical and operational

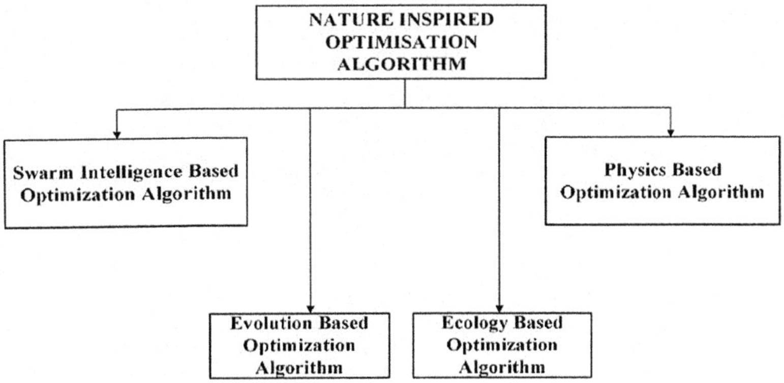

Fig. 1 Classification of nature-inspired algorithms

constraints. Figure 1 shows the classification of the algorithms. The nature-inspired algorithms applied [9] to the economic load dispatch problem were presented in Table 1.

A comprehensive review and brief survey of the application of the swarm intelligence algorithm under a nature-inspired algorithm to complex ELD problems of modern power systems is presented in this chapter.

2 Problem Formulation of ELD

The overall fuel cost for producing electric power is distributed among the generating units using the economic load dispatch issue. All participating generators should provide enough electricity to meet the system's demand and transmission losses [33]. First, the generator's cost function appears to be a spherical cost function. Because of valve point loading, the generating characteristics behave as non-convex and the cost function acts as a piecewise linear function, respectively. Furthermore, the input and spinning reserve limitations are defined with the practical implications of different fuels [54].

2.1 *Cost Function of Economic Dispatch*

The cost function without consideration of the valve point effect behaves as a quadratic function with cost characteristics convex in nature [65]. The cost function is represented in Eqs. (1) and (2) [10].

$$bj(F_{gr}) = \min(F_{gr}) \tag{1}$$

Table 1 Nature-inspired method applied to economic dispatch problem

	Methods with abbreviation	Reference
Swarm Intelligence Algorithm for solving Economic Dispatch Problem	Particle swarm optimization (PSO)	[10]
	Modified PSO (MPSO)	[11]
	Linearly decreasing particle swarm optimization (LPSO)	[12]
	New PSO-local random search (NPSO-LRS)	[13]
	Chaotic particle swarm optimization (CPSO)	[14]
	Anti-predatory PSO (APSO)	[15]
	Adaptive PSO (APSO)	[16]
	Timevarying acceleration coefficients-PSO (TVAC-PSO)	[17]
	Crazy PSO (CRAZYPSO)	[18]
	Self-organizing hierarchical PSO (SOH-PSO)	[19]
	Quantum-behaved particle swarm optimization with differential mutation (QPSO-DM)	[20]
	Improved coordinated aggregation-based PSO (ICA-PSO)	[21]
	Improved particle swarm optimization (IPSO)	[22]
	Fuzzy adaptive modified PSO (FAMPSO)	[23]
	Parallelized PSO using modified stochastic acceleration factors (PPSO-MSAF)	[24]
	Hybrid quantum mechanics-inspired PSO (HQPSO)	[25]
	Iteration PSO (IPSO)	[26]
	Quantum-behaved particle swarm optimization algorithm (QPSO)	[27]
	Iteration PSO with time-varying acceleration coefficients (IPSO-TVAC)	[28]
	Quantum-inspired PSO (QPSO)	[29]
	Hybrid multi-agent-based PSO (HMAPSO)	[30]
	θ-PSO	[32]
	Cuckoo search algorithm (CSA)	[33, 34]
	Modified CSA (MCSA)	[35]

(continued)

Table 1 (continued)

	Methods with abbreviation	Reference
	One rank CSA (ORCSA)	[36]
	Artificial bee colony algorithm (ABC)	[37, 38]
	Modified ABC (MABC)	[39]
	Incremental ABC with local search method (IABC-LS)	[40]
	Harvest season ABC (HSABC)	[41]
	Binary/real coded artificial bee colony algorithm (BRABC)	[42]
	ACO	[43]
	ACSA	[44]
	Continuous ant colony optimization (C-ANT)	[45]
	Firefly algorithm (FFA)	[46]
	Distance-based FFA (DFFA)	[47]
	Chaos mutation firefly algorithm (CMFA)	[48]
	Modified bat algorithm (MBA)	[49]
	Chaotic bat algorithm (CBA)	[50]
	Pseudo-inspired chaotic bat algorithm (PI-CBA)	[51]
	Continuous group search optimizer (CGSO)	[52]
	Modified Group search optimizer (MGSO)	[53]
	Social spider algorithm (SSA)	[54]
	Modified SSA (MSSA)	[55]
	Improved SSA (ISSA)	[56]
	Levy—SSA (LSSA)	[57]
	Orthogonal learning competitive swarm optimizer (OLCSO)	[58]
	Ant lion optimizer (ALO)	[59]
	Gray wolf optimization (GWO)	[60]
	Hybrid bacterial foraging with Nelder–Mead (HBF-NM)	[61]
	Hybrid chaotic PSO and sequential quadratic programming (CPSO-SQP)	[62]
	Hybrid differential evolution and PSO (DEPSO)	[63]
	Hybrid gray wolf optimization algorithm (HGWO)	[64]

$$\min(F_{gr}) = \sum_{gr=1}^{N} F_{gr}(P_{gr}) = \sum_{gr=1}^{N} a_{gr} P_{gr}^2 + b_{gr} P_{gr} + c_{gr} \tag{2}$$

The effect of the steam valve in the generating units prompts the non-linearities and non-convexity in the cost curve. Thus, the cost characteristics perform as a piecewise linear-quadratic function [66]. The precise mathematical design of the cost function is represented in Eq. (3) [13].

$$\min(F_{gr}) = \sum_{gr=1}^{N} F_{gr}(P_{gr}) = \sum_{gr=1}^{N} a_{gr} P_{gr}^2 + b_{gr} P_{gr} + c_{gr} + \left| e_{gr} * \sin\left(f_{gr}\left(P_{gr}^{min} - P_{gr}\right)\right)\right| \tag{3}$$

Here a_{gr}, b_{gr}, c_{gr}, e_{gr}, and f_{gr} are the input coefficient of cost for the generator, "F_{gr}" shows the cost to be optimized, and $(F_{gr}P_{gr})$ is the function of the cost of the generator for P_{gr} as the output. Figure 2 shows the characteristics curve of the cost.

The practical concept of the application of multiple fuels to the generating units is considered for the solution and the cost function is represented in Eq. (4) with the valve point effect [67].

$$\min(F_{gr}) = \sum_{gr=1} F_{gr}(P_{gr}) = \sum_{gr=1}^{N} \sum_{i=1} a_{gri} P_{gr}^2 + b_{gri} P_{gr} + c_{gri} + \left| e_{gri} * \sin\left(f_{gri}\left(P_{gr}^{min} - P_{gr}\right)\right)\right| \tag{4}$$

Here, a_{gri}, b_{gri}, c_{gri}, e_{gri}, and f_{gri} show the cost coefficient of different generating units for ith type of fuel.

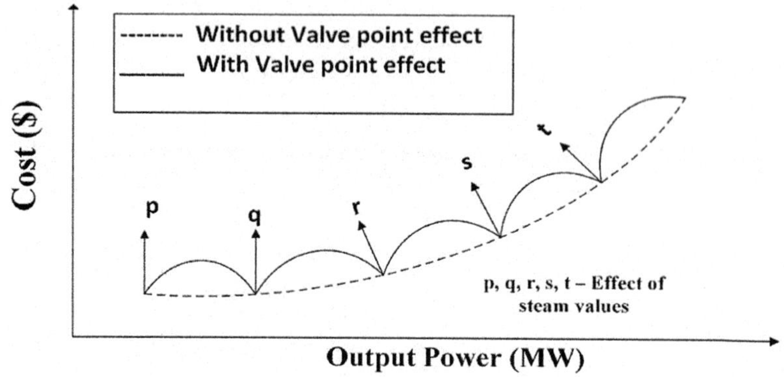

Fig. 2 Characteristics curve of fossil fuelbased generators

2.2 Constraints

2.2.1 Constraints for Real Power Balance

During any scheduled period, the overall generated power should satisfy the system demand and transmission losses in the power system for balancing power as shown in Eq. (5) [10].

$$\sum_{gr=1}^{N} P_{gr} - P_{Dgr} - P_{Lgr} = 0 \tag{5}$$

Here, P_{gr} represents the scheduled generated power, P_{Dgr} represents the system load demand and P_{Lgr} represents the losses that occur during the power transmission. P_{Lgr} is presented in terms of loss coefficient as "B" as shown in Eq. (6) [10].

$$P_{Lgr} = \sum_{x=1}^{N}\sum_{y=1}^{N} P_{grx} B_{xy} P_{gry} + \sum_{x=1}^{N} B_{0x} P_{grx} + B_{00} \tag{6}$$

where P_{Lgr} shows the transmission losses, B_{xy}, B_{Oy}, and B_{OO} are the transmission loss coefficients.

2.2.2 Constraints Regarding the Capacity of Generators

The power generated from each committed generator should be within the upper and lower limit capacity of the generators as represented in Eq. (7) [10].

$$P_{grmin} < P_{gr} < P_{grmax} \tag{7}$$

Here P_{grmin} and P_{grmax} are the minimum and maximum amount of power each generator is able to produce, respectively.

2.2.3 Generator Ramp Rate Limits

Ramp rate restricts the real-time and dynamic action for generating units, affecting the decisions for the operation of a power plant, which in turn affects the system economy. Due to ramp rate limits, the current scheduling may be disrupted when the generation increases. The ramp rate constraints of the thermal power plant are given in detail as follows in Eq. (8) [13]:

$$\max(P_{grx}^{min}, P_{grx}^{0} - DR_y) \le P_{grx} \le (P_{grx}^{max} + UR_y) \tag{8}$$

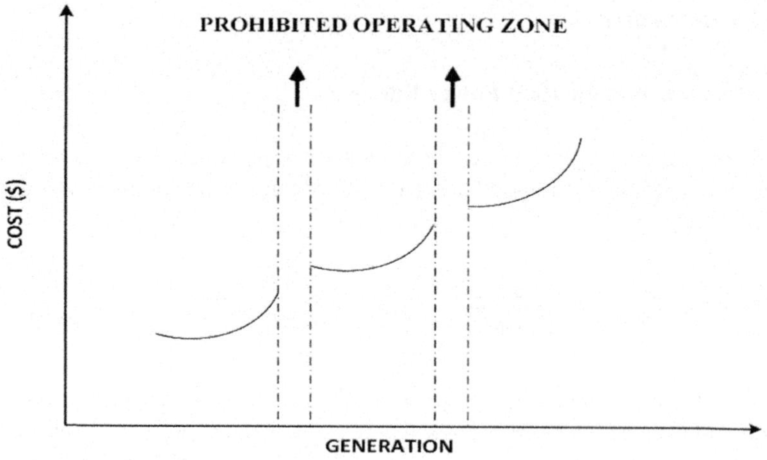

Fig. 3 Prohibited operating zones cost curve

Here ${P_{grx}}^{0}$ is the output power at the previous level and DR_y and UR_y are the down ramp rate limit and up ramp rate limit, respectively.

2.2.4 Prohibited Operating Zones (POZs)

The committed divisions of generators consist of specific units which cannot be operated as their operation is undesirable and unmanageable because of some particular physical limitations like steam valve operation and shaft bearing vibrations. These zones of operation are known as POZs and are shown in Fig. 3. The presence of POZs is a major reason for the introduction of non-convexity and non-linearity in the cost function. The POZs are characterized by some inequality constraints, which are listed below as per Eqs. (9), (10), and (11) [13].

$$P_{gry}^{min} < P_{gry} < P_{grx,1}^{l} \quad (9)$$

$$P_{gry,y-1}^{U} < P_{gry} < P_{gry,y}^{l} \quad (10)$$

$$P_{gry,n_x}^{U} < P_{gry} < P_{gry}^{max} \quad (11)$$

3 Swarm Intelligence-Based Optimization Algorithm

The swarm intelligence optimization algorithm is the problem-solving method inspired by the behavior of animal, insect, and bird societies. Various swarm intelligence methods were applied to get a quality solution in economic dispatch problems such as particle swarm optimization [10], cuckoo search (CS) [33], ant colony optimization (ACO) [43], Krill Herd (KH), artificial bee colony (ABC) [37], firefly algorithm (FA) [46], gray wolf optimization (GWO), ant lion optimization [60], and bacteria foraging optimization (BFA) [61] by many researchers. These methods are mostly applied to complex ELD problems with various complexities and dimensions.

3.1 Particle Swarm Optimization (PSO)

Kennedy et al. proposed this PSO method with the inspiration of bird flocking and fish schooling. Gaing applied the conventional PSO [10] to a practical constrained ELD problem. Here POZs, ramp rate limit, and line flow constraints are considered. The power generated from each generating unit is represented as a gene. The individual comprises numerous genes. Each individual under the randomly generated population indicates a solution aspirant for the solution of the ELD problem. This method was tested for 6 units, 15 units, and 40 units of test systems for 200 generations.

A modified PSO was applied by Park et al. [11]. In this chapter, the conventional PSO is modified to improve the acceleration of the optimization problem with the inclusion of a dynamic search-space reduction strategy. This method was validated for both valve point loading effects and multiple fuel economic dispatch problems.

Jeyakumar et al. adapted a linear declination of inertia weight to conventional PSO [12]. Economic dispatch with multiple fuels with the effect of the steam valve and multiple areas with tie line constraints was solved by the method. This method was also used to get the solution for the environmental impact economic dispatch problem. A new PSO was proposed by Selvakumar et al. [13] by splitting the cognitive nature by remembering the poorest position and integrating local random search in the original PSO. This method was proposed to solve three types of the test system with unitary fuel and one test system for multiple fuels.

Anti-predatory behavior of the swarm is adapted to the conventional PSO to raise the bar in terms of exploration and solution quality by Selvakumar et al. for solving of ELD problem [15]. This method is termed as anti-predatory particle swarm optimization (APSO). This method is tested for two test systems, one with VPE and the other with multiple fuel options.

An adaptive approach is applied to the original PSO by Panigrahi et al. [16]. The movement of swarm particles is controlled and the population with definite iteration is re-initialized. The particle with high fitness values moved slowly as compared to low fitness particles according to their rank. This method was tested for both static and dynamic economic dispatch (DED) problems. This method improves the convergence property and robustness. Chaturvedi et al. adapted time-varying acceleration coefficients for proper balancing between exploitation and exploration to avoid premature convergence and better exploration capability [17]. This technique was tested for convex and non-convex economic dispatch problems.

Chaturvedi et al. proposed an enhanced algorithm for original PSO [19] named self-organizing hierarchical PSO (SOH-PSO) applied to a three-test system with non-linearities and transmission losses. In this method, the velocity of the particle was again initialized, during the search process when stuck at local optima. Sun et al. proposed a modified version of PSO by applying quantum mechanics [20]. Again, the differential mutation was hybridized to quantum PSO for the application of three-test systems to reduce the optimum cost with lesser convergence time. Improved coordinated aggregation-based PSO (ICA-PSO) [21] is proposed by Vlachogiannis et al. to solve the economic dispatch problem. This type of aggregation helps the particle to get attracted to the places with maximum food. In this method, the particle memorizes the finest attained position to find the food and is attracted by other particles to find a better solution.

The application of chaotic sequence and crossover scheme is amended by Park et al. [22] to enhance constraint handling capacity. The chaotic sequence is replaced by the random number with the linear declination of inertia weight. Here the crossover operation was also assembled to find a more optimum solution as compared to other existing techniques with four diverse strategies. This method was tested for multiple fuel systems and for large power systems with 140 generating units. The original PSO was modified with the application of a fuzzy adaptive process by Niknam et al. [23]. The weight of inertia and learning factor were attuned dynamically using the IF/THEN rule of fuzzy logic. The performance of the proposed method was evaluated by the application to 13 and 40 units of test systems.

Subbaraj et al. proposed a parallelized PSO using modified stochastic acceleration factors (PSO-MSAF) [24] for the solution of the ELD problem. Multiple populations were created to work parallel with multiple workers. This method was used to update the learning parameter and particle velocity. The manager transmits the updated parameters to the parallel worker. These modifications made the proposed technique an effective one to get the quality solution for different generators. Hybrid quantum-based PSO [25] was suggested with the implementation of multiple populations by Chakraborty et al. The larger size of the power system could not be handled properly by a single population. The usage of a single population causes premature convergence as the local optima are stuck early. The multiple population schemes were adapted to overcome the difficulties. The proposed algorithm was validated for both single-type fuel and multiple fuel systems. Safari et al. suggested the usage of a best iteration index term to the particle to improvise the classical PSO [26]. The iteration index term shows the best fitness value, in every iteration by any particle. This process improvises the searching ability of the particle to get a superior solution with other techniques for two different test systems.

Time-varying acceleration coefficients were added to iterative PSO Mohammadi-Ivatloo et al. [28] to solve the economic dispatch problem. The integration of time-varying acceleration creates a suitable balance between the local search and global search process during all phases of iteration. This method uses three different tests for solving the complex ELD problem. Kumar et al. proposed a hybrid technique with the combination of the Nelder–Mead method to PSO technique named as multi-agent-based hybrid PSO [30]. The decision-making access is taken from Bee's decision-making approach for a better optimum solution. Basu proposed a modified PSO for solving of ELD problem [31]. Gaussian random variables were applied to the velocity term to improvise the efficiency of the search process. The convergence speed was increased with an improved optimum cost for large power systems and multiple linearities.

Hosseinnezhad proposed an enhanced version of PSO named as θ-PSO [32] to solve the ELD problem for four different test systems with diverse non-linearity constraints. The conventional PSO was revised with a phase angle vector as a replacement for the velocity vector. The new position of each solution particle was mapped with the phase angle.

3.2 Ant Colony Optimization (ACO)

This optimization algorithm considers the shortest route to reach the food source from the shelter or colony of the ant. This technique performs with two key steps. In the first step, the artificial ant creates the solution by moving to the nearby positions of a given problem following the transition rule for building the solutions. The updating process occurred considering the odor after every iteration, then the fragrance gets vanished in the second step. The ants in this step forgot the worst solution, which was initially learned. Song et al. applied ant colony optimization to the large-scale economic dispatch problem [43]. This method shows a minimum cost for two test systems as compared to the conventional genetic algorithm.

Pothiya et al. proposed an improved ant colony optimization algorithm [44] by including three approaches, i.e., variable reduction, priority list, and zoom feature. This method was used to solve for single fuels with non-convexity and multiple fuel test systems. Vlachos et al. proposed a continuous ant colony optimization algorithm to overcome the difficulties produced by the discrete method of solving the problem by conventional ACO [45]. This proposed algorithm improves the exploitation process for finding the improved solution. This was tested for four generating units with different types of fuels.

3.3 Cuckoo Search Algorithm (CSA)

This is an effective algorithm with lesser controlling parameters. Implementation of levy flight behavior efficiently processes the global search. The reproduction strategy of the cuckoo is the key inspiration to the algorithm as the cuckoo places its egg in the nest of another bird. This algorithm was proposed with three key rules a) One egg was laid by a cuckoo at another bird's nest randomly at a specific time, b) the eggs with better quality were considered as the solution which will be treated for the next generation, and c) the number of the nest of the host bird is fixed and the probability was learned by bird.

Vo et al. applied the original CSA to the economic dispatch problem with all the practical and operational constraints [34]. It was applied for multiple fuel systems with spinning reserve constraints. The CSA algorithm verified the optimum result as compared to other existing techniques. Basu and Chowdhury implemented CSA to the ELD problem as well as the microgrid problem [33]. In this chapter, the conventional cuckoo search was applied for a different test system and found an effective solution for the economic dispatch for single-fuel and multiple fuel systems as well as microgrid systems.

A modified strategy is adapted to the conventional CSA by Zhao et al. [35] to solve the non-smooth, non-linear economic dispatch problem. The self-adaptive step size was implemented with improved neighborhood strategy and generation of new solutions by improved lambda iteration methods. The algorithm was validated for four different benchmark functions. Nguyen & Vo. proposed one rank CSA [36] for the solution of the ELD problem with all the constraints and multiple fuel system. The conventional CSA is enhanced by two factors, a) the generated solution by the implementation of levy flight was merged with the extra egg together to assess the solution rank once, b) the best solution was bound to handle the inequality constraints and applied to ELD problem for finding the optimum cost with increased solution quality.

3.4 Artificial Bee Colony Algorithm

This algorithm was suggested by Karaboga for the application of many optimization problems. The inspiration for the algorithm was established, considering the intellectual behavior of honeybees. The collective and interacting behavior of honeybees with each other for the foraging process mainly encourages the algorithm. The algorithm shows a prominent result for the solution of the optimization problem for the uni-modal and multi-modal problems. The conventional version and modified version of the algorithm have been used for the ELD problem. Implementation of the ABC algorithm improves the quality of the solution. Hemamalini et al. applied the conventional artificial bee colony to the economic dispatch problem with practical constraints [37]. Certain behavioral approaches for specific honeybees were reserved for the constrained problem. The ABC algorithm was effectively skilled to trace the finest solutions. The quality of the solution was enhanced as compared to various techniques.

Labi et al. [38] also applied the conventional ABC for solving the ELD problem considering the valve point loading effect. This method was tested for 3 units, 13 units, and 40 units of test systems. Secui put forward a modified version of the ABC algorithm to improve the performance of the ELD problem and economic emission dispatch problem [39]. A novel relation was introduced to improvise the solution within the search area. Further mapping of cat and logistics chaotic mapping enhanced the performance of ABC. This modified method enhanced the capability of the algorithm to attain stable and quality solutions by avoiding convergence prematurely. The improved version of the ABC algorithm was applied by Aydin for solving the non-convex economic dispatch problem [40]. Aydin applied the incremental property as well as implemented the local search technique to the incremental property of the conventional ABC algorithm. During the process of optimization, the number of population was increased and a local search method was added to the method. The efficiency and quality of the solution increased as compared to the original ABC algorithm and other contributed methods.

The improvement of the ABC algorithm with the implementation of harvest season approaches improves the optimal cost and emission for a test system which was introduced by Afandi et al. [41]. The introduction of harvest search improves the search mechanism, by moving of honeybees in arbitrary directions for finding the food source. A harvest operator is also used to specify the availability of new food during random directional moves. Chandrasekaran et al. proposed real coded and binary coded of the ABC algorithm for the solution of the economic dispatch problem [42]. The original ABC algorithm is further improvised by a real coded version and then a binary coded version for improvement in optimal cost of economic dispatch considering the non-linear objective function and multiple fuel objectives.

3.5 Firefly Algorithm

This novel algorithm was anticipated by Yang with the inspiration of the blinking nature of fireflies. Fireflies are considered as agents for solving the optimization problem. The bioluminescent glowing of fireflies is considered as the fitness function for optimization. The firefly with brighter luminous attracts partners irrespective of sex to explore the search space. This FFA follows three rules for solving the problem, and they are

- Every firefly follows the movement toward the brighter fireflies irrespective of their sex.
- The attraction of fireflies depends upon the brightness. The brightness decreases with an increase in distance from other fireflies.
- The represented brightness of fireflies was due to the cost function of the economic dispatch problem.

Yang proposed and applied the optimization method to solve the ELD problem [46]. Chen et al. proposed an upgraded firefly algorithm for a constrained optimization problem [47]. Two enhancements were considered to conventional ABC for improvement in searchability. An adaptation strategy was assigned to the brightness depending on the distance and the reduction of the randomization parameter. Furthermore, a crossover operator was introduced for better solution quality. The proposed algorithm was also validated for different benchmark functions. The improvement in optimal solution is compared with other techniques enlisted. Yang et al. proposed a chaos mutation firefly algorithm (CMFA) [48] to solve the ELD problem with non-convexity and multiple fuel systems. The conventional firefly algorithm was modified with two new properties to enhance the optimization capability in a small CPU time. The fixed parameters used in the FFA were replaced by novel dynamic tuned parameters and the new mutation process adapts the self-adaptive mutation mechanism. This proposed algorithm enhances the speed and quality of the solution as compared to other applied techniques.

3.6 Bat Algorithm

This algorithm was proposed by Yang in the year 2010. The algorithm was inspired by the echolocation sensation within the bats for searching food. The generation of sound waves is another characteristic of the bat to move toward prey, while moving toward the prey the pulse rate rises with the decrease in loudness. Latif et al. proposed a modified bat algorithm to get the solution to the problem for two test systems [49]. The conventional bat algorithm was modified with two modifications: a) the anti-predatory approach was adopted in the BA algorithm, b) the inertia weight factor was introduced to calculate the velocity of the bat. This modified version of BA maintains a proper balance between exploration and exploitation, which increases the convergence rate. This algorithm provides a better solution as compared to a real coded genetic algorithm, particle swarm optimization.

Adarsh et al. introduce the chaotic sequence [50] for solving the economic dispatch problem and replace the random variables with different chaotic mapping to improve the performance for solving economic dispatch problem with non-linearities and multiple fuel systems. Pseudo-inspired chaotic bat algorithm (PI-CBA) was proposed by Shukla et al. [51] for five different test systems with practical constraints. The equality constraints were handled by the pseudo-code method and the random parameter was replaced by chaotic sequence mapping to enhance the optimal solution.

3.7 Other Swarm Intelligence Algorithms

Xiong et al. proposed an orthogonal learning competitive swarm optimizer [58] for solving the ELD problem. The competitive swarm intelligence method is a variant of PSO. The particles of the swarm in CSO were subdivided into two particles to move toward the food source. The particle with the best fitness value was considered as the winner and moved to the next iteration process and the other particle was looser, which will learn from the winner particle to update the velocity. The improvement was done for the learning process of the loser. An orthogonal learning strategy was introduced for the loser to learn. This technique was validated for 24 benchmark functions and three-test systems.

Kamboj et al. applied the ant lion optimizer (ALO) [59] algorithm to solve the problem. The stalking nature of ant lions inspires the algorithm. The hunting mechanism was processed by five steps, which constitute of random walk nature of ants, building the trap mechanism, setup of ants for the traps, grasping the prey, and starting the traps again. This algorithm was tested for four different small power systems and compared with other techniques to find efficiency. Gray wolf optimization (GWO) [60] algorithm was applied by Kamboj et al. for a better solution to the economic dispatch problem. This algorithm was mimicked by the hunting behavior of gray wolf toward the prey and comprised three steps: searching for prey, encircling the prey, and attacking the prey. The performance was tested for a small, medium, and large power system.

3.8 Hybrid Swarm Intelligence Algorithm

Hybrid methods were used to overcome the difficulties of conventional methods. The hybrid method combines two or more techniques to enhance the optimization capability. Some hybrid technique includes the swarm intelligence technique to enhance the convergence property. Panigrahi and Pandi proposed a hybrid technique with the combination of bacteria foraging algorithm to Nelder–Mead algorithm [61] for effective exploration of local optima. This hybrid technique has improved the search process which helps to improve the convergence property with optimum cost.

A hybrid chaotic PSO with the combination of the sequential quadratic programming (SQP) method was applied by Cai et al. [62] to the ELD problem. Here chaotic sequence applied to conventional PSO works as the main algorithm to optimize while the SQP method was used for fine-tuning the parameter to get optimal cost. Sayah et al. proposed a hybrid PSO and differential evolution algorithm to improve the solution for the ELD problem [63]. The differential information of DE was combined with the information of memory extracted from PSO for application to non-linear ELD problems. The combined approach was tested for four test systems with practical constraints.

Jayabarathi et al. [64] suggested a hybrid gray wolf optimization algorithm to solve complex economic dispatch problems. The conventional GWO was combined with a new mutation and crossover operator of the genetic algorithm to perform better for the non-linear problem. Chen et al. [68] suggested a biogeography learning strategy to the conventional PSO to solve the problem. This method maintains a perfect balance between local and global search to improve the quality of the solution. Five different test systems were tested by the proposed method to verify the efficacy.

4 Conclusions

A summarization and application of swarm intelligence techniques were presented for the ELD problem in the chapter. Many nature-inspired algorithms were applied directly and with a modified version to the complex and non-convex problem for optimum cost. Swarm intelligence methods solve effectually the non-continuous, non-convex, and non-smooth problems. With the no free lunch theorem, the motivation to solve the economic dispatch is still going on. Among the swarm intelligence approach, many researchers acquire the improved solution with various modifications and hybridization to the conventional PSO among the swarm intelligence approaches. Swarm intelligence approaches were also used to get the solution for the combined economic emission dispatch problem and economic dispatch to the microgrid.

The comprehensive review shows the effectiveness of swarm intelligence and nature-inspired algorithms in the economic dispatch problem in modern power systems. As no optimization techniques cannot be claimed as the best solution as per the no free lunch theorem, improvement to the application of the novel nature-inspired algorithm to the ELD problem can be realized. In addition, the nature-inspired algorithm is also used for many economic and emission remit problems with the integration of renewable energy resources.

Acknowledgements This research work was supported by "Woosong University's Academic Research Funding—2023".

References

1. Dhillon JS, Kothari DP (2010) Power system optimization. Preintce Hall of India Private Limited. https://toaz.info/doc-view
2. Gherbi YA, Lakdja F (2022) Dynamic economic emission dispatch considering valve point effect and variable active loss using new optimization method. IEEE International Conference on Electrical Sciences and Technologies in Maghreb (CISTEM) 4:3006–3020. https://doi.org/10.1109/CISTEM55808.2022.10043864
3. Liang ZX, Glover JD (1992) A zoom feature for a dynamic programming solution to economic dispatch including transmission losses. IEEE Trans Power Syst 7:544–550. https://doi.org/10.1109/59.141757
4. El KAA, Ma H, Hart JL (1994) Environmentally constrained economic dispatch using the Lagrangian relaxation method. IEEE Trans Power Syst 9(4):1723–1729. https://doi.org/10.1109/59.331423
5. Farag A, Al-Baiyat S, Cheng TC (1995) Economic load dispatch multiobjective optimization procedures using linear programming techniques. IEEE Trans Power Syst 10(2):731–738. https://doi.org/10.1109/59.387910
6. Reid GF, Hasdorff L (1973) Economic dispatch using quadratic programming. IEEE Trans Power Apparus Syst PAS-92(6):2015–2023. https://doi.org/10.1109/TPAS.1973.293582
7. Su CT, Chiou GJ (1997) A fast-computation Hopfield method to economic dispatch of power systems. IEEE Trans Power Syst 12(4):1759–1764. https://doi.org/10.1109/59.627888
8. Wood AJ, Wollenberg BF, Sheble BG (2013) Power generation, operation, and control. John Wiley & Sons. http://powerunit-ju.com/wp-content/uploads/2018

9. Dubey HM, Panigrahi BK, Pandit M (2014) Bio-inspired optimisation for economic load dispatch. A Rev Int J Bio-Inspired Comput 6(1):7–21. https://doi.org/10.1504/IJBIC.2014.059967
10. Gaing ZL (2003) Particle swarm optimization to solving the economic dispatch considering the generator constraints. IEEE Trans Power Syst 18(3):1187–1195. https://doi.org/10.1109/TPWRS.2003.814889
11. Park JB, Lee KS, Shin JR, Kwang Y (2005) Lee A particle swarm optimization for economic dispatch with nonsmooth cost functions. IEEE Trans Power Syst. 20(1):34–42 https://doi.org/10.1109/TPWRS.2004.831275
12. Jeyakumar DN, Jayabarathi T, Raghunathan, (2006) Particle swarm optimization for various types of economic dispatch problems. Int J Electr Power Energy Syst 28(1):36–42. https://doi.org/10.1016/j.ijepes.2005.09.004
13. Immanuel AS, Thanushkodi K (2007) A new particle swarm optimization solution to nonconvex economic dispatch problems. IEEE Trans Power Syst 22(1):42–51. https://doi.org/10.1109/TPWRS.2006.889132
14. Cai J, Ma X, Li L, Haipeng P (2007) Chaotic particle swarm optimization for economic dispatch considering the generator constraints. Energy Convers Manage 48(2):645–653. https://doi.org/10.1016/j.enconman.2006.05.020
15. Immanuel AS, Thanushkodi K (2008) Anti-predatory particle swarm optimization: solution to nonconvex economic dispatch problems. Electr Power Syst Res 78(1):2–10. https://doi.org/10.1016/j.epsr.2006.12.001
16. Panigrahi BK, Pandi VR, Das S (2008) Adaptive particle swarm optimization approach for static and dynamic economic load dispatch. Energy Convers Manag 49(6):1407–1415. https://doi.org/10.1016/j.enconman.2007.12.023
17. Chaturvedi KT, Pandit M, Srivastava L (2009) Particle swarm optimization with time varying acceleration coefficients for non-convex economic power dispatch. Int J Electr Power Energy Syst 31(6):249–257. https://doi.org/10.1016/j.ijepes.2009.01.010
18. Roy R, Ghoshal SP (2008) A novel crazy swarm optimized economic load dispatch for various types of cost functions. Int J Electr Power Energy Syst 30(4):242–253. https://doi.org/10.1016/j.ijepes.2007.07.007
19. Chaturvedi KT, Pandit M, Srivastava L (2008) Self-organizing hierarchical particle swarm optimization for nonconvex economic dispatch. IEEE Trans Power Syst 23(3):1079–1087. https://doi.org/10.1109/TPWRS.2008.926455
20. Sun J, Fang W, Wang D, Xu W (2009) Solving the economic dispatch problem with a modified quantum-behaved particle swarm optimization method. Energy Convers Manag 50(12):2967–2975. https://doi.org/10.1016/j.enconman.2009.07.015
21. Vlachogiannis JG, Lee KY (2009) Economic load dispatch—A comparative study on heuristic optimization techniques with an improved coordinated aggregation-based PSO. IEEE Trans Power Syst 24(2):991–1001. https://doi.org/10.1109/TPWRS.2009.2016524
22. Park JB, Jeong YW, Shin JR, Lee KY (2009) Closure to discussion of An improved particle swarm optimization for nonconvex economic dispatch problems. IEEE Trans Power Syst 25(4). https://doi.org/10.1109/TPWRS.2010.2069890
23. Niknam T, Mojarrad HD, Nayeripour M (2010) A new fuzzy adaptive particle swarm optimization for non-smooth economic dispatch. Energy 35(4):1764–1778. https://doi.org/10.1016/j.energy.2009.12.029
24. Subbaraj PR, Salivahanan S, Senthilkumar TR (2010) Parallel particle swarm optimization with modified stochastic acceleration factors for solving large scale economic dispatch problem. Int J Electr Power Energy Syst 32(9):1014–1023. https://doi.org/10.1016/j.ijepes.2010.02.003
25. Chakraborty ST, Yona A, Saber AY, Funabashi T (2011) Solving economic load dispatch problem with valve-point effects using a hybrid quantum mechanics inspired particle swarm optimisation. IET Gener Transm Distrib 5(10):1042–1052. https://doi.org/10.1049/iet-gtd.2011.0038
26. Safari A, Shayeghi H (2011) Iteration particle swarm optimization procedure for economic load dispatch with generator constraints. Expert Syst Appl 38(5):6043–6048. https://doi.org/10.1016/j.eswa.2010.11.015

27. Zhisheng Z (2010) Quantum-behaved particle swarm optimization algorithm for economic load dispatch of power system. Expert Syst Appl 37(2):1800–1803. https://doi.org/10.1016/j.eswa.2009.07.042
28. Ivatloo M, Rabiee BA, Soroudi A, Ehsan M (2012) Iteration PSO with time varying acceleration coefficients for solving non-convex economic dispatch problems. Int J Electr Power Energy Syst 42(1):508–516. https://doi.org/10.1016/j.ijepes.2012.04.060
29. Meng K, Wang HG, Dong HY, Wong KP (2009) Quantum-inspired particle swarm optimization for valve-point economic load dispatch. IEEE Trans Power Syst 25(1):215–222. https://doi.org/10.1109/TPWRS.2009.2030359
30. Kumar R, Sharma D, Sadu A (2011) A hybrid multi-agent based particle swarm optimization algorithm for economic power dispatch. Int J Electr Power Energy Syst 33(1):115–123. https://doi.org/10.1016/j.ijepes.2010.06.021
31. Basu M (2015) Modified particle swarm optimization for nonconvex economic dispatch problems. Int J Electr Power Energy Syst 69:304–312. https://doi.org/10.1016/j.ijepes.2015.01.015
32. Hosseinnezhad V, Babaei E (2013) Economic load dispatch using θ-PSO. Int J Electr Power Energy Syst 49:160–169. https://doi.org/10.1016/j.ijepes.2013.01.002
33. Basu M, Chowdhury A (2013) Cuckoo search algorithm for economic dispatch. Energy 60:99–108. https://doi.org/10.1016/j.energy.2013.07.011
34. Vo DN, Schegner P, Ongsakul, (2013) Cuckoo search algorithm for non-convex economic dispatch. IET Gener Transm Distrib 7(6):645–654. https://doi.org/10.1049/iet-gtd.2012.0142
35. Zhao J, Liu S, Zhou M, Guo X, Qi L (2018) Modified cuckoo search algorithm to solve economic power dispatch optimization problems. IEEE/CAA J Autom Sin 5(4):794–806. https://doi.org/10.1109/JAS.2018.7511138
36. Trung NG, Vo DN (2015) The application of one rank cuckoo search algorithm for solving economic load dispatch problems. Appl Soft Comput 37:763–773. https://doi.org/10.1016/j.asoc.2015.09.010
37. Hemamalini S, Simon PS (2010) Artificial bee colony algorithm for economic load dispatch problem with non-smooth cost functions. Electr Power Compon Syst 38(7):786–803. https://doi.org/10.1080/15325000903489710
38. Labbi Y, Attous DB, Mahdad B (2014) Artificial bee colony optimization for economic dispatch with valve point effect. Front Energy 8(4):449–458. https://doi.org/10.1007/s11708-014-0316-8
39. Secui DC (2015) A new modified artificial bee colony algorithm for the economic dispatch problem. Energy Convers Manag 89:43–62. https://doi.org/10.1016/j.enconman.2014.09.034
40. Aydın B, Ozyon S (2013) Solution to non-convex economic dispatch problem with valve point effects by incremental artificial bee colony with local search. Appl Soft Comput 13(5):2456–2466. https://doi.org/10.1016/j.asoc.2012.12.002
41. Afandin A, Miyauchi H (2014) Improved artificial bee colony algorithm considering harvest season for computing economic dispatch on power system. IEEJ Trans Electr Electron Eng 9(3):251–257. https://doi.org/10.1002/tee.21963
42. Chandrasekaran K, Hemamalini S, Simon SP, Padhy NP (2012) Thermal unit commitment using binary/real coded artificial bee colony algorithm. Electr Power Syst Res 84(1):109–119. https://doi.org/10.1016/j.epsr.2011.09.022
43. Pothiya S, Ngamroo I, Kongprawechnon W (2010) Ant colony optimisation for economic dispatch problem with non-smooth cost functions. Int J Electr Power Energy Syst 32(5):478–487. https://doi.org/10.1016/j.ijepes.2009.09.016
44. Song YH, Chou CSV, Min Y (1999) Large-scale economic dispatch by artificial ant colony search algorithms. Electr Mach & Power Syst 27(7):679–690. https://doi.org/10.1080/073135699268948
45. Vlachos A, Petikas I, Kyriakides S (2011) A Continuous Ant Colony (C-ANT) algorithm solving the Economic Load Dispatch (ELD) Problem. J Inf Optim Sci 32(1):1–13. https://doi.org/10.1080/02522667.2011.10700039

46. Yang XC, Hosseini SSS, Gandomi AH (2012) Firefly algorithm for solving non-convex economic dispatch problems with valve loading effect. Appl Soft Comput 12(3):1180–1186. https://doi.org/10.1016/j.asoc.2011.09.017
47. Chen G, Ding X (2015) Optimal economic dispatch with valve loading effect using self-adaptive firefly algorithm. Appl Intell 42(2):276–288. https://doi.org/10.1007/s10489-014-0593-2
48. Yang Y, Wei B, Liu H, Zhang Y, Zhao J, Manla E (2018) Chaos firefly algorithm with self-adaptation mutation mechanism for solving large-scale economic dispatch with valve-point effects and multiple fuel options. IEEE Access 6:45907–45922. https://doi.org/10.1109/ACCESS.2018.2865960
49. Latif A, Palensky P (2014) Economic dispatch using modified bat algorithm. Algorithms 7(3):328–338. https://doi.org/10.3390/a7030328
50. Adarsh BR, Raghunathan JT, Yang XC (2016) Economic dispatch using chaotic bat algorithm. Energy 96:666–675. https://doi.org/10.1016/j.energy.2015.12.096
51. Shukla A, Singh SN (2017) Pseudo-inspired CBA for ED of units with valve-point loading effects and multi-fuel options. IET Gener, Transm & Distrib 11(4):1039–1045. https://doi.org/10.1049/iet-gtd.2016.1360
52. Dalvand MM, Ivatloo BM, Najafi RA (2012) Continuous quick group search optimizer for solving non-convex economic dispatch problems. Electr Power Syst Res 93:93–105. https://doi.org/10.1016/j.epsr.2012.07.009
53. Zare K, Haque MT, Davoodi E (2012) Solving non-convex economic dispatch problem with valve point effects using modified group search optimizer method. Electr Power Syst Res 84(1):83–89. https://doi.org/10.1016/j.epsr.2011.10.004
54. James J, Yu K, Victor O, Li K (2016) A social spider algorithm for solving the non-convex economic load dispatch problem. Neurocomputing 171:955–965. https://doi.org/10.1016/j.neucom.2015.07.037
55. Elsayed WT, Hegazy YG, Bendary FM, El-bages MS (2016) Modified social spider algorithm for solving the economic dispatch problem. Eng Sci Technol, Int J 19(4):1672–1681. https://doi.org/10.1016/j.jestch.2016.09.002
56. Kien LC, Nguyen TT, Hien CT, Duong MQ (2019) A Novel Social Spider Optimization Algorithm for Large-Scale Economic Load Dispatch Problem. Energies 12(6):1–26. https://doi.org/10.3390/en12061075
57. Khodakhast E, Rasool H, Reza AZM, Jafar Z (2015) A new sufficent method based on levy-social spider technique for optimal economic dispatch of thermal power unit. J Intell & Fuzzy Syst 28(3):1137–1143. https://doi.org/10.3233/IFS-141397
58. Xiong G, Shi D (2018) Orthogonal learning competitive swarm optimizer for economic dispatch problems. Appl Soft Comput 66:134–148. https://doi.org/10.1016/j.asoc.2018.02.019
59. Kamboj VK, Bhadoria A, Bath SK (2017) Solution of non-convex economic load dispatch problem for small-scale power systems using ant lion optimizer. Neural Comput Appl 28(8): 2181–2192. https://link.springer.com/article/https://doi.org/10.1007/s00521-015-2148-9
60. Kamboj VK, Bath SK, Dhillon JS (2015) Solution of non-convex economic load dispatch problem using Grey Wolf Optimizer. Neural Comput Appl 27(5):1301–1316. https://doi.org/10.1007/s00521-015-1934-8
61. Panigrahi BK, Pandi VR (2008) Bacterial foraging optimisation: Nelder-Mead hybrid algorithm for economic load dispatch. IET Gener Transm Distrib 2(4):556–565. https://doi.org/10.1049/iet-gtd:20070422
62. Cai J, Li Q, Li L, Peng H, Yang Y (2012) A hybrid CPSO–SQP method for economic dispatch considering the valve-point effects. Energy Convers Manag 53(1):175–181. https://doi.org/10.1016/j.enconman.2011.08.023
63. Sayah S, Hamouda A (2013) A hybrid differential evolution algorithm based on particle swarm optimization for nonconvex economic dispatch problems. Appl Soft Comput 13(4):1608–1619. https://doi.org/10.1016/j.asoc.2012.12.014
64. Jayabarathi T, Raghunathan T, Adarsh BR, Suganthan PN (2016) Economic dispatch using hybrid grey wolf optimizer. Energy 111:630–641. https://doi.org/10.1016/j.energy.2016.05.105

65. Salkuti SR (2022) Emerging and advanced green energy technologies for sustainable and resilient future grid. Energies 15(18):6667. https://doi.org/10.3390/en15186667
66. *Salkuti SR (2020)* Multi-objective based economic environmental dispatch with stochastic solar-wind-thermal power system, Int J Electr Comput Eng 10(5): 4543–4551. http://doi.org/https://doi.org/10.11591/ijece.v10i5.pp4543-4551
67. Reddy SS (2017) Multi-objective based adaptive immune algorithm for solving the economic and environmental dispatch problem. Int J Appl Eng Res 12(6):1043–1048
68. Chen X, Xu B, Du W (2018) An improved particle swarm optimization with biogeography-based learning strategy for economic dispatch problems. Complexity. https://doi.org/10.1155/2018/7289674

Design of Observer-Based Robust Double Integral Sliding Mode Controller for Grid-Connected PV System

Raseswari Pradhan

Abstract Photovoltaic (PV) microgrid dynamics is always uncertain and nonlinear in nature due to parameter variation and environmental effects. There is a need for an appropriate mathematical model to represent it. This work focuses on this issue. Further, an optimal robust controller is required for inverter control of this PV microgrid. A Sliding-Mode-Control (SMC)-based controller can handle this problem. However, an SMC-based system needs full knowledge of system states. In absence of that information, these controllers can't be designed. An observer can be used to estimate the state variables. This chapter presents an extended observer-based double integral sliding mode controller (ESO-DISMC). The controller parameters are optimized using the water evaporation optimization (WEO) algorithm. Results from this work are compared with that of Genetic Algorithm (GA)- and Particle Swarm Optimization (PSO)-optimized controllers to verify the proposed controller's superiority over that of other controllers.

Keywords Photovoltaic microgrid · Inverter control · Uncertainty · ESO-DISMC · Robust control · Total harmonic distortion · Optimization

Nomenclature

PV	Photovoltaic
MPPT	Maximum power point tracking
ESO	Extended state observer
VSI	Voltage source inverter
WEO	Water evaporation optimization
THD	Total harmonic distortion
FFT	Fast Fourier transform

R. Pradhan (✉)
Department of Electrical Engineering, Veer Surendra Sai University of Technology, Burla, Odisha, India
e-mail: rpradhan_ee@vssut.ac.in

S. R. Salkuti et al. (eds.), *Power Quality in Microgrids: Issues, Challenges and Mitigation Techniques*, Lecture Notes in Electrical Engineering 1039,
https://doi.org/10.1007/978-981-99-2066-2_20

GA	Genetic algorithm
PSO	Particle swarm optimization
PWM	Pulse width modulation
SMC	Sliding mode controller
PI	Proportional–integral
FOPI	Fractional-order proportional–integral
PLL	Phase lock loop
$i_{d,ref}$	Reference d-axis inverter current
$i_{q,ref}$	Reference q-axis inverter current
i_d	D-axis inverter current
i_q	Q-axis inverter current
E_{gi}	Grid-fed voltage
L_{gf}	Filter inductance
r_{gf}	Filter resistance
V_i	Inverter output voltage
E_d, E_q	Dq-grid voltages
a_d, a_q	Dq-switching signals
d_1, d_2	External disturbance signals
MVP	Monolayer-evaporation phase
DVP	Droplet-evaporation phase
RVP	Regeneration of droplets phase
ITAE	Integral time absolute error
I_D	Diode current
I_{sh}	Shunt current
I	Output current
V_{sh}	Voltage across the shunt resistor and diode
I_0	Reverse bias saturation current
V_D	Voltage across the diode
V_T	Terminal voltage
n	Ideality or quality factor

1 Introduction

A grid-tied PV system has basically an inner and another outer loop of controlling currents. The inner one is meant to control inverter output to help in maintaining unity power factor as well as accurate waveforms. Similarly, the outer one helps in accomplishing the maximum power extraction from PV arrays. Various methods of control and switching pattern have been proposed in order to create the reference current [1]. Two control schemes widely used for the aforesaid purpose are the direct power control-based strategy and the vector control-based strategy. In the first control system, the load voltage and current are converted to the α-β reference frame to estimate the instantaneous values of the active and reactive power of the consumer load [2].

Using a suitable controller, the reference active power and reactive power are generated from the instantaneous active power and reactive power. These reference powers are utilized to give the reference grid current to the PWM controller to generate the switching signals for the inverter. In the vector control strategy [3, 4], the three-phase inverter currents are first transformed into a *d-q* reference frame using Park's transformation and then the currents i_d and i_q are controlled using various controllers. These controlled currents are again transformed into three-phase reference current which is fed to the PWM controller to control the gating pulse to the inverter. For the MPPT control, the real power control is adjusted for DC-link voltage regulation. The output of the DC-link voltage control $i_{d,ref}$ is used as a reference for a real power controller [5].

Voltage and current sensors provide the information for the phase-locked loop (PLL) and two *dqo* transformer blocks [6, 7]. Based on the direct and quadrature values of voltage and current, active and reactive power can be calculated as follows:

$$P = \frac{3}{2}(v_d i_d + v_q i_q) \tag{1}$$

$$Q = \frac{3}{2}(v_q i_d - v_d i_q) \tag{2}$$

2 Literature Review on Controllers for Grid-Tied PV Microgrid

The inverter control can further be classified as PV-side and grid-side control. MPPT operation is made using PV-side control. Similarly, power factor and harmonic current controlling are done at grid side. Voltage control and current control are employed in a PV microgrid. The voltage amplitude and frequency of the inverter need to be synchronized with each other for integrating it with the grid. The device PLL is used to measure the frequency and power angle. The inverter assembly circuit with the control strategy is shown in Fig. 1.

PI controller is considered as one of the popular controllers owing to its simplicity, accuracy, and ease to use. There are two references each for the dq-currents used. These signals are applied to produce the required switching signals for the inverter. It used PI control schemes [8]. However, this PI control algorithm with fixed gain fails to handle the dynamic operating conditions. This problem can be rectified using a fractional-order control scheme [9, 10].

A hybrid controller in which fuzzy logic plus a PID controller is presented in [11]. Here, the role of fuzzy logic is to tune PID controller gains. The controller is verified in different set power and loading conditions. Different variants of fuzzy

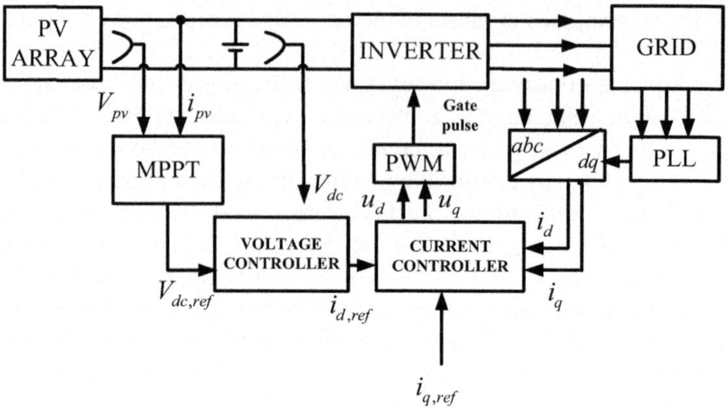

Fig. 1 Control scheme of inverter in single-stage grid-tied PV microgrid

logic controllers with DSP and DSPACE implementation on standalone and GCPVS have been proposed in the literature [11]. An intelligent fuzzy-neurocontroller is proposed for the inverter control of a PV system in presence of grid faults [12]. A hysteresis control for a PV system is presented that satisfies IEEE-519 regulations [13]. Compared with the case without the control scheme, it envisages that the system power quality improved significantly in the former case. An Artificial Neural Network (ANN)-based controller is presented in a grid-tied PV system [14].

Robust controllers for linear systems with uncertainties are very often designed by employing the linear matrix inequality (LMI) approach. A robust H_∞ controller for a single-phase grid-connected PV system using the LMI approach is proposed in [15]. An internal model control (IMC)-based state feedback control scheme is presented here. It is seen that it is unable to diminish higher ordered harmonics. Also, this system dynamics is very complex with many numbers of resonating factors.

Such an H_∞ controller is presented in [16]. It is designed for 1-ϕ PV dynamics with parameter uncertainty. A mathematical model of the said system was also formulated to test the controller performance [17] A structured singular value (μ) approach to design a robust controller for a boost converter with voltage-mode control is presented in [18]. Robust control of grid-tied parallel inverters using a nonlinear backstepping approach is presented in [19] in which the reference for active current component changes each time an inverter unit is connected to the system. A partial feedback linearization approach is employed to design a controller considering structured uncertainties within the PV system model. The upper bounds of the parameters and states of the PV system need to be known necessitating information on system operating points or the nature of the faults. But here capacitor banks are used for each node, which increases the system states and further makes the controller design complex.

A robust nonlinear sliding mode controller is more suitable for a system with nonlinearity such as a grid-connected PV system with nonlinear loads and inverters contributing to nonlinear behavior [20, 21]. SMC is found to be effective due to its robustness, system order reduction, and appropriateness to the ON–OFF behavior of power switches [20]. One of the most important features of the sliding mode regime, in variable structure systems (VSS), is the ability to achieve responses that are independent of the system parameters. A nonlinear sliding mode controller is proposed for a single-stage grid-connected PV system with a novel nonlinear sliding surface [22, 23]. Sliding mode controller is one of the popular robust control paradigms which helps in achieving the desired control objectives despite the presence of nonlinearities, parameter variations, and external disturbances. A Lyapunov-based finite time nonlinear controller is presented in [24] for reactive power and DC-link voltage control, where the Lyapunov approach is exploited to ensure stability and robustness. A reinforcement learning (RL) and sliding mode control (SMC)-based control scheme is proposed for a 3-ϕ PV system integrated into a grid [25]. An observer-based fuzzy-SMC is presented for inverter control of a 1-ϕ PV system [26]. Further, a Lyapunov-based fast terminal SMC is proposed for an uncertain hybrid system that is with PV and wind [27].

From all the above discussion, it is concluded that SMC-based controllers are very efficient and robust controllers. However, to design these, system state information is needed. While designing SMC-based controllers, it is assumed that all the states of the PVGCS are available. However, sometimes, the states of the system are unavailable. Therefore, it is necessary to design observers to estimate the unknown states [28]. An extended state observer is an efficient nonlinear observer that is suitable for the estimation of disturbances and uncertainties considering these as new state variables [29]. This observer has a success story of its performance in estimating uncertainty [30].

The objective of this chapter is to design a nonlinear extended state observer-based DISMC for an inverter current control of a grid-tied PV microgrid in presence of uncertainty. Further, there is no proper model available for the studied system in presence of uncertainty as these uncertain elements are not clearly defined. This work addresses this issue also.

3 Design of Grid-Tied PV Microgrid

The inverter control can further be classified as PV-side and grid-side control. MPPT operation is made using PV-side control. Similarly, power factor and harmonic current controlling are done at grid side. Voltage control and current control are employed in a PV microgrid. Voltage amplitude and frequency of the inverter control can further be classified into PV-side and grid-side control MPPT operation.

The equivalent circuit of the testing PV system is shown in Fig. 2. It is a single-stage grid-tied system. Only the 3-ϕ Voltage Source Inverter (VSI) is capable of maximum power extraction from the PV panel and power balancing phenomenon between PV and grid. The power factor should be unity also. Here, the output voltage and current of PV panels are V and I, respectively. All the elements used in Fig. 2 and Eqs. (1), (2) are having usual denotation [31].

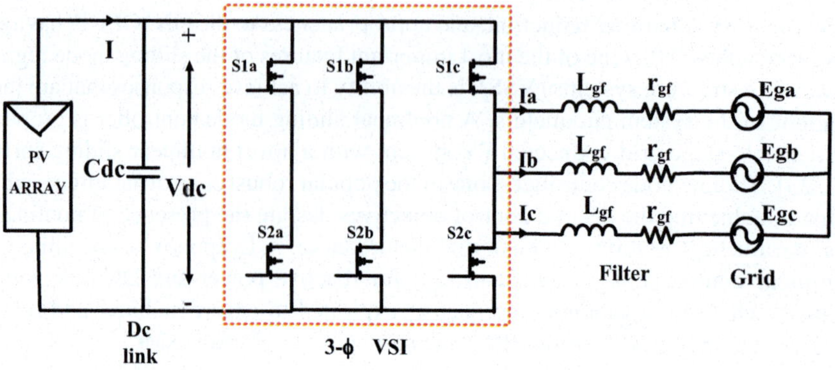

Fig. 2 Layout of 3-ϕ gridtied studied PV system

$$I = N_p I_{ph}\left[\exp\left\{\frac{q}{AKT}\left(\frac{V}{N_S}+\frac{IR_S}{N_P}\right)\right\}-1\right]-\frac{N_P}{R_P}\left(\frac{V}{N_S}+\frac{IR_S}{N_P}\right) \tag{3}$$

The voltage outputs of VSI are defined as follows:

$$V_a = L_{gf}\frac{di_a}{dt} + i_a r_{gf} + E_{ga} \tag{4}$$

$$V_b = L_{gf}\frac{di_b}{dt} + i_b r_{gf} + E_{gb} \tag{5}$$

$$V_c = L_{gf}\frac{di_c}{dt} + i_c r_{gf} + E_{gc} \tag{6}$$

Let the switching signals of the VSI are $S_{ia} = a_S$, $S_{ib} = b_S$, $S_{ic} = c_S$ where $i = 1$ or 2. Voltage equations can be rewritten in terms of gate signals as follows:

$$V_a = \frac{V_{dc}}{3}(2a_s - b_s - c_s) \tag{7}$$

$$V_b = \frac{V_{dc}}{3}(-a_s + 2b_s - c_s) \tag{8}$$

$$V_c = \frac{V_{dc}}{3}(-a_s - b_s + 2c_s) \tag{9}$$

Using Eqs. (4)–(6) in Eqs. (7)–(9), the equations as follows have resulted:

$$\frac{di_a}{dt} = \frac{-r_{gf}}{L_{gf}}i_a - \frac{1}{L_{gf}}E_{ga} + \frac{V_{dc}}{3L_{gf}}(2a_s - b_s - c_s) \tag{10}$$

$$\frac{di_b}{dt} = \frac{-r_{gf}}{L_{gf}} i_b - \frac{1}{L_{gf}} E_{gb} + \frac{V_{dc}}{3L_{gf}} (-a_s + 2b_s - c_s) \tag{11}$$

$$\frac{di_c}{dt} = \frac{-r_{gf}}{L_{gf}} i_c - \frac{1}{L_{gf}} E_{gc} + \frac{V_{dc}}{3L_{gf}} (-a_s - b_s + 2c_s) \tag{12}$$

Again, DC-link voltage (V_{dc}) can be calculated as follows:

$$\frac{dV_{dc}}{dt} = \frac{1}{C_{dc}} \left[i_{pv} - i_{inv} \right] \tag{13}$$

where the inverter input current is considered as follows:

$$i_{inv} = i_a a_s + i_b b_s + i_c c_s \tag{14}$$

It is observed that the equivalent mathematical model of the studied system is highly complex and nonlinear [32]. Finding a solution to this model is very tough. Simplification of these dynamics is necessary so as to find the solution easily. Such a simplified structure is defined as follows:

$$\frac{d}{dt}\begin{bmatrix} i_d \\ i_q \\ V_{dc} \end{bmatrix} = \begin{bmatrix} -\frac{r_{gf}}{L_{gf}} & w & \frac{a_d}{L_{gf}} \\ -w & -\frac{r_{gf}}{L_{gf}} & \frac{a_q}{L_{gf}} \\ -\frac{a_d}{C_{dc}} & -\frac{a_q}{C_{dc}} & \frac{1}{r_{pv} C_{dc}} \end{bmatrix} \begin{bmatrix} i_d \\ i_q \\ V_{dc} \end{bmatrix} + \begin{bmatrix} -\frac{1}{L_{gf}} & 0 \\ 0 & -\frac{1}{L_{gf}} \\ 0 & 0 \end{bmatrix} \begin{bmatrix} E_d \\ E_q \end{bmatrix} \tag{15}$$

where E_d, i_d, and a_d are the d-axis and E_{gq}, i_q, and a_q are the q-axis parameters of the grid-side voltage, current and gate signals for the inverter respectively. The term r_{pv} represents the dynamic resistance of the PV panel. The dynamics of the microgrid derived in above Eq. (15) do not involve uncertainties. However, the actual values of r_{gf}, L_{gf}, C_{dc} are unknown but they vary within known upper and lower bounds. Consider $\overline{r}_{gf}$, $\overline{L}_{gf}$, $\overline{C}_{dc}$ being the nominal values of grid filter resistance, grid filter inductance, and DC-link capacitance, respectively [33]. Then the actual parameter values with dynamic perturbation become

$$r_{gf} = \overline{r}_{gf} + \delta_{rgf} \tag{16}$$

$$L_{gf} = \overline{L}_{gf} + \delta_{Lgf} \tag{17}$$

$$C_{dc} = \overline{C}_{dc} + \delta_{C_{dc}} \tag{18}$$

where δ_{rgf}, δ_{cgf}, δ_{cdc} are the uncertainties of the nominal parameters in the range ± 1. The grid-connected PV system with the above-described uncertain parameters is defined as follows:

$$\frac{d}{dt}\begin{bmatrix} i_d \\ i_q \\ V_{dc} \end{bmatrix} = A\begin{bmatrix} i_d \\ i_q \\ V_{dc} \end{bmatrix} + B\begin{bmatrix} E_d \\ E_q \\ i_{pv} \end{bmatrix} \tag{19}$$

where

$$A = \begin{bmatrix} \frac{-(\overline{r}_{gf}+\delta_{rgf})}{(\overline{L}_{gf}+\delta_{Lgf})} & \omega & \frac{a_d}{(\overline{L}_{gf}+\delta_{Lgf})} \\ -\omega & \frac{-(\overline{r}_{gf}+\delta_{rgf})}{(\overline{L}_{gf}+\delta_{Lgf})} & \frac{a_q}{(\overline{L}_{gf}+\delta_{Lgf})} \\ \left(\frac{-a_d}{\overline{C}_{dc}+\delta_{C_{dc}}}\right) & \left(\frac{-a_q}{\overline{C}_{dc}+\delta_{C_{dc}}}\right) & 0 \end{bmatrix} \tag{20}$$

$$B = \begin{bmatrix} \frac{-1}{(\overline{L}_{gf}+\delta_{Lgf})} & 0 & 0 \\ 0 & \frac{-1}{(\overline{L}_{gf}+\delta_{Lgf})} & 0 \\ 0 & 0 & \frac{1}{(\overline{C}_{dc}+\delta_{C_{dc}})} \end{bmatrix} \tag{21}$$

Simplifying Eq. (19) gives

$$\dot{i}_d = \frac{-(\overline{r}_{gf}+\delta_{rgf})}{(\overline{L}_{gf}+\delta_{Lgf})} i_d + \omega i_q + \frac{a_d}{(\overline{L}_{gf}+\delta_{Lgf})} V_{dc} - \frac{1}{(\overline{L}_{gf}+\delta_{Lgf})} E_d \tag{22}$$

$$\dot{i}_q = -\omega i_d - \frac{(\overline{r}_{gf}+\delta_{rgf})}{(\overline{L}_{gf}+\delta_{Lgf})} i_q + \frac{a_q}{(\overline{L}_{gf}+\delta_{Lgf})} V_{dc} - \frac{1}{(\overline{L}_{gf}+\delta_{Lgf})} E_q \tag{23}$$

$$\dot{V}_{dc} = \frac{-a_d}{(\overline{C}_{dc}+\delta_{Cdc})} i_d - \frac{a_q}{(\overline{C}_{dc}+\delta_{Cdc})} i_q \tag{24}$$

where $i_d, i_q,$ and V_{dc} are chosen as the state variables; E_d and E_q as the input variables; and i_d, i_q, V_{pv} as the output variables, i.e.,

$$\begin{bmatrix} x_1 \\ x_2 \\ x_3 \end{bmatrix} = \begin{bmatrix} i_d \\ i_q \\ V_{pv} \end{bmatrix} \tag{25}$$

$$\begin{bmatrix} y_1 \\ y_2 \\ y_3 \end{bmatrix} = \begin{bmatrix} i_d \\ i_q \\ V_{pv} \end{bmatrix} \tag{26}$$

$$\begin{bmatrix} u_1 \\ u_2 \end{bmatrix} = \begin{bmatrix} E_d \\ E_q \end{bmatrix} \tag{27}$$

Equations (22)–(24) can be represented in standard state-space form as follows:

$$\dot{x}_1 = \frac{1}{\overline{L}_{gf}+\delta_{\lg f}}[(-\overline{r}_{gf}+\delta_{rgf})x_1 + a_d x_3 - u_1)] + \omega x_2 \tag{28}$$

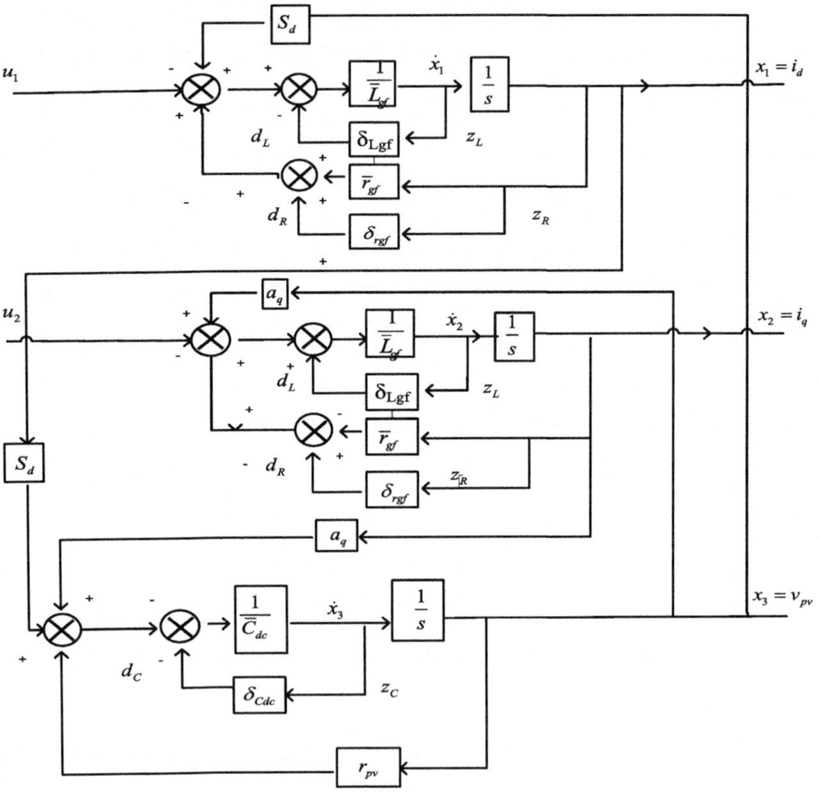

Fig. 3 Structured uncertainties block diagram of grid-connected PV system

$$\dot{x}_2 = \frac{1}{\overline{L}_{gf} + \delta_{\lg f}}[(-\overline{r}_{gf} + \delta_{rgf})x_2 + a_q x_3 - u_2)] - \omega x_1 \tag{29}$$

$$\dot{x}_3 = \frac{1}{\overline{C}_{dc} + \delta_{cdc}}(-a_d x_1 - a_q x_2 + r_{pv} x_3) \tag{30}$$

Figure 3 shows the block diagram of the grid-connected PV system including the uncertainties.

4 Design of Grid-Tied PV Microgrid

The PV system model is remodeled considering two disturbing signals that are unknown like d_1 and d_2. It is assumed that none of the system states are known. The controller can only be designed after all states are either known or measurable. To find these states, a modified version of the observer called ESO is formulated.

The ESO is calculated by taking a hyperbolic tangent function. Let the system has two states x_1 and x_2 that are fully unknown. To solve this issue, two other state variables are introduced like Z_1 and Z_2. The modified state equations are represented as follows:

$$\dot{Z}_1 = -a_1 \tanh(b_1(Z_1 - x_1) \tag{31}$$

$$\dot{Z}_2 = -a_2 \tanh(b_2(Z_1 - x_2) \tag{32}$$

where x_1 and x_2 denote the states of the system which are not fully known. Figure 4 presents the overall control structure using ESO-DISMC for a grid-connected PV system. The estimated errors can be determined as follows:

$$e_1 = d_1 - Z_1 \tag{33}$$

$$e_2 = d_2 - Z_2 \tag{34}$$

$$d_1 = e_1 + Z_1 \tag{35}$$

$$d_2 = e_2 + Z_2 \tag{36}$$

These estimated errors are also assumed to be bounded. The design steps of the ESO-DISMC are as follows. The reference current of i_d is estimated from the MPPT output and takes the reference current of i_q as null. Then, the dq-sliding surfaces S_d and S_q are designed as follows:

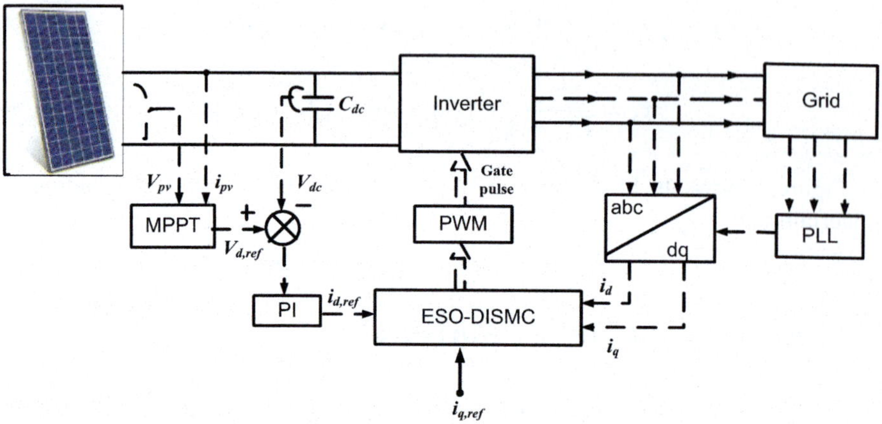

Fig. 4 Layout of PV system with ESO-DISMC-based current controller

$$\begin{array}{l} S_d = a_d X_{id} + d_1 \\ S_q = a_q X_{iq} + d_2 \end{array} ; \ (i = 1, 2, 3) \tag{37}$$

where

$$X_{1d} = i_d - i_{dref} \tag{38}$$

$$X_{2d} = \int X_{1d} dt \tag{39}$$

$$X_{3d} = \int X_{2d} dt \tag{40}$$

$$X_{1q} = i_q - i_{qref} \tag{41}$$

$$X_{2q} = \int X_{1q} dt \tag{42}$$

$$X_{3q} = \int X_{2q} dt \tag{43}$$

Here, a_d, and a_q are the coefficients of S_d and S_q, respectively. Figure 5 shows the structure of the proposed controller. Next, the control signal u is computed using the sliding mode existence rule. This signal u is calculated by adding its equivalent term u_{eq} and nonlinear term U_n. The equivalent values of control inputs $u_{d,equ}$ and $u_{q,equ}$ can be calculated by solving Eqs. (44) and (45).

$$\begin{array}{l} \dot{S}_d = 0 \text{ at } u_{d,ref} \\ \dot{S}_q = 0 \text{ at } u_{q,ref} \end{array} \tag{44}$$

On solving the above equations, one obtains

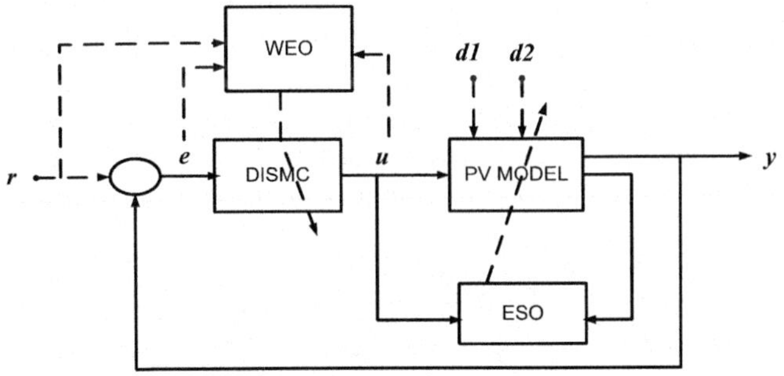

Fig. 5 Structure of the ESO-DISMC

$$u_{d,equ} = \left[L_{gf}\frac{d}{dt}\left(i_{d,ref}\right) + R_{gf}i_d - L_{gf}\omega i_q\right] + E_d + \alpha_d L_{gf}X_{1d} + \beta_d L_{gf}X_{2d}dt \tag{45}$$

$$u_{q,equ} = \left[L_{gf}\frac{d}{dt}\left(i_{q,ref}\right) + R_{gf}i_q + L_{gf}\omega i_d\right] + E_q + \alpha_q L_{gf}X_{1q} + \beta_q L_{gf}X_{2q}dt \tag{46}$$

The discontinuous control terms are calculated as follows:

$$u_{dn} = K_d \text{ sgn}(S_d) \tag{47}$$

$$u_{qn} = K_q \text{ sgn}\left(S_q\right) \tag{48}$$

Eventually, the required control signal is calculated as follows:

$$u_d = u_{deq} + u_{dn} \tag{49}$$

$$u_q = u_{qeq} + u_{qn} \tag{50}$$

5 Water Droplet Optimization (WEO) Algorithm

It is also a population-based optimization algorithm that is capable of solving the complex problem to give optimal solutions. The scheme is basically formulated inspired by natural phenomena like the evaporation of tiny water droplets on a solid surface. The surface wettability is also known as the charge holding capability (q) which varies between 0 and 0.7 times than that of an electron. Based on the charge holding capacity, there are three phases in this optimization scheme such as monolayer-evaporation phase (MVP), droplet-evaporation phase (DVP), and regeneration of droplets phase (RVP). The value of the charge of the substrate at MVP is more than 0.4 times than that of an electron whereas it is less than 0.4 times than that of an electron for DVP. DISMC controller uses the error between the desired currents and the reference currents and is based on the Integral Time Absolute Error (ITAE) index. The fitness function used in this algorithm is as follows:

$$J_k^t = \int_0^t \delta\left|V_{dc}\left(K_p, K_i\right) + S_d(K_d, \alpha_d, \beta_d) + S_q\left(K_q, \alpha_q, \beta_q\right)\right|d\delta \tag{51}$$

Where K_p, K_i are parameters of the PI controller and (K_d, α_d, β_d) and (K_q, α_q, β_q) are parameters of proposed optimized ESO-DISMC for dq components of currents respectively.

5.1 MVP

The amount of q is amplified depending on how much the surface is wet. At first, all the water droplets are uniformly distributed on the observed surface. Now, the parameters are updated such that the monolayer-evaporation probability (P_{ij}^t) is optimized. Here, defined j counts the variable number and i counts the iteration number of the procedure. The scaling function (E_{sub}) is further calculated such that it remains between $E_{\max}$ and $E_{\min}$. The scaling function is computed as follows:

$$E_{sub}(k)^t = \frac{(E_{\max} - E_{\min}) \times \left(J_k^t - \text{Min}(J)\right)}{(\text{Max}(J) - \text{Min}(J))} + E_{\min} \tag{52}$$

$$if\ rand\left(X_{ij}\right)\begin{cases} < E\left(E_{sub}(k)^t\right);\ \sum (P_{ME})_{ij}^t = 1 \\ \geq E\left(E_{sub}(k)^t\right);\ \sum (P_{ME})_{ij}^t = 0 \end{cases} \tag{53}$$

Here, $(rand\ (X_{ij}))$ is any random set of parameters that need to be tuned. When the value is equal to 1 then it is a fit case else it is an unfit case.

5.2 DVP

In this case, the vapor forming becomes slower as there is condensing of the contact angle (θ) between the droplet and surface. This angle is a very important component of this algorithm as it controls the optimization convergence rate. Usually, the best convergence is found when this angle lies $-45°$ and $-25°$. The fitness function formulated in terms of contact angle is shown as follows:

$$\theta_i^t = \frac{(\theta_{\max} - \theta_{\min}) \times \left(J_i^t - \text{Min}(J)\right)}{(\text{Max}(J) - \text{Min}(J))} + \theta_{\min} \tag{54}$$

After calculating θ, the probability of droplet-evaporation matrix (P_{DEM}) is formulated as follows:

$$if\ \ rand(X)_{ij}\begin{cases} < J\left(\theta_k^t\right);\ \sum (P_{DEM})_{ij}^t = 1 \\ \geq J\left(\theta_i^t\right);\ \sum (P_{DEM})_{ij}^t = 0 \end{cases} \tag{55}$$

5.3 RVP

A fresh droplet is expected to replace an evaporated one continuously. This phase evaluates that generation probability. This probability function is incorporated into the formulated optimization cost function. The parameter's value (X^{t+1}) is updated following the same mechanism as in the case of the fresh droplet generation.

$$X(k+1) = X(k) + \mu \times \begin{cases} (P_{ME})_{ij}^{k};\ k \leq \frac{k_{\max}}{2} \\ (P_{DE})_{ij}^{k};\ k > \frac{k_{\max}}{2} \end{cases} \tag{56}$$

where μ is the perturbed size of droplet selection to calculate $(k+1)$th set from that of the kth set. The term $k_{\max}$ depicts the overall time spent in each iteration.

6 Results and Discussion

The studied PV system is analyzed and controlled from the obtained simulation results [31]. The purpose of the application of the WEO algorithm is to calculate all the optimized values of dq-axis controller gains. For correct and quick convergence, it is necessary to select the upper and lower limits of each gain properly. Before selecting the upper and lower limits, an analysis is done on system outcomes for different sets of controller gains as shown in Table 1. Following the data of this table, it becomes easier to select the upper and lower bounds of controller gains. Table 2 reports the inequality constraints of all the reported dq-axis controller gains used in this system. In this work, five samples of water droplets were for 50 iterations to accomplish the tuning performance of the WEO-DISMC current controller.

The above five samples of water droplets are having different substrate energy levels $E_{\max}$. The convergence behaviors of the droplets are shown in Fig. 6. This figure delivers that the sample with substrate energy-3 converges faster with a lower objective function after convergence. The convergence behaviors for the proposed

Table 1 Performance analysis in case of selecting different sets of gains for the proposed WEO-optimized ESO-DISMC controller

Controller components						THD (%)	Settling time of i_d (ms)	Settling time of i_q (ms)
α_d	β_d	α_q	β_q	K_d	K_q			
1	1	1	1	1	1	4.25	70	>100
1	100	1	100	0.1	0.1	3.87	65	80
20	100	10	100	0.5	0.5	2.74	72	140
5	100	90	15	0.9	0.4	2.17	0.07	1.19
10	80	80	20	0.7	0.99	2.19	1.01	1.28
12	50	100	20	0.88	0.9	2.14	3.2	1.09

Table 2 Set optimization constraints in WEO algorithm for tuning ESO-DISMC

Parameter	Inequality constraint limits	
	Lower	Upper
α_d	0.1	100
β_d	0.1	100
α_q	0.1	100
β_q	0.1	100
K_d	0	1
K_q	0	1
K_p	0.1	20
K_i	0.1	1000
Substrate interaction energy, E	−0.5	−3.5
Contact angle, θ (°C)	−45	−25

current controllers tuned with tuning algorithms such as PSO, GA, and the proposed WEO are compared in Fig. 7. From this figure, it can be observed that the value of the cost function in the case of the proposed WEO-tuned ESO-DISMC is lower than that of PSO-tuned ESO-DISMC and GA-tuned ESO-DISMC.

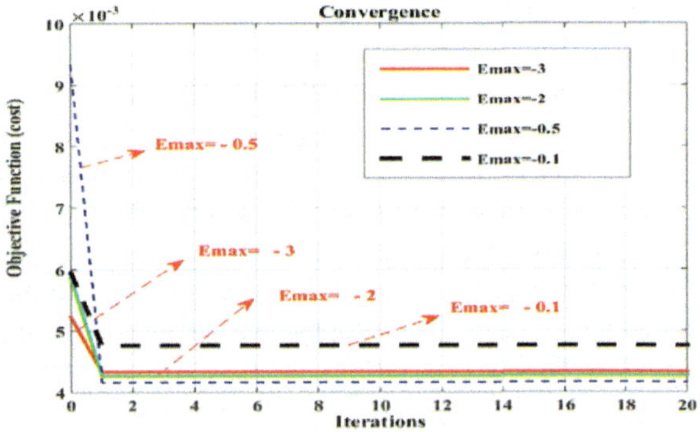

Fig. 6 Convergence characteristics for different maximum values of substrate energies ($E_{\max}$)

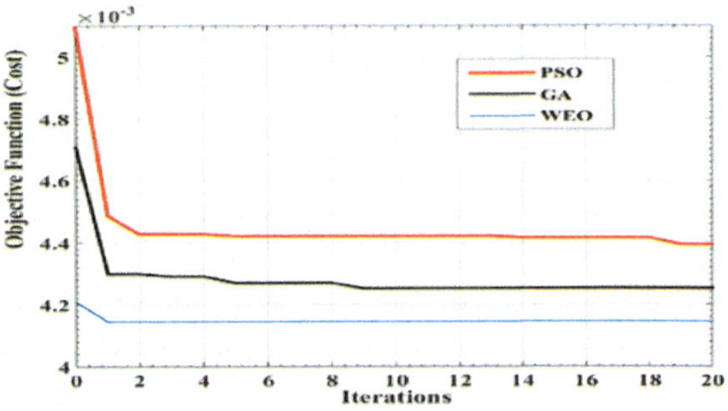

Fig. 7 Comparison of convergence curve (GA, PSO, and WEO applied to design DISMC)

Further, FFT analysis is done in the case of all the three optimization algorithms tuned proposed ESO-DISMC on the studied system as shown in Fig. 8a–c. From this analysis, it is noted that THD in the case of WEO-based ESO-DISMC is lower than that of GA and PSO algorithms. The tuned controller parameters using GA, PSO, and WEO along with their corresponding fitness function values are given in Table 3.

The performance of the proposed extended state observer-based DISMC is evaluated using a model designed in MATLAB/Simulink. To evaluate the tracking performance of the proposed extended state observer-based DISMC, the reference set points for i_d and i_q were changed from 0 to 1 at 1 s and again from 1 to 0.8 at 2 s. Figure 9a, b shows the plots for i_d and i_q for different values of r_{gf}, respectively. Similarly, Fig. 10a, b shows the plots for i_d and i_q for different values of L_{gf}, respectively.

Also, the THD of the grid current was evaluated as shown in Fig. 11. The THD was found to be 2.15%. A comparative analysis of the proposed ESO-DISMC was made with that of the DISMC controller comparing robustness, THD of grid current and settling time, and peak overshoot amplitude. The overall comparative study on results found from PV systems with normal DISMC and ESO-DISMC is shown in Table 4.

7 Conclusions

This chapter presents a simplified model of a PV system with uncertainty. It also proposed the design, implementation, and evaluation of a WEO-optimized ESO-DISMC for a grid-tied PV microgrid. The proposed controller is capable of enhancing the power quality of the grid-fed power as per the grid code. UPF and MPPT operations are also ensured. Results obtained also confirm that both the inverter currents are tracked accurately. Further, the controller is found to be robust against random reference changes. The proposed controller is compared with ESO-DISMC tuned with GA and PSO algorithms. Also, results are compared with that of DISMC tuned with the WEO algorithm. Comparison factors are system modeling, robustness, THD of grid current and settling time, and peak overshoot amplitude. It was found that among all the aforesaid controllers, the proposed WEO-tuned ESO-DISMC provides superior performance even when all the states of the system are not known.

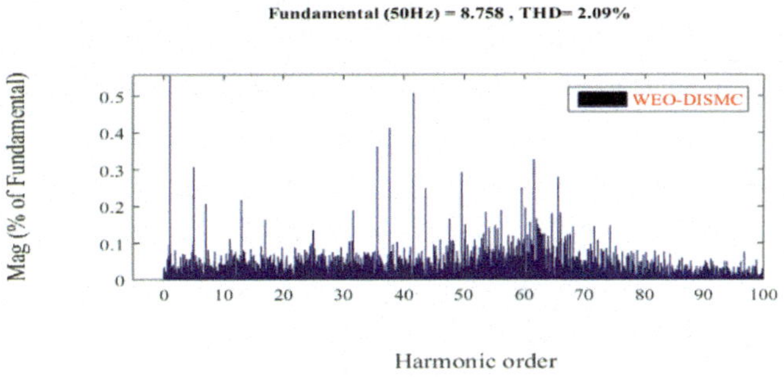

(a)

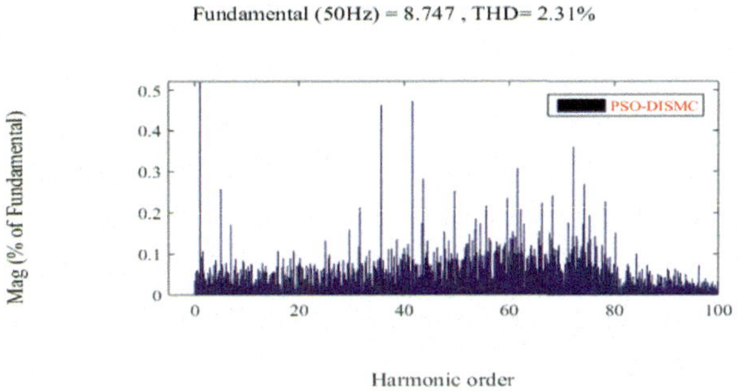

(b)

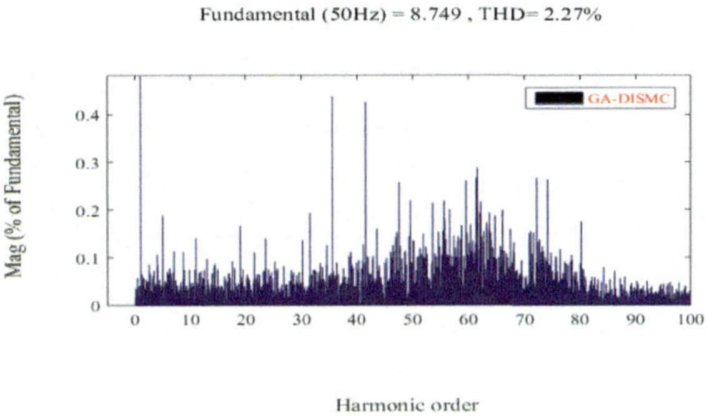

(c)

Fig. 8 THD of phase-a grid current, I_a for **a** WEO-DISMC controller, **b** PSO-DISMC and **c** GA-DISMC

Table 3 Comparison of ESO-DISMC controller tuned with different optimization algorithms

Parameters	WEO	PSO	GA
α_d	29.83	35.21	12.15
β_d	86.42	73.24	57.81
K_d	0.892	0.677	0.921
α_q	16.06	79.87	26.08
β_q	18.89	6.17	37.09
K_q	0.98	0.42	0.49
K_p	9.12	6.02	13.88
K_i	455.72	365.26	201.17
Fitness Function	2.1e-3	3.2e-3	4.4e-3
Total generations	50	50	50

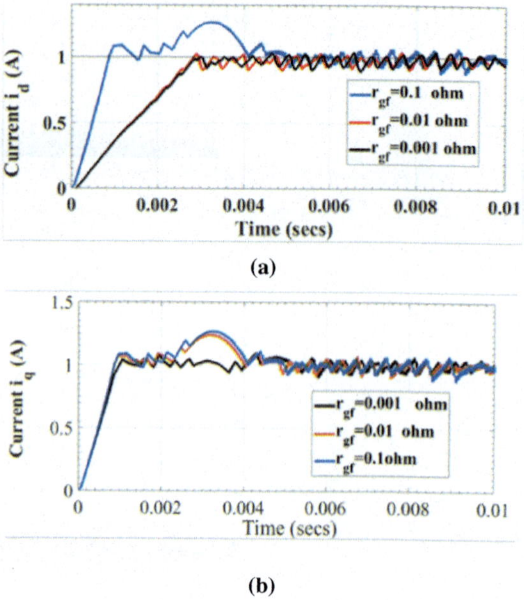

Fig. 9 Step response of **a** current i_d and **b** current i_q with multivariable H_∞ controller for perturbation in line resistance r_{gf}

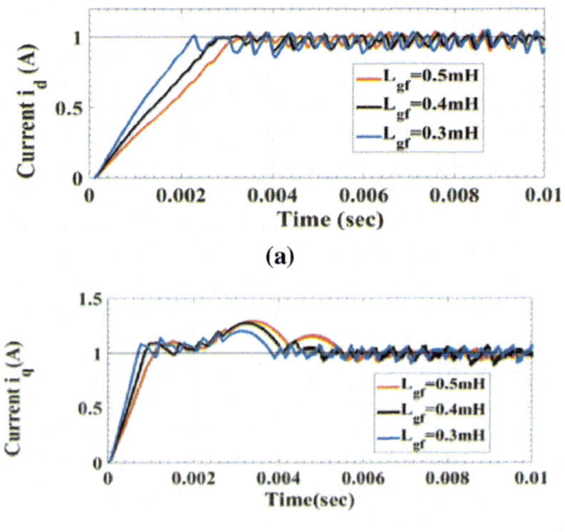

Fig. 10 Step response of **a** current i_d and **b** current i_q proposed controller for perturbation in line inductance L_{gf}

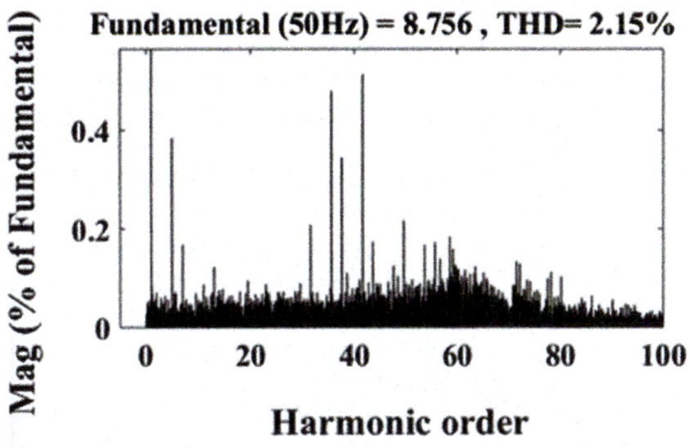

Fig. 11 THD of grid current of phase-a in case of proposed ESO-DISMC

Table 4 Comparison of performance of WEO-tuned ESO-DISMC with WEO-tuned DISMC

Parameter		DISMC	ESO-DISMC
System model		Nonlinear	Nonlinear
Robust stability		Yes	Yes
THD (%)		2.09	2.14
Sensitivity to Change in i_d	Settling time(ms)	2.8	4.2
	Peak-to-peak value	0.05	0.09
Sensitivity to Change in i_q	Settling time(ms)	2	1.19
	Peak-to-peak value	0.04	0.09

References

1. Williams DO, Megahed TF, Abdelkader SM (2022) Increasing green energy penetration and efficient utilization through a finite control set model predictive-based virtual inductance droop control. Electr Power Syst Res 210:108176. https://doi.org/10.1016/j.epsr.2022.108176
2. Pradhan R (2022) Role of advanced control technologies in the evolution of smart. Next generation smart grids: modeling, control and optimization. Springer Nature, Singapore, pp 377–401. https://doi.org/10.1007/978-981-16-7794-6_16
3. Khalid M, Mohan A (2020) Performance analysis of vector controlled PMSM drive modulated with sinusoidal PWM and space vector PWM. In: IEEE international power and renewable energy conference, pp 1–6. https://doi.org/10.1109/IPRECON49514.2020.9315210
4. Rao SNVB, Kumar YVP, Pradeep DJ, Reddy CP, Flah A, Kraiem H, Al-Asad JF (2022) Power quality improvement in renewable-energy-based microgrid clusters using fuzzy space vector PWM controlled inverter. Sustainability 14(8):4663. https://doi.org/10.3390/su14084663
5. Pradhan SS, Pradhan R, Subudhi B (2020) A hybrid incremental conductance and sliding mode controller for a PV microgrid system. Int J Intell Syst Des Comput 3(2):145–160. https://doi.org/10.1504/IJISDC.2020.115170
6. Bharatee A, Ray PK, Ghosh A (2021) A power management scheme for grid-connected PV integrated with hybrid energy storage system. J Mod Power Syst Clean Energy 10(4):954–963. https://doi.org/10.35833/MPCE.2021.000023
7. Arfeen ZA, Kermadi M, Azam MK, Siddiqui TA, Akhtar ZU, Ado M, Abdullah MP (2020) Insights and trends of optimal voltage-frequency control DG-based inverter for autonomous microgrid: state-of-the-art review. Int Trans Electr Energy Syst 30(10):e12555. https://doi.org/10.1002/2050-7038.12555
8. Ramadan HS, Padmanaban S, Mosaad MI (2022) Metaheuristic-based near-optimal fractional order PI controller for on-grid fuel cell dynamic performance enhancement. Electr Power Syst Res 208:107897. https://doi.org/10.1016/j.epsr.2022.107897
9. Pradhan SS, Pradhan R, Subudhi B (2021) An optimal fractional-order-proportional-integral controller for a grid-tied photovoltaic system. In: IEEE international symposium of Asian control association on intelligent robotics and industrial automation (IRIA), pp 166–171. https://doi.org/10.1109/IRIA53009.2021.9588694
10. Ayachi AY, Hamoudi F, Kheldoun A, Didier G, Rabiai Z (2021) Fuzzy logic enhanced control for a single-stage grid-tied photovoltaic system with shunt active filtering capability. Int Trans Electr Energy Syst 31(10):e13008. https://doi.org/10.1002/2050-7038.13008
11. Priyadarshi N, Sanjeevikumar P, Bhaskar MS, Azam F, Taha IB, Hussien MG (2022) An adaptive TS-fuzzy model based RBF neural network learning for grid integrated photovoltaic applications. IET Renew Power Gener 16(14):3149–3160. https://doi.org/10.1049/rpg2.12505
12. Douiri MR (2019) Particle swarm optimized neuro-fuzzy system for photovoltaic power forecasting model. Sol Energy 184:91–104. https://doi.org/10.1016/j.solener.2019.03.098
13. Kumar K, Alam M, Verma S, Dutta V (2020) Effect of hysteresis band control strategy on energy efficiency and durability of solar-hydrogen storage based microgrid in partial cloudy condition. J Energy Storage 32:101936. https://doi.org/10.1016/j.est.2020.101936
14. Bhattacharyya S, Singh B, Chandra A, Al-Haddad K (2019) Artificial neural network based advanced current control for grid-tied photovoltaic system. In: IEEE national power electronics conference (NPEC), pp 1–6. https://doi.org/10.1109/NPEC47332.2019.9034761
15. Bouzid AEM, Chaoui H, Zerrougui M, Elghali SB, Benbouzid M (2021) Robust control based on linear matrix inequalities criterion of single phase distributed electrical energy systems operating in islanded and grid-connected modes. Appl Energy 292:116776. https://doi.org/10.1016/j.apenergy.2021.116776
16. Pradhan SS, Pradhan R, Subudhi B (2019) Design and analysis of an H_∞ controller for a single phase grid connected photovoltaic system with parametric uncertainties. In: IEEE 2019 second international conference on advanced computational and communication paradigms (ICACCP), pp 1–6. https://doi.org/10.1109/ICACCP.2019.8882898

17. Koch GG, Osorio CR, Pinheiro H, Oliveira RC, Montagner VF (2019) Design procedure combining linear matrix inequalities and genetic algorithm for robust control of grid-connected converters. IEEE Trans Ind Appl 56(2):1896–1906. https://doi.org/10.1109/TIA.2019.2959604
18. Sabir A, Ibrir S (2020) A robust control scheme for grid-connected photovoltaic converters with low-voltage ride-through ability without phase-locked loop. ISA Trans 96:287–298. https://doi.org/10.1016/j.isatra.2019.05.027
19. Bhunia M, Subudhi B, Ray PK (2021) Design and real-time implementation of cascaded model reference adaptive controllers for a three-phase grid-connected PV system. IEEE J Photovolt 11(5):1319–1331. https://doi.org/10.1109/JPHOTOV.2021.3093047
20. Menaga D, Sankaranarayanan V (2020) A novel nonlinear sliding mode controller for a single stage grid-connected photovoltaic system. ISA Trans 107:329–339. https://doi.org/10.1016/j.isatra.2020.07.021
21. Armghan H, Yang M, Armghan A, Ali N (2021) Double integral action based sliding mode controller design for the back-to-back converters in grid-connected hybrid wind-PV system. Int J Electr Power Energy Syst 127:106655. https://doi.org/10.1016/j.ijepes.2020.106655
22. Bag A, Subudhi B, Ray PK (2019) A combined reinforcement learning and sliding mode control scheme for grid integration of a PV system. CSEE J Power Energy Syst 5(4):498–506. https://doi.org/10.17775/CSEEJPES.2017.01000
23. Zhang Y, Wang J, Li H, Zheng TQ, Lai JS, Li J, Chen Q (2020) Dynamic performance improving sliding-mode control-based feedback linearization for PV system under LVRT condition. IEEE Trans Power Electron 35(11):11745–11757. https://doi.org/10.1109/TPEL.2020.2983315
24. Kumar N, Sharma A (2022) Design and analysis of nonlinear controller for a standalone photovoltaic system using Lyapunov stability theory. J Solar Energy Eng 144(1). https://doi.org/10.1115/1.4051584
25. El Aroudi A, Haroun R, Cid-Pastor A, Kouzou A, Martinez-Salamero L (2013) A comparison between static and dynamic performances of a Z-source and a dual-stage boost converter under SMC for PV energy applications. Energy Procedia 42:587–596. https://doi.org/10.1016/j.egypro.2013.11.060
26. Yuan W, Gao G (2019) Disturbance observer-based adaptive integral sliding mode control for the hybrid automobile electro-coating conveying mechanism. Int J Adv Robot Syst 16(3). https://doi.org/10.1177/1729881419844126
27. Heydari E, Varjani AY, Diallo D (2020) Fast terminal sliding mode control-based direct power control for single-stage single-phase PV system. Control Eng Pract 104:104635. https://doi.org/10.1016/j.conengprac.2020.104635
28. Fallahzadeh SAA, Abjadi NR, Kargar A, Blaabjerg F (2021) Applying sliding-mode control to a double-stage single-phase grid-connected PV system. J Renew Energy Environ 18(1):1–12. https://doi.org/10.30501/JREE.2020.233358.1114
29. Apte A, Thakar U, Joshi V (2019) Disturbance observer based speed control of PMSM using fractional order PI controller. IEEE/CAA J Autom Sin 6(1):316–326. https://doi.org/10.1109/JAS.2019.1911354
30. Nerkar SS, Londhe PS, Patre BM (2022) Design of super twisting disturbance observer based control for autonomous underwater vehicle. Int J Dyn Control 10(1):306–322. https://doi.org/10.1007/s40435-021-00797-1
31. Salkuti SR (2022) Emerging and advanced green energy technologies for sustainable and resilient future grid. Energies 15(18):6667. https://doi.org/10.3390/en15186667
32. Punna S, Mailugundla R, Salkuti SR (2022) Design, analysis and implementation of bidirectional DC–DC converters for HESS in DC microgrid applications. Smart Cities 5(2):433–454. https://doi.org/10.3390/smartcities5020024
33. Pradhan SS, Pradhan R (2020) A modified double-integral-sliding-mode-controller for a micro-grid system with uncertainty. Distrib Gener Altern Energy J 35(1):19–46. https://doi.org/10.13052/dgaej2156-3306.3512

An Introduction to Demand Response in the Microgrid

Krishna Mohan Reddy Pothireddy, Sandeep Vuddanti, and Surender Reddy Salkuti

Abstract This chapter focuses on a basic introduction to conventional energy sources, renewable energy sources (RESs), the need for energy management, demand response (DR), advantages in employing DR, types of DR, and issues related to the application of DR in the microgrid (MG) and their impacts on the consumers and the generating companies. Moreover, issues related to load curtailment, load growth, errors related to the forecasting and its impact, the effect of rebound effect on market clearing price (MCP), and on total consumer tariff (TCT). Besides, curtailing load demand reduces the operational cost (OC), and might lift the TCT and MCP, whereas load growth increases the OC but may not increase the MCP and TCT. Load redistribution might create a local peak at the low-price scheduling horizon which is termed as the rebound effect, TCT and OC of the MG will increase to a large value. A case study in which an IEEE 33-bus system was considered to show the impact of congestion on a few buses and the role of the independent system operator (ISO) in alleviating this congestion by generating false locational marginal price (LMP) or system marginal price (SMP).

Keywords Conventional energy sources · Renewable energy sources · Energy management · Microgrid · Demand response · Market clearing price · Total consumer tariff · Operational cost

---

K. M. R. Pothireddy · S. Vuddanti
Department of Electrical Engineering, National Institute of Technology Andhra Pradesh, Tadepalligudem, Andhra Pradesh, India
e-mail: sandeep@nitandhra.ac.in

S. R. Salkuti (✉)
Department of Railroad and Electrical Engineering, Woosong University, Daejeon 34606, Republic of Korea
e-mail: surender@wsu.ac.kr

S. R. Salkuti et al. (eds.), *Power Quality in Microgrids: Issues, Challenges and Mitigation Techniques*, Lecture Notes in Electrical Engineering 1039,
https://doi.org/10.1007/978-981-99-2066-2_21

Nomenclature

MG	Microgrid
DR	Demand response
MCP	Market clearing price
EMS	Energy management system
CESs	Conventional energy sources
RESs	Renewable energy sources
MPPT	Maximum power point tracking
OC	Operational cost of the MG
MCP	Market clearing price
TCT	Total consumer tariff
PV	Photovoltaic
WT	Wind turbine
EV	Electric vehicle
BESS	Battery energy storage system
PHEV	Plug-in hybrid electric vehicle
SOC	State of charge
HEMS	Home energy management system
PBDR	Price-based demand response
IBDR	Incentive-based demand response
RTP	Real-time pricing
LMP	Locational marginal price
SMP	System marginal price
ISO	Independent system operator
MILP	Mixed integer linear programming
MINLP	Mixed integer non-linear programming
NLP	Non-linear programming
QP	Quadratic programming
MCP_t	Market clearing price at time t
P_l	Load demand
P_i^t	Power generation by the ith generator at time t

1 Introduction

Conventional energy sources (CESs) play an important role in balancing the energy required by the electrical loads by using coal, gas, nuclear energy, or a mix of all as fuel for producing electricity [1]. CESs generate emissions and the amount of emission generated is a function of the quality of fuel used [2]. Due to the growing concerns about environmental pollution, power generation through the CESs got reduced since it is considered as one of the major sources of polluting the environment [3]. Various methods employed for reducing the emissions from the CESs are deploying a post-combustion filter, swapping aged fuel burners, switching to low-emission fuels, and generating a schedule considering emission objectives but the methods incur additional capital [4, 5]. To meet the power balance condition, decarbonize the grid, and reduce pollution, the power system engineers introduced renewable energy sources (RESs) as an alternative to the CESs [6]. RESs are free from pollution, abundant in nature, and incur the less operational cost. RESs, loads, and storage are unitedly termed as microgrid (MG). However, the uncertainty and uncontrollability introduced by the RESs are the derating factors for effective and efficient energy management in the MG [7]. RESs are basically non-dispatchable sources of energy.

Generation blend, i.e., combining dispatchable (diesel generator, micro-turbine, fuel cell, and storage) and non-dispatchable sources for power generation increases the reliability of the system. Moreover, the power generation from RESs depends on climatic conditions which hinder the complete generation shift from CESs to RESs. MG can be operated in grid-connected mode or in isolated or autonomous mode. In the grid-connected mode, all the RESs, i.e., wind, PV, biomass, etc. operate in maximum power point condition whereas in an isolated mode based on the availability of battery SOC the RESs may be operated in MPPT or not. If the SOC of the storage element is higher and the load demand is lower, then RESs derate their output power, i.e., PV operated in voltage-controlled mode and wind derates its output by the pitch control mechanism. The wind is more uncertain and less variable whereas the PV is less uncertain and highly variable. The load demand itself is unpredictable and volatile in nature, the penetration of large capacity RESs will aggravate the power imbalance. Volatile power production from RESs, volatile load demand, and volatile electricity price necessitate energy management in an MG. Demand response (DR) plays an eminent role in energy management in an MG in which changes have been introduced in the consumer's consumption patterns of electricity with the change in price [8].

Figure 1 shows the functions of EMS. The function of the energy management system is to set constraints on voltage limits, the maximum amount of energy consumption, controlling thermostats to maintain room temperature, etc. by doing so the EMS [9] helps in reducing the OC of the MG and TCT of the consumer. Loads are classified into curtailable, non-curtailable, and shiftable loads [10, 11]. In a home energy management system (HEMS), the user must specify their loads as curtailable and non-curtailable. Based on the user specification the HEMS manages the energy consumption in accordance with the price of electricity and reduces the TCT. Electricity price information will be received through the local area network and electricity through service mains. A survey on the significance of DR has been presented in [12].

Figure 2 represents the advantages of employing DR programs [13]. The peak load on the system occurs occasionally, the majority of times the peak load generator is kept idle; therefore, if the peak demand on the system is reduced by peak shifting, curtailing, or conservation, the peak-to-average ratio of the system reduces and results in increased load factor.

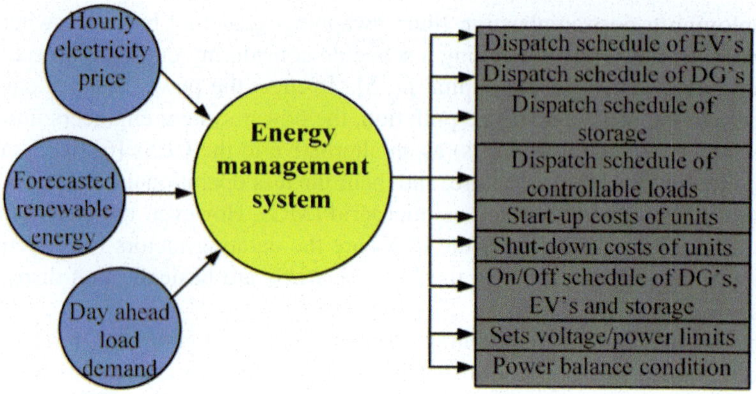

Fig. 1 Functions of the energy management system

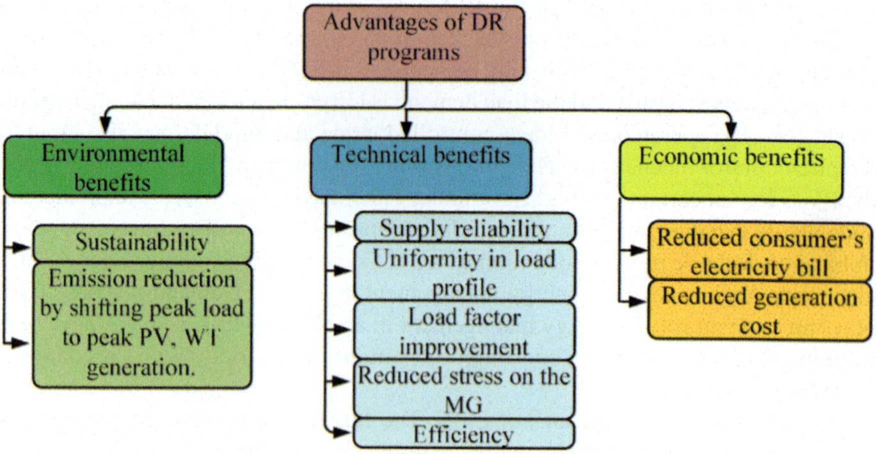

Fig. 2 Advantages of employing demand response programs

Further, DR improves the reliability of supply by adjusting the load demand as per the generation. If the load demand is less than the generation, the load growth technique in DR programs increases the load demand. If the load demand is more than the supply through load conservation or direct load curtailment excess load will be curtailed. Therefore, the stress or burden on the system will be mitigated. Further, the efficiency of the power system increases since the load reduction changes the generation schedule where the peak generator is removed from the dispatch schedule. The function of a peak generator is to deliver power during peak load demand and the drawbacks of this generator are hostile to the environment and ineffective. Moreover, the OC of this generator is the highest, it will be turned on during the peak instant of load demand. A sudden outage on the system might bring the entire power system to halt, to overcome this curtailing the load is mandatory to overcome the burden on the committed generator resulting in increased reliability on the system. The OC and emission reduction can be reduced by simply removing this peak generator from committing to meet the demand by applying for DR programs.

In an isolated MG, the RESs control their output based on the SOC of the battery energy storage system (BESS). This is because there is no provision for frequency regulation in the absence of utility, the MG resources should control the generation depending on the load demand. Shifting the peak load demand to the scheduling horizon where the RESs generate peak power. Therefore, this method greatly enhances the utilization factor of RESs. Further, the absence of a peak generator doesn't produce harmful substances that are unfriendly to the environment, therefore global warming or its impact on the environment greatly reduces.

To introduce flexibility in consumers and alleviate them to curtail the load demand, the operators introduce two programs as shown in Fig. 3, namely, price-based DR (PBDR) [14] and incentive-based DR (IBDR). In the PBDR, the variation in price induces a change in the user electricity consumption whereas in IBDR, incentives have been distributed for load curtailment. In PBDR, the volatility or variation in the hourly price is more in real-time pricing, whereas, in the time of use and critical peak pricing, the price variation is less. However, user satisfaction and comfort should not be constrained. DR can be applied during normal operation and during an emergency when the system reliability is jeopardized. A. Abdollahi et al. proposed an efficient smart infrastructure in [15], by employing RTP for EM. IBDR is classified into four types based on whether the operator controls the load or the user controls the load. Direct load curtailment (DLC) is one of the EM techniques in which the independent system operator (ISO) curtails the load demand during an emergency. Figure 4 shows the price elasticity, the expected LMP is a function of forecasted load demand.

As shown in Fig. 4, the error in demand forecasting should be limited to minimize the expected LMP. Uncertainties in load, PV, and WT power generation and

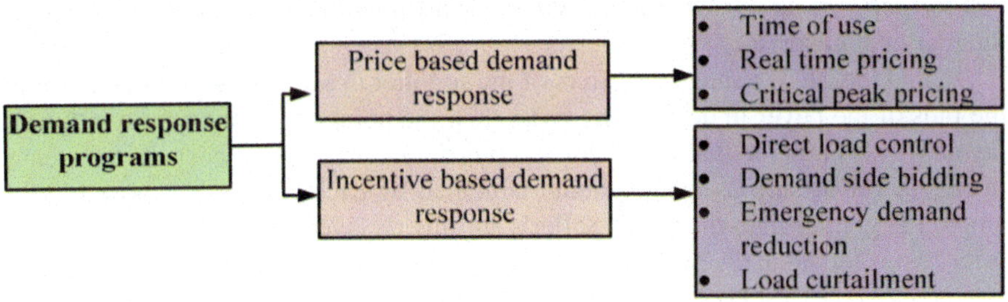

Fig. 3 Demand response programs

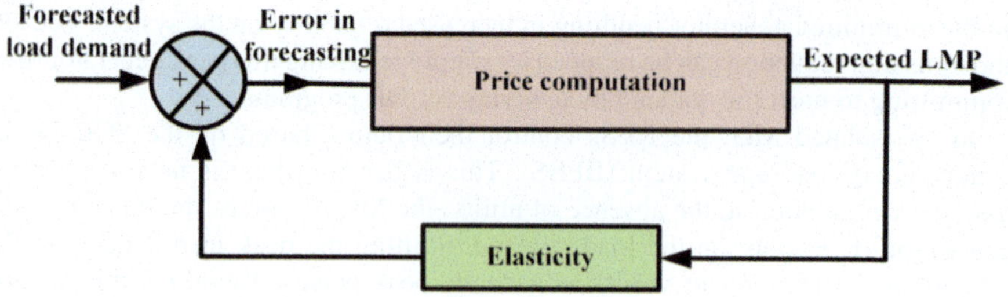

Fig. 4 Price elasticity

uncertainties in price forecasting can be modeled using the following distribution functions:

- Uncertainty in wind power can be modeled as Weibull distribution.
- Uncertainty in load demand can be modeled as the normal distribution or Gaussian distribution.
- Price uncertainty can be modeled as the normal distribution.

There are two types of auction mechanisms available [16, 17], they are operational cost minimization and payment cost minimization mechanisms. In the operational cost minimization mechanism, MCP is determined after optimizing the objective whereas, in the payment cost minimization model, the MCP itself is a decision variable and the optimization technique has to find it first. Therefore, the payment cost minimization model is complex and time-consuming [18–21]. There are two types of settlement rules available, they are pay-as-bid or pay-as-offer and pay-at-MCP. In general, the latter one is mostly used and all the markets follow this settlement rule [22]. In pay-as-bid, all the generators receive the money based on their bidding value, whereas, in pay-at-MCP, all the loads will pay a uniform tariff that too depends on the MCP. MCP will be decided by the highest bid generator i.e., the marginal generator.

To find the MCP arrange the offers of the supplier in ascending order and arrange the bids of the buyer in descending order then the point where the supply offer and demand bids coincide that point can be considered as marginal cost as represented in Fig. 5. Buyer surplus is the amount benefited by each consumer whose bidding amount is greater than the MCP. Seller's surplus received by the generators whose offer cost is less than the MCP. Figure 5 shows the surplus costs when the settlement rule employed by ISO is pay-at-MCP. If the pay-at-offer rule is employed no surplus will be received, all the generators receive based on their bidding price.

Figure 6 shows the available resources in the MG for EM. It is categorized into supply-side resources and demand-side resources. Supply-side resources offer less flexibility and demand-side resources offer the highest flexibility. Further, supply-side energy management is time-consuming. The loads have to be clustered into elastic and inelastic loads for applying for the DR programs. Further, the elastic loads are categorized into self-elastic loads and cross-elastic loads. Self-elasticity is

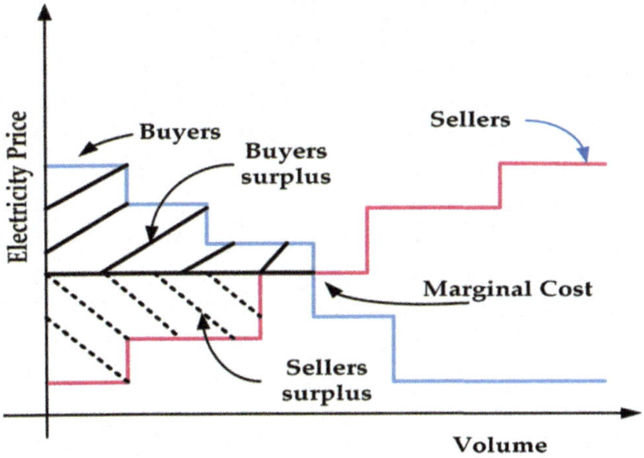

Fig. 5 Supply versus demand curve

defined as changes in electricity consumption with a change in the electricity price on the same scheduling horizon, whereas cross elasticity is defined as the changes in the consumer electricity consumption with changes in the electricity price at different scheduling hours. In general, the self-elastic loads are on/off type loads such as lighting loads whose elasticity is bounded on one scheduling horizon.

Lighting loads are generally classified under self-elastic loads whose period for curtailment is one. The majority of EM techniques rely on HVAC systems for energy management since the flexibility offered by these sources is more. P100 represents the load either can be on/off and elasticity extends to only one period. P75 represents the highest elasticity. Plug-in hybrid electric vehicle (PHEV) charges when the load demand on the system is low and discharges when the load demand on the system is more. Therefore, PHEV offers some flexibility for EM. Figure 7 shows the various types of clustering mechanisms for clustering the load demand. Further, the elbow method is a popular method, which is used to find the best number of clusters.

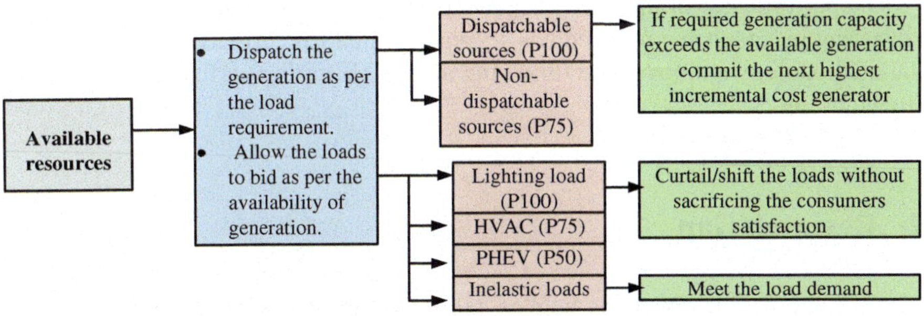

Fig. 6 Resources of microgrid for dispatch

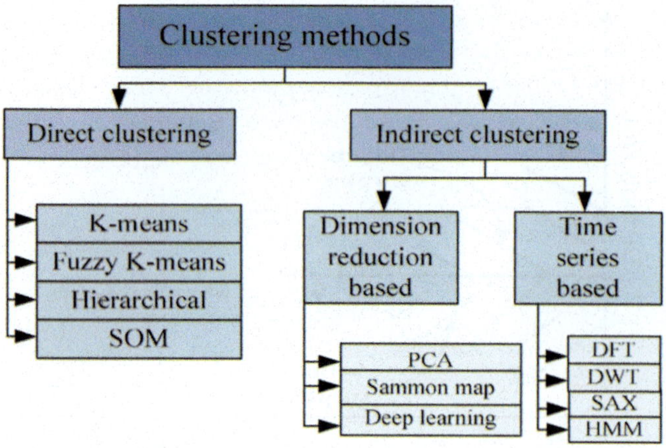

Fig. 7 Various types of clustering methods

Fig. 8 Clustering performance indicators

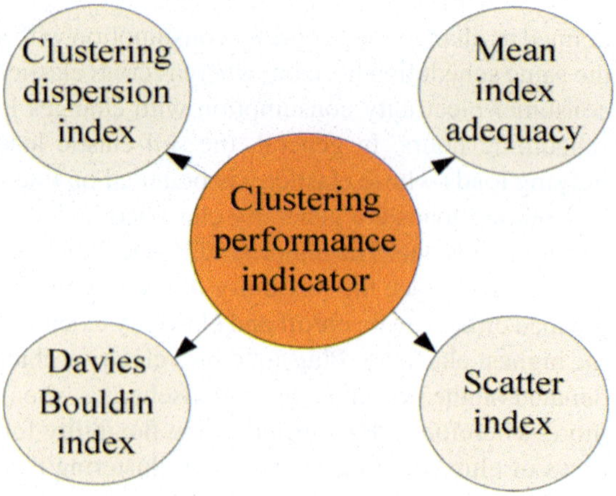

The performance of the clustering mechanisms plays a vital role in whether the clustering mechanism employed has segregated the load correctly or not. To assess the methodology employed the following are the clustering performance indicators shown in Fig. 8.

2 Research Gap

As said above, DR programs help in reducing the OC of the MG but there are challenges in the application to the MG. This section presents some of the research gaps in the application of DR as shown in Fig. 9.

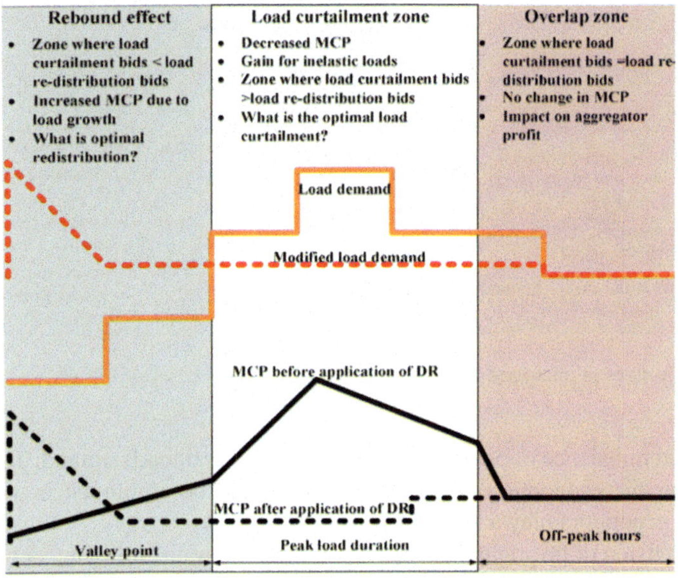

Fig. 9 Research gaps in the application of demand response

- A gap in mitigating the imbalance of power in the MG due to stochastic and variable sources.
- Impact of load redistribution on MCP and TCT.
- Impact of load curtailment on MCP and TCT.
- Impact of overlap between the demand curtailed loads and demand redistributed loads.
- Impact of rebound effect on inelastic loads.
- Lagging in incentive provision methods.
- The incentives are to be distributed to the demand-responsive loads.
- A glitch on load shaping methods.

3 Research Objectives

Different types of objectives are considered in a power system as shown in Fig. 10. Conventional unit commitment [23], and economic dispatch can solve the first objective function, i.e., the reduction of operational expenditure modeled as a single-objective function. If the objective is to reduce both emission and operational costs then the problem is considered as a multi-objective function. As the two objectives are conflicting, therefore, majority of researchers use a trade-off method such as Pareto optimal and Pareto dominance for solving the multi-objective functions. Static economic dispatch only solves the problems with certainty, if the system is having stochastic or uncertain sources then the static economic dispatch fails and

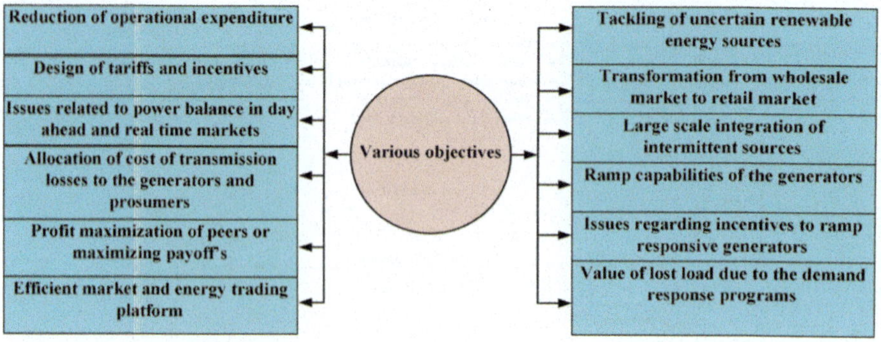

Fig. 10 Objectives of microgrid

results in an imbalance of power between generation and load demand. To overcome this, a dynamic economic dispatch schedule should be employed in which ramp capability generators play a significant role.

The provision of fast ramp generators, i.e., DR programs mitigates the imbalance of power greatly but incentives should be distributed to the demand-responsive loads. The main function of the aggregator is to collect the bids from the consumers and aggregate the individual consumer load demand and place them as a single load demand. Aggregator is a profit organization that always tries to maximize its profit whereas the ISO is a non-profit organization where function is to maintain the system reliability by matching the demand and supply throughout the scheduling horizon.

4 Problem Modeling

The problem can be modeled as a convex problem (linear, integer), non-convex problem (quadratic), single objective, multi-objective, etc. as shown in Fig. 11. The presence of local constraints introduces the non-convexity, further, the presence of prohibited operating zones, valve point effects, and multi-fuel mix divide the solution space into non-convex with more small convex regions. In the convex modeling, objective function and inequality constraint considered should be convex and equality constraints must be affine.

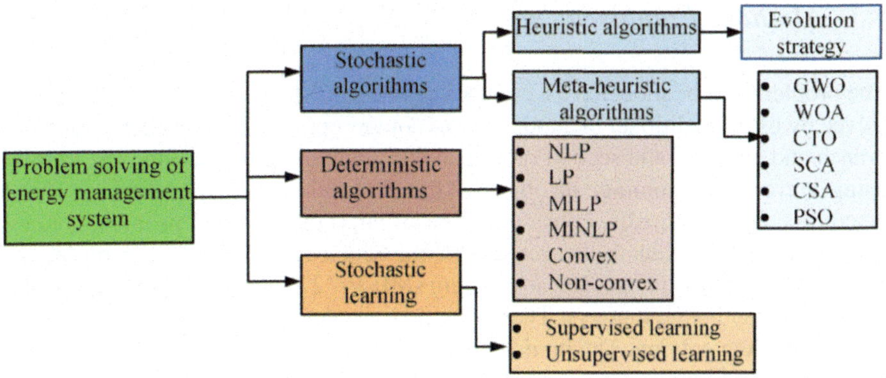

Fig. 11 Problem-solving methodologies

4.1 Decision Variables

The possible decision variables in solving the optimal scheduling problems may be as follows:

- Commands related to the power exchange between the main grid and microgrid and the amount of power to be exchanged [24].
- Commands related to the power exchange between the two community microgrids and the amount of power to be exchanged.
- Scheduling status of storage, electric vehicle (EV), and charge/discharge commands.
- Commands related to the market clearing price of the microgrid.
- State of charge maintenance, start-up, and shut-down decisions.

Among the above decision variables, a few are binary variables for example, ON/OFF state of generators and storage, etc.

4.2 Constraints

Constraints are classified into equality and inequality constraints.

- The basic power balance condition, i.e., generation should match the load demand which is termed as equality constraint or coupling constraint.
- Minimum and maximum generation limits, ramp-up, ramp-down, start-up, shut-down, SOC of storage, minimum up, minimum down time, battery energy limits, EV storage capacity, charge/discharge capacity, etc. are considered as an inequality constraint or local constraint.

4.3 Methods Employed

The problem can be modeled as a convex problem [25] (linear, integer) and can be solved by using traditional methods such as convex optimization, branch and bound, branch and cut [26], and second-order cone programming. For solving the problem using convex programming, the objective function and inequality constraints should be convex, and the equality constraint to be affine. If the modeled problem is linear, it is sufficient to evaluate the vertices of the relaxed problem. If the problem is modeled as a mixed integer linear problem (MILP) [27], the vertices of the relaxed constraint may not be integers. Therefore, first, we need to obtain an integer set for further problem-solving. The branch-and-cut method helps in obtaining the vertices as integer, as it cuts the major portion of the outside the feasible region, and the solution converges much faster than the main problem. Branching will be done from the vertices whichever is non-integer. Moreover, the branch-and-cut method has limitations in solving large-size problems.

The Lagrangian relaxation [28] method finds a feasible primal solution and reduces the dual gap. Primal feasibility suggests that all the constraints in the problem must be satisfied, and dual feasibility suggests that the dual variables associated with inequality constraint must be non-negative. The presence of local constraints introduces non-convexity [29], therefore the usage of above-said solution methods may not be feasible. Dynamic programming can be used for non-convex problems but the limitation is problem dimensionality. Priority scheduling is one more method for committing the generators but the OC increases. If the problem is modeled as a multi-objective problem, then the objectives need to be solved independently and a trade-off has to be taken.

Meta-heuristic algorithms are population based, easy to implement, convergence speed, and robust; however, there is a possibility of the algorithm getting trapped in the local optima. Further, these algorithms are stochastic in nature, parameter dependent and each time when the algorithm runs it results in a different optimum. To overcome this stochasticity all the meta-heuristic algorithms should be run a number of times and the mean, median, and standard deviation of the result have to be taken.

5 Theoretical Case Study

Consider an IEEE 33-bus system as shown in Fig. 12, the LMP for each bus varies as per the marginal cost of the unit that is committed or based on the transmission loss and line congestion [30]. Suppose that the bus nearer to the slack bus is having lowest LMP (since it is nearer to the reference bus), therefore, all the EV owners get converged toward this bus for charging their EVs. This increases the congestion of that transmission line and the independent system operator has to buy extra power from the grid or other microgrids. This can be overcome by generating a fake LMP

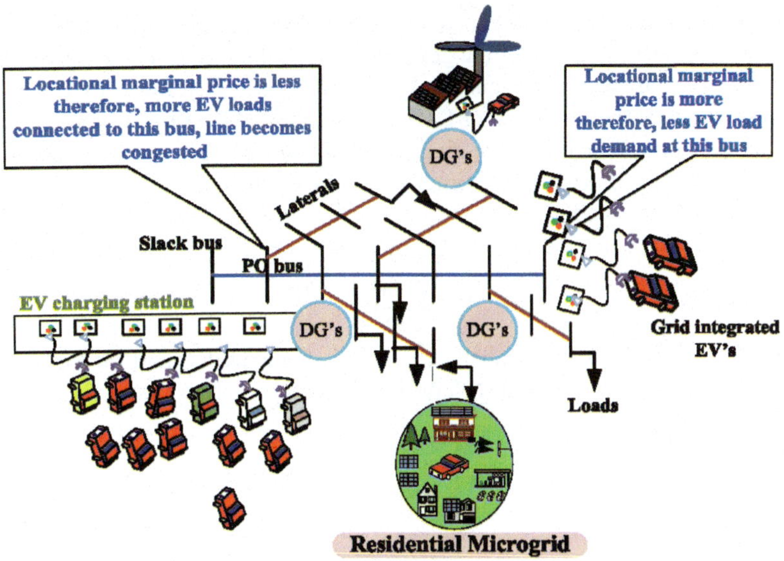

Fig. 12 Case study of IEEE 33-bus system

that is higher than the previous LMP such that all the vehicles get diverged to other buses where the LMP is less. The ISO must know how to generate fake LMP by changing the scheduling status of the dispatchable generators. EV charging stations were located at bus 2 and bus 7, whereas three DGs and one local residential MG are connected. The resources available in the test system for EM in an MG are EVs, WT, PV, and a utility which is not shown in Fig. 12 to avoid brevity.

Further, from the same Fig. 12, the thick line in the charging station represents isolation which means no EV is charging, and the thin line indicates the charging of the EV. The distribution network is of low voltage network which is radial in form and is complex to perform the load flow analysis. Further, the characteristics of the LV distribution network are

- Weakly meshed, i.e., ill network.
- The R/X ratio is too high compared to the transmission network.
- Standard load flow analysis techniques used for transmission networks, i.e., Gauss–Seidel, Newton–Raphson, and fast-decoupled cannot be employed for the distribution network.
- Presence of shunt capacitors for voltage profile improvement.

MCP at a time "t" can be defined as the incremental fuel cost of the highest marginal generator.

$$\text{MCP}_t = \begin{cases} A\,during\,normal\,load \\ B + \frac{x}{y}\,during\,peak\,load \end{cases} \tag{1}$$

where "A" is the incremental fuel cost of the base load marginal generator in Eq. (1), "B" is the IFC of the peak marginal generator, "x" is the no-load and start-up cost of the MG, and "y" is the power generated by the marginal generator. From the above equation, it is clear that the MCP or LMP is a function of the power generated by the marginal generator hence by changing the dispatch schedule of the generators the MCP can be changed. TCT at the time "t" can be defined as

$$\mathrm{TCT}_t = \sum_{t=1}^{24} \sum_{l=1}^{L} P_l * MCP_t \tag{2}$$

where "L" stands for load number, "P_l" stands for load demand on the system as shown in Eq. (2). OC at the time "t" of the MG can be defined as

$$\mathrm{OC}_t = \sum_{t=1}^{24} \sum_{i=1}^{N} P_i^t * MCP_t \tag{3}$$

P_i is the amount of power generated by the generator "i" at time "t" in Eq. (3).

5.1 Queries and Conclusions at Peak Scheduling Hours

There are so many queries while applying for DR programs those are: At what scheduling instant the load demand should be curtailed, how much to be curtailed, and where the curtailed load should be redistributed. Whether the consumer gets benefits for load curtailment and how much compensation is to be provided to the consumer. There is a chance of load curtailment and load redistribution at the same scheduling hour which is termed as an overlap between demand-responsive loads and demand redistribution bids. Due to the occurrence of overlap the profit margin of the aggregator gets reduced. Moreover, the load reduction causes a change in the market clearing price (MCP) or locational marginal price (LMP). Does a change in load demand with a change in price create a change in the generator commitment status? Does a change in the load demand create a change in the scheduling status of the already committed generator? Do these changes impact the total consumer tariff?

If the change in load demand necessitates the curtailment of an already committed generator, then the MCP reduces, if the change in load demand impacts only the generator schedule, then the consumer tariff will increase, this is due to the increase in MCP. If the curtailed load demand is less than the capacity of the marginal generator then the MCP increases. Consider the curtailed load demand is 25% of the capacity of the marginal generator, then the MCP is "P", if it is 50, 75, and 100% then the MCP will be "Q", "R", and "S", respectively. Then the relation between each MCP is as follows: S < P < Q < R. All these conclusions are valid at peak duration only.

To change the MCP at the off-peak hours, curtailment of load should turn off the committed generators. Similarly, the load growth doesn't change the MCP until and unless the scheduling of the uncommitted generator takes place. However, the OC of the MG reduces during load curtailment and increases during load growth. This is valid under peak load and under off-peak load conditions.

5.2 Forecasting and Its Issues

Price forecasting, renewable power generation forecasting, and load forecasting play important role in the EMS of the MG. Forecasting all these and scheduling the day ahead is mandatory for the smooth operation of the MG. Time series forecasting models, such as autoregressive integrated moving average (ARIMA) [31] and long short-term memory (LSTM) [32], are popularly used forecasting techniques. The problem with recurrent neural networks (RNN) [33] is, they cannot handle long-term dependence on data, LSTM is a form of RNN with storage. The problem associated with artificial intelligence-based forecasting is data dependent, problems associated with the presence of outliers in the data, and cyberthreats, therefore pre-processing of data is mandatory for forecasting. These days researchers have developed so many neural network models by hybridization, they are convolutional neural network (CNN) [34]-based LSTM, multi-layer perceptron (MLP) [35]-based CNN, LSTM-based autoencoder [36, 37], etc.

Errors in forecasting the load demand create an imbalance in the real-time market. It causes a change in the dispatch schedule, necessitates the turning on of fast ramp generators, and necessitates the purchase of electricity in the spot market at a price higher than the day-ahead price which will increase the OC of the MG. Errors in forecasting can be avoided by using a proper forecasting technique that gives less errors i.e., mean absolute error, mean absolute percentage error, etc.

5.3 Rebound Effect and Its Impacts on MCP and TCT

Another issue related to the application of demand response programs is due to the occurrence of the rebound effect or payback effect [38–40], as shown in Fig. 13.

The shifting of the peak from the high-price scheduling horizon "*t*" to the low-price scheduling horizon "*h*" is termed as the rebound effect. The shifting of the peak necessitates the commitment of the peak load generator which increases the MCP at this time instant "*h*" [41–43]. The inelastic consumers scheduled at "*h*" have to pay extra tariffs because of the increase in MCP. Moreover, the demand recovery is a function of weather, economic conditions, the number of demand reduction periods, and on the amount of load curtailment.

The RE can be avoided by setting a constraint on the amount of load shift on a particular scheduling horizon [44–46]. Schedule the curtailed loads to more

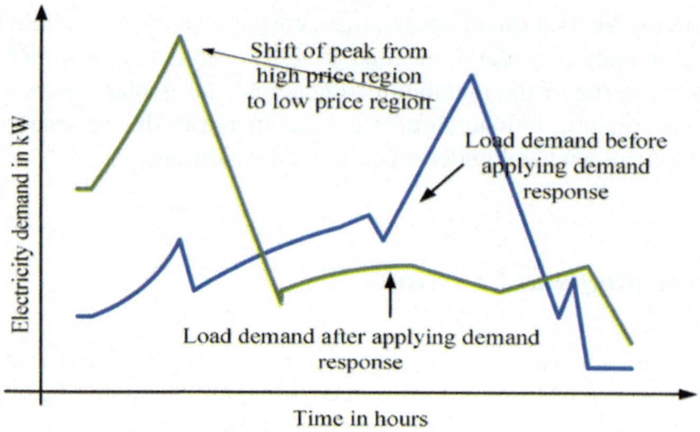

Fig. 13 Rebound or payback effect

scheduling hours. Set the ramp-up constraint on the redistributed load. Majorly, inelastic consumers suffer due to the occurrence of RE [47–49].

6 Conclusion

This chapter discusses the issues that are present in the MG. Majorly, issues related to load curtailment, load redistribution, and the overlap between the demand reduction bids and demand redistribution bids, issues related to the reduction of aggregator profit, various objectives considered in the MG, and solution methodologies employed for solving the objectives. Various impacts of demand response programs on MCP, TCT, and OC are also presented. The generation mix with the deployment of MG increases the reliability of the system, which further leads to an increased share of RESs supply. The curtailed load may be redistributed to the scheduling hour where the peak RESs occur instantly and this results in a reduction in OC. While participating in grid-connected mode or in the dispatch of multiple microgrids the OC incurred by the microgrid should not exceed the OC incurred when operating independently. A brief introduction to artificial intelligence-based time series forecasting models has been presented.

Acknowledgements This research work was supported by "Woosong University's Academic Research Funding—2023".

References

1. Talaq JH, El-Hawary F, El-Hawary ME (1994) A summary of environmental/economic dispatch algorithms. IEEE Trans Power Syst 9(3):1508–1516. https://doi.org/10.1109/59.336110
2. Ramanathan R (1994) Emission constrained economic dispatch. IEEE Trans Power Syst 9(4):1994–2000. https://doi.org/10.1109/59.331461
3. Akbari-Dibavar A, Mohammadi-Ivatloo B, Zare K, Khalili T, Bidram A (2021) Economic-emission dispatch problem in power systems with carbon capture power plants. IEEE Trans Ind Appl 57(4):3341–3351. https://doi.org/10.1109/TIA.2021.3079329
4. Delson JK (1974) Controlled emission dispatch. IEEE Trans Power Appar Syst 93(5):1359–1366. https://doi.org/10.1109/TPAS.1974.293861
5. Lamont O (1995) Emission dispatch models and algorithms for the 1990s. IEEE Trans Power Syst 10(2):941–947. https://doi.org/10.1109/59.387937
6. Mohan Reddy PK, Prakash M (2020) Optimal dispatch of energy resources in an isolated micro-grid with battery energy storage system. In: 4th international conference on intelligent computing and control systems (ICICCS), pp 730–735. https://doi.org/10.1109/ICICCS48265.2020.9121176
7. Pothireddy KM, Vuddanti S, Salkuti SR (2022) Impact of demand response on optimal sizing of distributed generation and customer tariff. Energies. https://doi.org/10.3390/en15010190
8. Conejo AJ, Morales JM, Baringo L (2010) Real-time demand response model. IEEE Trans Smart Grid 1(3):236–242. https://doi.org/10.1109/TSG.2010.2078843
9. Su C, Kirschen D (2009) Quantifying the effect of demand response on electricity markets. IEEE Trans Power Syst 24(3):1199–1207. https://doi.org/10.1109/TPWRS.2009.2023259
10. Strbac G, Kirschen D (1999) Assessing the competitiveness of demand-side bidding. IEEE Trans Power Syst 14(1):120–125. https://doi.org/10.1109/59.744498
11. Strbac G, Farmer G, Cory ED (1996) Framework for the incorporation of demand-side in a competitive electricity market. IEEE Proceed Gener, Transm Distrib 143(3): 232–237. https://doi.org/10.1049/ip-gtd:19960290
12. Deng R, Yang Y, Chow MY, Chen J (2015) A Survey on demand response in smart grids: mathematical models and approaches. IEEE Trans Ind Inf 11(3):570–582. https://doi.org/10.1109/TII.2015.2414719
13. Rahimi F, Ipakchi A (2010) Demand response as a market resource under the smart grid paradigm. IEEE Trans Smart Grid 1(1):82–88. https://doi.org/10.1109/TSG.2010.2045906
14. Liu N, Yu X, Wang C, Li C, Ma L, Lei J (2017) Energy-sharing model with price-based demand response for microgrids of peer-to-peer prosumers. IEEE Trans Power Syst 32(5):3569–3583. https://doi.org/10.1109/TPWRS.2017.2649558
15. Abdollahi A, Parsa Moghaddam M, Rashidinejad M, Sheikh-El-Eslami MK (2012) Investigation of economic and environmental-driven demand response measures incorporating UC. IEEE Trans Smart Grid 3(1):12–25. https://doi.org/10.1109/TSG.2011.2172996
16. Fernández-Blanco R, Arroyo JM, Alguacil N, Guan X (2016) Incorporating price-responsive demand in energy scheduling based on consumer payment minimization. IEEE Trans Smart Grid 7(2):817–826. https://doi.org/10.1109/TSG.2015.2427584
17. Fernández-Blanco R, Arroyo JM, Alguacil N (2014) Network-constrained day-ahead auction for consumer payment minimization. IEEE Trans Power Syst 29(2):526–536. https://doi.org/10.1109/TPWRS.2013.2284283
18. Alonso J, Trias A, Gaitan V, Alba AA (1999) Thermal plant bids and market clearing in an electricity pool. Minimization of costs versus minimization of consumer payments. IEEE Trans Power Syst 14(4):1327–1334. https://doi.org/10.1109/59.801892
19. Vazquez C, Rivier M, Perez-Arriaga IJ (2002) Production cost minimization versus consumer payment minimization in electricity pools. IEEE Trans Power Syst 17(1):119–127. https://doi.org/10.1109/59.982202
20. Hao S, Angelidis GA, Singh H, Papalexopoulos AD (1998) Consumer payment minimization in power pool auctions. IEEE Trans Power Syst 13(3):986–991. https://doi.org/10.1109/59.709087

21. Li H, Lin X, Yang H, Liang X, Lu R, Shen X (2014) EPPDR: an efficient privacy-preserving demand response scheme with adaptive key evolution in smart grid. IEEE Trans Parallel Distrib Syst 25(8):2053–2064. https://doi.org/10.1109/TPDS.2013.124
22. De Jonghe C, Hobbs B, Belmans R (2012) optimal generation mix with short-term demand response and wind penetration. IEEE Trans Power Syst 27(2):830–839. https://doi.org/10.1109/TPWRS.2011.2174257
23. Palensky P, Dietrich D (2011) Demand side management: demand response, intelligent energy systems, and smart loads. IEEE Trans Ind Inf 7(3):381–388. https://doi.org/10.1109/TII.2011.2158841
24. Deng R, Yang Z, Chow MY, Chen J (2015) A survey on demand response in smart grids: mathematical models and approaches. IEEE Trans Ind Inf 11(3):570–582. https://doi.org/10.1109/TII.2015.2414719
25. Bragin MA, Luh PB, Yan JH, Stern GA (2015) Novel exploitation of convex hull invariance for solving unit commitment by using surrogate Lagrangian relaxation and branch-and-cut. IEEE Power Energy Soc Gen Meet 1–5. https://doi.org/10.1109/PESGM.2015.7286409
26. Bragin MA, Luh PB, Yan JH, Stern GA (2014) Surrogate Lagrangian relaxation and branch-and-cut for unit commitment with combined cycle units. In: 2014 IEEE PES general meeting conference & exposition, pp 1–5. https://doi.org/10.1109/PESGM.2014.6939901
27. Han X, Luh PB, Yan JH, Stern GA (2011) Energy and spinning reserve payment cost co-optimization. In: 2011 IEEE power and energy society general meeting, pp 1–8. https://doi.org/10.1109/PES.2011.6039741
28. Bragin MA, Han X, Luh PB, Yan JH (2011) Payment cost minimization using Lagrangian relaxation and modified surrogate optimization approach. IEEE Power Energy Soc Gen Meet 1–7. https://doi.org/10.1109/PES.2011.6039191
29. Bragin MA, Luh PB, Yan JH (2012) An efficient surrogate optimization method for solving linear mixed-integer problems with cross-coupling constraints. In: Proceedings of the 10th world congress on intelligent control and automation, pp 4055–4060. https://doi.org/10.1109/WCICA.2012.6359153
30. Bragin MA, Luh PB, Yan JH, Yu N, Stern GA (2013) Efficient surrogate optimization for payment cost co-optimization with transmission capacity constraints. IEEE Power Energy Soc Gen Meet 1–5. https://doi.org/10.1109/PESMG.2013.6672447
31. Wei L, Zhen-gang Z (2009) Based on time sequence of arima model in the application of short-term electricity load forecasting. In: International conference on research challenges in computer science, pp 11–14. https://doi.org/10.1109/ICRCCS.2009.12
32. Tokgöz G, Ünal A (2018) A RNN based time series approach for forecasting turkish electricity load. In: 2018 26th signal processing and communications applications conference, pp 1–4. https://doi.org/10.1109/SIU.2018.8404313
33. Hua Y, Zhao Z, Li R, Chen X, Liu Z, Zhang H (2019) deep learning with long short-term memory for time series prediction. IEEE Commun Mag 57(6):114–119. https://doi.org/10.1109/MCOM.2019.1800155
34. Koprinska I, Wu D, Wang Z (2018) Convolutional neural networks for energy time series forecasting. In: 2018 international joint conference on neural networks (IJCNN), pp 1–8. https://doi.org/10.1109/IJCNN.2018.8489399
35. Li L, Ota K, Dong M (2017) Everything is image: CNN-based short-term electrical load forecasting for smart grid. In: 14th international symposium on pervasive systems, algorithms and networks & 2017 11th international conference on frontier of computer science and technology & 2017 third international symposium of creative computing (ISPAN-FCST-ISCC), Exeter, pp 344–351. https://doi.org/10.1109/ISPAN-FCST-ISCC.2017.78
36. Essien A, Giannetti C (2019) A deep learning framework for univariate time series prediction using convolutional LSTM stacked autoencoders. In: IEEE international symposium on innovations in intelligent systems and applications, pp 1–6. https://doi.org/10.1109/INISTA.2019.8778417
37. Park K, Jeong J, Kim D, Kim H (2020) Missing-insensitive short-term load forecasting leveraging autoencoder and LSTM. IEEE Access 8:206039–206048. https://doi.org/10.1109/ACCESS.2020.3036885

38. Bitaraf H, Rahman S (2018) Reducing curtailed wind energy through energy storage and demand response. IEEE Trans Sustain Energy 9(1):228–236. https://doi.org/10.1109/TSTE.2017.2724546
39. Shi W, Li N, Xie X, Chu C, Gadh R (2014) Optimal residential demand response in distribution networks. IEEE J Select Areas Commun 32(7):1441–1450. https://doi.org/10.1109/JSAC.2014.2332131
40. Zhou X (2022) Exploiting integrated demand response for operating reserve provision considering rebound effects. IEEE Access 10:15151–15162. https://doi.org/10.1109/ACCESS.2022.3148398
41. Halbe S, Chowdhury B, Abbas A (2019) Mitigating rebound effect of demand response using battery energy storage and electric water heaters. In: IEEE 16th international conference on smart cities: improving quality of life using ICT & IoT and AI (HONET-ICT), pp 095–099. https://doi.org/10.1109/HONET.2019.8908081
42. Zhang W, Lian J, Chang C, Kalsi K (2013) Aggregated modeling and control of air conditioning loads for demand response. IEEE Trans Power Syst 28(4):4655–4664. https://doi.org/10.1109/TPWRS.2013.2266121
43. Samad T, Koch E, Stluka P (2016) Automated demand response for smart buildings and microgrids: the state of the practice and research challenges. Proceed IEEE 104(4):726–744. https://doi.org/10.1109/JPROC.2016.2520639
44. Talebi A, Hatami A (2020) Optimal scheduling of hvac set-point for electricity peak reduction and avoid rebound phenomenon. In: 28th Iranian conference on electrical engineering (ICEE), pp 1–5. https://doi.org/10.1109/ICEE50131.2020.9260721
45. Faddel S, Mohammed OA (2019) Automated distributed electric vehicle controller for residential demand side management. IEEE Trans Ind Appl 55(1):16–25. https://doi.org/10.1109/TIA.2018.2866255
46. Chen C, Kishore S, Snyder LV (2011) An innovative RTP-based residential power scheduling scheme for smart grids. In: IEEE international conference on acoustics, speech and signal processing, pp 5956–5959. https://doi.org/10.1109/ICASSP.2011.5947718
47. Ludwig P, Winzer C (2022) Tariff menus to avoid rebound peaks: results from a discrete choice experiment with Swiss customers. Energies 15(17). https://www.mdpi.com/1996-1073/15/17/6354
48. Battula AR, Vuddanti S, Salkuti SR (2021) Review of energy management system approaches in microgrids. Energies 14(17). https://doi.org/10.3390/en14175459
49. Battula AR, Vuddanti S, Salkuti SR (2023) A day ahead demand schedule strategy for optimal operation of microgrid with uncertainty. Smart Cities 6(1):491–509. https://doi.org/10.3390/smartcities6010023

Active Power Load Data Dimensionality Reduction Using Autoencoder

Venkataramana Veeramsetty, Prabhu Kiran, Munjampally Sushma, Amuda Mahesh Babu, Rathlavath Rakesh, Kunchala Raju, and Surender Reddy Salkuti

Abstract Dimensionality reduction is a machine learning based technique used to convert the data from higher dimensionality space to lower dimensionality space. This technique helps to build lighter version machine learning based predictive models. In this paper, a deep learning model i.e.. autoencoders is used to reduce the dimensionality of active power load data from higher dimensionality space consists 14 input features to lower dimensionality space consists 7 input features. Original active power load data is prepared based on data collected from 33/11KV substation located in Godishala village, Telangana State, India, from 01.01.2021 to 31.12.2021. Autoencoder model is trained and tested with python program using visual studio. From the simulation results, observed that autoencoder model reduces the dimensionality space of load data with almost same variance that original data exist.

Keywords Dimensionality reduction · Auto-encoder · Deep leaning · Load forecasting · Error metrics

Nomenclature

W_{ih} Weights between input and latent space
W_{hk} Weights between output and latent space
b_h Bias parameters at latent space
b_k Bias parameters at output layer

V. Veeramsetty
Center for AI and Deep Learning, SR University, Warangal, India

P. Kiran · M. Sushma · A. M. Babu · R. Rakesh · K. Raju
Department of Electrical and Electronics Engineering, SR University, Warangal, India

S. R. Salkuti (✉)
Department of Railroad and Electrical Engineering, Woosong University, Daejeon 34606, Republic of Korea
e-mail: surender@wsu.ac.kr

S. R. Salkuti et al. (eds.), *Power Quality in Microgrids: Issues, Challenges and Mitigation Techniques*, Lecture Notes in Electrical Engineering 1039,
https://doi.org/10.1007/978-981-99-2066-2_22

net_h Net input to latent space
net_k Net input to output layer
O_h Output of latent space
W^n_{hk} Weights connected to nth neuron in output layer
ΔW^n_{hk} Change in weights connected to nth neuron in output layer
Δb_k Change in bias connected to neurons in output layer
ΔW^n_{ih} Change in weights connected to nth neuron in latent space
Δb_h Change in bias connected to neurons in latent space
L(T-1) Load on substation one hour before the time of forecasting
L(T-2) Load on substation two hours before the time of forecasting
L(T-3) Load on substation three hours before the time of forecasting
L(T-4) Load on substation four hours before the time of forecasting
L(T-24) Load on substation one day before the time of forecasting
L(T-48) Load on substation two days before the time of forecasting
L(T-72) Load on substation three days before the time of forecasting
L(T-96) Load on substation four days before the time of forecasting

1 Introduction

Active power load forecasting is an essential task for proper and optimal operation of electric power distribution system. Forecasting of load on an electric power distribution system is done with various machine learning models with huge amount of data and input features. Few literature on electric power load forecasting with various number of features are discussed in this section as follows: electric power load at particular time of the day has been forecast based on last three hours and load at same hour for the last 4 days in [40]. The active power load on a 33/11 kV substation is forecast one hour ahead based on input features L(T-1), L(T-2), L(T-24), L(T-48), day, season, temperature, and humidity in [44]. Load forecasting using LSTM and XGBOOST models with region, industry, and capacity features is implemented in [25]. Electric power load forecasting with LSTM model using features like ambient temperature, humidity, and wind speed is discussed in [3]. The enormous number of dimensions for high-dimensional data, distance, and proximity estimations in high-dimensional space are unreliable [21]. It is often desirable to reduce the dimension of data to a smaller dimension when undertaking data analysis with a high number of attributes [4]. It is necessary to execute the Dimensionality Reduction (DR), which is intended to project the original high-dimensional data into a low-dimensional feature subspace while retaining some desired information [22].

DR is key machine learning technique used to convert data from higher dimensional space to lower dimensionality space [7, 47] in order to build a predictive machine learning models with less number of model parameters. This process leads to use less memory space to deploy deep learning models on edge devices and less computation time [31]. DR is a method of extracting enough information from data to determine the prediction by removing noise and redundant information that causes over-fitting and improves discriminating.

DR techniques classified into two categories i.e., feature selection[1] and feature extraction [21]. Feature extraction and feature selection are widely used to remove redundant and unnecessary features in order to improve performance of deep learning model [46]. In feature selection, dimensionality of the data space is reduced by selecting the most appropriate features from the existing features i.e. correlation[15], random forest, etc.[2, 28]. Whereas in feature extraction, dimensionality of the data space is reduced by extracting the less number of new features from the existing features i.e., factor analysis, principal component analysis, etc. Existing DR approaches have some shortcomings, such as cluster over-positioning and the lack of an angular reference [45]. Furthermore, they can all be thought of as applying a set of weights to a collection of feature variables in order to generate transformed variables with training. The distinction is that the weight used for feature selection is binary, and only features with weights equal to one are maintained, but the weight used for feature extraction is continuous, and all features are eventually retained [14].

DR of an electric power load dataset using principal component analysis is proposed in [13, 38]. In this papers, principal component analysis is used to build recurrent neural network model with less number of model parameters by reducing the dimensionality of the electric power dataset space. DR of an electric power load dataset using random forest is proposed in [43]. In this paper, random forest is used to build gated recurrent unit model with less number of model parameters to forecast the active power load on a distribution substation. A deep learning model i.e., long short-term memory model with lighter version by reducing the dimensionlaity space of dataset using factor analysis is proposed in [39].

DR for a electric power load dataset by identifying the correlation between various input features is proposed in [41, 42]. In this paper, DR based on correlation techniques is used to build artificial neural network model with less number of model parameters to forecast the load on a 33/11KV substation located in Warangal, India. Autoencoder based DR approach is proposed in [17] to analyze, design, and optimize the electromagnetic nanostructures. DR technique by combining the information gain (IG) and principal component analysis (PCA) is proposed in [34]. A DR algorithm locally linear embedding is applied in [9] for image feature selection.

Performed a comparative analysis of three broad techniques to feature selection with real-world applications [18]. Feature selection for clustering is discussed in [35] using filter models, wrapper models, hybrid models, and embedded models. Review of nonlinear methods for reducing dimensionality and their applications in data visualization is provided in [12]. Discussion about deep learning approaches for text feature extraction is presented in [24].

In this paper, a deep learning model i.e., autoencoder [5, 29] is used to reduce the original active power load dataset consists of 14 features like temperature, humidity, season, day status etc., into a dataset consists of less number of features with almost same variance that original data have. In this paper, best number of features that can reduced from original 14 features is identified based on autoencoder loss, number of null values and variance difference from original data variance.

Major contributions of this paper are as follows:

- Practical active power load data is collected from a 33/11 KV substation which is located in Godishala, Telangana State, India, and made publicly available at https://data.mendeley.com/datasets/tj54nv46hj
- Autoencoder model is used for the first time to reduce the dimensionality of the active power load data.
- New objective function is proposed to identify the optimal autoencoder architecture

The remaining part of the paper is organized as follows: Sect. 2 describes methodology, Sect. 3 presents analysis of simulation results, and Sect. 3 provides conclusions of this work. Appendix sections demonstrates autoencoder training and data reconstruction with a small sample data.

2 Materials and Methods

This section presents functioning of autoencoder, details of active power load dataset and error metrics that are used to obtain the performance of the autoencoder model.

2.1 Autoencoder

Autoencoder is a neural network that is used to reconstruct its input [8, 16, 26]. It has two parts i.e., encoder and decoder [33]. Encoder is used to get compressed features at latent space [10, 11, 23, 48], whereas decoder is used to reconstruct the original inputs from the compressed latent space. In autoencoders, the learning algorithm performs back propagation by assigning input data as target values [6]. In each hidden layer and in latent space ReLu activation function [19, 27, 30] is considered and that is mathematically modelled as shown in Eq. (1). The gradient of ReLu activation function is zero for z ≤ 0 and whereas for $z > 0$ the gradient is equal to 1. Linear activation function is considered for output layer of autoencoder that is mathematically modelled as shown in Eq. (2). In Eq. (2), “w” represents weight matrix between latent space and output layer, “f(z)” represents output of latent space, and “b” represents bias parameters in output layer.

$$f(z) = \max(0, z) \tag{1}$$

$$g(w, f(z)) = w^T f(z) + b \tag{2}$$

Architecture of the autoencoder is shown in Fig. 1. In Fig. 1, L(T-1), L(T-2), L(T-3), L(T-4), L(T-24), L(T-48), L(T-72), L(T-96),Temp, Humidity, Day, and season

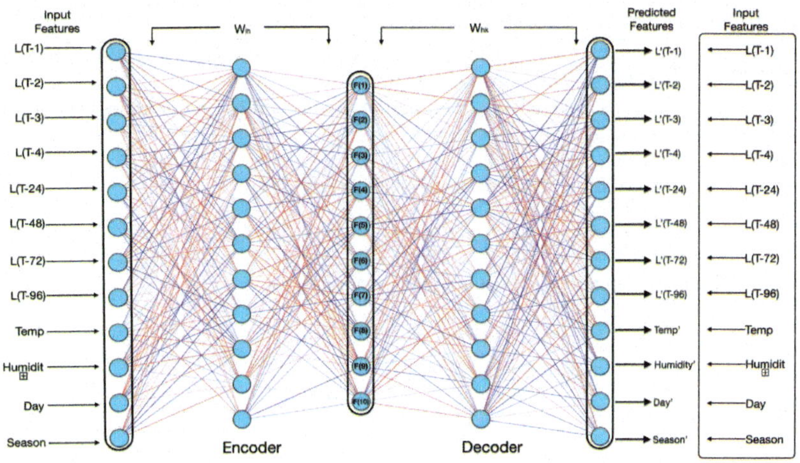

Fig. 1 Autoencoder architecture

represents original input features in the load dataset. Whereas L'(T-1), L'(T-2), L'(T-3), L'(T-4), L'(T-24), L'(T-48), L'(T-72), L'(T-96), Temp', Humidity', Day', and season' are the output at output layer while expecting same input features which are given to input layer. F1, F2, F3, F4, F5, F6, and F7 are the reduced extracted features from input features in the original load data. Training algorithm shown in Algorithm 1 helps to update the weights and bias parameters of autoencoder in such a way that autoencoder provides output which is almost same as int input. Performance of the autoencoder is observed in terms of training and validation loss i.e., mean square error [20, 32, 36] as shown in Eq. (9).

$$\text{loss} = \frac{1}{n}\sum_{i=1}^{n_s ample}\sum_{k=1}^{n_o}(L_k - L_k')^2 \tag{3}$$

Different autoencoder architectures are trained with training data and validated with testing data. Optimal autoencoder architecture is identified based on the objection function show in Eq. (11).

$$\text{Min Obj} = 0.5 * \text{Norm\_Validation\_loss} + 0.5 * \text{Norm\_Variance\_Difference} \tag{11}$$

Encoder part of optimal autoencoder architecture is shown in Fig. 2. It consists two dense layers. First layer consists 132 weight parameters and 11 bias parameters. Whereas, second layer has 110 weight parameters and 10 bias parameters. So total number of model parameters that the optimal autoencoder is 263. These 263 model parameters helps to reconstruct the data from original data that consists 12 input features.

Algorithm 1 Autoencoder training algorithm

Step 1: Read input features of original load data ▷ L=L(T-1),L(T-2),L(T-3),L(T-4),L(T-24),L(T-48),L(T-72),L(T-96),Temp, Humidity, Day status, season
Step 2: Initialize weights W_{ih} and W_{ho}, bias parameters b_h and b_o, and learning rate η
Step 3: Assume del_W as max(W_{ih})
while $del_W \leq 0.001$ **do**
Step 4: Calculate net input to latent space using equation (4)

$$net_h = W_{ih}^T L + b_h \tag{4}$$

Step 5: Calculate output of latent space using equation (5)

$$O_h = max(0, net_h) \tag{5}$$

Step 6: Calculate output of output layer using equation (6)

$$O_k = W_{ho}^T O_h + b_o \tag{6}$$

Step 7: update W_{ho} using equation (7) and b_o using (8)

$$W_{ho} = W_{ho} + \eta(L_k - L_k')O_h \tag{7}$$

$$b_o = b_o + \eta(L_k - L_k') \tag{8}$$

Step 8: update W_{ih} using equation (9) and b_h using (10) if total input to hidden neuron "h" is positive, otherwise no need to update W_{ih} and b_h.

$$W_{ih} = W_{ih} + \eta L \sum_{k=1}^{n_k}(L_k - L_k')W_{ho_k} \tag{9}$$

$$b_h = b_h + \eta \sum_{k=1}^{n_k}(L_k - L_k')W_{ho_k} \tag{10}$$

find del_W as maximum change among W_{ih}, W_{ho}, b_h and b_o
end while
Step 9: Store model in terms of model paramneters W_{ih}, W_{ho}, b_h and b_o, and architecture

2.2 *Active Power Load Dataset*

Active power load data of 33/11 KV substation located in godishala, India and weather data of this region is taken from [37]. This is a online mendeley data repository can be accessed through web link https://data.mendeley.com/datasets/tj54nv46hj/1. From this data, new dataset reconstructed to predict load L(T) from the input features like L(T-1), L(T-2), L(T-3), L(T-4), L(T-24), L(T-48), L(T-72), L(T-96), Temp, Humidity, Day, and season. This dataset consists total 8664 samples with 12 input features and one predicted variable i.e. L(T). In this paper, dimensionality of the dataset is reduced by compressing 12 input features into 7 features which are extracted from latent space of the autoencoder.

Fig. 2 Autoencoder architecture: encoder part

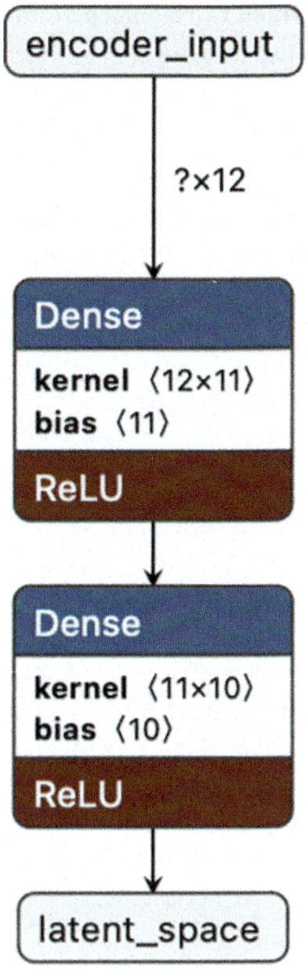

3 Results and Discussion

Autoencoder is developed based on data available in [37] using Google Colab. This section presents training and testing performance of autoencoder model. Out of 8664 samples 70% of samples i.e., 5804 samples are used for training and remaining 30% of samples i.e., 2614 samples are used for testing. Stochastic gradient descent optimizer has been used train the autoencoder model as shown in Algorithm 1.

3.1 Optimal Batch Size

Autoencoder with different number of neurons in latent space i.e., 11, 10, 9, 8, 7, 6, 5 and 4 is trained and validated with various batch sizes i.e., 8, 16, 32 and 64. For each autoencoder architecture optimal batch size is identified based on lowest validation loss. Training and validation errors for different latent space with different batch size are presented in Table 1. From the Table 1, it is observed that autoencoder

Table 1 Training performance of various autoencoder architectures with respect to batch sizes

Latent space	Batch sizes			
	8		16	
	Training loss	Validation loss	Training loss	Validation loss
11	0.0180	0.0198	0.0091	0.0090
10	0.0117	0.0109	0.0082	0.0077
9	0.0116	0.0125	0.0125	0.0120
8	0.0140	0.0144	0.0123	0.0125
7	0.0228	0.0222	0.0173	0.0164
6	0.0285	0.0266	0.0336	0.0328
5	0.0345	0.0373	0.0369	0.0368
4	0.0349	0.0371	0.0392	0.0394
Latent space	Batch sizes			
	32		64	
	Training loss	Validation loss	Training loss	Validation loss
11	0.0077	0.0078	0.0076	**0.0075**
10	0.0061	0.0064	0.0061	**0.0062**
9	0.0097	0.0096	0.0062	**0.0062**
8	0.0093	0.0090	0.0086	**0.0083**
7	0.0152	0.0146	0.0131	**0.0130**
6	0.0185	**0.0182**	0.0216	0.0212
5	0.0350	0.0368	0.0330	**0.0316**
4	0.0338	0.0351	0.0323	**0.0311**

architerures with 11, 10, 9, 8, 7, 5 and 4 have lowest mean square error with batch size 64, and for latent space with 6 neurons lowest mean square error with batch size 32. Among all the 8 autoencoder architectures, autoencoder with 10 neurons in latent space has minimum training and validation errors i.e., 0.0061 and 0.0062 respectively.

The converging characteristics of autoencoder with 10 neurons in latent space with batch size of 64 for training and validation data is shown in Fig. 3. From the Fig. 3, it is observed that the both validation and training losses are almost following and reducing as iteration progress. It shows that autoencoder architectures with 10 neurons in latent space is trained perfectly without either overfit or underfit problems.

The variation between actual input features and output features after reconstruction for a latent space with 10 neurons is shown in Fig. 4 for batch size 64. From the Fig. 4, it is observed that the reconstructed features almost close to original input features.

Table 2 shows the variance of data extracted from each neuron in the latent space. Table 3 shows the comparison between average variance of reconstructed data and

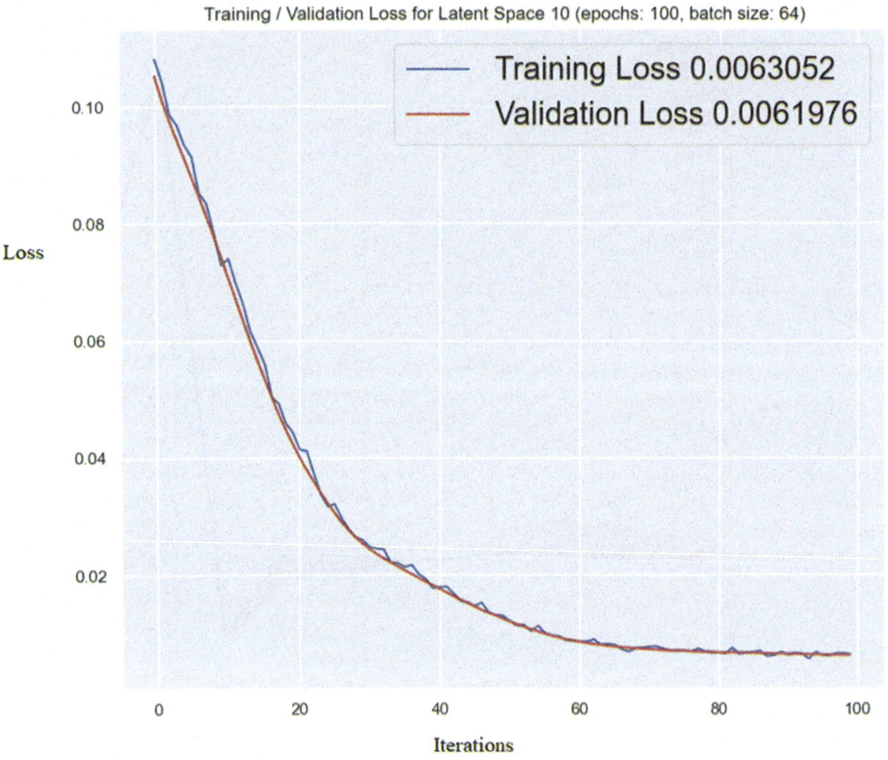

Fig. 3 Converging characteristics

original data. From the Table 3, it is observed that the variance difference between reconstructed data and original data is low in case of latent space with 10 neurons.

The value of the objective function mentioned in Eq. (11) for various autoencoder architectures is shown in Table 4. From the Table 4, it is observed that autoencoder with 10 neurons in the latent space has minimum objective function value hence it is considered as an optimal model.

Table 5 presents information about zero values in each feature column with training data for various autoencoder architectures. Training data consists total 5804 samples. The autoencoder with 10 neurons in the latent space is considered as optimal model, and in this architecture the value of the fourth feature for all the samples of training data is zero.

Table 6 presents information about the number of zero values for each feature column in the compressed data for various autoencoder architectures. Training data consists total 2860 samples. The autoencoder with 10 neurons in the latent space is considered as optimal model, and in this architecture the value of the fourth feature in the compressed data for all the samples of testing data is zero. As the all the values against fourth feature in the compressed data is zero with both training and

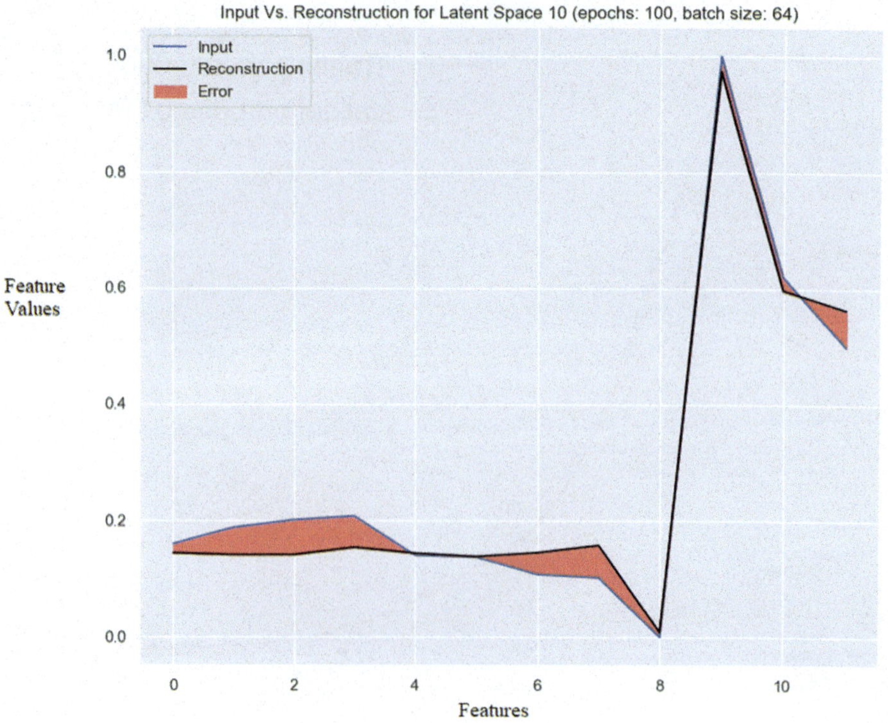

Fig. 4 Original versus reconstructed features

Table 2 Variance of each feature corresponding to neuron in latent space

Latent Space	Variance information for each feature										
	1	2	3	4	5	6	7	8	9	10	11
11	0.46	0.28	0.00	0.22	0.34	0.48	0.00	0.33	1.04	0.10	0.16
10	0.35	0.05	0.29	0.00	0.78	0.30	0.08	0.30	0.28	0.54	NA
9	0.42	0.02	0.35	0.00	0.26	0.97	1.11	0.37	0.56	NA	NA
8	0.78	1.07	0.00	0.96	1.25	0.00	1.42	0.60	NA	NA	NA
7	0.19	1.43	0.31	0.25	0.14	0.32	0.61	NA	NA	NA	NA
6	0.00	0.09	0.08	0.00	0.00	2.79	NA	NA	NA	NA	NA
5	0.00	0.00	0.04	0.59	0.07	NA	NA	NA	NA	NA	NA
4	0.00	0.11	0.16	0.00	NA	NA	NA	NA	NA	NA	NA

Table 3 Comparison between reconstructed and original data variance

Latent space	Variance of reconstructed data	Original Data Variance	Variance Difference
11	0.3092	0.0630	−0.2463
10	0.2974	0.0630	**−0.2345**
9	0.4507	0.0630	−0.3877
8	0.7597	0.0630	−0.6967
7	0.4626	0.0630	−0.3997
6	0.4945	0.0630	−0.4315
5	0.1415	0.0630	−0.0786
4	0.0685	0.0630	−0.0055

Table 4 Objective function value for various autoencoder architectures

Latent space	Objective
11	0.200547
10	0.16609
9	0.276458
8	0.542518
7	0.419536
6	0.543784
5	0.552839
4	0.490237

Table 5 Information about zero values in each feature column with training data

Latent Space	Number of zeros in each feature column										
	1	2	3	4	5	6	7	8	9	10	11
11	129	96	5776	177	312	106	5804	481	197	0	236
10	317	1483	123	5804	0	782	9	75	0	38	...
9	0	4254	854	5804	0	0	15	27	69	...	...
8	0	466	5804	68	0	5804	0	11	...	...	...
7	2685	253	0	4787	0	533	16	...	...	...	...
6	5804	0	295	5053	5804	412	...	...	...	...	...
5	5804	5804	3887	0	0	...	...	...	...	...	...
4	5144	0	908	5804	...	...	...	...	...	...	...

Table 6 Information about zero values in each feature column with testing data

Latent space	Number of zeros in each feature column										
	1	2	3	4	5	6	7	8	9	10	11
11	73	48	2847	112	143	58	2860	29	98	0	108
10	158	744	57	2860	0	418	6	37	0	12	
9	0	2122	463	2860	0	0	8	10	37		
8	0	238	2860	25	0	2860	0	2			
7	1304	131	0	2359	0	281	6				
6	2860	1	152	2459	2860	212					
5	2860	2860	1898	0	0						
4	2498	0	479	2860							

Table 7 Statistical information of original dataset

Parameter	L(T-1)	L(T-2)	L(T-3)	L(T-4)	L(T-24)	L(T-48)
Count	5804	5804	5804	5804	5804	5804
Mean	0.283	0.283	0.283	0.283	0.283	0.286
Std	0.220	0.220	0.220	0.220	0.220	0.219
Min	0.000	0.003	0.000	0.000	0.003	0.000
25%	0.103	0.103	0.103	0.104	0.103	0.106
50%	0.206	0.206	0.207	0.208	0.207	0.209
75%	0.431	0.430	0.431	0.431	0.437	0.433
Max	1.000	1.000	1.000	0.981	0.981	1.000
Parameter	L(T-72)	L(T-96)	DAY	SEASON	TEMP	HUMIDITY
Count	5804	5804	5804	5804	5804	5804
Mean	0.285	0.289	0.140	0.503	0.540	0.611
Std	0.218	0.218	0.347	0.413	0.153	0.235
Min	0.000	0.000	0.000	0.000	0.000	0.000
25%	0.106	0.110	0.000	0.000	0.466	0.443
50%	0.210	0.215	0.000	0.500	0.534	0.659
75%	0.433	0.437	0.000	1.000	0.621	0.830
Max	0.999	1.000	1.000	1.000	1.000	1.000

testing data, fourth feature is discarded from compressed data. Now the size of the compressed data is $(5804 + 2860 = 8644) \times 9$ instead of 8644×10.

The original dataset that is prepared based on data collected from 33/11 KV substation in Godidhala, Telangana State, India with features like L(T-1), L(T-2), L(T-3), L(T-4), L(T-24), L(T-48), L(T-72), L(T-96), Temperature, Humidity, Season and Day status is available at public repository at https://data.mendeley.com/datasets/7vdt5rz47x/1. Table 7 presents statistical information of the original dataset.

Table 8 Statistical information of compressed dataset

Parameter	F1	F2	F3	F4	F5	F6	F7	F8	F9	F10
Count	5804	5804	5804	5804	5804	5804	5804	5804	5804	5804
Mean	0.870	0.289	1.529	0.000	2.772	0.758	0.923	1.362	1.042	2.148
Std	0.588	0.228	0.539	0.000	0.886	0.546	0.286	0.552	0.528	0.733
Min	0.000	0.000	0.000	0.000	0.780	0.000	0.000	0.000	0.029	0.000
25%	0.338	0.000	1.324	0.000	2.131	0.377	0.838	1.021	0.662	1.740
50%	0.809	0.315	1.679	0.000	2.570	0.720	1.005	1.414	0.886	2.210
75%	1.428	0.478	1.926	0.000	3.288	1.166	1.116	1.746	1.333	2.637
Max	2.537	0.809	2.371	0.000	5.518	2.532	1.469	2.638	3.036	4.096

The compressed dataset that is developed from autoencoder is available at public repository at https://data.mendeley.com/datasets/7vdt5rz47x/1. Table 8 presents statistical information of the compressed dataset.

4 Conclusions

This study proposed a data-driven autoencoder model to reconstruct the data with less dimensions to train the deep learning models in order to forecast the load. Autoencoder architectures with various neurons in the latent space is trained and tested, and the optimal model is identified based on validation loss and variance. Based on the simulation results, autoencoder with 10 neurons in latent space is identified as optimal model with minimum validation loss 0.0062 and variance difference 0.2344. With this trained autoencoder, original data with shape 8644×12 is converted into 8644×10. From the simulation results, it is observed that the output of the fourth neuron in the latent space which represents feature four in the compressed data has a constant value i.e., zero. Hence, fourth feature from the compressed data is removed and the new shape of reconstructed data is 8644×9.

Acknowledgements This research work was supported by "Woosong University's Academic Research Funding—2023."

Appendix

Autoencoder

This section presents step by step procedure for training of autoencoder and also explains how trained encoder is used to reconstruct the data from original dataset.

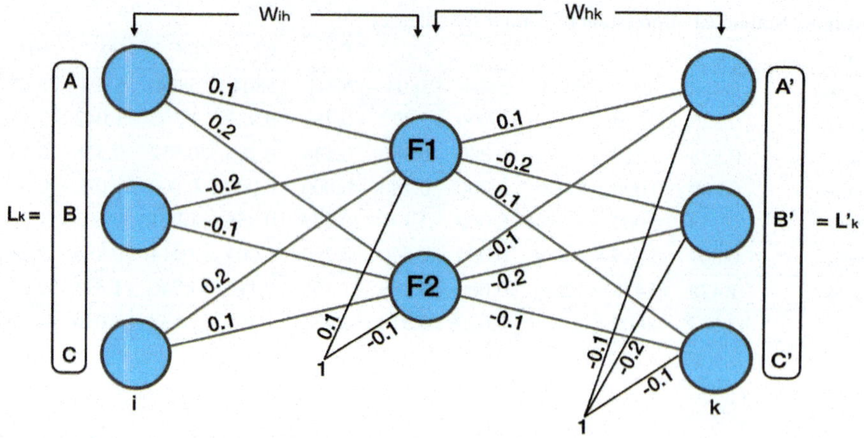

Fig. 5 Sample autoencoder architecture

Table 9 Sample training data

Sample	A	B	C
1	0.1	0.2	0.3
2	0.3	0.2	0.1

The sample architecture of autoencoder that used in this section to explain the training and data reconstruction procedure is shown in Fig. 5.

The initial weight and bias matrices are shown in equations (12–15). The data samples that are used to train the autoencoder is shown in Table 9.

$$W_{ih} = \begin{bmatrix} 0.1 & 0.2 \\ -0.2 & -0.1 \\ 0.2 & 0.1 \end{bmatrix} \tag{12}$$

$$W_{hk} = \begin{bmatrix} 0.1 & -0.2 & 0.1 \\ -0.1 & 0.2 & -0.1 \end{bmatrix} \tag{13}$$

$$b_h = \begin{bmatrix} 0.1 \\ -0.1 \end{bmatrix} \tag{14}$$

$$b_k = \begin{bmatrix} 0.1 \\ -0.2 \\ 0.1 \end{bmatrix} \tag{15}$$

Auto Encoder Training

In this section, update of weights and bias parameters of autoencoder based on given training data for one iteration is explained.

- Iteration 01
  - Sample 01 = [0.1 0.2 0.3]

Calculate net input to hidden layer/latent space using Eq. (4) and the result for the first sample is shown below

$$net_h = \begin{bmatrix} 0.1 & -0.2 & 0.2 \\ 0.2 & -0.1 & 0.1 \end{bmatrix} * \begin{bmatrix} 0.1 \\ 0.2 \\ 0.3 \end{bmatrix} + \begin{bmatrix} 0.1 \\ -0.1 \end{bmatrix} = \begin{bmatrix} 0.13 \\ -0.07 \end{bmatrix} \tag{16}$$

Calculate the output of hidden layer/latent space using Eq. (5) and the result for the first sample is shown below

$$O_h = \max\left(0, \begin{bmatrix} 0.13 \\ -0.07 \end{bmatrix}\right) = \begin{bmatrix} 0.13 \\ 0 \end{bmatrix} \tag{17}$$

Calculate the input of output layer using Eq. (6) and the result for the first sample is shown below

$$net_k = \begin{bmatrix} 0.1 & -0.1 \\ -0.2 & 0.2 \\ 0.1 & -0.1 \end{bmatrix} * \begin{bmatrix} 0.13 \\ 0 \end{bmatrix} + \begin{bmatrix} 0.1 \\ -0.2 \\ 0.1 \end{bmatrix} = \begin{bmatrix} 0.11 \\ -0.23 \\ 0.11 \end{bmatrix} \tag{18}$$

As linear activation function is used in the output layer, the output and input for output layer remains same and that is

$$net_k = L'_k = \begin{bmatrix} A' \\ B' \\ C' \end{bmatrix} = \begin{bmatrix} 0.11 \\ -0.23 \\ 0.11 \end{bmatrix} \tag{19}$$

Calculate the change in weights connected to each neuron in the output layer using Eq. (20). Where n is a neuron in output layer.

$$\Delta W^n_{hk} = \eta * (L(n) - L'(n)) * o_j \tag{20}$$

$$\Delta W^1_{hk} = 0.1 * (0.1 - 0.11) * \begin{bmatrix} 0.13 \\ 0 \end{bmatrix} = \begin{bmatrix} -0.0001 \\ 0 \end{bmatrix} \tag{21}$$

$$\Delta W_{hk}^{2} = 0.1 * (0.2 + 0.23) * \begin{bmatrix} 0.13 \\ 0 \end{bmatrix} = \begin{bmatrix} 0.0056 \\ 0 \end{bmatrix} \tag{22}$$

$$\Delta W_{hk}^{3} = 0.1 * (0.3 - 0.11) * \begin{bmatrix} 0.13 \\ 0 \end{bmatrix} = \begin{bmatrix} 0.0025 \\ 0 \end{bmatrix} \tag{23}$$

Finally, change in weights between latent space and output layer is given by

$$\Delta W_{hk} = \begin{bmatrix} \Delta W_{hk}^{1} & \Delta W_{hk}^{2} & \Delta W_{hk}^{3} \end{bmatrix} = \begin{bmatrix} -0.0001 & 0.0056 & 0.0025 \\ 0 & 0 & 0 \end{bmatrix} \tag{24}$$

Now update the weights between latent space and output layer using equation (25).

$$W_{hk} = W_{hk} + \Delta W_{hk} \tag{25}$$

$$W_{hk} = \begin{bmatrix} 0.1 & -0.2 & 0.1 \\ -0.1 & 0.2 & -0.1 \end{bmatrix} + \begin{bmatrix} -0.0001 & 0.0056 & 0.0025 \\ 0 & 0 & 0 \end{bmatrix} = \begin{bmatrix} 0.0999 & -0.1944 & 0.1025 \\ -0.1 & 0.2 & -0.1 \end{bmatrix} \tag{26}$$

Calculate the change in bias connected to each neuron in the output layer using Eq. (27) and update bias parameters using Eq. (29)

$$\Delta b_k = \eta * \left(\begin{bmatrix} A \\ B \\ C \end{bmatrix} - \begin{bmatrix} A' \\ B' \\ C' \end{bmatrix} \right) \tag{27}$$

$$\Delta b_k = 0.1 * \left(\begin{bmatrix} 0.1 \\ 0.2 \\ 0.3 \end{bmatrix} - \begin{bmatrix} 0.11 \\ -0.23 \\ 0.11 \end{bmatrix} \right) = \begin{bmatrix} -0.001 \\ 0.043 \\ 0.019 \end{bmatrix} \tag{28}$$

$$b_k = b_k + \Delta b_k \tag{29}$$

$$b_k = \begin{bmatrix} 0.1 \\ -0.2 \\ 0.1 \end{bmatrix} + \begin{bmatrix} -0.001 \\ 0.043 \\ 0.019 \end{bmatrix} = \begin{bmatrix} 0.099 \\ -0.157 \\ 0.119 \end{bmatrix} \tag{30}$$

Calculate the change in weights connected to each neuron in the latent space using Eq. (31). Where n is a neuron in output layer and j is neuron in latent space.

$$\Delta W_{ih}^{j} = \eta * L * \sum (L(n) - L'(n)) * W_{hk}^{n} \tag{31}$$

$$\Delta W_{ih}^{1} = 0.1 * \begin{bmatrix} 0.1 \\ 0.2 \\ 0.3 \end{bmatrix} * [(0.1 - 0.11) * 0.0999 + (0.2 + 0.23) * -0.1944 + (0.3 - 0.11) * 0.1025] \quad (32)$$

$$\Delta W_{ih}^{1} = \begin{bmatrix} -0.0007 \\ -0.0013 \\ -0.002 \end{bmatrix} \quad (33)$$

$$\Delta W_{ih}^{2} = 0.1 * \begin{bmatrix} 0.1 \\ 0.2 \\ 0.3 \end{bmatrix} * [(0.1 - 0.11) * -0.1 + (0.2 + 0.23) * 0.2 + (0.3 - 0.11) * -0.1] \quad (34)$$

$$\Delta W_{ih}^{2} = \begin{bmatrix} 0.0007 \\ 0.0014 \\ 0.002 \end{bmatrix} \quad (35)$$

Finally, change in weights between latent space and input layer is given by

$$\Delta W_{ih} = \left[\Delta W_{ih}^{1}\ \Delta W_{ih}^{2}\right] = \begin{bmatrix} -0.0007 & 0.0007 \\ -0.0013 & 0.0014 \\ -0.002 & 0.002 \end{bmatrix} \quad (36)$$

Now update the weights between latent space and input layer using Eq. (37).

$$W_{ih} = W_{ih} + \Delta W_{ih} \quad (37)$$

$$W_{ih} = \begin{bmatrix} 0.1 & 0.2 \\ -0.2 & -0.1 \\ 0.2 & 0.1 \end{bmatrix} + \begin{bmatrix} -0.0007 & 0.0007 \\ -0.0013 & 0.0014 \\ -0.002 & 0.002 \end{bmatrix} = \begin{bmatrix} 0.0993 & 0.2007 \\ -0.2013 & 0.1014 \\ 0.198/ & 0.102 \end{bmatrix} \quad (38)$$

Calculate the change in bias connected to each neuron "j" in the latent space using Eq. (39) and update bias parameters using equation (??).

$$\Delta b_{h}^{j} = \eta * \sum (L(n) - L'(n)) * W_{hk}^{n} \quad (39)$$

$$\Delta b_{h}^{1} = 0.1 * [(0.1 - 0.11) * 0.0999 + (0.2 + 0.23) * -0.1944 + (0.3 - 0.11) * 0.1025] = -0.0065 \quad (40)$$

$$\Delta b_h^2 = 0.1 * [(0.1 - 0.11) * -0.1 + (0.2 + 0.23) * 0.2 + (0.3 - 0.11) * -0.1] = 0.0068 \quad (41)$$

Finally, change in bias parameter for the neurons in latent space are given by

$$\Delta b_h = \begin{bmatrix} \Delta b_h^1 & \Delta b_h^2 \end{bmatrix} = \begin{bmatrix} -0.0065 \\ 0.0068 \end{bmatrix} \quad (42)$$

$$b_h = b_h + \Delta b_h \quad (43)$$

$$b_h = \begin{bmatrix} 0.1 \\ -0.1 \end{bmatrix} + \begin{bmatrix} -0.0065 \\ 0.0068 \end{bmatrix} = \begin{bmatrix} 0.0935 \\ -0.0932 \end{bmatrix} \quad (44)$$

- Iteration 01
  - Sample 02 = [0.3 0.2 0.1]

Calculate net input to hidden layer/latent space using Eq. (4) and the result for the second sample is shown below

$$net_h = \begin{bmatrix} 0.0993 & -0.2013 & 0.198 \\ 0.2007 & 0.1014 & 0.102 \end{bmatrix} * \begin{bmatrix} 0.3 \\ 0.2 \\ 0.1 \end{bmatrix} + \begin{bmatrix} 0.0935 \\ -0.0932 \end{bmatrix} = \begin{bmatrix} 0.1028 \\ -0.0025 \end{bmatrix} \quad (45)$$

Calculate the output of hidden layer/latent space using Eq. (5) and the result for the first sample is shown below

$$O_h = max(0, \begin{bmatrix} 0.1028 \\ -0.0025 \end{bmatrix}) = \begin{bmatrix} 0.1028 \\ 0 \end{bmatrix} \quad (46)$$

Calculate the input of output layer using Eq. (6) and the result for the first sample is shown below

$$net_k = \begin{bmatrix} 0.0999 & -0.1 \\ -0.1944 & 0.2 \\ 0.1025 & -0.1 \end{bmatrix} * \begin{bmatrix} 0.1028 \\ 0 \end{bmatrix} + \begin{bmatrix} 0.099 \\ -0.157 \\ 0.119 \end{bmatrix} = \begin{bmatrix} 0.1093 \\ -0.177 \\ 0.1295 \end{bmatrix} \quad (47)$$

As linear activation function is used in the output layer, the output and input for output layer remains same and that is

$$net_k = L_k' = \begin{bmatrix} A' \\ B' \\ C' \end{bmatrix} = \begin{bmatrix} 0.1093 \\ -0.177 \\ 0.1295 \end{bmatrix} \quad (48)$$

Calculate the change in weights connected to each neuron in the output layer using Eq. (20).

$$\Delta W_{hk}^{1} = 0.1 * (0.3 - 0.1093) * \begin{bmatrix} 0.1028 \\ 0 \end{bmatrix} = \begin{bmatrix} -0.002 \\ 0 \end{bmatrix} \tag{49}$$

$$\Delta W_{hk}^{2} = 0.1 * (0.2 + 0.177) * \begin{bmatrix} 0.1028 \\ 0 \end{bmatrix} = \begin{bmatrix} 0.0039 \\ 0 \end{bmatrix} \tag{50}$$

$$\Delta W_{hk}^{3} = 0.1 * (0.1 - 0.1295) * \begin{bmatrix} 0.1028 \\ 0 \end{bmatrix} = \begin{bmatrix} -0.0003 \\ 0 \end{bmatrix} \tag{51}$$

Finally, change in weights between latent space and output layer is given by

$$\Delta W_{hk} = \left[\Delta W_{hk}^{1}\ \Delta W_{hk}^{2}\ \Delta W_{hk}^{3}\right] = \begin{bmatrix} 0.00196 & 0.00388 & -0.0003 \\ 0 & 0 & 0 \end{bmatrix} \tag{52}$$

Now update the weights between latent space and output layer using Eq. (25).

$$W_{hk} = \begin{bmatrix} 0.0999 & -0.1944 & 0.1025 \\ -0.1 & 0.2 & -0.1 \end{bmatrix} + \begin{bmatrix} 0.00196 & 0.00388 & -0.0003 \\ 0 & 0 & 0 \end{bmatrix} = \begin{bmatrix} 0.1019 & -0.1905 & 0.1022 \\ -0.1 & 0.2 & -0.1 \end{bmatrix} \tag{53}$$

Calculate the change in bias connected to each neuron in the output layer using Eq. (27) and update bias parameters using Eq. (29)

$$\Delta b_k = 0.1 * \left(\begin{bmatrix} 0.3 \\ 0.2 \\ 0.1 \end{bmatrix} - \begin{bmatrix} 0.1093 \\ -0.177 \\ 0.1295 \end{bmatrix}\right) = \begin{bmatrix} 0.019 \\ 0.038 \\ 0.0103 \end{bmatrix} \tag{54}$$

$$b_k = \begin{bmatrix} 0.099 \\ -0.157 \\ 0.119 \end{bmatrix} + \begin{bmatrix} 0.019 \\ 0.038 \\ 0.0103 \end{bmatrix} = \begin{bmatrix} 0.118 \\ -0.119 \\ 0.1293 \end{bmatrix} \tag{55}$$

Calculate the change in weights connected to each neuron in the latent space using Eq. (31). Where n is a neuron in output layer and j is neuron in latent space.

$$\Delta W_{ih}^{1} = 0.1 * \begin{bmatrix} 0.3 \\ 0.2 \\ 0.1 \end{bmatrix} * [(0.3 - 0.1093) * 0.1019 + (0.2 + 0.177) * -0.1905 + (0.1 - 0.1295) * 0.1022] \tag{56}$$

$$\Delta W_{ih}^{1} = \begin{bmatrix} -0.0017 \\ -0.0011 \\ -0.0006 \end{bmatrix} \tag{57}$$

$$\Delta W_{ih}^{2} = 0.1 * \begin{bmatrix} 0.3 \\ 0.2 \\ 0.1 \end{bmatrix} * [(0.3 - 0.1093) * -0.1 + (0.2 + 0.177) * 0.2 + (0.1 - 0.1295) * -0.1] \tag{58}$$

$$\Delta W_{ih}^{2} = \begin{bmatrix} 0.0018 \\ 0.0012 \\ 0.0006 \end{bmatrix} \tag{59}$$

Finally, change in weights between latent space and input layer is given by

$$\Delta W_{ih} = \left[\Delta W_{ih}^{1}\ \Delta W_{ih}^{2}\right] = \begin{bmatrix} -0.0017\ 0.0018 \\ -0.0011\ 0.0012 \\ -0.0006\ 0.0006 \end{bmatrix} \tag{60}$$

Now update the weights between latent space and input layer using Eq. (37).

$$W_{ih} = \begin{bmatrix} 0.0993 & 0.2007 \\ -0.2013 & 0.1014 \\ 0.198/ & 0.102 \end{bmatrix} + \begin{bmatrix} -0.0017\ 0.0018 \\ -0.0011\ 0.0012 \\ -0.0006\ 0.0006 \end{bmatrix} = \begin{bmatrix} 0.0976 & 0.2025 \\ -0.2024 & 0.1026 \\ 0.1974/ & 0.1026 \end{bmatrix} \tag{61}$$

Calculate the change in bias connected to each neuron "j" in the latent space using Eq. (39) and update bias parameters using Eq. (29).

$$\Delta b_{h}^{1} = 0.1 * [(0.3 - 0.1093) * 0.1019 + (0.2 + 0.177) * -0.1905 + (0.1 - 0.1295) * 0.1022] = -0.0055 \tag{62}$$

$$\Delta b_{h}^{2} = 0.1 * [(0.3 - 0.1093) * -0.1 + (0.2 + 0.177) * 0.2 + (0.1 - 0.1295) * -0.1] = 0.0059 \tag{63}$$

Finally, change in bias parameter for the neurons in latent space are given by

$$\Delta b_{h} = \left[\Delta b_{h}^{1}\ \Delta b_{h}^{2}\right] = \begin{bmatrix} -0.0055 \\ 0.0059 \end{bmatrix} \tag{64}$$

$$b_{h} = \begin{bmatrix} 0.0935 \\ -0.0932 \end{bmatrix} + \begin{bmatrix} -0.0055 \\ 0.0059 \end{bmatrix} = \begin{bmatrix} 0.088 \\ -0.0873 \end{bmatrix} \tag{65}$$

$$net_{h} = \begin{bmatrix} 0.0993 & -0.2013 & 0.198 \\ 0.2007 & 0.1014 & 0.102 \end{bmatrix} * \begin{bmatrix} 0.3 \\ 0.2 \\ 0.1 \end{bmatrix} + \begin{bmatrix} 0.0935 \\ -0.0932 \end{bmatrix} = \begin{bmatrix} 0.1028 \\ -0.0025 \end{bmatrix} \tag{66}$$

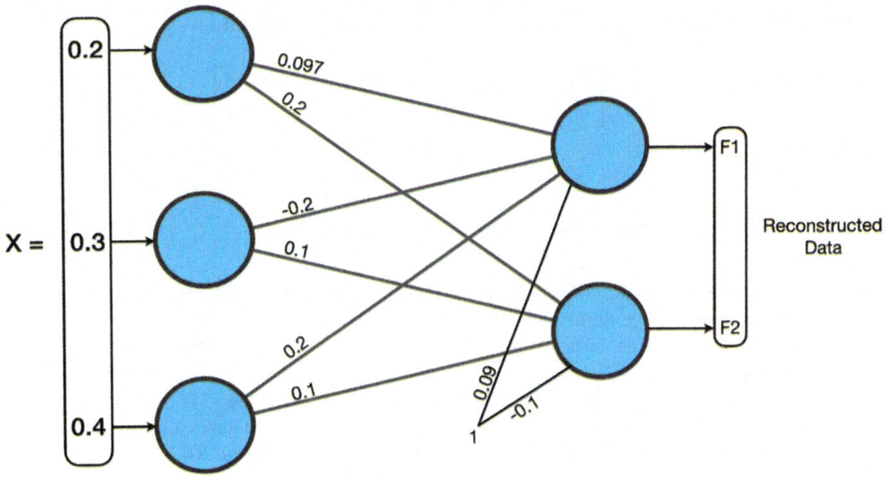

Fig. 6 Encoder part of sample autoencoder architecture

Calculate the output of hidden layerlatent space using Eq. (5) and the result for the first sample is shown below

$$O_h = \max\left(0, \begin{bmatrix} 0.1028 \\ -0.0025 \end{bmatrix}\right) = \begin{bmatrix} 0.1028 \\ 0 \end{bmatrix} \tag{67}$$

Reconstruction of Data Using Trained Autoencoder

Encoder part of trained autoencoder which is shown in Fig. 6 is used to reconstruct the data. Weight and bias matrices of encoder part are shown below.

$$W_{ih} = \begin{bmatrix} 0.0976 & 0.2025 \\ -0.2024 & 0.1026 \\ 0.1974/ & 0.1026 \end{bmatrix} \tag{68}$$

$$b_h = \begin{bmatrix} 0.088 \\ -0.0873 \end{bmatrix} \tag{69}$$

A sample data [0.2,0.3,0.4] is reconstructed into a data sample consists only two features as shown below

$$net_h = \begin{bmatrix} 0.0976 & -0.2024 & 0.1974 \\ 0.2025 & 0.1026 & 0.1026 \end{bmatrix} * \begin{bmatrix} 0.2 \\ 0.3 \\ 0.4 \end{bmatrix} + \begin{bmatrix} 0.088 \\ -0.0873 \end{bmatrix} = \begin{bmatrix} 0.13 \\ 0.01 \end{bmatrix} \tag{70}$$

The output of hidden layer/latent space is reconstructed data.

$$O_h = \begin{bmatrix} F_1 \\ F_2 \end{bmatrix} = \max\left(0, \begin{bmatrix} 0.13 \\ 0.01 \end{bmatrix}\right) = \begin{bmatrix} 0.13 \\ 0.01 \end{bmatrix} \quad (71)$$

References

1. Alimoussa M, Porebski A, Vandenbroucke N, El Fkihi S, Oulad Haj Thami R (2022) Compact hybrid multi-color space descriptor using clustering-based feature selection for texture classification. J Imaging 8(8). https://doi.org/10.3390/jimaging8080217. https://www.mdpi.com/2313-433X/8/8/217
2. Alomari ES, Nuiaa RR, Alyasseri ZAA, Mohammed HJ, Sani NS, Esa MI, Musawi BA (2023) Malware detection using deep learning and correlation-based feature selection. Symmetry 15(1). https://doi.org/10.3390/sym15010123. https://www.mdpi.com/2073-8994/15/1/123
3. Bacanin N, Stoean C, Zivkovic M, Rakic M, Strulak-Wójcikiewicz R, Stoean R (2023) On the benefits of using metaheuristics in the hyperparameter tuning of deep learning models for energy load forecasting. Energies 16(3). https://doi.org/10.3390/en16031434. https://www.mdpi.com/1996-1073/16/3/1434
4. Buatoom U, Jamil MU (2023) Improving classification performance with statistically weighted dimensions and dimensionality reduction. Appl Sci 13(3). https://doi.org/10.3390/app13032005. https://www.mdpi.com/2076-3417/13/3/2005
5. de Castro-Cros M, Velasco M, Angulo C (2023) Analysis of gas turbine compressor performance after a major maintenance operation using an autoencoder architecture. Sensors 23(3). https://doi.org/10.3390/s23031236. https://www.mdpi.com/1424-8220/23/3/1236
6. Chikkankod AV, Longo L (2022) On the dimensionality and utility of convolutional autoencoder& rsquo;s latent space trained with topology-preserving spectral EEG head-maps. Mach Learn Knowl Extract 4(4):1042–1064
7. Dessureault JS, Massicotte D (2022) Dpdrc, a novel machine learning method about the decision process for dimensionality reduction before clustering. AI 3(1):1–21. https://doi.org/10.3390/ai3010001. https://www.mdpi.com/2673-2688/3/1/1
8. Do JS, Kareem AB, Hur JW (2023) Lstm-autoencoder for vibration anomaly detection in vertical carousel storage and retrieval system (vcsrs). Sensors 23(2). https://doi.org/10.3390/s23021009. https://www.mdpi.com/1424-8220/23/2/1009
9. Fan L (2020) Dimensionality reduction of image feature based on geometric parameter adaptive lle algorithm. J Intell Fuzzy Syst 38(2):1569–1577. https://doi.org/10.3233/JIFS-179520
10. Ferner C, Wegenkittl S (2022) Benefits from variational regularization in language models. Mach Learn Knowl Extract 4(2):542–555
11. Fiorini S, Ciavotta M, Maurino A (2022) Listening to the city, attentively: A spatio-temporal attention-boosted autoencoder for the short-term flow prediction problem. Algorithms 15(10). https://doi.org/10.3390/a15100376. https://www.mdpi.com/1999-4893/15/10/376
12. Gisbrecht A, Hammer B (2015) Data visualization by nonlinear dimensionality reduction. Wiley Interdisc Rev Data Mining Knowl Discov 5(2):51–73. https://doi.org/10.1002/widm.1147
13. Grimaccia F, Mussetta M, Niccolai A, Veeramsetty V, Chandra DR (2022) Short-term load forecasting in dso substation networks with dimensionality reduction techniques. In: 2022 IEEE international conference on environment and electrical engineering and 2022 IEEE industrial and commercial power systems Europe (EEEIC/I&CPS Europe). IEEE, pp 1–6. https://doi.org/10.1109/EEEIC/ICPSEurope54979.2022.9854749
14. Huang X, Wu L, Ye Y (2019) A review on dimensionality reduction techniques. Int J Pattern Recogn Artif Intell 33(10):1950017. https://doi.org/10.1142/S0218001419500174

15. Ibrahim S, Nazir S, Velastin SA (2021) Feature selection using correlation analysis and principal component analysis for accurate breast cancer diagnosis. J Imaging 7(11). https://doi.org/10.3390/jimaging7110225. https://www.mdpi.com/2313-433X/7/11/225
16. Katsamenis I, Bakalos N, Karolou EE, Doulamis A, Doulamis N (2022) Fall detection using multi-property spatiotemporal autoencoders in maritime environments. Technologies 10(2). https://doi.org/10.3390/technologies10020047. https://www.mdpi.com/2227-7080/10/2/47
17. Kiarashinejad Y, Abdollahramezani S, Adibi A (2020) Deep learning approach based on dimensionality reduction for designing electromagnetic nanostructures. NPJ Comput Mater 6(1):1–12. https://doi.org/10.1038/s41524-020-0276-y
18. Kumar V, Minz S (2014) Feature selection: a literature review. SmartCR 4(3):211–229. https://doi.org/10.6029/smartcr.2014.03.007
19. Lee J, Ryu S, Chung W, Kim S, Kang YJ (2023) Estimates of internal forces in torsionally braced steel i-girder bridges using deep neural networks. Appl Sci 13(3). https://doi.org/10.3390/app13031499. https://www.mdpi.com/2076-3417/13/3/1499
20. Li D, Tang Z, Kang Q, Zhang X, Li Y (2023) Machine learning-based method for predicting compressive strength of concrete. Processes 11(2). https://doi.org/10.3390/pr11020390. https://www.mdpi.com/2227-9717/11/2/390
21. Li J, Zhang J, Bah MJ, Wang J, Zhu Y, Yang G, Li L, Zhang K (2022) An auto-encoder with genetic algorithm for high dimensional data: towards accurate and interpretable outlier detection. Algorithms 15(11). https://doi.org/10.3390/a15110429. https://www.mdpi.com/1999-4893/15/11/429
22. Li X, Zhang L, You J (2019) Locally weighted discriminant analysis for hyperspectral image classification. Remote Sens 11(2):109. https://doi.org/10.3390/rs11020109
23. Li Y, Yan Y (2023) Training autoencoders using relative entropy constraints. Appl Sci 13(1). https://doi.org/10.3390/app13010287. https://www.mdpi.com/2076-3417/13/1/287
24. Liang H, Sun X, Sun Y, Gao Y (2017) Text feature extraction based on deep learning: a review. EURASIP J Wireless Commun Netw 2017(1):1–12. https://doi.org/10.1186/s13638-017-0993-1
25. Ma H, Yang P, Wang F, Wang X, Yang D, Feng B (2023) Short-term heavy overload forecasting of public transformers based on combined lstm-xgboost model. Energies 16(3). https://doi.org/10.3390/en16031507. https://www.mdpi.com/1996-1073/16/3/1507
26. McClelland JL, Rumelhart DE, Group PR et al (1987) Parallel distributed processing: explorations in the microstructure of cognition: psychological and biological models, vol 2. MIT press. https://doi.org/10.7551/mitpress/5236.001.0001
27. Nayak SR, Nayak DR, Sinha U, Arora V, Pachori RB (2023) An efficient deep learning method for detection of covid-19 infection using chest x-ray images. Diagnostics 13(1). https://doi.org/10.3390/diagnostics13010131. https://www.mdpi.com/2075-4418/13/1/131
28. Oppel H, Munz M (2021) Analysis of feature dimension reduction techniques applied on the prediction of impact force in sports climbing based on IMU data. AI 2(4):662–683. https://doi.org/10.3390/ai2040040. https://www.mdpi.com/2673-2688/2/4/40
29. Peralta B, Soria R, Nicolis O, Ruggeri F, Caro L, Bronfman A (2023) Outlier vehicle trajectory detection using deep autoencoders in Santiago, Chile. Sensors **23**(3). https://doi.org/10.3390/s23031440. https://www.mdpi.com/1424-8220/23/3/1440
30. Podder P, Das SR, Mondal MRH, Bharati S, Maliha A, Hasan MJ, Piltan F (2023) Lddnet: a deep learning framework for the diagnosis of infectious lung diseases. Sensors 23(1). https://doi.org/10.3390/s23010480. https://www.mdpi.com/1424-8220/23/1/480
31. Reddy GT, Reddy MPK, Lakshmanna K, Kaluri R, Rajput DS, Srivastava G, Baker T (2020) Analysis of dimensionality reduction techniques on big data. IEEE Access 8:54776–54788. https://doi.org/10.1109/ACCESS.2020.2980942
32. Sadek AH, Fahmy OM, Nasr M, Mostafa MK (2023) Predicting Cu(ii) adsorption from aqueous solutions onto nano zero-valent aluminum (nzval) by machine learning and artificial intelligence techniques. Sustainability 15(3). https://doi.org/10.3390/su15032081. https://www.mdpi.com/2071-1050/15/3/2081

33. Sahoo AK, Pradhan C, Barik RK, Dubey H (2019) Deepreco: Deep learning based health recommender system using collaborative filtering. Computation 7(2). https://doi.org/10.3390/computation7020025. https://www.mdpi.com/2079-3197/7/2/25
34. Salo F, Nassif AB, Essex A (2019) Dimensionality reduction with ig-pca and ensemble classifier for network intrusion detection. Comput Netw 148:164–175. https://doi.org/10.1016/j.comnet.2018.11.010
35. Tang J, Alelyani S, Liu H (2014) Feature selection for classification: a review. Data Classif Algorithms Appl 37. https://doi.org/10.1201/b17320
36. Vardhan BVS, Khedkar M, Srivastava I, Thakre P, Bokde ND (1996) A comparative analysis of hyperparameter tuned stochastic short term load forecasting for power system operator. Energies 16(3). https://doi.org/10.3390/en16031243. https://www.mdpi.com/1996-1073/16/3/1243
37. Veeramsetty V (2022) Electric power load dataset. https://data.mendeley.com/datasets/tj54nv46hj/1
38. Veeramsetty V, Chandra DR, Grimaccia F, Mussetta M (2022) Short term electric power load forecasting using principal component analysis and recurrent neural networks. Forecasting 4(1):149–164. https://doi.org/10.3390/forecast4010008
39. Veeramsetty V, Chandra DR, Salkuti SR (2021) Short-term electric power load forecasting using factor analysis and long short-term memory for smart cities. Int J Circ Theory Appl 49(6):1678–1703. https://doi.org/10.1002/cta.2928
40. Veeramsetty V, Deshmukh R (2020) Electric power load forecasting on a 33/11 kv substation using artificial neural networks. SN Appl Sci 2(5):855. https://doi.org/10.1007/s42452-020-2601-y
41. Veeramsetty V, Mohnot A, Singal G, Salkuti SR (2021) Short term active power load prediction on a 33/11 kv substation using regression models. Energies 14(11):2981. https://doi.org/10.3390/en14112981
42. Veeramsetty V, Rakesh Chandra D, Salkuti SR (2022) Short term active power load forecasting using machine learning with feature selection. In: Next generation smart grids: modeling, control and optimization. Springer, pp 103–124. https://doi.org/10.1007/978-981-16-7794-6_5
43. Veeramsetty V, Reddy KR, Santhosh M, Mohnot A, Singal G (2022) Short-term electric power load forecasting using random forest and gated recurrent unit. Electr Eng 104(1):307–329. https://doi.org/10.1007/s00202-021-01376-5
44. Veeramsetty V, Sai Pavan Kumar M, Salkuti SR (2022) Platform-independent web application for short-term electric power load forecasting on 33/11 kv substation using regression tree. Computers 11(8):119. https://doi.org/10.3390/computers11080119
45. Viale L, Daga AP, Fasana A, Garibaldi L (2023) Dimensionality reduction methods of a clustered dataset for the diagnosis of a scada-equipped complex machine. Machines 11(1). https://doi.org/10.3390/machines11010036. https://www.mdpi.com/2075-1702/11/1/36
46. Xing H, Chen B, Feng Y, Ni Y, Hou D, Wang X, Kong Y (2022) Mapping irrigated, rainfed and paddy croplands from time-series sentinel-2 images by integrating pixel-based classification and image segmentation on google earth engine. Geocarto Int 1–20. https://doi.org/10.1080/10106049.2022.2076923
47. You K, Qiu G, Gu Y (2022) Rolling bearing fault diagnosis using hybrid neural network with principal component analysis. Sensors 22(22). https://doi.org/10.3390/s22228906. https://www.mdpi.com/1424-8220/22/22/8906
48. Zhang Z, Huang R, Han F, Wang Z (2019) Image error concealment based on deep neural network. Algorithms 12(4). https://doi.org/10.3390/a12040082. https://www.mdpi.com/1999-4893/12/4/82

Design of Misalignment-Tolerant Orthogonal Wireless Power Transfer Coils for Unmanned Aerial Vehicles

Y. Satyavani, Phaneendra Babu Bobba, and V. Sandeep

Abstract This chapter presents a Single Transmitter and Multi-Receiver (STMR)-based Wireless Power Transfer (WPT) technique for charging the battery of Unmanned Aerial Vehicles (UAVs). STMR-based WPT systems are having the flexibility to power multiple loads at a time but have lower system efficiency in long-distance power transfer due to cross coupling and misalignment of the coils. To overcome the limitations of STMR systems, a new design is proposed which has a transmitter that consists of two orthogonal circular coils which increase the efficiency of the WPT system by increasing flux density over a long distance. The proposed multiple receivers are placed in such a way that UAV receives the power optimally. The concepts referred for the design, simulation, and equivalent circuit model for the WPT system have been analyzed. The transfer efficiency for different cases of misalignment has been obtained with an operating frequency of 85 kHz. The Inductor–Capacitor–Capacitor (LCC)-compensation circuit is being used at the transmitter side while the series capacitor is connected at the receiver side. The performance comparison of Orthogonal Two Circular Transmitter Coils Multi-receivers (OTCTC-MR)- and STMRbased WPT system have been presented with possible misalignment of the transmitter and receiver coils for the output power of 360 W and current 7.5 A at receiver coils to charge the battery of 48 V. It has been shown that for the same power output, the number of coils at the received side is reduced by 69 turns in OTCTC-MR-based WPT system compared to STMRbased WPT system which is having 156 turns. Hence, the weight of the UAV, losses are reduced, and the payload can be increased with OTCTC-MR-based WPT system for transferring power to charge the battery of the UAV.

Y. Satyavani · V. Sandeep
Department of Electrical and Electronics Engineering, National Institute of Technology, Tadepalligudem, Andhra Pradesh 534101, India
e-mail: sandeep@nitandhra.ac.in

P. B. Bobba (✉)
Department of Electrical and Electronics Engineering, GRIET, Hyderabad, Telangana 500095, India
e-mail: bobbaphani@griet.ac.in

S. R. Salkuti et al. (eds.), *Power Quality in Microgrids: Issues, Challenges and Mitigation Techniques*, Lecture Notes in Electrical Engineering 1039,
https://doi.org/10.1007/978-981-99-2066-2_23

Keywords LCC compensation · Misalignment · Multi-receivers · Single transmitter and multi-receiver · Unmanned aerial vehicles · Wireless power transfer

Nomenclature

STMR	Single transmitter and multi-receiver
WPT	Wireless power transfer
UAV	Unmanned aerial vehicles
OTCTC-MR	Orthogonal two circular transmitter coils multi-receivers
LCC	Inductor–capacitor–capacitor

1 Introduction

Unmanned Aerial Vehicles (UAVs) are evolving not only for being used in the military but also being used in surveillance, delivery, health care, agriculture, journalism, and many more [1, 2]. They have the advantage of low maintenance, stealth capabilities, and ease to use. They can be used as charging stations which can be used to power other electronic devices and to charge the battery [3] of other UAVs by transmitting power to them. Depending upon the range, operating height, altitude, weight, and size, UAVs are classified into different types [4]. The major drawbacks due to which the UAVs are used for limited applications are low battery life and short-range operation. The commercially available UAVs are powered by a lithium battery with a flight time of 20–40 min [5, 6]. The flight time can be increased by using the high-rating battery, but this decreases the payload or by replacing the battery automatically at base stations immediately after landing [7] which is impractical due to its complexity and cost. The alternative to increasing the flight time is to create base stations (charging stations) for charging the battery [8]. The charging can be done in two ways, one way is by charging the battery through an electrical connection between UAV and the charging station and the second way is to charge the battery wirelessly. The former way has higher transfer efficiency but due to the contacts being exposed to the atmosphere reduces its efficiency. The later one which is charging wirelessly can be one by using Wireless Power Transfer (WPT). WPT can be done with magnetic coupling and resonance coupling. The WPT has two coils, a transmitter coil and a receiver coil. Both coils are magnetically coupled with a coupling coefficient (k). Power transfers from the transmitter coil to the receiver coil based on the magnetic resonance power transfer. The transmitter coil and receiver coil and all other components of the WPT circuit are set to operate at same resonant frequency. The transmitter coil is placed on the charging station and the receiver coil is placed in UAV. The geometrical dimension of the coil affects the power transfer efficiency and mutual inductance [9]. The coils are designed such that they should

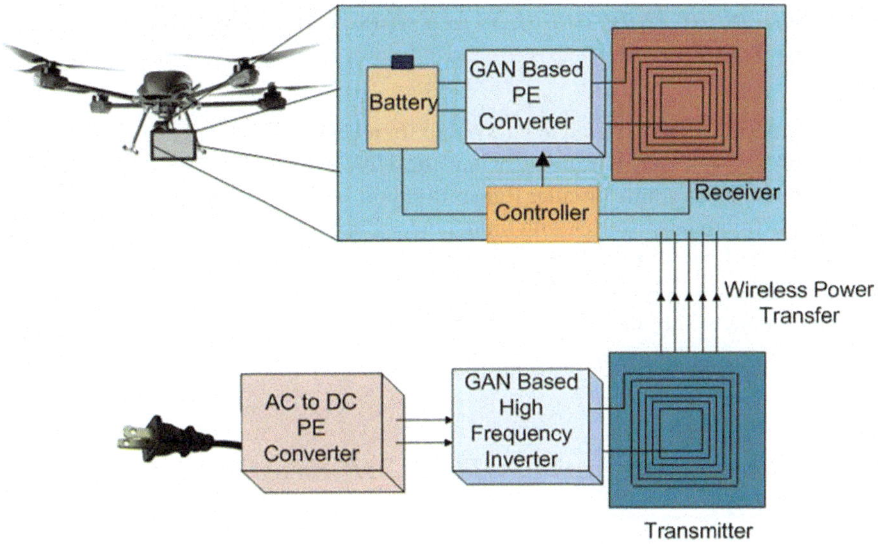

Fig. 1 WPT system for UAV charging

overcome the misalignment effect [10] and not interfere with UAV equipment like sensors, and cameras. The schematic for the WPT system for UAV charging is shown in Fig. 1.

The Single Transmitter and Multi-receiver (STMR)-based [11] WPT has one circular [12] transmitter coil which is placed on the charging station and three planar receiver coils [13] connected to UAV. When the transmitter coil is excited with a high-frequency power supply it produces flux due to which an emf is induced in the transmitter coil. As the transmitter is short-circuited the current flows through the resonator coil and produces flux.

The flux links with the receiver resonator coil. For improving efficiency and misalignment [14, 15] tolerant, a new STMRbased WPT system with two orthogonal [16] circular coils as a transmitter has been designed. In this Orthogonal Two Circular Transmitter Coils Multi-receivers design (OTCTC-MR), improved efficiency has been observed compared to the other STMR considering different cases of misalignment.

2 Design of STMR-Based WPT System

In this section, the schematic diagram of STMR and OTCTC-MR, expressions for WPT system parameters has been presented.

2.1 Schematic Diagram of STMR-Based WPT System for UAV

The schematic of STMRbased WPT system for a UAV is shown in Fig. 2. The battery of the UAV is charged wirelessly when the UAV lands on the charging pad. The charging pad has a single circular transmitter coil with few turns which is connected to the high-frequency inverter with Inductance–Capacitor–Capacitor (LCC) compensation [17, 18] circuit to reduce the reactive power and to maintain the unity power factor. The receiver has three planar circular coils fitted to the UAV. The receiver coils are connected to the rectifier circuit and the output of the rectifier is connected to the battery for charging.

2.2 Schematic Diagram of OTCTC-MR-Based WPT System for UAV

The schematic diagram for the transmitter has two orthogonal circular transmitter coils which are placed on the charging station and three receivers out of which two receivers are connected to the legs of the UAV and the third one is connected to UAV.

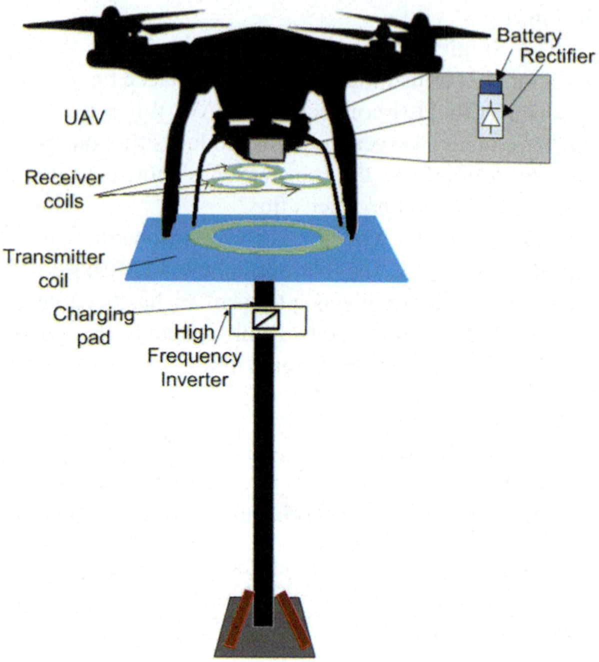

Fig. 2 Schematic diagram of STMR

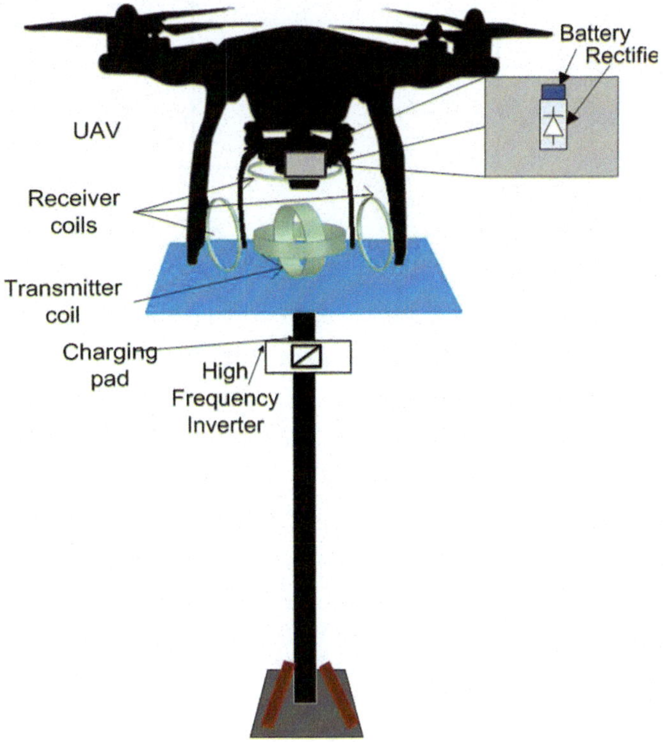

Fig. 3 Schematic diagram of OTCTC-MR

The schematic of two orthogonal circular transmitter coils and multi-receiver coil WPT system is shown in Fig. 3.

2.3 *Expressions for WPT System Circuit Parameters*

Self-inductance of circular coils is given by

$$L = \mu_0 * N^2 * R * \left(In\frac{16 * R}{d} - \frac{7}{8} \right). \tag{1}$$

Considering the two coils, one is as a transmitter coil with a center as O, the radius of R_p and another is a receiver coil with a center as C, a radius of R_S, which is given in Fig. 4a. Figure 4a shows angle misalignment of transmitter and receiver coils ($\theta = 0$, $\psi = 0$), Fig. 4b shows lateral angle misalignment ($\psi = 0$), and Fig. 4c shows arbitrary lateral and angular misalignment.

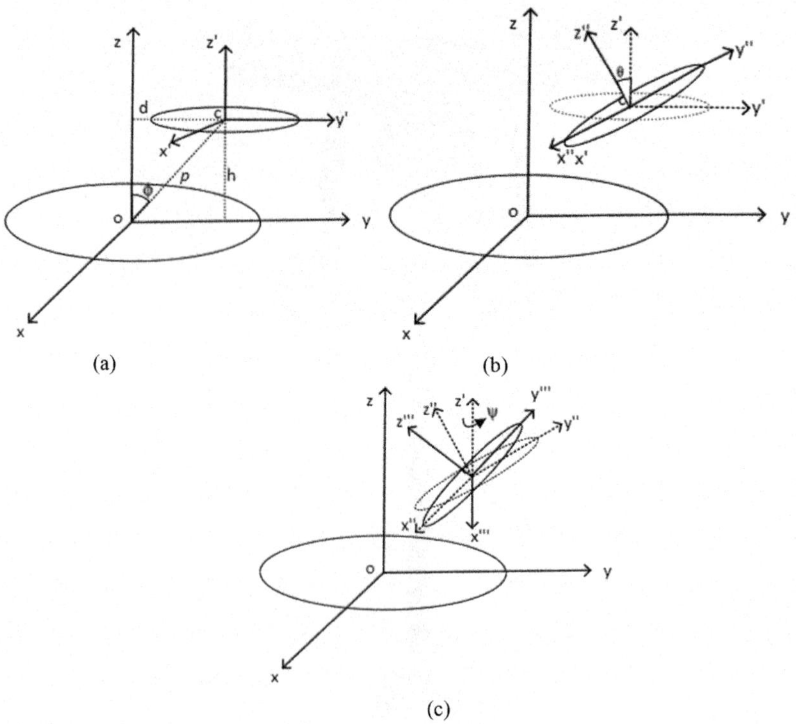

Fig. 4 **a** Angle misalignment of transmitter and receiver coils ($\theta = 0$, $\psi = 0$); **b** lateral angle misalignment ($\psi = 0$); **c** arbitrary lateral and angular misalignment

The receiver coil is in plane $\lambda = ax + by + cz + d = 0$, the center of the receiver coil C = (x_c, y_c, z_c), and the arbitrary point on the secondary coil is denoted as D (x_0, y_0, z_0). The arbitrary points [19] on the receiver coil are expressed as

$$D_{1,2}\left(x_C \mp R_s\frac{ab}{lL}, y_c \pm R_s\frac{l}{L}, z_C \mp R_s\frac{bc}{lL}\right), \tag{2}$$

where $L = \sqrt{(a^2 + b^2 + c^2)}$ and $l = \sqrt{(a^2 + c^2)}$.

The arbitrary point on the transmitter coil is expressed as

$$x_p = R_p\cos(t), \quad y_p = R_p\sin(t), \quad z_p = 0 \tag{3}$$

The parametric coordinates of an arbitrary point $E_s(x_s, y_s, z_s)$ on the secondary coil are given by

$$x_s = x_c + R_s u_x \cos\varphi + R_s v_x \sin\varphi, \tag{4}$$

$$y_s = y_c + R_s u_y \cos\varphi + R_s v_y \sin\varphi, \tag{5}$$

$$z_s = z_c + R_s u_z \cos\varphi + R_s v_z \sin\varphi, \tag{6}$$

where $\varphi\epsilon(0, 2\pi)$. The mutual inductance of circular coils is given by

$$M = \frac{\mu_0 R_S}{\pi}\int_0^{2\pi} \frac{[P_1 cos\varphi + P_2 sin\varphi + P_3]\psi(k)}{k\sqrt{V_0^3}} d\varphi, \tag{7}$$

where

$$\alpha = \frac{R_s}{R_P},\quad \beta = \frac{x_C}{R_p},\quad \gamma = \frac{x_C}{R_P},\quad \delta = \frac{z_c}{R_p} \tag{8}$$

$$p_1 = \pm\frac{\gamma c}{l},\quad P_2 = \mp\frac{\beta L^2 + \gamma ab}{lL},\quad P_3 = \frac{\alpha C}{L},\quad P_4 = \mp\frac{\beta ab + \gamma l^2 + \delta bc}{lL},$$
$$P_5 = \mp\frac{\beta c - \delta a}{l} \tag{9}$$

$$A_0 = 1 + \alpha^2 + \beta^2 + \gamma^2 + \delta^2 + 2\alpha(P_4 \cos\varphi + P_5 \sin\varphi) \tag{10}$$

$$V_0^2 = \alpha^2\left[\left(1 - \frac{b^2c^2}{l^2L^2}\right)\cos^2\varphi + \frac{c^2}{l^2}\sin^2\varphi + \frac{abc}{l^2L}\sin 2\varphi\right] + \beta^2 + \gamma^2 \mp 2\alpha\frac{\beta ab - \gamma l^2}{lL}\cos\varphi \mp \frac{2\alpha\beta c}{l}\sin\varphi \tag{11}$$

$$k = \sqrt{\frac{4V_0}{A_0 + 2V_0}},\quad \psi(k) = \left(1 - \frac{k^2}{2}\right)K(k) - E(k). \tag{12}$$

To maintain the circuit at a resonance state, the following three equations must be satisfied:

$$j\omega L_f - \frac{1}{j\omega C_f} = 0 \tag{13}$$

$$j\omega L_f - j\omega L_1 - \frac{1}{j\omega C_f} = 0 \tag{14}$$

$$j\omega L_2 - \frac{1}{j\omega C_2} = 0. \tag{15}$$

Applying KVL to the circuit shown in Fig. 5,

$$V_1 = \left(R_{Lf} + j\omega L_f\right) I_{Lf} + \frac{I_{Cf}}{j\omega C_f} \tag{16}$$

$$V_1 = \left(R_{Lf} + j\omega L_f\right) I_{Lf} + \left(j\omega L_1 + j\omega L_1' + \frac{1}{j\omega C_1} + R_1 + R_1'\right) I_1 + j\omega M_{12} I_2 + j\omega M_{1'3} I_3 + j\omega M_{1'4} I_4, \tag{17}$$

where V_1 is the input voltage to the LCC compensation at the transmitter side. R_{Lf}, L_f are the resistance to the inductance of a compensation inductor at the transmitter side. L_1 and $L_1^{'}$ are inductances of a horizontal transmitter coil and vertical transmitter coil respectively. M_{12} is the mutual inductance between the horizontal transmitter coil and horizontal receiver coil; $M_{1'3}$, $M_{1'4}$ are mutual inductances between the vertical transmitter coil and the two vertical receiver coils.

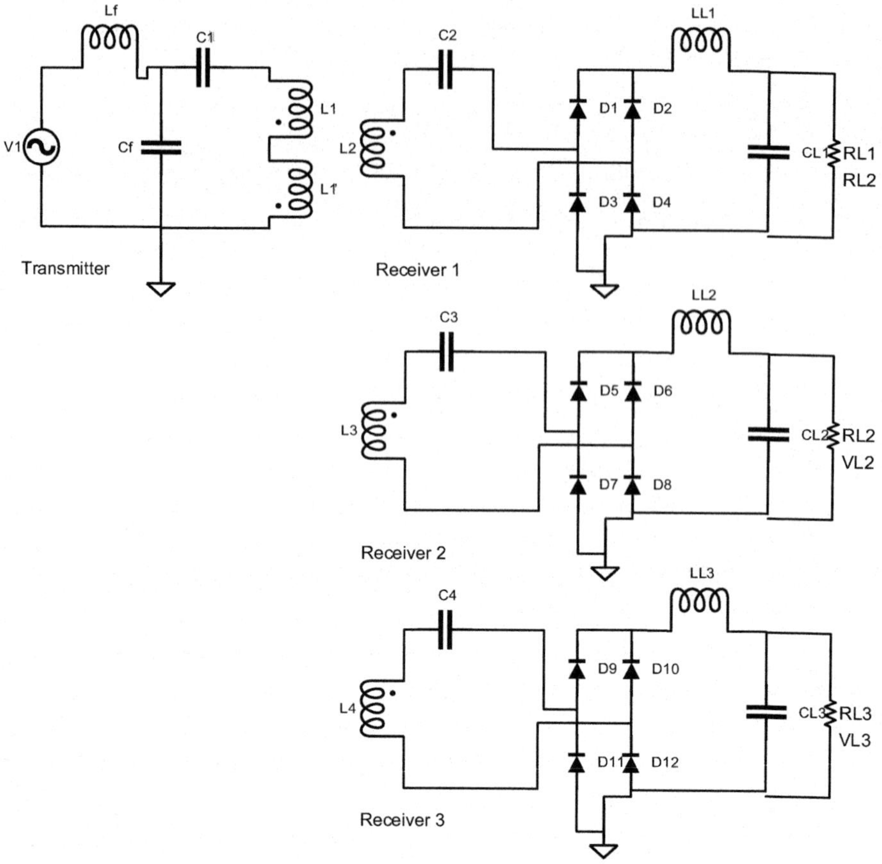

Fig. 5 WPT system circuit diagram

Assumptions

The mutual inductance between the transmitter coil 1 (L_1) and the receiver coil 1 is equal to mutual inductance between the transmitter coil 2 ($L_1^{'}$) and receiver coils 2 and 3 as M. Self-inductance, current, and voltage of all the receiver coils are equal.

$$V_1 = \left(R_{Lf} + j\omega L_f\right)I_{Lf} + \left(j\omega L_1 + j\omega L_1^{'} + \frac{1}{j\omega C_1} + R_1 + R_1^{'}\right)I_1 + 3j\omega M I_2 \tag{18}$$

$$V_L = (j\omega L_2 + R_2)I_2 + \frac{1}{j\omega C_2} + j\omega M I_1 \tag{19}$$

$$I_{Lf} = I_{Cf} + I_1, \tag{20}$$

$$I_L = I_2 = I_{C_2}. \tag{21}$$

When AC power supply is applied to the circuit at resonance frequency than the input voltage V_1 will appear across the Capacitor C_f and the currents $I_1 = I_{Cf}$:

$$V_1 = \frac{I_1}{j\omega C_f}. \tag{22}$$

From Eqs. (14) and (20), we can conclude that

$$I_1 = \frac{V_1}{j\omega_0 L_f}. \tag{23}$$

3 Design of STMR and OTCTC-MR WPT Coils

The design of the WPT system was done in Ansys. An excitation of 9 A is given to the transmitter coil. The WPT system was operated at 85 kHz. The battery voltage is taken as 48 V.

3.1 Design of STMR-Based WPT System

The representation of transmitter and receiver coils for STMRbased WPT system is shown in Fig. 6. It has three planar receiver coils connected in series placed in the UAV. The transmitter coil and the receiver coils are placed 10 cm apart. The transmitter coil has 20 turns, and the receiver has 52 turns, and the physical parameters of the coils are shown in Table 1.

The coils are designed in Ansys 3D design and are simulated to obtain the flux density plot for zero misalignments shown in Fig. 7. The flux flows from the transmitter coil to the receiver coils. The flux linkages are shown in different colors and the flux density value for each color is indicated at the top left of the flux density plot. The flux leakage is more if the misalignment is more and the flux linking the receiver coils is also less. The traditional design for WPT for UAV application was given a schematic diagram in Fig. 6, 9 A of current is given to the transmitter coil and 2.5 A of current is excited to receiver coils. The system is operated at 85 KHz of frequency and obtained 120 W of power from each receiver. A total power of 360 W of power is obtained. The parameters of the coils are given in detailed information in Table 2 by displacing the receiver coils vertically. For 10 cm of distance between the transmitter and receiver coils, the coefficient of coupling between transmitter and receiver coil 1 (KTR1) is 0.0549, between transmitter and receiver coil 2 (KTR2) is 0.0556, and between transmitter and receiver coil 3 (KTR3) is 0.0501. The system is simulated by considering the displacement of the receiver coils center to the center of the transmitter coil in (mm). The self- and mutual inductances are obtained from the

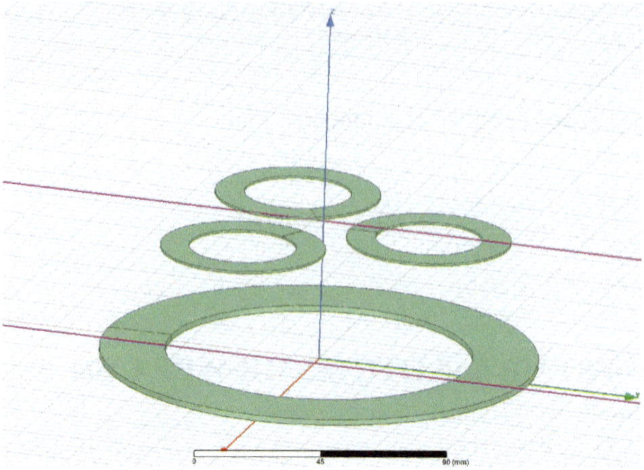

Fig. 6 Representation of transmitter and three planar receiver coils

Table 1 Transmitter and receiver coil parameters

	Number of turns	Inner diameter (mm)	Outer diameter (mm)	Thickness of coil (mm)
Transmitter coil	20	55.943	80	2.187
Receiver coil 1	52	19.47	30	0.81
Receiver coil 2	52	19.47	30	0.81
Receiver coil 3	52	19.47	30	0.81

simulation and the values of inductances and capacitances of compensation circuit are calculated using (15) and (16) and are tabulated in Table 3.

The values of the coupling [20] coefficient effect between the transmitter and the receiver coils are obtained from the simulation of coils in Ansys for each case of gap or displacement between transmitter and receiver coils. For each case, the values of the coefficient of coupling between the transmitter and the receiver coils between the receiver coils are given in Table 4. The coefficient of coupling is high in case-1, i.e., for the less displacement between transmitter and receiver coils.

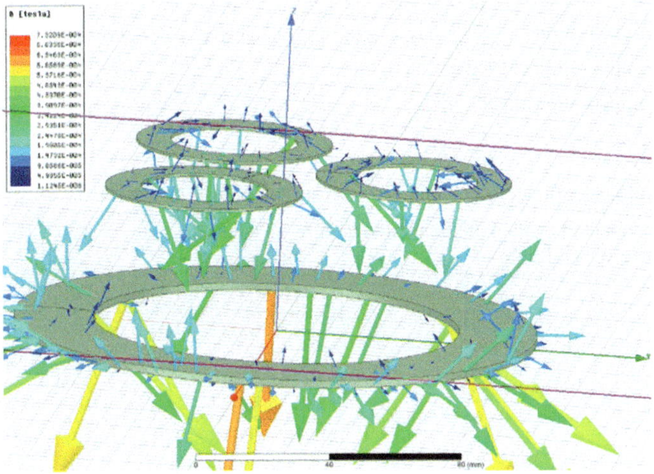

Fig. 7 Flux density plot of WPT coils obtained from FEM simulation

Table 2 Coupling coefficient of the transmitter coil and receiver coil for angle misalignment between them

Case	Displacement of receiver coils center to the center of transmitter coil (mm)	K_{TR1}	K_{TR2}	K_{TR3}
1	0	0.05396	0.053933	0.0538
2	10	0.0549	0.0556	0.0501
3	30	0.05706	0.058016	0.03437
4	50	0.048606	0.048459	0.006835
5	−10	0.04888	0.0489	0.06104
6	−20	0.033285	0.033211	0.0684
7	−50	0.0156	0.015672	0.068395

Table 3 WPT system parameters for LTspice simulation

Parameters	Values
Self-inductance of transmitter coilL_T	24.85 μH
Self-inductance of receiver coil 1 L_{R1}	195.91 μH
Self-inductance of receiver coil 2 L_{R2}	195.91 μH
Self-inductance of receiver coil 3 L_{R3}	195.91 μH
Compensation circuit inductance at transmitter side L_{f1}	5.1 μH
Compensation circuit shunt capacitance at transmitter side C_{f1}	0.688 μF
Compensation circuit series capacitor at transmitter side C_1	0.178 μF
Receiver 1 compensation capacitor C_{R1}	1.79×10^{-8}
Receiver 2 compensation capacitor C_{R2}	1.79×10^{-8}
Receiver 3 compensation capacitor C_{R3}	1.79×10^{-8}

Table 4 Coupling coefficient effect between transmitter and receiver coils for displacement

Case	Displacement between the transmitter and receiver coil (mm)	K_{TR1}	K_{TR2}	K_{TR3}	K_{R1R2}	K_{R1R3}	K_{R2R3}
1	60	0.042	0.042	0.042	0.019	0.019	0.0197
2	70	0.033	0.033	0.033	0.019	0.019	0.0196
3	80	0.027	0.027	0.027	0.019	0.019	0.0195
4	90	0.022	0.022	0.022	0.019	0.019	0.0194
5	100	0.018	0.018	0.018	0.019	0.019	0.0195

3.2 Design of OTCTC-MR-Based WPT System

To increase the power efficiency of the WPT system and to transfer the power to the receivers placed on the UAV, a new design is proposed which is shown in Fig. 8. Two of the three receiver coils receiver coils 1 and 3 are placed vertically, attached to the legs of the UAV and a horizontal receiver coil 3 is fitted to the UAV. Two circular vertical and horizontal transmitter coils placed orthogonally and connected in series are placed on the charging station. Power is transferred to all receivers through the two orthogonal circular transmitter coils.

The OTCTC-MR coils are designed in Ansys with an excitation of 9 A of current given to both vertical and horizontal transmitter coils. 2.5 A was obtained at each receiver with a total of 7.5 A for three receivers. The system is operated at 85 kHz of frequency and is simulated in Ansys. The flux density plot is as shown in Fig. 9. It shows the flux linkages from the transmitter coil to the receiver coils.

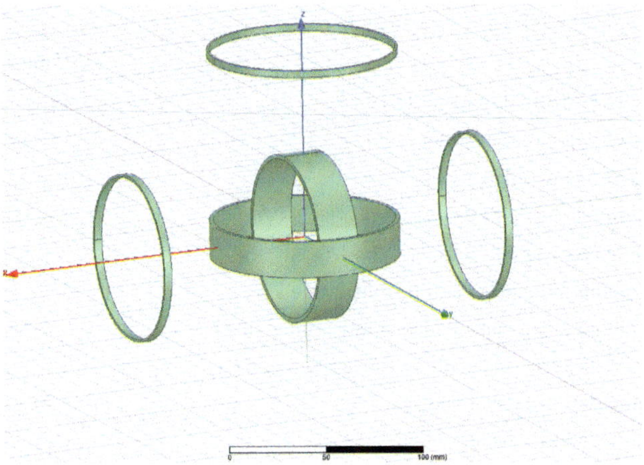

Fig. 8 Representation of OTCTC-MR coils in FEM simulation

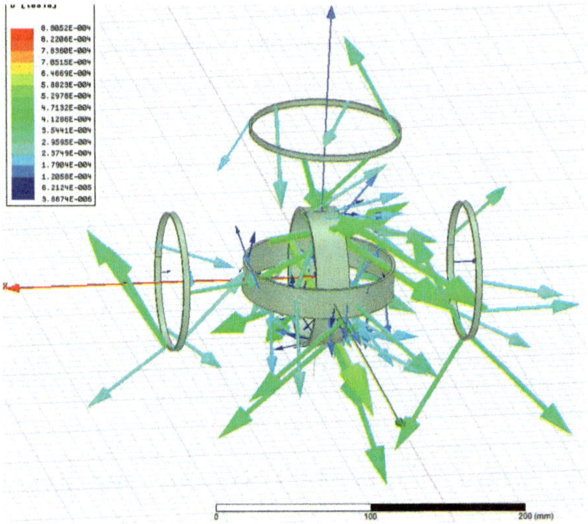

Fig. 9 Flux density plot of OTCTC-MR coils obtained from FEM simulation

The self- and mutual inductances are obtained from the simulation and the values of inductances and capacitances of the compensation circuit are calculated using (15) and (16) and the WPT coil parameters from the Ansys simulation are given in Table 5. The coil's physical parameters are given in Table 6. The vertical transmitter has 10 turns and the horizontal transmitter has 10 turns. Receivers 1 and 2 have 30 turns and receiver 3 has 27 turns. The inner and outer diameters of coils and the thickness of the coils are also mentioned.

Table 5 OTCTC-MR WPT system parameters

Parameters	Values
Self-inductance of horizontal transmitter coil L_T	11.245 μH
Self-inductance of vertical transmitter coil $L_T^{'}$	11.569 μH
Self-inductance of receiver coil 1 L_{R1}	189.24 μH
Self-inductance of receiver coil 2 L_{R2}	199.341 μH
Self-inductance of receiver coil 3 L_{R3}	203.86 μH
Compensation circuit inductance at transmitter side L_{f1}	5.7 μH
Compensation circuit shunt capacitance at transmitter side C_{f1}	0.6156 μF
Compensation circuit series capacitor at transmitter side C_1	0.183 μF
Receiver 1 compensation capacitor C_{R1}	1.92×10^{-8}
Receiver 2 compensation capacitor C_{R2}	1.76×10^{-8}
Receiver 3 compensation capacitor C_{R3}	1.72×10^{-8}

Table 6 Transmitter and receiver coil parameters

	Turns	Inner diameter (mm)	Outer diameter (mm)	Thickness of coil (mm)
Transmitter coil (horizontal coil)	9	50	47.813	19.683
Transmitter coil (vertical coil)	10	45	42.813	21.87
Receiver coil 1 (horizontal coil)	27	50	48.785	1.215
Receiver coil 2 (vertical coil 1)	30	45	43.785	1.215
Receiver coil 3 (vertical coil 2)	30	45	43.785	1.215

4 Simulation of WPT Systems and Results

The inductance values obtained from the Ansys simulation are used to design the WPT system circuit in LT-Spice. The coupling coefficient between the transmitter and the receiver coils for different cases of misalignments [21] is found using the Ansys simulation.

4.1 Simulation of STMR-Based WPT System

The inductance values obtained from the Ansys simulation are used to design the WPT system circuit in LT-Spice. The LT-Spice simulation circuit is shown in Fig. 10.

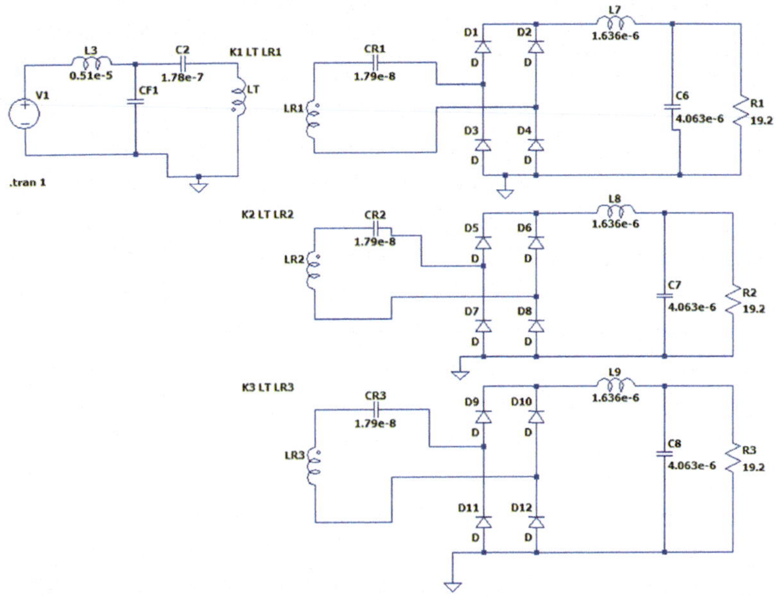

Fig. 10 LT-Spice simulation circuit of STMR WPT system

LCC compensation circuit is used on the transmitter side in order to reduce the reactive power and improve the power factor at the input side. The input voltage of 48 V is given to the transmitter circuit. A current of 9 A flows in the transmitter circuit and 2.5 A flows in each receiver circuit and the power of 120 W is observed at the load resistance. The operating frequency is 85 kHz. The output voltage obtained at the load resistances is 48 V, the plots obtained from the simulation are shown in Fig. 11.

The output power and efficiency [22, 23] for various misalignment cases between the transmitter and receiver coil are mentioned in Table 2 and are given in Table 7. Due to misalignment between receiver coils and transmitter coils [24], the total power to the battery from all the receivers is reduced as shown in the graph given in Fig. 12.

4.2 Simulation of OTCTC-MR-Based WPT System

The inductance values obtained from the Ansys simulation are used to design the WPT system circuit in LT-Spice. The LT-Spice simulation circuit is shown in Fig. 13. LCC compensation circuit is used on the transmitter side to reduce the reactive power and improve the power factor at the input side. The input voltage of 48 V is given to the transmitter circuit. A current of 9 A flows in the transmitter circuit and 2.5 A flows in each receiver circuit and the power of 120 W is observed at the load resistance

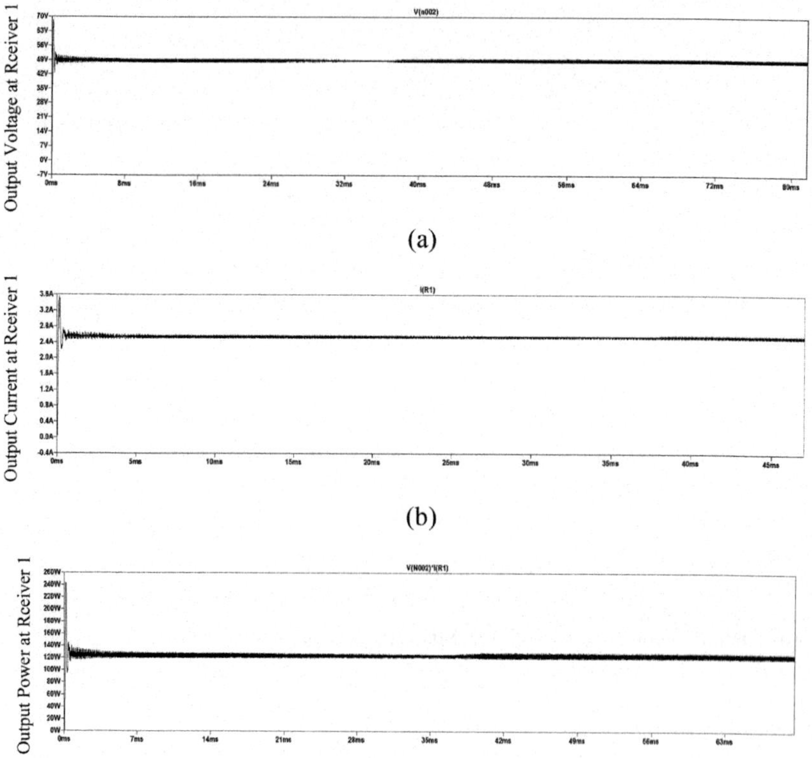

(c)

Fig. 11 Output parameters of receiver coil 1 obtained from LTspice simulation of WPT system **a** output voltage **b** output current **c** output power

Table 7 Output power and efficiency of WPT system under different cases of misalignment

Case	P_{in}	P_{OUTR1}	P_{OUTR2}	P_{OUTR3}	% Efficiency
1	382.4	123.9	123.9	123.9	97.20
2	376.4	122.25	122.25	122.25	97.423
3	383.4	130.37	130.63	86.62	90.648
4	244.9	114.72	114.59	9.67	97.55
5	235.9	115.74	117.2	0.721	93.348
6	318.2	82.515	82.998	145.55	97.745
7	220.0	34.248	35.7	147.04	98.532

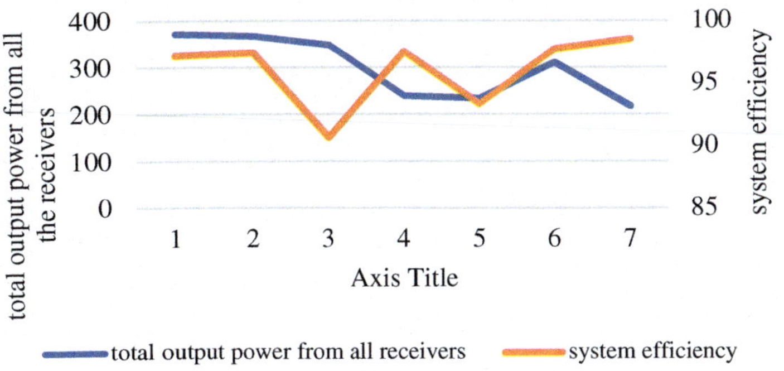

Fig. 12 Graphical representation output power and efficiency

[25]. The operating frequency is 85 kHz. The output voltage obtained at the load resistances is 48 V, power at receiver 1 is 120 W, and the current is 2.5 A for receiver 1, the plots obtained from the simulation are shown in Fig. 14.

Design parameters of transmitter and receiver coils for Ansys design are given in Table 6. In order to consider the misalignment effect, different cases of displacement of receiver coils w.r.t transmitter coil by varying the values of *a*, *b*, *h* are considered. In case-1, the horizontal displacement w.r.t center of the transmitter coil, the value displacement *a* is 100 mm and *b* is 100 m, i.e., the horizontal displacement of the receiver coil from the center of the transmitter coil is zero. Similarly, the other cases

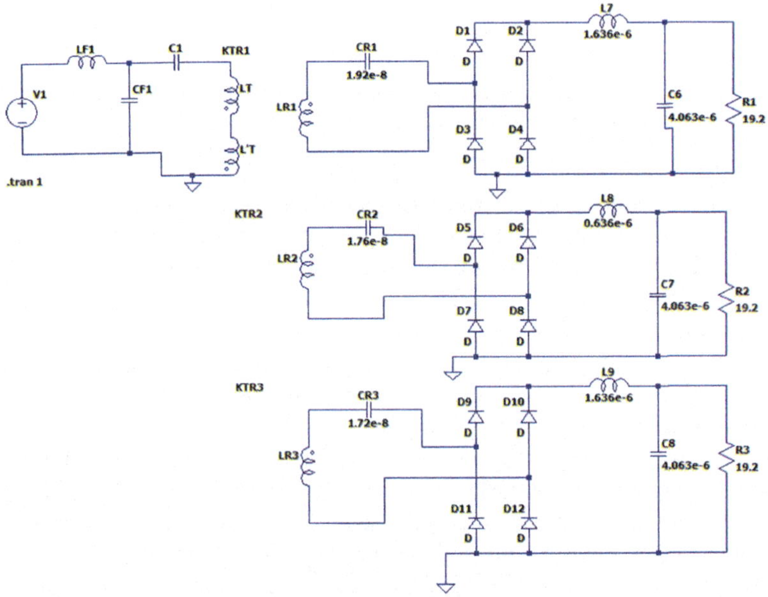

Fig. 13 LT-Spice simulation circuit of OTCTC-MR WPT system

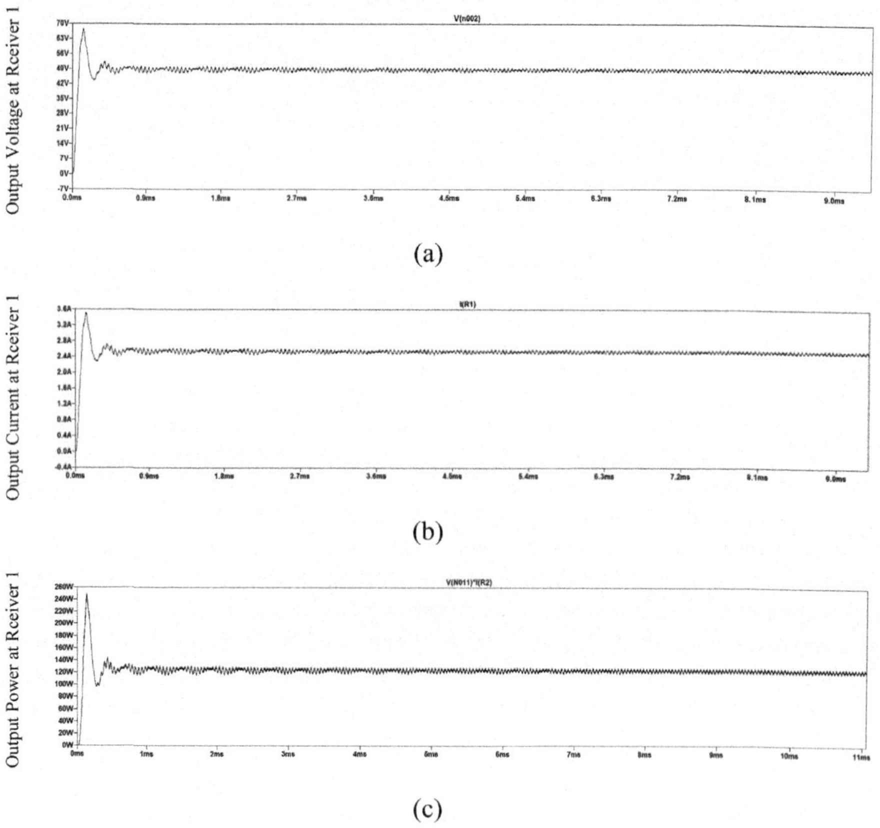

Fig. 14 Output parameters of receiver coil 1 obtained from LTspice simulation of WPT system **a** output voltage **b** output current **c** output power

are taken by considering the different vertical and horizontal displacements, i.e., for different values of *a*, *b*, and *h*. The cases for different displacements are given in Table 8. The simulation is carried out for different cases to obtain a flux density plot and the coefficient of coupling for various misalignments obtained from the simulation is given in Table 9.

For each case of misalignment, the values of self-inductances and mutual inductances are obtained from the simulation. For each case, the LTspice simulation is carried out by replacing the values of mutual and self-inductances. The output power for each receiver is obtained from the simulation and is tabulated in Table 10 for different misalignment cases. The maximum output power of 1852.18 W and the efficiency of 99.121% are obtained in case-10, where the gap (*h*) between transmitter and receiver coil is 50 mm and the horizontal misalignment is zero, i.e., *a* and *b* are 100 mm as shown in Fig. 15. Since the leakage flux is less because of less gap (*h*) and zero misalignments the power obtained in this case is maximum. When the misalignment is more in case-6, i.e., the gap (*h*) is 100 mm and the receivers are

Table 8 Cases considered for the displacement of receiver coils w.r.t to transmitter coil with different values of *a*, *b*, and *h*

Case number	*a* (mm)	*b* (mm)	*h* (mm)
1	100	100	100
2	90	110	100
3	80	120	100
4	70	130	100
5	60	140	100
6	50	150	100
7	100	100	90
8	100	100	80
9	100	100	60
10	100	100	50

Table 9 Coupling coefficient values for different cases mentioned in Table 8

Case number	K_{TR1}	K_{TR2}	K_{TR3}
1	0.0314	0.03026	0.03026
2	0.0314	0.0374	0.0234
3	0.03026	0.0487	0.01916
4	0.02827	0.0632	0.0156
5	0.0254	0.08496	0.01303
6	0.02258	0.1179	0.0108
7	0.03989	0.02953	0.029778
8	0.04977	0.02946	0.029381
9	0.084166	0.02978	0.02965
10	0.1136	0.02977	0.0298

displaced horizontally toward the left side by 50 mm, i.e., *a* is 50 mm and *b* is 150 mm, the output power obtained at the receivers is 2160 W and the efficiency is 99%. It has been observed that the output power obtained at receiver coils is more even if there is misalignment i.e., the OTCTC-MR-based WPT design is misalignment tolerant.

Due to misalignment between the receiver coils and transmitter coil, the total power to the battery from all the receivers is reduced as shown in the graph given in Fig. 16.

5 Conclusions

In this chapter, the STMR and OTCTC-MR-based WPT systems were designed and analyzed, and the results were presented. From the simulations, it has been shown that OTCTC-MR-based WPT system design has a smaller number of turns on the receiver

Table 10 Output power and efficiency of WPT system under different cases of misalignment

Case	Input power (W)	Receiver 1 output power (W)	Receiver 2 output power (W)	Receiver 3 output power (W)	System efficiency (%)
1	409.79	122.45	123.96	123.96	90.38
2	434.23	118	202.89	78	91.861
3	524.94	108.64	349.32	51.417	97.035
4	737.77	94.31	591.71	33.405	97.51
5	1191.4	74	1077	22.592	98.5
6	2162.2	59	2086	15	99
7	476.5	191.87	124.9	124.16	92.53
8	857.79	303.09	126.79	125.74	64.77
9	1143.7	874.9	124.48	124.08	98.230
10	1868.6	1603.7	124.4	124.08	99.121

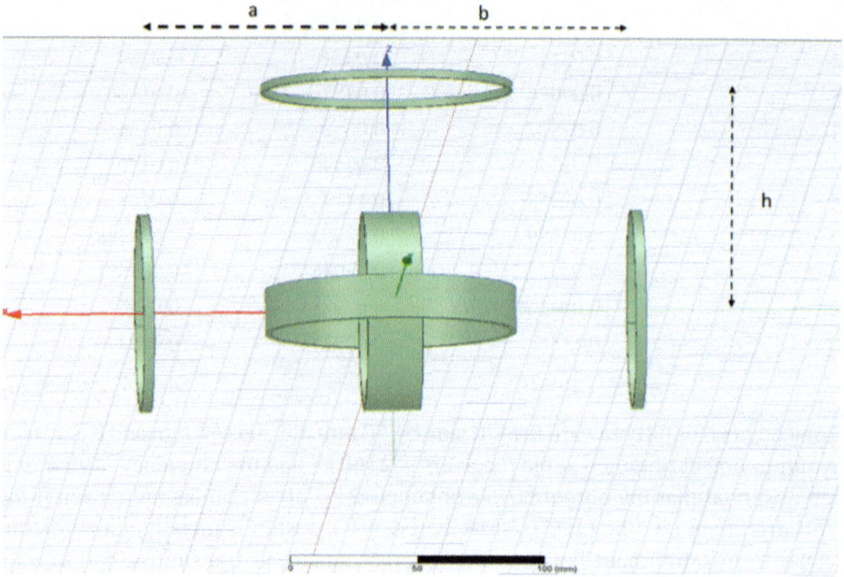

Fig. 15 OTCTC-MR coils with horizontal and vertical displacement parameters *a*, *b,* and *h*

side (87 turns) than the STMR (156 turns) for the same power output of 120 W at each receiver coil and a total of 360 W at three receiver coils. Therefore, the weight of the receiver coils for OTCTC-MR-based WPT is reduced as the number of coils at the receiver side is reduced and the payload of the UAV can be increased. An increment in total output power and system efficiency in the case of the misalignment between the transmitter coil and the receiver coils is observed in OTCTC-MR-based WPT system.

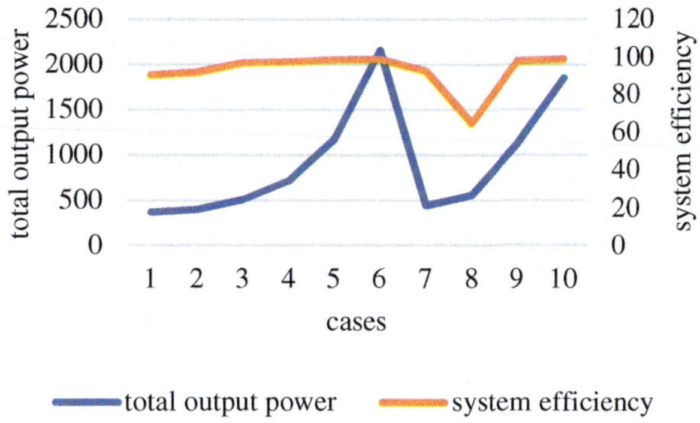

Fig. 16 Output power and efficiency for different cases of misalignment

Hence long distance of power transfer is obtained. The mutual inductance between the three receiver coils can be avoided due to the presence of the cross-coupling effect between the receiver coils. With OTCTC-MR, a good efficient design is obtained for the wireless power transfer for UAV by reducing the coupling effect between the receiver coils and reducing the misalignment problem.

References

1. Chittoor PK, Chokkalingam B, Mihet-Popa L (2021) A review on UAV wireless charging: fundamentals, applications, charging techniques and standards. IEEE Access 9:69235–69266. https://doi.org/10.1109/ACCESS.2021.3077041
2. Shakhatreh H, Sawalmeh AH, Al-Fuqaha A, Dou Z, Almaita E, Khalil I et al (2019) Unmanned aerial vehicles (UAVs): a survey on civil applications and key research challenges. IEEE Access 7:48572–48634. https://doi.org/10.1109/ACCESS.2019.2909530
3. Griffin B, Detweiler C (2012) Resonant wireless power transfer to ground sensors from UAV. In: IEEE international conference on robotic automation, pp 2660–2665. https://doi.org/10.1109/ICRA.2012.6225205
4. Paiva E, Rodas J, Kali Y, Lesme F, Lesme JL, Rodríguez-Piñeiro J (2021) A Review of UAVs topologies and control techniques. In: IEEE international conference on automation, pp 1–6. https://doi.org/10.1109/ICAACCA51523.2021.9465186
5. Lee B, Kwon S, Park P, Kim K (2014) Active power management system for an unmanned aerial vehicle powered by solar cells, a fuel cell, and batteries. IEEE Trans Aerosp Electron Sys 50(4):3167–3177. https://doi.org/10.1109/TAES.2014.130468
6. Inspire 2 - Product Information - DJI. (n.d.). DJI Official. https://www.dji.com/inspire-2/info#specs
7. Lee D, Zhou J, Lin WT (2015) Autonomous battery swapping system for quadcopter. In: International conference on unmanned aircraft system, pp 118–124. https://doi.org/10.1109/ICUAS.2015.7152282
8. Satyavani Y, Bobba PB, Sandeep V (2021) Design and development of wireless power transfer system for UAV. In: International conference on sustainable energy and future electric transportation, pp 1–6. https://doi.org/10.1109/SeFet48154.2021.9375735
9. Kavitha M, Bobba PB, Prasad D (2016) Effect of coil geometry and shielding on wireless power transfer system. In: IEEE power India international conference, pp 1–6. https://doi.org/10.1109/POWERI.2016.8077154

10. Kavitha M, Prasad D, Bobba PB (2019) Methods for overcoming misalignment effects and charging control of a dynamic wireless electric vehicle charging system. IET Electr Power Appl 13(8):1184–1192. https://doi.org/10.1049/iet-epa.2018.5509
11. Lee S, Kim S, Seo C (2013) Design of multiple receiver for wireless power transfer using metamaterial. In: Asia-pacific microwave conference proceedings, pp 1036–1038. https://doi.org/10.1109/APMC.2013.6695016
12. Kuragayala A, Bobba PB, Satyavani Y (2021) Design and development of wireless power transfer system for implantable cardioverter defibrillator. Int J Appl Electromagn Mech 65(4):703–714. https://doi.org/10.3233/JAE-201524
13. Zhang J, Yuan X, Wang C, He Y (2017) Comparative analysis of two-coil and three-coil structures for wireless power transfer. IEEE Trans Power Electron 32(1):341–352. https://doi.org/10.1109/TPEL.2016.2526780
14. Dai Z, Fang Z, Huang H, He Y, Wang J (2018) Selective omnidirectional magnetic resonant coupling wireless power transfer with multiple-receiver system. IEEE Access 6:19287–19294. https://doi.org/10.1109/ACCESS.2018.2809797
15. Campi T, Cruciani S, Maradei F, Feliziani M (2019) Wireless charging system integrated in a small unmanned aerial vehicle (UAV) with high tolerance to planar coil misalignment. In: Joint IEEE international symposium EMC & APEMC, pp 601–604. https://doi.org/10.23919/EMCTokyo.2019.8893934
16. Ng WM, Zhang C, Lin D, Hui SYR (2014) Two- and three-dimensional omnidirectional wireless power transfer. IEEE Trans Power Electron 29(9):4470–4474. https://doi.org/10.1109/TPEL.2014.2300866
17. Kan T, Nguyen T, White JC, Malhan RK, Mi CC (2017) A new integration method for an electric vehicle wireless charging system using LCC compensation topology: analysis and design. IEEE Trans Power Electron 32(2):1638–1650. https://doi.org/10.1109/TPEL.2016.2552060
18. Houran MA, Yang X, Chen W (2018) Magnetically coupled resonance WPT: review of compensation topologies, resonator structures with misalignment, and EMI diagnostics. Electronics 7(11):296. https://doi.org/10.3390/electronics7110296
19. Babic S, Sirois F, Akyel C, Girardi C (2010) Mutual inductance calculation between circular filaments arbitrarily positioned in space: alternative to Grover's formula. IEEE Trans Magn 46(9):3591–3600. https://doi.org/10.1109/TMAG.2010.2047651
20. Ahn D, Hong S (2013) Effect of coupling between multiple transmitters or multiple receivers on wireless power transfer. IEEE Trans Ind Electron 60(7):2602–2613. https://doi.org/10.1109/TIE.2012.2196902
21. Kavitha M, Prasad D, Bobba PB (2019) Methods for overcoming misalignment effects and charging control of a dynamic wireless EV charging system. IET Electr Power Appl 13(8):1184–1192. https://doi.org/10.1049/iet-epa.2018.5509
22. Salkuti SR (2022) Emerging and advanced green energy technologies for sustainable and resilient future grid. Energies 15(18):6667. https://doi.org/10.3390/en15186667
23. Vuddanti S, Shivanand MN, Reddy SS (2021) Design of a one kilowatt wireless charging system for electric vehicle in line with Bharath EV standards. Int J Emerg Electr Power Syst 22(3):255–267. https://doi.org/10.1515/ijeeps-2020-0178
24. Oh H, Oh S, Koo K, Choi W, Shin J, Hwang KC et al (2021) Mid-range wireless power transfer system for various types of multiple receivers using power customized resonator. IEEE Access 9:45230–45241. https://doi.org/10.1109/ACCESS.2021.3067023
25. Sandeep V, Shastri S, Sardar A, Salkuti SR (2020) Modeling of battery pack sizing for electric vehicles. Int J Power Electron Drive Syst 11(4):1987–1994. https://doi.org/10.11591/ijpeds.v11.i4.pp1987-1994

Energy Storage Technologies for Next-Generation Electrical Power Systems

Seong-Cheol Kim, Sravanthi Pagidipala, and Surender Reddy Salkuti

Abstract This chapter aims to present the current practices, challenges, and opportunities for various energy storage technologies for utilization in electrical networks. Renewable energy sources (RERs) power output cannot meet the immediate demand of customers due to their intermittent behavior and hence there is a requirement for energy storage systems (ESSs). The ESS plays a deciding role in the effective utilization of electric vehicles (EVs). In comparison with the existing technologies in ESSs such as ultracapacitors, fuel cells, and battery packs, the most economically viable option is considered as the battery packs. There are different types of ESSs such as supercapacitors, fuel cells, flywheels, and various electrochemical and non-electrochemical storages. This chapter discusses and analyzes the characteristics, size, functions, applications, stability, cost, etc. of various ESSs.

Keywords Renewable energy · Energy storage · Electric vehicles · Power systems · Microgrids · Battery · Supercapacitors

Nomenclature

PQ	Power quality
ES	Energy storage
KE	Kinetic energy
PE	Potential energy

S.-C. Kim · S. R. Salkuti (✉)
Department of Railroad and Electrical Engineering, Woosong University, Daejeon 34606, Republic of Korea
e-mail: surender@wsu.ac.kr

S.-C. Kim
e-mail: kmin@wsu.ac.kr

S. Pagidipala
Department of Electrical Engineering, National Institute of Technology Andhra Pradesh, Tadepalligudem, Andhra Pradesh, India

S. R. Salkuti et al. (eds.), *Power Quality in Microgrids: Issues, Challenges and Mitigation Techniques*, Lecture Notes in Electrical Engineering 1039,
https://doi.org/10.1007/978-981-99-2066-2_24

MGs	Microgrids
PHS	Pumped hydroelectric storage
SMES	Superconducting magnetic energy storage
SOC	State of charge
EDLCs	Electrochemical double-layer capacitors
ESR	Equivalent series resistance

1 Introduction

The storage of electrical energy has become an inevitable component in the modern hybrid power network due to the large-scale deployment of renewable energy resources (RERs) and electric vehicles (EVs) [1, 2]. This energy storage (ES) can solve several operational problems in power networks due to intermittent characteristics of the RERs and EVs while providing various other benefits such as numerous ancillary services support, peak shaving, valley filling, demandside management, and reliability improvement. Storage provides ancillary services support to provide frequency regulation as well as the power balance between the system load and power generated, and to provide a reliable power supply to the customers. Therefore, energy storage can hold substantial promise for transforming the electric power industry [3, 4]. The large-scale integration of RERs energy storage systems (ESSs) for the generation of electrical power has been a most common practice throughout the world as everyone is caring for the environment to reduce the issues of imminent depletion of fossil fuels, global warming, and rising fuel prices. However, there are several challenges to implementing large-scale penetration of RERs and EVs [5]. Among them, the most important challenge is the uncertain and non-dispatchable nature of RERs, which causes the increased cost of power generation as well as the power quality (PQ) concerns [6]. To address these issues one of the solutions could be using energy storage devices, which address this issue by absorbing/supplying load demand when load power is surplus/deficit [7].

Typical ESS applications include improved PQ, transmission and distribution facility deferral, load leveling, peak shaving, voltage regulation, spinning reserve, and integration with renewable energy (RE) generation plants [8]. And also, the online assessment of reliability indices and other system parameters measures the health of the system and protects the system from major unexpected contingencies [9]. Electrical energy can be stored in various forms such as kinetic energy (KE), potential energy (PE), and electrochemical energy. This stored energy is then converted back to electrical form but while conversion, some losses may be present. Because of the self-sufficiency provided by smart grid (SG) technologies, microgrids are becoming very popular [10]. ESSs can be considered as their essence for providing continuous, reliable, and sustainable electrical energy. Modern microgrids (MGs) are groups with many interconnected sources and sinks of power, which can operate on their own as well as in synchronism with the main utility grid. This autonomous operation of MGs is called electrical islanding.

The present electrical network is undergoing a continuous revolution due to various concerns such as global climate change, energy crisis as well as grid reliability [11, 12]. Hence, the world energy scenario has been moving toward large-scale integration of EVs and RERs into the existing electrical grid. The main aim goal of the entire effort is to suppress the impact of power generation and utilization on the environment. Therefore, the distributed and RE-based generation facilities will gradually replace centralized and fossil fuel-based conventional power generation [13]. Furthermore, the concepts of the microgrid and SGs will alleviate the hassle of grid management systems and improve the grid's flexibility and reliability in line with the concept of decentralized distributed power generation. Various RESs such as solar, wind, and wave energies have been prominent these days and implemented in practice. Among various RERs, solar and wind are the most promising. However, without combining with ESSs, these RERs cannot be used as long-term and reliable electrical solutions [14]. The ESSs serve as energy backup, which stores energy during excess power consumption and releases it when the grid experiences a power deficit. The ESSs act as uninterruptable power supplies (UPS), providing a steady voltage and frequency during power outages [15]. ESSs are also very important key components for smart grid and MG systems.

Energy storage systems (ESSs) play a crucial role in maintaining power balance in renewable power generation and isolated power supply systems. However, in recent years, the single ESS combines energy storage with complementary characteristics to get ideal storage characteristics. A control and operation strategy for ESSs is developed in [16] for the application in hybrid microgrids with AC coupling and a detailed review is carried out for the existing applications. An energy storage capacity planning methodology for enhancing offshore wind power consumption has been proposed in [17] by considering the uncertainty of offshore wind power, the annual load demand, line structure as well as various types of power sources.

2 Classification of Energy Storage Systems (ESSs)

In the modern electric grid system, the integration of RERs, ESSs, and electric load is growing rapidly in recent years. However, the intermittent nature is the main issue regarding these RESs and EVs as by varying environmental conditions power generation from these RERs changes [18]. The use of RERs in global electrical grids has been growing rapidly since the beginning of the year 2000. The maturity in RERs significantly reduced installation and maintenance costs. In recent days, the majority of the developing countries in South Asia, Latin America, and South Africa have been moving toward RERs. Due to the reduced cost of RERs and land availability countries like the United States and Australia have been attracted to the RES investments [19, 20]. The ESSs act as uninterruptable power supplies (UPS), providing a steady voltage and frequency during power outages.

The ESSs are required to mitigate the power mismatch between demand and supply in the electrical system [21]. The criteria for the selection of the ES technologies depend on application, energy and power ratings, size, lifetime, response time, capacity, and cost. The broad classifications of ES technologies include mechanical energy storage, magnetic energy storage, chemical energy storage, electrochemical energy storage, and thermal energy storage [22]. The major mechanical ESSs include pumped hydro storage, flywheel ES, etc. Magnetic ESSs include Superconducting Magnetic Energy Storage (SMES). Chemical ESSs include alkaline fuel cells (AFC) and proton-exchange membrane fuel cells (PEMFC). Electrochemical ESSs include Li-ion batteries, Ni-Cd batteries, lead–acid batteries, etc. The thermal ESSs include sensible heat. The power rating of this sensible heat is in the range of 0.1–300 MW, efficiency is between (30–60)%, the cost of this battery is (200–300) $/kW, and the life is about 10–40 years.

3 Battery Energy Storage System (ESS)

3.1 Modeling of Battery ESS

The discharge time, capacity, and current relationship for a lead–acid battery is given by

$$t = \frac{Q_P}{I^k}, \tag{1}$$

where k is a constant, I is the current drawn from the battery, Q_P is capacity when discharged at a rate of 1 A, and t is the amount of time that a battery can sustain. There are three types of battery models, and they are electrical circuit-based, electrochemical-based, and experimental circuit-based battery models. Figure 1 depicts the basic electrical circuit-based model by the series combination of voltage source and resistance [23]. The internal resistance (R) of the battery is constant, and it is independent of the charging/discharging current magnitude. The terminal voltage of the battery depends on the electric charge of the battery.

In this electrical circuit model [24, 25], the terminal voltage of the battery is expressed by using

$$E = E_o - K\frac{Q}{(Q-t)} + Ae^{-Bi_t}, \tag{2}$$

where E_o is the nominal voltage, K is polarization voltage, E is the no-load voltage, Q is charge capacity of the battery, A is amplitude of exponential zone, and B is the exponential time zone constant [26]. Terminal current (i_t) in terms of battery current (i_b) is expressed as

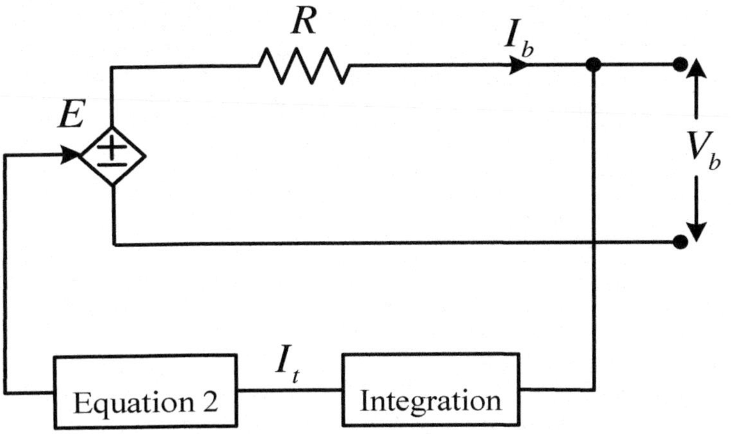

Fig. 1 Electrical equivalent circuit of lead–acid battery

$$i_t = \int_0^t i_b dt. \quad (3)$$

The terminal voltage of the battery is expressed as

$$V_b = E - Ri_b. \quad (4)$$

The capacity model of the battery describes the capacity as a function of current ($Q_{\max}(i)$) is given by

$$Q_{\max}(i) = \frac{Q^0_{\max} kcT}{1 - e^{-kT} + c(kT - 1 + e^{-kT})}. \quad (5)$$

This capacity model has three constants, $Q^0_{\max}$, k and c, where $Q^0_{\max}$ is the maximum capacity.

The lifetime model of the battery can be expressed as

$$C_F = a_1 + a_2 e^{-a_3 R} + a_4 e^{-a_5 R}, \quad (6)$$

where C_F is cycles to failures, a_i are fitting constants, and R is the range of the cycle.

3.2 *State of Charge (SOC) of a Battery*

The state of charge (SOC) of a battery cell gives the average available lithium concentration in the cell compared to the maximum concentration of lithium that can be achieved in the cell [27, 28]. Considering the maximum theoretical lithium concentration in the electrode to be C_s and the average concentration of lithium at a time t to be $C_{s,t,\text{avg}}$. Then, the lithium stoichiometry (θ_t) is given by

$$\theta_t = \frac{C_{s,t,\text{avg}}}{C_{s,\max}}. \tag{7}$$

Cell SOC (Z_t) is given by

$$Z_t = (\theta_t^{\text{neg}} - \theta_{0\%}^{\text{neg}})/(\theta_{100}^{\text{neg}} - \theta_{0\%}^{\text{neg}}) = (\theta_t^{\text{pos}} - \theta_{0\%}^{\text{pos}})/(\theta_{100}^{\text{pos}} - \theta_{0\%}^{\text{pos}}). \tag{8}$$

As nominal cell voltage depends on the amount of Li-ions in the electrode current collector surface and SOC depends on the average concentration over the entire surface of electrodes. As average value doesn't depend on temperature, resting of cells, etc. [29, 30], SOC remains independent while as cell voltage changes accordingly. Hence voltage can only be an indirect measuring unit of SOC, while as the change in current shows a direct relationship with the same. SOC relationship with current is mathematical as a function of cell efficiency (η), the total capacity of the cell (Q), and the state of charging (charging/discharging of the cell) [31]. It is expressed by using

$$z(t) = z(0) - \frac{1}{Q}\int_0^t \eta i(\tau)d\tau. \tag{9}$$

State of health (SOH) of a battery is a quantity that reflects the aging process of a battery and the life span of the battery depending upon its performance analysis. Major after-effects of cell aging include a drastic capacity decrease of the battery to around 20–30% and a cell internal resistance increase up to 50–100%. The state of life (SOL) of a battery depicts the remaining life/percentage of life left for a battery in percentage [32, 33]. It changes according to the number of faults and aging effects a battery has gone through.

3.3 Total Energy and Power of a Battery

The total energy of a battery is amount of energy present in a battery which is independent of temperature and rate of charging or discharging of a cell [34]. Mathematically, it can be expressed as

$$W(t) = Q \int_{\text{Zmin}}^{z(t)} OCV(\tau)d\tau \approx QV_{\text{nom}}\Delta Z. \tag{10}$$

The power of a battery is the rate of energy movement in limits to its design aspects [35]. Normally, it is calculated with a high pulse power characterization (HPPC) test measuring cell resistance for different SOC and temperature values [36]. The cell model for HPPC discharge power gives

$$V(t) = OCV(Z(t)) - I(t)R \tag{11}$$

$$I(t) = \frac{OCV(Z(t)) - V(t)}{R}. \tag{12}$$

The voltage change in between minimum limit (V_{min}) and maximum limit (V_{max}) during charging and discharging is given by

$$P_{\text{discharge}} = V(t)I(t) = V_{\text{min}}\frac{OCV(Z(t)) - V(t)}{R_{\text{dis}}} \tag{13}$$

$$P_{\text{charge}} = V(t)I(t) = V_{\text{max}}\frac{OCV(Z(t)) - V(t)}{R_{chg}}. \tag{14}$$

During the calculation of charge power and discharge power, voltage is clamped at V_{max} and V_{min}, respectively.

4 Classification of Energy Storage Batteries

4.1 Lead–Acid Batteries

These are one of the most common and widely used rechargeable electrochemical storage devices. Due to their low cost and reliability, they have been used in RE-based power generation and HEV applications [37, 38]. The lead–acid battery is quite a mature technology and hasn't really changed much apart from its physical structure. When a lead–acid battery is discharged less than 20% of its rated capacity, its lifetime is seriously affected. When charged and discharged at a high rate of current, its lifetime is reduced. Due to heavyweight lead collectors, the lead–acid battery has low power and energy densities. The cycle life of the lead–acid battery is in the range of 1200–1800 cycles, with an efficiency range of 85–90%. They have energy and power densities of 30–50 Wh/kg and 75–300 W/kg, respectively. The nominal cell voltage is about 2.105 V [39, 40]. Generally, they are used in automobiles, uninterruptible power supply (UPS), etc.

In general, the cathode plates are made of lead dioxide (PbO_2), and the anode plates of porous lead (Pb) [26]. Both plates are immersed in an electrolyte, sulfuric acid (H_2SO_4). During the discharging process, lead over the anode reacts with H_2SO_4

making lead sulfate and electrons. These batteries are inexpensive, ready for availability, and have mature technology. At a temperature of 25 °C, its self-discharge rate is less than 2% of its rated capacity. Thus, it is most suitable for longer term power storage applications [41, 42]. The temperature range of operation is from −30 to +40 °C range. The life of lead–acid batteries is about 6–100 years, the cost is about 250–300 $/kW, and the power rating is about 100 MW. The detailed research in these fields has culminated in collectors that are non-corrosive and have a high energy and power density. Its lifetime improved with proper control methods and energy management strategy.

4.2 Lithium–Ion (Li-Ion) Batteries

These batteries are used in utility applications, automobiles, consumer electronics, etc. The cost of this battery is around 1000–2000 $/kW. They have high energy and power densities as well as high efficiency [43]. The life of a Li-ion battery is about 30 years, its efficiency is about 85–90%, and the power rating is 100 MW. These batteries have a high cost, require power electronics, and have safety issues.

4.3 Nickel–Metal Hydride (Ni-MH) Batteries

The applications of Ni-MH batteries include electronic devices, EVs, electrical tools, UPS, etc. In Ni-MH batteries, positive and negative electrodes are made with nickel hydroxide and metal alloy, respectively [44, 45]. The Ni-MH battery technology was invented around 100 years ago. Their cycle life depends on the depth of discharge (DOD). For 100% DOD, it has a life cycle of more than 1000, and for 10% DOD, it has a life of over 10,000,000 cycles. It has an efficiency of 65–70% range. Their energy and power density are (70–110) Wh/kg and (150–300) W/kg, respectively. They have operating temperatures ranging from −30 to +70 °C. But the desired performance can be obtained at a temperature between 0 and 40 °C. The Ni-MH battery's rated voltage is around 1.2 V [46]. They are used in consumer electronics and telecom backup, etc. These batteries have a high energy density, mature technology, and better cycle life than lead–acid batteries. The Ni-MH batteries required less maintenance and are environmentally friendly [47]. They have a wide range of operating temperatures and less memory effect. The operation of Ni-MH batteries generates heat and complex charging technology is necessary for safe operation. For larger scale, cost of operation will increase.

4.4 *Nickel–Cadmium (Ni-Cd) Batteries*

In Ni-Cd batteries, an aqueous alkali solution is used as an electrolyte. The Ni-Cd and Ni-MH batteries are used in EVs because they have high energy density and can be discharged for a very low SOC range and long life cycle. They are used in utility/telecom backup, consumer electronics, etc. The Ni-Cd battery cycle life depends on DOD. Its cycle life is in the range of 2000–2500 cycles. The range of charge/discharge efficiency is 60–70%. Its energy and power density are 70–110 Wh/kg and 150–300 W/kg. The life of Ni-Cd battery is about 40 years, the cost is in the range of 1500–2500 $/kW, and the power rating is in the range of 10–100 kW. The operating temperature of Ni-Cd batteries ranges from −40 to +60 °C. The rated voltage of Ni-Cd battery is 1.2 V, the cut-off voltage is 1 V, and the self-discharge rate is 10%/month. The negative electrodes are made of nickel species and the positive electrodes are made of cadmium species active materials in Ni-Cd batteries. Ni-Cd batteries have a high specific energy, but they have a high memory effect, low performance, and high price.

4.5 *Nickel–Zinc (NiZn) Batteries*

In NiZn batteries, the positive electrode is made with nickel, and the negative electrodes are made with zinc hydroxide. The electrolyte in NiZn batteries is a potassium peroxide aqueous solution [48]. NiZn batteries have a life cycle of 100–300 cycles. NiZn batteries have a 75% efficiency rating. It has a power density of 150–300 W/kg and an energy density of 60–65 Wh/kg. The operating temperature of NiZn batteries is −10 to +50 °C. The nominal voltage of the NiZn battery is 1.65 V.

5 Supercapacitors and Fuel Cells

5.1 *Supercapacitors (SCs)*

The electrochemical double-layer capacitors (EDLCs) or ultracapacitors are termed as supercapacitors (SCs) and they have high energy density as compared to general capacitors. The design of EDLCs is such that the surface area of the electrode is more with a thin high-permittivity dielectric between them so as to have a higher value of capacitance [43]. The material of the electrode and electrolyte affects the performance of SCs. The SCs have capacitances in the range of 1000–5000 F, and energy densities in the range of 3.5–4.9 Wh/kg.

The battery and SC have complementary properties. Both have their benefits and drawbacks. The system accuracy is improved when these two energy storage devices are combined [49]. The battery life cycle is greatly enhanced with adding an SC storage device. This will also minimize replacement cost of battery. As a result, the combination of battery and SC is an appealing option for wind power, PV generation, and EV applications. SCs are used mostly where fast charging/discharging processes are to be done and it provides a high current pulse. High energy density and high capacitances make activated carbon electrodes as the most commonly used electrode material. The carbon could be in different forms such as graphene and carbon nanotubes. Their capacities are about a thousand times compared to those of conventional capacitors making a large amount of energy to be stored in them.

SCs are highly effective in reducing power gaps lasting from a few seconds to a few minutes, and they can be recharged quickly [50]. Because of the rapid charging/discharging processes, supercapacitors are used as peak load boosters for plug-in hybrid electric vehicles (PHEV), in UPSs, automobiles with regenerative braking, etc. The energy and power density of SCs in the range of 2.5–15 Wh/kg and 500–5000 W/kg, respectively. The efficiency of SC is more than 90%. The major demerit of SC is the high self-discharge rate, which averages more than 20% per day. The cost of SCs is relatively high compared to other storage devices. SCs when operated singly can only store so much power. As a result, practical implementations of SCs involve connecting various cells in parallel strings, in order to maximize the storage capacity.

5.1.1 Modeling of SC

SCs store the energy in the form of charge differences appearing on their positive and negative plates separated by some distance. They are also called as double-layer capacitors and are similar to conventional capacitors except that it has a high capacitance value due to their bigger plate area and less distance between plates. There are three electrical equivalent circuit models of SC, i.e., a series combination of resistance and capacitance (series RC) model, a parallel combination of resistance and capacitance (parallel RC) model, and a series–parallel combination of resistance and capacitance (series–parallel RC) model [51]. The dynamic behavior of SC is determined by using resistance and capacitance. In general, series resistance in the equivalent circuit of SC determines the equivalent series resistance (ESR). It reflects the losses during the charging/discharging process of SC. If the charging current abruptly drops to zero in a series RC model, the SC voltage gradually decreases and takes some time to achieve a steady state. This behavior of SC is not discussed in the series RC model.

This problem is somewhat discussed in the parallel RC model of SC. The parallel RC model is made up of two or more series RC circuit models that are connected in parallel. There are various time constants for various RC models. As a result, during the charging/discharging operation, every RC branch in a parallel RC circuit behaves differently. As compared to the series RC model, the dynamic behavior of the parallel RC model of SC is more accurate. A series–parallel model of SC is shown in Fig. 2. The ESR in Fig. 2 represents series resistance which represents losses due to resistors in the charging/discharging operation of SC, where R_p and C_p represent

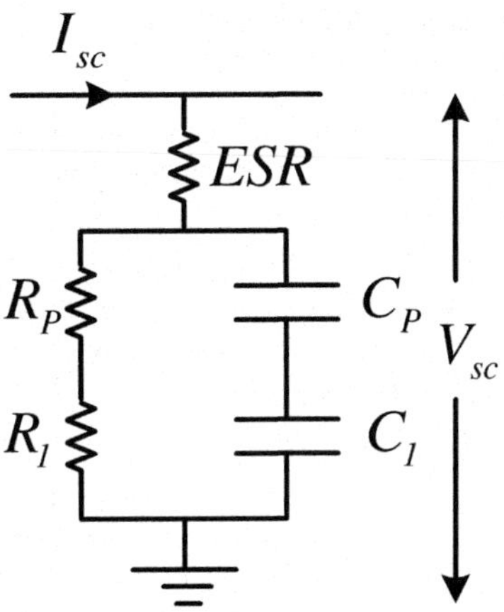

Fig. 2 Equivalent circuit representation of SC

parallel resistance and capacitance, respectively. R_1 represents the self-discharge rate of SC and C_1 is actual storage capacitance. V_{sc} and i_{sc} represent terminal voltage and output current from SC.

The capacitance of *SC* can be represented by

$$C = \varepsilon_R \varepsilon_0 \frac{A}{d}. \tag{15}$$

The stored energy of a capacitor (W_{stored}) is given by

$$W_{\text{stored}} = \frac{1}{2} C V^2. \tag{16}$$

The *SOC* of a supercapacitor is given by

$$SOC_{\min} \leq SOC \leq SOC_{\max} \tag{17}$$

$$SOC = SOC_i + \frac{1}{Q_{SC}} \int_0^t i_{CH} dt, \tag{18}$$

where SOC_i is the initial *SOC* of the supercapacitor, Q_{SC} is rated supercapacitor charge, and i_{CH} is supercapacitor charging current.

5.2 Fuel Cells (FCs)

The FC is an electrochemical device that converts chemical energy into electrical by making and breaking molecular bonds. When the hydrogen is fed at the anode and oxygen is fed at the cathode, then a potential difference occurs between the two electrodes [52]. The hydrogen here is the fuel and oxygen is the oxidant causing a chemical reaction between them. FC can be of the alkaline fuel cell (AFC) or proton-exchange membrane FC (PEMFC). The AFC has a power rating in the range of 10–100 kW, an efficiency of 60–70%, a cost is about 500 $/kW, and a life of 5–15 years. The PEMFC has a power rating in the range of 1–250 kW, an efficiency of 50–60%, a cost of 200 $/kW, and a life cycle of about 10 years. Mostly FC has a power rating of up to 50 MW and good energy density from 800 to 10000 Wh/kg but its efficiency is less with high cost. FC can be used for medium as well as large-scale applications from kW to MW ratings. The advantages of FC over other ESSs are higher efficiency, less noise, low carbon emission, longer lifespan, and high reliability. However, the efficiency of FC depends on the extraction of power from it. It is relatively more expensive than other energy storage devices and compared to lead–acid batteries and SC, the reaction time is slower.

6 Non-electrochemical Storage

6.1 Compressed Air Energy Storage (CAES)

Some of the main costs cover the costs of setting up the underground storage, intercoolers for dissipating the heat, and compressors and expanders for the air (i.e., compressors for the air to be stored and expand the air feeding the turbine). The CAES has huge energy and power capacity. However, the major disadvantages are large scale, require fuel input, long construction time, and geographically limited.

6.2 Flywheel

Flywheel technology is based on a spinning wheel with a fixed axle. Common applications of the flywheel are used to provide continuous energy when the main energy source is disconnected. Flywheel also has disadvantages. One of the most known limits to flywheel conception is the tensile of the material used to build the rotor. Flywheels have high cycle life, quick recharge, high power density, and independent power and energy sizing. The life of the flywheel is about 20 years, efficiency is in the range of 70–95%, the cost is about 250–300 $/kW, and the power rating is in the range of 2–20 MW.

Pumped Hydroelectric Storage (PHS) has huge energy and power capacity. The life of PHS is 70 years and the power rating is 10 MW–3 GW. The efficiency of PHS is about 75%, and the cost is in the range of 40–50 $/kW. Superconducting Magnetic Energy Storage (SMES) is used in emergency bridging power and power quality applications. However, it is expensive, has large parasitic losses, and has low energy density. The SMES has a power rating of 1 MW–3 GW, a cost is about 10,000 $/kW, an efficiency of around 95%, and a life of about 20 years. SMES is used in emergency bridging power and improved power quality. However, SMES is expensive, has large parasitic losses, and has low energy density.

7 Hybrid Energy Storage System (HESS)

Different energy storage systems are in use now, but no device can provide a quick response and longer power supply. Two key terms are helpful when designing hybrid energy storage systems, namely, energy density and energy density. The difference between the two is that a high-energy-density device can store more energy, while a higher energy density describes the dynamics of the device delivering energy, i.e., its response. Energy density is defined in watts-hours/kg, whereas energy density is defined in watts/kg. Various challenges in the ES technologies include developing charging and discharging models of traditional storage devices of interest (flywheel, supercapacitor, and battery), evaluating the impacts of storage for standalone and grid-connected distribution networks with RERs, optimizing the power of the integrated RERs with storage under different conditions, regulatory framework to promote ES, ES for optimal operation of distribution network, ES for demand response management, the role of fuel cells, hydrogen storage in ES technology, innovations, trends, key issues, and future technological scope.

8 Conclusions

This chapter discussed various types of energy storage technologies for next-generation electrical power systems including microgrids and smart grids. Classification of storage systems based on electrochemical and non-electrochemical is discussed in this chapter. CAES and pumped hydro energy storage are the most reliable sources of energy storage systems when compared to other sources of energy storage to meet the high power demand. The round-trip efficiency of these ESSs is about 75%. Every ES system has various advantages and at the same time some issues. A combination of ESSs can provide several advantages over using a single ESS. Hybrid ESSs (HESSs) incorporate the characteristics of various ES elements to increase the stability and reliability of the system.

Acknowledgements This research work was funded by "Woosong University's Academic Research Funding—2023".

References

1. Wei L, Wu M, Yan M, Liu S, Cao Q, Wang H (2019) A review on electrothermal modeling of supercapacitors for energy storage applications. IEEE J Emerg Select Topics in Power Electron 7(3):1677–1690. https://doi.org/10.1109/JESTPE.2019.2925336
2. Perdana YS, Muyeen SM, Al-Durra A, Morales-Paredes HK, Simões NG (2019) Direct connection of supercapacitor-battery hybrid storage system to the grid-tied photovoltaic system. IEEE Trans Sustain Energy 10(3):1370–1379. https://doi.org/10.1109/TSTE.2018.2868073
3. Podder AK, Chakraborty O, Islam S, Manoj Kumar N, Alhelou HH (2021) Control strategies of different hybrid energy storage systems for electric vehicles applications. IEEE Access 9:51865–51895. https://doi.org/10.1109/ACCESS.2021.3069593
4. Acar C (2018) A comprehensive evaluation of energy storage options for better sustainability. Int J Energy Res 42(12):3732–3746. https://doi.org/10.1002/er.4102
5. Duan Y, Harley RG (2009) Present and future trends in wind turbine generator designs. In: IEEE power electronics and machines in wind applications, Lincoln, NE, USA, pp 1–6. https://doi.org/10.1109/PEMWA.2009.5208401
6. Chen H, Cong TN, Yang W, Tan C, Li Y, Ding Y (2009) Progress in electrical energy storage system: a critical review. Prog Nat Sci 19(3):291–312. https://doi.org/10.1016/j.pnsc.2008.07.014
7. Chakraborty MR, Dawn S, Saha PK, Basu JB, Ustun TS (2022) A comparative review on energy storage systems and their application in deregulated systems. Batteries 8(9):124. https://doi.org/10.3390/batteries8090124
8. Wei QS, Zhang X, Kim BH, Oh BS (2022) Experimental investigation of supercapacitor based regenerative energy storage for a fuel cell vehicle equipped with an alternator. Int J Hydrogen Energy 47(3):1954–1964. https://doi.org/10.1016/j.ijhydene.2021.10.102
9. Şahİn ME, Blaabjerg F, Sangwongwanİch A (2021) Modelling of supercapacitors based on simplified equivalent circuit. CPSS Trans Power Electron Appl 6(1):31–39. https://doi.org/10.24295/CPSSTPEA.2021.00003
10. Hadjipaschalis I, Poullikkas A, Efthimiou V (2009) Overview of current and future energy storage technologies for electric power applications. Renew Sustain Energy Rev 13(6–7):1513–1522. https://doi.org/10.1016/j.rser.2008.09.028
11. Patil SB, Pawar AS (2020) Stability analysis of supercapacitor based DC microgrid system with energy storage. In: IEEE international conference for innovation in technology (INOCON), Bengaluru, India, pp 1–4. https://doi.org/10.1109/INOCON50539.2020.9298267
12. Salkuti SR (2021) Energy storage and electric vehicles: technology, operation, challenges, and cost-benefit analysis. Int J Adv Comput Sci Appl 12(4). https://doi.org/10.14569/IJACSA.2021.0120406
13. Le TS, Nguyen TN, Bui DK, Ngo TD (2023) Optimal sizing of renewable energy storage: a techno-economic analysis of hydrogen, battery and hybrid systems considering degradation and seasonal storage. Appl Energy 336:120817. https://doi.org/10.1016/j.apenergy.2023.120817
14. Rocabert J, Capó-Misut R, Muñoz-Aguilar RS, Candela JI, Rodriguez P (2019) Control of energy storage system integrating electrochemical batteries and supercapacitors for grid-connected applications. IEEE Trans Ind Appl 55(2):1853–1862. https://doi.org/10.1109/TIA.2018.2873534
15. Lashway CR, Mohammed OA (2016) Adaptive battery management and parameter estimation through physics-based modeling and experimental verification. IEEE Trans Transp Electrification 2(4):454–464. https://doi.org/10.1109/TTE.2016.2558843

16. Ramos F, Pinheiro A, Nascimento R, Araujo Silva Junior W, Mohamed MA, Annuk A, Marinho MHN (2022) Development of operation strategy for battery energy storage system into hybrid AC microgrids. Sustainability 14:13765. https://doi.org/10.3390/su142113765
17. Yu H, Yang X, Chen H, Lou S, Lin Y (2022) Energy storage capacity planning method for improving offshore wind power consumption. Sustainability 14:14589. https://doi.org/10.3390/su142114589
18. Brekken TKA, Ozkan-Haller HT, Simmons A (2012) A methodology for large-scale ocean wave power time-series generation. IEEE J Oceanic Eng 37(2):294–300. https://doi.org/10.1109/JOE.2012.2187393
19. Rahim AHMA (2022) Fuzzy logic control of supercapacitor energy storage system in a microgrid. In: International conference on innovations in science, engineering and technology (ICISET), Chittagong, Bangladesh, pp 516–521. https://doi.org/10.1109/ICISET54810.2022.9775843
20. May GJ, Davidson A, Monahov B (2018) Lead batteries for utility energy storage: a review. J Energy Storage 15:145–157. https://doi.org/10.1016/j.est.2017.11.008
21. Akinyele DO, Rayudu RK (2014) Review of energy storage technologies for sustainable power networks. Sustain Energy Technol Assess 8:74–91. https://doi.org/10.1016/j.seta.2014.07.004
22. Panhwar IH et al (2020) Mitigating power fluctuations for energy storage in wind energy conversion system using supercapacitors. IEEE Access 8:189747–189760. https://doi.org/10.1109/ACCESS.2020.3031446
23. Tian S, Hong M, Ouyang M (2009) An experimental study and nonlinear modeling of discharge I-V behavior of valve-regulated lead-acid batteries. IEEE Trans Energy Convers 24(2):452–458. https://doi.org/10.1109/TEC.2008.2001447
24. Reddy SS (2018) Emergency reserve activation considering demand-side resources and battery storage in a hybrid power system. Electr Eng 100:1589–1599. https://doi.org/10.1007/s00202-017-0637-9
25. Frate GF, Ferrari L, Sdringola P, Desideri U, Sciacovelli A (2023) Thermally integrated pumped thermal energy storage for multi-energy districts: integrated modelling, assessment and comparison with batteries. J Energy Storage 61:106734. https://doi.org/10.1016/j.est.2023.106734
26. Zhang Q, Li G (2020) Experimental study on a semi-active battery-supercapacitor hybrid energy storage system for electric vehicle application. IEEE Trans Power Electron 35(1):1014–1021. https://doi.org/10.1109/TPEL.2019.2912425
27. Tan KM, Babu TS, Ramachandaramurthy VK, Kasinathan P, Solanki SG, Raveendran SK (2021) Empowering smart grid: a comprehensive review of energy storage technology and application with renewable energy integration. J Energy Storage 39:102591. https://doi.org/10.1016/j.est.2021.102591
28. Buller S, Karden E, Kok D, Doncker RWD (2002) Modeling the dynamic behavior of supercapacitors using impedance spectroscopy. IEEE Trans Ind Appl 38(6):1622–1626. https://doi.org/10.1109/TIA.2002.804762
29. Wang X, Qu G, Hao S, Zhao G, Zhang X, Ma W, Xu X (2021) Improvement of nickel-cobalt-based supercapacitors energy storage performance by modification of elements. J Colloid Interface Sci 602:712–720. https://doi.org/10.1016/j.jcis.2021.06.063
30. Sridhar S, Salkuti SR (2022) Development and future scope of renewable energy and energy storage systems. Smart Cities 5(2):668–699. https://doi.org/10.3390/smartcities5020035
31. Ding X, Duan L, Zhou Y, Yang L, Li M, Tian F, Gao C (2023) Thermodynamic analysis and economic assessment of a novel multi-generation liquid air energy storage system coupled with thermochemical energy storage and gas turbine combined cycle. J Energy Storage 60:106614. https://doi.org/10.1016/j.est.2023.106614
32. Zhu C, Li X, Lu R, Xu J, Fan G (2017) A comparative study of three electrochemical battery models with different operating conditions. In: IEEE transportation electrification conference and expo, Asia-Pacific (ITEC Asia-Pacific), Harbin, China, pp 1–7. https://doi.org/10.1109/ITEC-AP.2017.8080786

33. Hannan MA, Hoque MM, Mohamed A, Ayob A (2017) Review of energy storage systems for electric vehicle applications: issues and challenges. Renew Sustain Energy Rev 69:771–789. https://doi.org/10.1016/j.rser.2016.11.171
34. Khaligh A, Li Z (2010) Battery, ultracapacitor, fuel cell, and hybrid energy storage systems for electric, hybrid electric, fuel cell, and plug-in hybrid electric vehicles: state of the art. IEEE Trans Vehicular Technol 59(6):2806–2814. https://doi.org/10.1109/TVT.2010.2047877
35. Malhotra A, Battke B, Beuse M, Stephan A, Schmidt T (2016) Use cases for stationary battery technologies: a review of the literature and existing projects. Renew Sustain Energy Rev 56:705–721. https://doi.org/10.1016/j.rser.2015.11.085
36. Ram V, Infantraj SSR (2022) Modelling and simulation of a hydrogen-based hybrid energy storage system with a switching algorithm. World Electric Vehicle J 13(10):188. https://doi.org/10.3390/wevj13100188
37. Zhang J, Liu J, Chen L, Zhang L, Zeng P, Li Y (2023) Reliability evaluation of high permeability renewable energy distribution network considering energy storage charge and discharge strategy. Energy Rep 9(3):361–368. https://doi.org/10.1016/j.egyr.2023.01.006
38. Aneke M, Wang M (2016) Energy storage technologies and real life applications—a state of the art review. Appl Energy 179:350–377. https://doi.org/10.1016/j.apenergy.2016.06.097
39. Reddy SS (2017) Optimal scheduling of thermal-wind-solar power system with storage. Renew Energy 101:1357–1368. https://doi.org/10.1016/j.renene.2016.10.022
40. Ali H (2022) A hybrid energy storage system based on supercapacitor and electric vehicle batteries for frequency stability improvement of islanded microgrids. In: 23rd international Middle East power systems conference (MEPCON), Cairo, Egypt, pp 1–6. https://doi.org/10.1109/MEPCON55441.2022.10021708
41. Olabi AG, Abbas Q, Makky AA, Abdelkareem MA (2022) Supercapacitors as next generation energy storage devices: properties and applications. Energy 248:123617. https://doi.org/10.1016/j.energy.2022.123617
42. Sharma P, Salkuti SR, Kim SC (2022) Advancements in energy storage technologies for smart grid development. Int J Electr Comput Eng 12(4):3421–3429. https://doi.org/10.11591/ijece.v12i4.pp3421-3429
43. Benoy SM, Pandey M, Bhattacharjya D, Saikia BK (2022) Recent trends in supercapacitor-battery hybrid energy storage devices based on carbon materials. J Energy Storage 52(B):104938. https://doi.org/10.1016/j.est.2022.104938
44. Reddy SS (2017) Optimal power flow with renewable energy resources including storage. Electr Eng 99:685–695. https://doi.org/10.1007/s00202-016-0402-5
45. Park S, Salkuti SR (2019) Optimal energy management of railroad electrical systems with renewable energy and energy storage systems. Sustainability 11(22):6293. https://doi.org/10.3390/su11226293
46. Fagiolari L, Sampò M, Lamberti A, Amici J, Francia C, Bodoardo S, Bella F (2022) Integrated energy conversion and storage devices: interfacing solar cells, batteries and supercapacitors. Energy Storage Mater 51:400–434. https://doi.org/10.1016/j.ensm.2022.06.051
47. Matos CR, Silva PP, Carneiro JF (2022) Overview of compressed air energy storage projects and regulatory framework for energy storage. J Energy Storage 55(D):105862. https://doi.org/10.1016/j.est.2022.105862
48. Salkuti SR (2020) Comparative analysis of electrochemical energy storage technologies for smart grid. TELKOMNIKA Telecommun Comput Electron Control18(4):2118–2124. https://doi.org/10.12928/telkomnika.v18i4.14039
49. Huang WC, Zhang Q, You F (2023) Impacts of battery energy storage technologies and renewable integration on the energy transition in the New York State. Adv Appl Energy 9:100126. https://doi.org/10.1016/j.adapen.2023.100126
50. Selleri G, Poli F, Neri R, Gasperini L, Gualandi C, Soavi F, Fabiani D (2023) Energy harvesting and storage with ceramic piezoelectric transducers coupled with an ionic liquid-based supercapacitor. J Energy Storage 60:106660. https://doi.org/10.1016/j.est.2023.106660
51. Luo L, Lan Y, Zhang Q, Deng J, Luo L, Zeng Q, Gao H, Zhao W (2022) A review on biomass-derived activated carbon as electrode materials for energy storage supercapacitors. J Energy Storage 55(D):105839. https://doi.org/10.1016/j.est.2022.105839

52. Saikia BK, Benoy SM, Bora M, Tamuly J, Pandey M, Bhattacharya D (2020) A brief review on supercapacitor energy storage devices and utilization of natural carbon resources as their electrode materials. Fuel 282:118796. https://doi.org/10.1016/j.fuel.2020.118796

Modeling and Sizing of the Hybrid Renewable System Opting Genetic Algorithm

Kumari Namrata, Nishant Kumar, Ch Sekhar, Ramjee Prasad Gupta, and Surender Reddy Salkuti

Abstract Generally speaking, renewable energy is energy derived from naturally replenishable resources. It exists compared to other energy sources, this one may be used across large geographic regions with the advantage that it cannot be exhausted and is constantly renewed. The objective of this work presented here highlights the availability and potential resources of renewable energy like sun photovoltaic (SPV) and wind turbines. Mathematical models of system components of the designed hybrid energy systems are necessary before optimizing size because they give information about the performance of system components in different operating conditions. The proposed hybrid system for techno-economic analysis was carried out through genetic algorithm optimization. The above cases are implemented and analyzed using MATLAB software. The analysis results in this chapter prove that the designed model of a hybrid energy system where solar photovoltaic sun plate, a storage unit, and a diesel or micro turbine generator as a backup and stand-by source is the most economical option for remote locations.

Keywords Solar photovoltaic · Genetic algorithm · Hybrid system · Optimization · Cost of energy

---

K. Namrata · C. Sekhar
Department of Electrical Engineering, NIT Jamshedpur, Jamshedpur, India
e-mail: namrata.ee@nitjsr.ac.in

C. Sekhar
e-mail: 2019pgeepe18@nitjsr.ac.in

N. Kumar
Department of Electrical Engineering, BK Birla Institute of Engineering and Technology, Pilani, India

R. P. Gupta
Department of Electrical Engineering, MIT Muzaffarpur, Muzaffarpur, India

S. R. Salkuti (✉)
Department of Railroad and Electrical Engineering, Woosong University, Daejeon, South Korea
e-mail: surender@wsu.ac.kr

S. R. Salkuti et al. (eds.), *Power Quality in Microgrids: Issues, Challenges and Mitigation Techniques*, Lecture Notes in Electrical Engineering 1039,
https://doi.org/10.1007/978-981-99-2066-2_25

Nomenclature

PV	Photovoltaic
COE	Cost of energy
LLP	Loss of load probability
TAC	Total annual cost
TALE	Total annual load energy
HOMER	Hybrid optimization of multiple energy resources

1 Introduction

The thirst for getting an uninterruptable power supply is increasing with the growing population and technology. This thirst cannot be fulfilled with conventional energy sources; as a result, the need for nonconventional energy sources is increasing in an exponential manner [1]. To meet the demand in a practical manner, we have to overcome several factors and conditions. Hence, the concept of using renewable energy systems is getting more interest [2]. Using one source of a renewable energy system does not fulfill the need for an uninterruptable power supply, therefore hybrid energy systems are in widespread usage [3]. After the installation of the renewable energy system to the existing grid system, the problem of optimization, stability, reliability, and controllability occurs and this can be minimized by the introduction of a genetic algorithm [4]. For improved performance and intelligence of smart systems involving renewable energy requires optimization algorithms [5]. The genetic algorithm is one of the biological computation methods used as a search algorithm based on the principles of biological evolution. The problem of scheduling in power systems after the integration of renewable energy systems can be solved by genetic algorithms.

Nowadays hybrid energy systems are in more use, especially the systems that contain solar photovoltaic systems because they are widely used in energy supply for remote locations, especially for small isolated regions [6]. Here, in this chapter, genetic algorithm is going to be used for the optimization of a hybrid renewable energy system. The hybrid renewable energy system discussed here consists of a solar PV system and a diesel or micro turbine generator, a combination of the renewable energy system and non-renewable system [7]. The cost of producing energy is significantly decreased by the sizing optimization of the hybrid system or the optimization of the hybrid system as defined. Shubhangi et al. [8] proposed a grid-tied hybrid system for rural electrification. Nishant et al. [9] proposed a hybrid renewable-based system for techno-socio-economic analysis and electrifying the rural community in India. A complete model of this hybrid system with sizing optimization is designed to supply power to small communities where insufficient power is present [10, 11]. Apart from working with both sources together, the analysis of the standalone PV system and analysis of the backup source alone is being carried out here. The genetic algorithm is used here for the optimization process [12]. The objective function does the work of minimizing energy production cost and also covers the load demand through a specified value of loss of load probability. Taking the consideration of global warming emissions into account energy production cost minimization is possible in this optimization process [13]. The analysis of solar radiation data is considered as the first step in this optimization process [14, 15]. After the analysis, the solar tilt angle is to be optimized for further calculations and optimization process and hence the optimization of the photovoltaic panels' tilt angle is carried out. Inspite of extending the main utility grid to different remote areas for power supply, the installation of the hybrid system in those remote areas with sizing optimization is found to be more cost-effective and more beneficial.

2 Introduction to Genetic Algorithm-Based Model

The use of hybrid energy systems, which rely on various renewable sources, is becoming more popular. The effectiveness of these systems was proven when they provided power to various locations where small isolated loads are present. The effects of greenhouse gases can be minimized or can be mitigated by the use of hybrid energy systems because it reduces the emissions of different gases such as carbon dioxide, sulfur dioxide, nitrous oxide, etc. The main advantages of hybrid energy systems are the reduced cost of maintenance and low emission levels of pollutants. To achieve such advantages, the genetic algorithm is opted to optimize and size the system [16]. Further renewable-based hybrid energy systems are approached where the hybrid system may include a nonconventional energy source like a solar photovoltaic system and a conventional energy source like a battery or a diesel generator [17]. A diesel generator is preferred for emergency situations to act as an aid source. A simple design of a hybrid energy system includes the connection of the DC bus and AC bus through a bidirectional inverter. Solar panels deliver the DC output almost at a constant level by the use of a solar charge converter and a battery bank and this DC output is carried through the DC bus. The diesel generator on the other side delivers AC output which is carried through the AC bus and also a load is connected to the AC side. As more than one type of energy source supplies power to the load in this system, so it is referred as a hybrid energy system. A hybrid energy system can be used for certain applications to increase the rate of reliability and also to secure the energy produced and consumed. Factors affecting the hybrid energy system are:

- The nature of renewable energy system.
- The efficiency of the renewable energy system.
- Availability of the energy source.
- Reliability and flexibility of the system with the conventional system.

Several projects using solar PV hybrid energy systems are being implemented in different localities for power supply to remote regions and minor communities. As per some projects, microturbines are used instead of diesel generators because the operational costs and maintenance of microturbines are lower compared to diesel generators [18]. Other attractive features of microturbines are lower noise, lesser pollutant emissions, and higher reliability. Nowadays, the better choice for standby sources is given to microturbines as compared to diesel generators. The primary goal of these initiatives is the optimization of the sizing of hybrid energy system components in order to reduce energy production costs, maximize the use of solar components, and cut back on pollution emissions [19].

Calculation of energy balance per hour by each energy source throughout the year is a process to be carried out before performing the optimization process. The calculation purpose demands mathematical modeling for components present in the designed system, which further requires the collection and analysis of data related to climate changes and other data. In the case's sun PV data analysis on solar radiation and hybrid energy systems and the mathematical model of every component present in the described hybrid system are to be developed and it is a prior step for optimization of the energy system with respect to components sizing [20].

In the field of solar hybrid energy systems, many suitable approaches have been developed for analyzing and predicting solar radiation on a particular day, month, year, calculating the appropriate tilt angles of PV panels, surface azimuth angles, and other necessary angles for optimum performance, developing suitable mathematical models for the solar panels. Those developed approaches recommend iterative or graphical techniques for the optimization purpose of the developing system [21]. In the study of optimization cases, linear programming and iterative attempts are much easier theoretically but practically it is found to be difficult for using the same iterative attempts to solve optimization problems. By the use of novel approaches, optimization problems can be solved more efficiently. There are very few novel approaches or algorithms developed by a few researchers that can be utilized for optimum solutions. The novel approaches include novel searching techniques for selecting an optimum possible solution from the available possible solutions in the design space for the optimum performance of the system or components. For selecting the optimum solution, a search method is developed.

The method is used for choosing the most ideal configuration that meets the given objective functions after the implementation of an appropriate searching technique. This type of approach is best suitable for satisfying multi-objective functions and also when optimization of various variables or parameters for constructing the decision vector is needed. Similar searching approaches are recommended for the optimization of the operating principles of the hybrid system. Some well-known novel searching techniques are particle swamp optimization, genetic algorithm, and simulated annealing approach. All of the aforementioned methods have distinctive characteristics with regard to optimization, nonetheless, an analysis of the literature reveals that the genetic algorithm is the best option for hybrid systems. Its exceptional efficiency makes utilizing a genetic algorithm to get the global optimal solution the best option, and it is also suggested for optimization issues requiring a large number of optimized parameters or variables [22]. The disadvantages of using genetic algorithms are the complexity of coding and the requirement for more computation time for optimization problems.

Installation of this hybrid system decreases the cost of operation and also minimizes pollutant emissions. The effects of greenhouse gases that include carbon dioxide (CO_2), nitrous oxide (NO_2), sulfur dioxide (SO_2), and nitric oxide (NO) emissions can be mitigated by the use of this optimization to satisfy the Kyoto Protocol requirements. India is the 80th country for accepting the amendment related to the second commitment period of the Kyoto Protocol. The Kyoto Protocol is an international agreement linked to the United Nations Framework convention on climate change (UNFCCC) that binds emission reduction targets internationally by its parties [23]. Here, reduced pollution emissions and low maintenance costs are the key benefits. A hybrid energy system employing a sun photovoltaic system like the source of renewable and a diesel engine or a microturbine as standby energy is being tried out using genetic algorithm optimization. Figure 1 shows a block schematic of the proposed solution.

In the block diagram of the proposed hybrid energy system, the bidirectional inverter is used to link the two buses viz. DC bus and AC bus. For preserving consistency in the power supply, the DC bus transports the outputs of both the DC battery bank and the solar PV panels that have gone through the solar charger converter. The AC bus consists of the input of the diesel generator or the microturbine and output to the load through the bidirectional inverter. Here more than one energy source is utilized for power supply to the AC load or the utility grid, hence termed as hybrid energy system [23, 24]. When compared to energy systems that rely on a single energy source, this approach can increase energy dependability and security.

A similar model is designed in the Palestine territory and is analyzed. It is a Mediterranean region and suitable for hybrid systems containing Solar PV systems in it. Palestine has a significant potential for solar radiation thanks to its high annual and monthly solar hours. Between 5.5 and 6 kWh/m^2 are the typical values for annual daily solar radiation on a horizontal surface on average.

In the territories of Palestine, many small communities are present in isolated regions and remote regions. Those small communities mainly depend on diesel generators for power supply to their homes. The territories of Palestine or Palestinian territories utilize external sources of power with certain restrictions over required

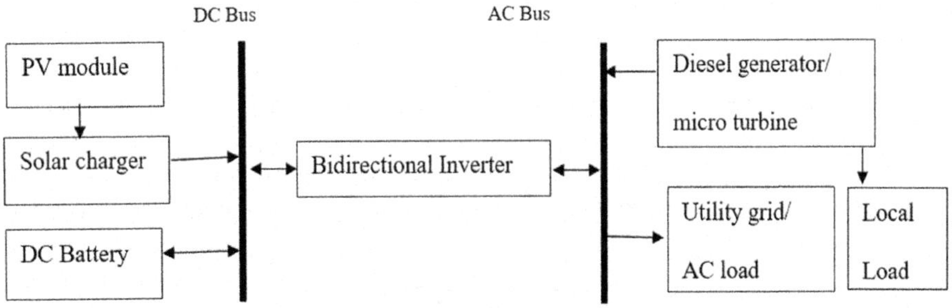

Fig. 1 Block diagram of the designed hybrid energy system

capability [25] like output, more than 30% of Palestinian families currently experience power outages. Many institutions operating in Palestine have made delivering electricity to these isolated areas and addressing the issue of inadequate power supplies their top concerns. In order to provide power to these small towns, a few projects using hybrid energy systems (solar panels and diesel generators) have been constructed. Compared to diesel generators in hybrid energy systems, microturbines are having more attractive features like low operational costs, lower maintenance costs, high reliability, less noise, fuel flexibility, and low pollutant emissions [26]. These features enhance the use of micro turbines instead of diesel generators as standby sources. The main objectives of using a genetic algorithm in this model design are:

- Sizing optimization of components present in the hybrid energy system.
- Minimization of cost of energy production.
- Maximum utilization of solar PV panels.
- Minimization of pollutant emissions.

Prior to executing the optimization or while carrying it out, it is essential to complete an energy balance per hour over the course of the whole year. The computation of the energy produced by various energy sources is done for this energy balance. Each and every component of the proposed system must be mathematically modeled for this computation, and the study of climatic data from local meteorological sources is also necessary. In addition, before creating an optimization method for the size optimization of hybrid system components, the study of solar radiation data and the building of a numerical simulation of the elements that make up the hybrid system used today must be finished.

Analysis and estimation of solar radiation data are carried out by Angstrom-type models, higher-order correlations, and on the basis of ambient temperature. The mathematical model of solar panels also requires the determination of optimal tilt angles and surface azimuth angles. These methods are utilized in this chapter to construct new models by improvising on already existing ones.

Several scholars advocate iterative strategies and/or graphical techniques to aim for size hybrid optimization of energy elements of the system, but only a few researchers really use the method based on unique algorithms. With the unique algorithmic technique, a conceptual area of potential output is initially built to ideal component dimensions of the hybrid system. The best option or configuration that fulfills the given goal function is then chosen using a search strategy. These cutting-edge methods are advised for solving multi-objective problems as well as for decision vector optimization, which consists of numerous variables or parameters [27]. These novel algorithms are recommended for the optimization of the operating strategy of hybrid systems.

For solving such optimization problems, using linear programming or iterative approaches makes the system more difficult in reality. Some examples of these novel searching techniques are genetic algorithm, simulated annealing approach, and particle swarm optimization, etc. these are some common approaches for optimization problems. The best strategy for the optimization of hybrid systems, according to a thorough analysis of all these approaches, is a genetic algorithm because of its advantages over other algorithms. The usage of the genetic algorithm is appropriate for optimization problems with a lot of optimized parameters since it is effective at finding the global optimal solutions. The problems with the genetic algorithm are the longer computation time for solving optimization problems and coding is relatively harder compared to other algorithms.

3 System Modeling

3.1 Modeling of Components

Firstly, the optimization of the tilt angle of the PV module is carried out by maximizing energy production annually. Taking the reference of measured solar radiation on a horizontal surface, the solar radiation on a tilted surface is calculated. For calculating the solar radiation on a tilted surface, the anisotropic model is utilized, and for calculating the diffused component of solar radiation on a tilted surface, the correlation of Orgill and Holand is employed. The following are the equations that were used to calculate this model:

$$P_{\mathrm{PV-gen}} = V_{\mathrm{PV}} I_{\mathrm{PV}} = V_{\mathrm{PV}} I_0 \left[e\left(\frac{V_{\mathrm{PV}} + I_{\mathrm{PV}} * R_S}{a * V_T} \right) - 1 \right] - V_{\mathrm{PV}} \left(\frac{V_{\mathrm{PV}} + I_{\mathrm{PV}} * R_S}{R_P} \right) \tag{1}$$

$$I_{\mathrm{PV}}(G, T) = I_{\mathrm{mp\text{-}STC}} * \left({}^{G}\!/_{G_{\mathrm{STC}}} \right) \tag{2}$$

This type of approach is suitable for the surface present near the equator or oriented towards the equator. A genetic algorithm has been used to carry out this tilt angle optimization and a one kWh rating PV panel was selected for the calculation of annual solar energy production. Maximizing the yearly solar energy output is the algorithm's goal function. The Nablus location was taken into account while using the hourly solar temperature and solar radiation data for this modeling. The lower constraint of the tilt angle for optimization was 0°, and the upper constraint was 90°.

For the calculation of generated energy of solar PV panels, a mathematical model of the solar PV panel is designed accurately and the model should define its operation effectively. The model calculation should consider the effect of variations in solar temperature and solar radiation. The functioning of solar PV panels and their capacity to produce electricity are impacted by variables such as humidity, wind speed, and other environmental changes in addition to temperature and radiation fluctuations. As these elements have an indirect impact, they are still not considered in the mathematical model of solar photovoltaic panels. Their consideration is taken in the techno-economic analysis of PV systems. The effects on solar irradiance level include the presence of water vapor particles, dust accumulation due to humidity, and the humidity on the encapsulating material of the solar cell. These effects should be taken into consideration as they affect the radiation levels. Water vapor particles on solar cells have been taken into account in the model because they have an impact on how solar radiation varies.

The annual maintenance cost of the PV system includes the cost of regular cleaning of the PV panel surfaces and also it removes dust accumulation on the panel surface. The efficiency and performance of solar PV panels are degraded due to the exposure of panels to the humid climate and this type of degradation is considered in the analysis of this model and is discussed later. In order to determine the energy in the model, various approaches are available for the calculation of solar PV parameters like the series resistance (RS), the shunt resistance (RP), and the ideality factor of the diode (a). Apart from this model, the single-diode model or two-diode model of solar PV cells can be used for mathematical models for this purpose. Assuming that the above-mentioned parameters of the PV cell model are calculated or are available, then Eq. (1) can be used for the calculation of generated power by PV panels. According to the equation, the current and voltage of solar panels must both be previously understood, but because the parameters for both voltage and current are interdependent, a numerical method must be developed for determining the current and voltage of solar panels given that IPV (G, T) is the output current, and calculated or recognized. The resulting current of the PV panel may be estimated using Eq. (2). Peak power tip current during routine testing, which may be found on nameplate data or provided by the manufacturer, is the Imp-STC utilized in Eq. (2). This equation may be utilized for the aforementioned purpose and it meets the requirement since most charger regulators often employ a maximum power point tracker to maintain the operating area of solar PV output around the maximum power point and prevent unneeded changes in output power. In Eq. (2), the calculated radiation on a tilted surface is denoted by G and its units are W/m^2 and the standard test for radiation is denoted by GSTC and the standard value of GSTC is 1000 W/m^2.

Here the optimization of various components used in the hybrid energy system with respect to their sizes is carried by the genetic algorithm. The selection of the types of components is made from the specified set of assigned types or brand-named components. The objective function required for the genetic algorithm is the minimization of the cost of energy. Also, the tilt angles of solar PV panels and surface azimuth angles are also included in the optimized parameters. Possibility cases are discussed in the results and discussion section. Apart from these optimized parameters, the mounting structure type of PV panels is also included. The case where the diesel generator is used instead of a microturbine as a standby source is also analyzed and studied. Also, the cases where the load reliability is considered to be 100% and loss of load probability is considered with a specified limit are also analyzed and studied.

As was already said, the usage of genetic algorithms determines the computational values of the sun PV model parameters. The selected parameters cover the available range of radiation of solar and the range of temperature of the solar cell. The decision vector includes the model parameters as discussed before the ideality factor, the shunt resistance, and the series resistance of the diode (a). The decision vector needed to take into account additional characteristics like saturation current and photon current since the comparable path was utilized to represent the photovoltaic circuit. If the two-diode model is preferred then the saturation current and ideality factor of the second diode are also to be included in decision vectors as parameters. So many research people have tested various alternatives for optimum calculations and a three parameters single diode model is recommended for this optimization because the three parameters single diode model provided the most accurate results than the other models.

The lower constraint of series resistance is 0.01 Ω, and the upper constraint is 1.2 Ω. The lower constraint of shunt resistance is 50 Ω, and the upper constraint is 1000 Ω. The diode ideality constant has an upper limitation of 2, and a lower limit of 1. For concision, the equations for the calculation of photon current (I_{Ph}), the diode saturation current (I_0), and the thermal voltage (VT) are not considered here.

The performance of the PV system does not remain the same for the whole year, and there is a degradation in its performance by a certain ratio. The internal connection between the PV cell and module, outside moisture that has accumulated on it, the kind of PV panels, the substances used in wrapping are the causes of this deterioration. This PV system's deterioration rate should typically be one percent. This degradation rate might also take the losses resulting from mismatch into account. The incompatibility damages are the difference in the total power of the mismatches and all panels added together and the total energy for the entire group of arrays. A portion of these losses is attributable to panel manufacturing flaws, while the rest portion of the losses is attributable to the distributional losses of solar PV arrays made up of many strings and derived from the electrical properties of the solar PV modules. Many methods are used to account for the deterioration of this model, but one of them involves compensating the impact on the solar PV system with increases in the size of the PV system by a certain % that is the same as the % provided to the percentage rate of degradation. This states that the size of solar photovoltaic module circuit at the nth annual is equal to the sum solar-power system size in the $(n-1)$th year and the % rate of deterioration as provided in Eq. (3).

$$P_{\text{PV system}(n)} = P_{\text{PV system}(n-1)} * DPR * P_{\text{PV system}(n-1)}; \quad n = 2, 3, 4, \ldots, \text{project lifetime} \tag{3}$$

where $P_{\text{PV system}(n)}$ is the photovoltaic module circuit size at nth annual, $P_{\text{PV system}(n-1)}$ is the PV system size at (n−1)th year and DPR denotes the degradation percentage rate.

The consumption of fuel by the microturbine depends on the actual power generated for a specific interval of time. For various loads, the amount of flow of natural gas as fuel in the microturbine is provided by the manufacturer, usually in plot data or in table data. Equation (4) gives the first-order relation between the output power of the 30 kW Capstone microturbine and the amount of natural gas flow in it. The natural gas flow is represented by FFMT and its units are m^3/h, and the microturbine output power is represented by P_{MTmod} and its units are kW.

$$FF_{\mathrm{MT}} = 0.314 * \left(P_{\mathrm{MT}_{\mathrm{MOD}}}\right) + 1.548 \quad (4)$$

The output power is also affected by the effects of the altitude and ambient conditions over the sea level. These effects alter the amount of generated power and must be taken into consideration. So, modified output power by microturbine is given by Eq. (5).

$$P_{MT\_MOD} = P_{\mathrm{MT\_RATINGS}}(1 - 0.0001 * H) * (0.911 - 0.0051 * T_{\mathrm{amb}}) \quad (5)$$

where P_{MTrating} is the microturbine-rated power, and T_{amb} (°C) indicates the atmospheric temperature and h denotes the height in (m) above sea level.

3.2 Modeling Based on Economic Analysis

The economic analysis is the consideration of different costs that are necessary for modeling the designed model. The different types of costs mostly include the capital costs of different components used, the cost of their installation, maintenance cost, operational costs, and also replacement costs for any component [24]. Life cycle costing is utilized in this economic analysis and is necessary for the analysis and comparison of different models or scenarios, which indeed helps to find the least cost model. This strategy depends on optimizing a solar PV system such that the energy generated by the panels and the stored energy in the battery bank is accessible before the source is filled. Calculating the cost of each component utilized in the model throughout the course of the component's maximum useful life is known as life cycle costing. Also, it specifies how long the given model or project will last. The maximum lifespan of a solar PV panel is typically 25 years, that of a microturbine is specified by the manufacturer in terms of operating time before overhaul and replacement, and that of a battery depends on usage or on the quantity of discharge and charge cycles the battery has undergone as well as the depth of discharge value.

4 Simulation and Optimization of Sizes of Components

4.1 Simulation Attempt

Energy is balanced for every hour throughout the year and this is possible by a simulation program approach. The created simulation program will simulate the functioning of the desired system that integrates many energy sources according to power flow management in the designed system (solar PV energy and microturbine). The flow of power between the sources present in the hybrid energy system and the priorities which are used to calculate this power flow is determined as per this strategy. This plan is based on making the most of a solar PV system, which means that the energy produced by the panels and the energy that has been stored in the batteries comes before the supply is loaded. In case this energy does not meet the requirement of load power, then the microturbine can be used to run as a backup source or standby source.

There are some cases where the energy stored by the battery is charged fully and the generated energy by solar PV panels exceeds the requirement of load power. The excess energy is to be utilized, so a dump load is introduced. The dump load consumes the excess energy generated. If the battery tank discharges to its maximum allowable depth of discharge level and the energy generated by the solar PV system is not sufficient to supply the required load, the microturbine is allowed to operate so as to charge the battery and supply the load. During this charging process, the bidirectional inverter acts as a rectifier and allows the power to flow to the battery bank for charging the battery.

In this chapter, the energy balance is conducted by assuming that there is no interruption in the power flow in the entire year or by assuming that a specific value for the loss of load probability is fixed. The calculation of loss of load probability is done by Eq. (6).

$$\text{LLP} = \frac{\sum_{1}^{h=8760}\text{Energy deficit}(h)}{\sum_{1}^{8760}\text{Load demand}(h)} \tag{6}$$

The quantity of load energy required for a specific hour that cannot be satisfied by the energy provided by various generating sources or storage sources is referred to as the energy deficit (h) when the loss of load probability is stated as LLP. The energy deficit can be calculated by using Eq. (7).

$$\text{Energy}_{\text{deficit}}(h) = \text{Load\_demand}(h) - [E_{\text{PV}}(h) + E_{\text{MTE}}(h) + E_{\text{MTH}}(h) * E_B(h-1)] \tag{7}$$

The energy produced by the solar panels for an hour is denoted by E_{PV} (h), the electrical energy generated by the microturbine for an hour is denoted by E_{MTE} (h), the thermal energy produced by the microturbine for an hour is denoted by E_{MTH} (h), and the energy stored in the battery at the end of the prior hour is denoted by E_B (h−1).

4.2 *Optimization of Component Sizes Based on Genetic Algorithm*

For solving optimization problems, the genetic algorithm plays a useful novel algorithm in various aspects of life. At each step, the individual solutions are selected, which constitute the population. The genetic algorithm follows three rules to form the next generation or population from the current or existing population. Those three rules are, namely, selection rules, crossover rules, and mutation rules.

As was previously said, the primary goal of optimizing the parameters for a hybrid energy system is to ensure that the sizes of the components utilized in the system fulfill the various preset objective functions. To obtain the optimized parameters, in this problem genetic algorithm is approached and suitable MATLAB code has been developed for running the optimization problem. The developed MATLAB code consists of two programs and those are genetic algorithm programming and a program related to fitness function for optimization. The first population generation is formed by choosing random population members or by producing potential outcomes or solutions in the genetic algorithm's MATLAB code. Each potential answer is converted into a code for the choice vector by taking into account the lower and higher bounds. This initial population generation undergoes successive iterations, and each generation of population members is evaluated for calculating the cost of energy, and the cost of energy is the required fitness function. The member evaluated from the generation selects the optimum fitness function or the minimum cost of energy. The flow chart of a genetic algorithm with the optimization procedure is shown in Fig. 2. After so many iterations and many executions through this genetic algorithm optimization process, it was seen that the population size of 50 members was suitable and sufficient for the optimization but the required population size for the most optimum solution is found to be 80. In many cases, the population size was found to be less than 60. The selected crossover factor for this purpose is 0.8 but the selected mutation factor is 0.2.

The selection function approach was the function given by Roulette and the crossover function followed was the arithmetic one. The Gaussian function was selected as the mutation function, and the number of specified generations is considered as the stopping criteria. The quantity of defined generations was chosen as the terminating criterion, and the Gaussian mutation function was chosen as the mutation function. In reality, since it depends on the optimization issue itself, it is possible to approach the other sorts of functions adequately. However, depending on the optimization challenge, alternative function types may be employed successfully. Regardless of whether the supplied issue is limited or not, the number of parameters that need to be optimized must fall inside the specification factors in order for the function to be employed. The MATLAB optimization toolbox is used for a comparable optimization issue in certain research articles. It was found that the functions that are defined before or before produce outcomes that are almost accurate with fewer generation numbers. If the reader is interested in finding more information about these specified functions is advised to check from the Help menu found in the MATLAB optimization toolbox.

For every member of the population in each generation, the calculation of loss of load probability is carried out. After calculation, the member who is not satisfying the required load for a certain percentage value is taken out from the population and is not included for further iteration. This is the selection process. After this, the process of crossover, mating, and mutation is carried on the members who successfully got selected for the next generation. To keep the population size constant at the

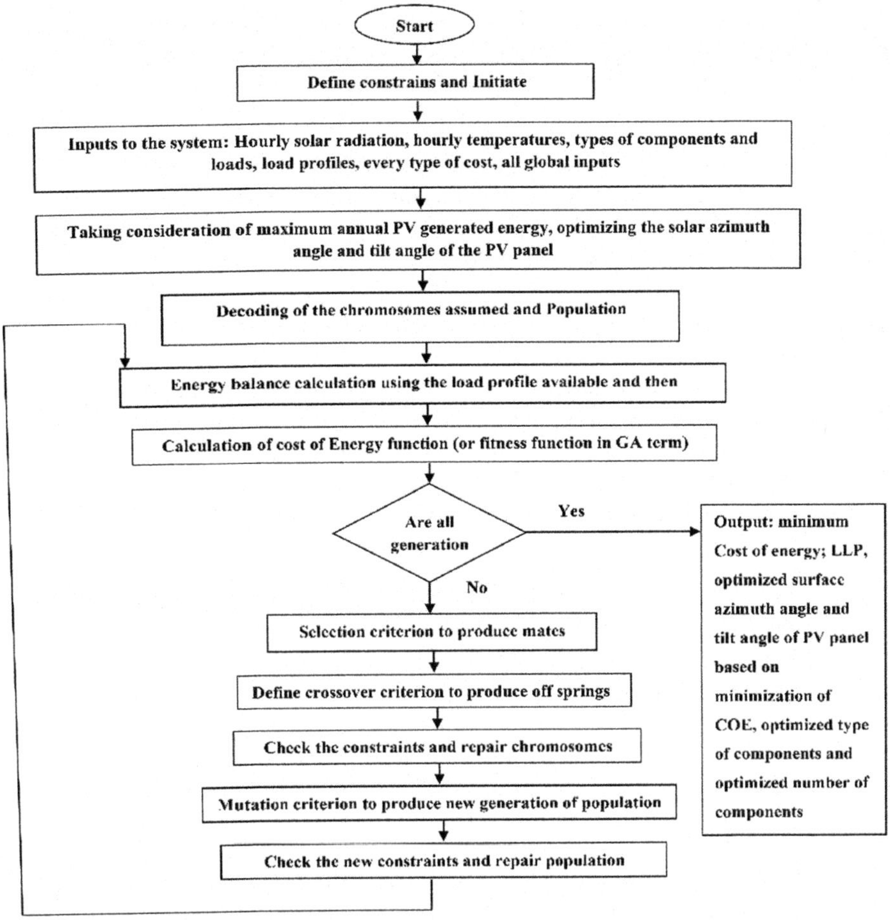

Fig. 2 Genetic algorithm flow diagram

beginning of each generation, the excluded members are replaced by new excluded members that are acquired when the mutation process is complete. Until all iterations or generations have been completed, or until the halting requirement has been met, this process of substitution and mutation continues. Each generation chooses a member from the population based on how satisfied they are with the energy function's lowest cost. The parameters to be optimized and included in the decision vector are stated below:

- The type of brand of solar PV panel (T_{PV}).
- The type of charging battery used (T_{B}).
- The microturbine type utilized in the hybrid energy system (T_{MT}).
- The solar PV mounting fixture type (T_{F}).
- The quantity of solar PV panels for the model (N_{PV}).

- The quantity of microturbines to be required for the model (N_{MT}).
- The solar panel tilt angle for minimization of cost of energy (β).
- The surface azimuth angle of the panels for the minimization of cost of energy (λ).

In the optimization sizing of components, priority is given to the parameters related to the number of each component rather than the type and brand of the components to be used. Equation 8 provides the decision vector equation with mentioned parameters and the decision vector is expressed as DV.

$$\text{DV} = \left[T_{\text{PV}}, T_B, T_{\text{MT}}, T_F, N_{\text{PV}}, N_B, N_{\text{MT}}, \beta, \gamma\right] \tag{8}$$

The cost of the energy function is the objective function that is to be optimized in this model of a hybrid energy system. The goal is the minimization of the fitness function present in the algorithm. The definition of cost of energy can be given as the cost of generating energy of 1 kWh to supply to the load and being used by it. The equation for calculating the cost of energy is given as,

$$\text{COE} = \frac{\text{TAC}}{\text{TALE}} \tag{9}$$

where the total annual cost and total annual load energy are denoted by TAC and TALE, respectively. The total annual cost includes different types of costs of components that are being used in the construction of a hybrid energy system. The different types of costs are discussed in the previous section and apart from them the TAC calculation also includes the cost of gaseous emissions like emission of CO_2, NOx, and SOx. The various types of costs that are related to different components are displayed in Tables 1, 2, 3, and 4.

While solving the optimized function by genetic algorithm, the constraints should also be considered. The value of the upper constraint and lower constraint for various parameters that are used in the optimization process depends on the parameter itself. The parameters to be optimized are the types and numbers of the components. The lower limitation for component types should be 0 or larger, and the upper restriction for types is the number of accessible kinds for this function. The lower constraint for the numbered parameters should be equal to 0 or greater than 0, and the upper

Table 1 Features of different kinds of batteries utilized in the study of the system

Battery parameters	Type of battery used			
	Type 1	Type 2	Type 3	Type 4
Rating of battery (Ah)	2430	1700	1215	648
Battery Voltage (V)	2	2	2	2
Cost of battery (Rs/unit)	88,900	58,450	42,700	21,350
Maintenance cost (Rs/unit)	1050	770	560	280

Table 2 Features of two microturbines used in the study of the system

Microturbine features	Type of microturbine used	
	Type 1	Type 2
Rating of the microturbine (Ah)	65	30
Capital expenses (Rs)	174,300	207,900
Cost of maintenance (Rs/unit)	1.23	1.4

Table 3 Specifications of different PV panel types and mounting structures applied in the system studies

PV panel parameters	Photovoltaic type			
	Type 1	Type 2	Type 3	Type 4
PV panel rating (W)	2430	1700	1215	648
Number of panels for each mounting structure of type 1	2	2	2	2
Number of panels for each mounting structure of type 1	11	12	9	9
Cost of panel (Rs/Watt)	889	584.50	427	213.50
Installation cost (Rs/Watt)	49	49	49	49
Maintenance cost (Rs/Watt)	1.75/yr	1.75/yr	1.75/yr	1.75/yr
Cost of type 1 mounting structure (Rs/unit)	18,200	20,300	27,790	27,790
Cost of type 2 mounting structure (Rs/unit)	171,640	254,520	254,520	254,520

constraint has no limitation theoretically and practically it depends on the designer of the model.

5 Inputs for Simulation

Ramallah (31.8 N latitude; 35.23 E longitude), where a surrounding electrified neighborhood was chosen, is located in the Mediterranean area of Palestine territory, where the hybrid energy system was developed to give power to outlying small communities. A little bit of survey regarding the measurements and other features was conducted in this region by some researchers. It was assumed to be that this community may have consumption similar to the nearby un-electrified community. The climate in the territories of Palestine is a seasonal type climate as a result the consumption of load changes from season to season. This consumption of the load is divided into two categories and they are load consumption in the summer season or hot period and load consumption in the winter season or cold period. In each category of load consumption, the days are divided into working days and weekend days. The load consumption measurement study was conducted over the course of a full week, and the hourly measured results have been recorded. The survey report details the heat load profile and the kind of heat load that is in use in this specific neighborhood for

Table 4 Remaining inputs of the simulation

Objects	Cost/value
Expenditure cost of bidirectional inverter (Rs/kW)	50,050
(%) Efficiency of bidirectional inverter used	91
Expenditure cost of photovoltaic solar charger converter (Rs/kW)	31,500
(%) Efficiency of the PV charger converter	94
Expenditure cost of the additional diesel generator	38,500
(%) Watt hour efficiency of the battery	85
Lifetime of the solar panels (years)	23–25
Lifetime of the battery used (years)	4–6
Lifetime of the designed system (years)	23–25
Fuel cost of natural gas (Rs/m^3)	23.1
Cost of diesel (Rs/l)	125
Replacement time period for battery of microturbine (h)	7500–8000
Replacement time period for igniters, thermocouples, and injectors of fuel (h)	17,000–20,000
Replacement time period for air filters and fuel (h)	19,000–20,000
Replacement time period for DG (diesel generator) (h)	22,000–24,000
Replacement time period for microturbine (h)	76,000–80,000
% rate of inflation of fuel	5
Carbon dioxide emissions from microturbine (kg/MW hour)	788
Carbon dioxide emissions from DG (kg/MW hour)	650.1
Cardin dioxide cost (Rs/Kg)	0.98–1.00
Nitrous oxide emissions from microturbines (kg/MW hour)	0.238
Nitrous oxide emissions from DG (kg/MW hour)	9.79
Nitrous oxides cost (Rs/Kg)	294.00
Sulfur oxide emissions from microturbines (kg/MW hour)	0.0078
Sulfur oxide emissions from DG (kg/MW hour)	0.21
Sulfur oxide cost (Rs/Kg)	70.00

each hour. The heat load that would be provided directly through the microturbine was determined using this research. The hourly load profiles for various categories for both the electric load and the heat load are shown in Fig. 3. Solar radiation and hourly average temperatures are needed to calculate the electricity produced by solar PV. The average annual solar radiation on a horizontal surface at the place chosen in the region is 5.94 kW h/m^2/day, yet there are more than 3000 h of sunshine each year.

Four different types of solar PV panels have been chosen for the purpose of size optimization, along with four different types of battery chargers, two different types of PV panel mounting fixtures, and two different types of microturbines. It is mentioned earlier that different cases of this model are carried out in different manners. One

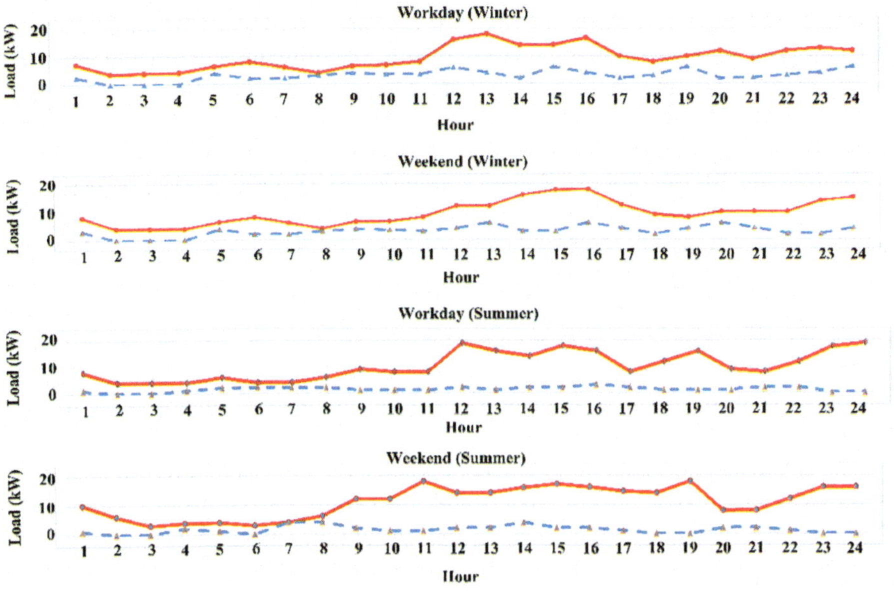

Fig. 3 Hourly typical load profiles (solid red for total load and dashed blue for heat load)

such manner is by assigning a type and then optimization of the number of units from the assigned type. Another possibility is the optimization of both the types and numbers of the components. In Tables 1, 2, 3, the ratings and different costs of various selected types of unit batteries, solar PV panels, and microturbines are given.

In Table 4, the required simulation program inputs are shown. Also, the gas emissions of CO_2, NO_X, and SO_X from the microturbines and diesel generators are displayed. The gas emission units are kg/MW h, and the cost of each emission is expressed in the units of $/kg. The types of mounting fixtures on PV panels include the fixing type of PV panel on a roof or ground with an inclination of tilt angle by a certain angle and the single axis fixture type with a solar tracker that absolutely follow the movements of the sun. The movement of this single-axis fixture is designed to keep the solar panel's surface azimuth angle equal to that of the sun. With each form of the fixture, the number of solar panels needed only relies on the mounting style and the kind of panels being used.

6 Development of Optimization

As per the survey of different research papers, software based on MATLAB has been developed for the analysis and optimization purpose of the genetic algorithm-based renewable energy system. This software is used for the optimization of the off-grid hybrid renewable energy system, which is based on solar PV renewable energy source. The tilt angle, component sizes, and component type are all optimized by this program. The PV panel mounting fixture type is one of the optimal parameters. For the simulation and optimization of the hybrid energy system, different commercial tools of software have been developed. Solar Design Tool, Solar Pro, RET Screen, PVSYST 4.33, and HOMER software are the most often used commercial software programs for the optimization of hybrid energy systems. Every software tool mentioned here are having specific features and capabilities of its own in order to simulate and optimize the hybrid energy system. The most popular and frequently used of these programs is HOMER. The HOMER program can simulate various PV energy system designs and can do sensitivity and optimization analyses. The built software in this chapter may be used to deal with numerous instances, calculate various parameters, and analyze the impact of changing any parameter on the outcomes. HOMER or other simulation and optimization tools can also be extremely helpful.

The characterization of the various components for various real-world conditions is carried by the mathematical models and it accounts for the different conditions that affect the operation of components present in the system. Such as in the software developed for this system the way of calculation for the generated power from solar panels is specified and also the angle of tiltness of the solar panel and surface azimuthal angle of the panel are specified. Those specified values and the consideration way of derating and degradation of the solar panel varies from different software such as HOMER. In the model created for the microturbine's fuel consumption, the impact of the surrounding temperature and the site's elevation above sea level must be taken into account. These effects are not considered in the software we use such as HOMER. There is the possibility for optimization of the type and size of the components along with the optimization of the tracking system type of the PV panel, also there is the possibility of inclusion of the optimization of tilt angles and surface azimuth angles.

The developed software is used to find and calculate the results on the basis of hourly, monthly, and yearly calculations. The software code is sometimes modified to address issues like the system's annual hours not being able to handle the load requirement. It is possible to plot any variable to make the results more obvious.

7 Discussion Based on Results

7.1 Mathematical Modeling Results of Solar Panels

The method to be utilized for modeling solar PV panels depends on the extraction of the parameters and/or variables using a genetic algorithm. Similar to Table 2, Table 5 shows the values for the three parameters for each kind of solar panel that will be utilized in the solar PV model. The average absolute values of the currents are compared to the corresponding currents at the greatest power point in order to

Table 5 PV model parameters (extracted using a genetic algorithm)

Parameter	PV type			
	Type 1	Type 2	Type 3	Type 4
Ideality factor (a)	1.175	1.014	1.08	1.037
Series resistance (R_S) in Ω	0.228	0.715	0.371	0.338
Shunt resistance (R_S) in Ω	921.9	987.6	972.6	935.9
Percentage of average absolute error in current (%)	0.18	0.25	0.19	0.19

calculate percentage values of average absolute errors in the currents. The difference between the value of output current from the PV panels using the extraction process by genetic algorithm and the value of current from the nameplate details is termed as a calculated error.

7.2 Results Obtained on Analysis of Solar Data

For optimization of tilt angle of the PV panel and the surface azimuth angles, a genetic algorithm has been utilized. The PV tilt angle has a value of 32.8° in terms of optimizing the yearly solar PV energy production. The optimization of the surface azimuth angle of the PV panel is also performed. In the northern hemisphere, solar panels are often oriented towards the south, but for the yearly energy output to be maximized, they must be angled to the west or east by a certain value. In this model study, the certain angle value obtained is +16° towards the west of south and it is an optimized value for greater energy production annually. It is clear that the determined surface azimuth angle, which is thought to be the ideal value for maximum yearly solar energy output, may not be the same as the ideal angle thought to be the best for minimizing energy costs. To maximize production, the ideal surface azimuth angle should be given more importance.

7.3 Results Based on Sizing Optimization of Hybrid System

In this chapter, various cases and scenarios have been studied and analyzed for selecting the most optimal solution or configuration which is suitable for satisfying the load demand with the minimum cost of energy production and with the noted value of loss of load probability. Some of the selected scenarios/case results have been displayed in Table 6.

Table 6 Obtained results from simulation considering different cases

Studied cases	Type of PV panel	Number of solar panels	Type of mounting-fixture of solar panel	Surface azimuth angle (^{0})	Tilt angle (^{0})	Type of micro-turbine used	Number of micro-turbines	Type of Battery used	Number of units of battery	COE (Rs/ kWh)	Operating hours of microturbine	LLP (%)
1	4	96	1	14	30	1	1	4	20	18.11	2690	0
2	4	96	1	14	32.8	1	1	4	20	18.11	2695	0
3	4	96	1	16	30	1	1	4	20	18.11	2693	0
4	4	96	1	0	30	1	1	4	20	18.16	2710	0
5	4	96	1	0	32.8	1	1	4	20	18.17	2713	0
6	4	84	2	NA	36	1	1	4	19	18.63	2777	0
7	2	117	1	14	29	1	1	2	8	18.39	2765	0
8	4	87	1	17	29	1	1	2	8	18.20	2760	0
9	2	120	1	15	25	1	1	4	20	18.28	2725	0
10	1	149	1	14	30	1	1	1	6	19.51	2780	0
11	3	96	1	12	30	1	1	3	11	18.61	2778	0
12	1	150	1	12	29	1	1	3	12	19.54	2774	0
13	1	137	1	14	30	1	1	4	23	18.60	2864	0
14	4	88	1	14	29	1	1	4	19	18.52	2921	0

(continued)

Table 6 (continued)

Studied cases	Type of PV panel	Number of solar panels	Type of mounting-fixture of solar panel	Surface azimuth angle $(^0)$	Tilt angle $(^0)$	Type of micro-turbine used	Number of micro-turbines	Type of Battery used	Number of units of battery	COE (Rs/kWh)	Operating hours of microturbine	LLP (%)
15	4	80	1	16	29	1	1	4	12	21.21	6365	0
16	NA	0	NA	NA	NA	1	2	NA	0	22.89	8761	0
17	NA	0	NA	NA	NA	1	2	NA	0	24.57	8760	0
18	4	490	1	13	40	NA	0	4	337	35.7	NA	2
19	4	507	1	14	30	NA	0	4	415	39.9	NA	0
20	4	459	1	15	30	NA	0	4	302	32.9	NA	4.9

* NA means not applicable

7.4 Discussion of Result of Best Optimized Case Study

Case 1: This is a best-optimized case with optimization of a number of components of each type and respective component type. In this case, the values of tilt angles and surface azimuthal angles are being optimized in order to minimize the cost of energy production instead of optimizing the angles with respect to the maximization of annual solar energy production. As mentioned earlier in the previous section that the optimum value for the tilt angle was 32.8 degrees and the optimum value for the surface azimuth angle was 16 degrees in order to obtain maximum annual solar energy production. It is also clear that the optimum value of tilt angle and the surface azimuth angle is for maximizing the annual solar energy production but does not ensure the minimization of the cost of energy production and the optimum value might be different. In reality, the difference between the optimum values for maximum solar energy production and the optimum values for the minimum cost of energy production is relatively smaller. Here in case 1, the working mode of the microturbine is set to cogeneration mode so that the electrical load is supplied electrical energy by the microturbine electrical output, and the local heat load is supplied the heat energy directly from the generated heat energy through a heat exchanger during the working of microturbine. Here the microturbine operates at its rated power in order to supply the electric power to the electric load and the additional amount of power generated is used for charging a battery via a bidirectional inverter. In case number 14, the electrical output from the microturbine is utilized to supply both the electrical load and heat loads. In case number 15, the microturbine operation depends on the pattern of load changing as per demand. In these cases, no additional power generated is available for charging the batteries. From Table 5, it can be seen that the annual runtime for the microturbine is about 2692 h. It indicates that daily, the microturbine is allowed to operate and function for a specific number of hours. According to the simulation findings, the microturbine generates roughly 57% of the total energy produced by the solar panels and the microturbine itself. According to the results of the simulation, just around 4% of the energy created overall has been wasted, while 14% of energy is lost annually. The energy efficiency in this scenario is around 86% annually, and the microturbine's capital cost appears to be higher than the capital cost of the other system components. The whole intended solar system (PV panels, a PV regulator, and a battery bank) accounts for about 41% of the total initial costs, the microturbine accounts for about 49% of the total initial expenses, and the remaining 10% is accounted for the bidirectional inverter and rest arrangements.

In Fig. 4, the monthly energy contribution of the solar PV system and microturbine is shown along with the load energy and dumped energy of every month. It is clear that the microturbine produces more energy during periods of decreased solar photovoltaic energy. During the same time period, the amount of dump energy is also high. It results from the direct usage of heat energy by the heat load from the produced heat energy of the microturbine through the heat exchanger when the microturbine is operating at its rated power.

Case 2: This situation is comparable to Example 1, however, neither the tilt degrees nor the surface azimuth angles are tuned to maximize yearly energy output in this instance. The surface azimuth angle is tuned to reduce the cost of energy production, whereas only the tilt angle is targeted to maximize yearly energy output.

Case 3: This situation is comparable to Example 1, however, neither the tilt degrees nor the surface azimuth angles are tuned to maximize yearly energy output in this instance. The only angle that is designed for maximum yearly energy output is the surface azimuth angle, whereas the tilt angle is adjusted for lowering the cost of energy production.

Case 4: It has a similarity to case number 1 but here both the tilt angles and surface azimuth angles are not optimized according to the maximization of annual energy production. Here the surface azimuth angle is being directed towards the south direction for annual energy production and the tilt angle is being optimized for the minimization of the cost of energy production.

Case 5: This situation is comparable to Example 1, however, neither the tilt degrees nor the surface azimuth angles are tuned to maximize yearly energy output in this instance. Here, the tilt angle has been tuned to maximize the yearly energy output while the surface azimuth angle has been set to face south for annual energy production.

It can be clearly noticed that from case 1 to case 5, the difference between the cost values of energy production is comparatively smaller. So it indicates that the optimization of tilt angle for maximization of annual energy production of solar PV

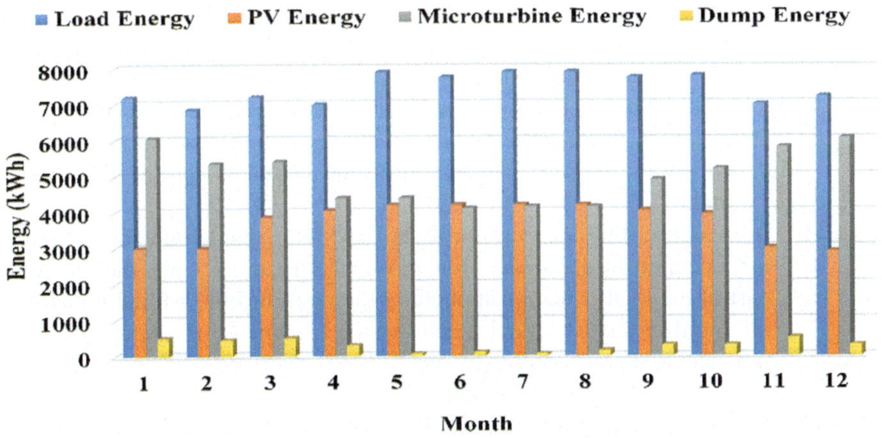

Fig. 4 Monthly PV panels generated energy, microturbine generated energy, and dump energy

system is more effective with the surface azimuth angle being directed towards the south. Also, it suggests that there is no need for calculating the surface azimuth angles for optimized value and no additional mathematical manipulations and evaluations are required for calculating the optimized value of tilt angle for minimization of cost of energy production.

Case 6: From here the evaluation of other cases is carried out which are not the main optimized cases but their analysis helps for further development. Case 6 is identical to Example 1, but here the yearly energy production is taken into account using the photovoltaic mounting fixture type 2. The yearly energy output from fixture type 2 is significantly higher than that of fixture type 1, and fewer solar panels are needed than with type 1 as well. That is feasible because the type 2 fixture tracks the sun's path; as a result, the solar panel's surface azimuth angle is equal to the solar azimuth angle. This speeds up the energy harvesting process.

The selected "cases" for minimizing the cost of electricity by taking into consideration various types of solar panels and various battery units range similarly from case 7 to case 13. In these situations, many component types are provided, and the optimization process is carried out based on a variety of components.

Case 14: It is very much similar to case 1 but here the microturbine is used to work in cogeneration mode. The electrical output of the microturbine is utilized to supply both the electric loads and heat loads. In this case, it may be seen that the operation of the microturbine in cogeneration mode decreases the cost of energy.

Case 15: It is very much similar to case 1 but here the microturbine is used to operate in accordance with the load flow. The electrical output of the microturbine changes according to the changes in the electrical part of the load. If the number of operation hours for a year, in this case, is compared to that of case 1, then it is observed to be double but the total consumption of natural gas for case 15 is less than that of case 1. In this case, the amount of consumption of natural gas is 30,244 m^3 annually and for case 1, it is 32,029 m^3 annually. The reason for this is that the microturbine is operated at full rated power in case 1 but in this case, it is operated as per the load connected to it. Because the battery charging in this situation is entirely dependent on the solar panels, more solar PV panels are needed than in example 1.

Case 16: In this case, the microturbine is utilized as a standalone source and is not operated in a cogeneration mode of operation. Only the electrical part of the load is being supplied by the microturbine but the number of microturbines is two. Each microturbine is operated for half of the total time and fuel consumption by one turbine was 18,102 m^3 and by another turbine was 27,252 m^3 annually.

Case 17: This situation is quite similar to Example 16, except in this instance, both microturbines are used for cogeneration, supplying both the electrical and thermal loads. Each microturbine operates for half of the total period, using 21,843 m^3 of fuel for one and 32,382 m^3 for the other each year.

Till now in all the cases, the load is assumed to be covered completely for the whole year without any load interruption. In a hybrid system of PV and microturbine, this is achieved easily without an increase in the cost of energy, particularly for cases from 1 to 13. For these cases, the turn on and turn off of the microturbine can be easily done through signals from the battery.

Case 18: In cases 16 and 17, the microturbine is used as a standalone source but in cases 18–20, the PV system is assumed to be in standalone mode, i.e. the microturbine is OFF and the PV and battery are ON. Cases 18–20 are used to evaluate the effect of different values of loss of load probability on the cost of energy. In case 18, the assumed value for the loss of load probability is 2%, and the load rejection is not zero. The value of the cost of energy for this case is comparatively higher than that of case 1 where a hybrid system is considered. Since the microturbine is off here, so there are no releases of pollutant emissions.

Case 19: This case is comparable to Case 18, however, since there is no load rejection, there is 100% dependability in this scenario. Due to the increased number of solar PV panels and battery units needed to offer a full load, the cost of energy is now more expensive than in the prior scenario, increasing by around 12%. This issue is being looked at because important loads like medical facilities and outlying telecom stations that require a continuous power supply but are located distant from the main grid may experience a similar situation. Examples of this situation are hospitals and phone booths located in rural areas without access to fuels like natural gas.

Case 20: As stated earlier it is similar to cases 18 and 19 but the assumed value for the loss of load probability is 5% and the cost of energy is comparatively lower than cases 18 and 19. Energy cost is down by around 17.6% in comparison to example 19, where 0% load rejection is expected.

8 Generalization of Results

The generalization of using microturbines as a backup source or a standby source in the described design of a hybrid energy system is discussed here. Basically, many research works on hybrid energy systems have proven the usefulness of using diesel generators as standby sources but in this chapter, the practicality and usefulness of using microturbines instead of diesel generators have been shown and proven for remote site applications.

This is a generalized model designed here by a few research workers for the region in Palestine and as stated before the hybrid energy system of solar PV, and microturbine is the best-optimized scenario. Perhaps this chapter that analyzed the detailed study of such type of hybrid system with microturbines as standby sources is one of the recently published studies in this field. Till now this type of hybrid model is not installed on large scale in any part of the world, and the results of this can be used for further research and as an experimental mark for the purpose of comparing the simulation results with other studies. The same was applied to the case of the Palestine territorial region when solar PV with microturbine is considered to be an optimized scenario. It can hope that this study may encourage the installation of this optimized scenario model of solar PV with microturbine as a trial run project so that it can be analyzed for future development.

According to research study reports, a hybrid energy system combining solar PV with a diesel generator has been installed to provide needed power in Palestine's distant districts. While the majority of these examples of hybrid systems were executed under the supervision of the Energy Research Center at An-Najah National University, some of them were carried out under the supervision of non-governmental organizations (NGOs). In each case, the solar PV system rating is within the range of 5–30 kWh. Such type of implementation of these projects proves the feasibility and possibility of conducting this type of project in spite of the many obstacles it faces. Such type of support is expected from every government.

Atouf village electrification, located in Atouf, the northern section of the West bank of Palestine, is one of these initiatives. On December 2007, this project has been established. In this project, the solar PV panels have a rating of 11.7 kWh and a battery bank with a rating of 120 kWh, and the rating of the diesel generator used is 20 KVA. For a duration of 6 months, the performance of this system was observed and monitored. After monitoring for 6 months, the economic analysis was carried and it indicated that the cost of energy production was near 0.65 \$/kWh and the same economical analysis has been carried out for the scenario of using a diesel generator only, indicating the cost of energy production near to 0.75 \$/kWh. In this chapter, the cost of energy production for the scenario of a hybrid system of solar PV with microturbine is 0.398 \$/kWh as shown in Table 6. When compared to the Atouf project, it is evident that the cost of energy generation is lower in the hybrid energy system scenario. That is a result of recent decreases in the capital expenses of various components. The cost of electricity for the hybrid system of solar PV with microturbine is lower when compared to other scenarios, as shown in the same table.

9 Design of the Optimized Case of Hybrid Energy System

As seen from Table 5 and from the discussion of cases, case 1 seems to be the best-optimized case with 92 PV panels and 20 battery units. For the system in this case, the recommended DC voltage is 48 V. An optimized battery unit has a rated voltage of 2 V, so the number of battery units required is 24 and be connected in series. For these, an additional four battery units are required to raise the cost of energy to 0.2624 \$/kWh. The maximum power point voltage for the optimum solar PV panel is equal to 29.8 V. It shows that 10 of these panels of this kind must be linked together in a series to make a string. As 92 PV panels are needed for the optimal situation, as was already indicated, a total of nine strings (each consisting of 10 PV panels) must be linked in parallel. In order to complete this installation, 90 panels are needed. Hence, the cost of energy is calculated as 0.2617 $/kWh for 24 battery units instead of 20 battery units and for 90 panels instead of 92 panels.

The charger regulator to be installed in this PV system must have an output-rated voltage as 48 V but the input voltage of the regulator must withstand the open circuit voltage of PV panels. The open circuit voltage is equal to 369 V (= 10 panels * 36.9 V). The solar PV system's maximum output is 25 kW, hence the charger regulator's power rating should not be less than 25 kW and is advised to be 25 kW. For improved performance and efficiency, a maximum power point circuit that has been pre-modeled has to be included in the developed model for such charger regulator ratings.

The input of the bidirectional inverter is from the DC output of panels, and the output is connected to the AC electrical load. So the input voltage rating of the inverter is selected to be 48 V in DC and the output voltage ratings are three phase, 400 V, and 50 Hz. The power rating of the bidirectional inverter is set at 25 kW since the maximum peak power in the obtained load curve is determined to be 19 kW. The microturbine's suggested power rating is 30 kW, and its recommended voltage range is three phases, 360–480 V, and 10–60 Hz.

10 Conclusion

The objective of this project was to create the best possible model of a renewable energy system using a genetic algorithm to provide electricity to rural areas in the Palestine region. The examination of system size optimization to lower the cost of energy production also considered the cost of emissions that contribute to global warming. The analytical results in this chapter demonstrate that the best cost-effective alternative for distant sites is a hybrid energy system with solar PV panels, a battery unit, and a diesel or micro turbine generator as a backup and standby source.

For supplying the mentioned load with zero loss of load probability, it was found that a number of 92 solar panels generating energy of 21.62 kWh and 20 battery units generating 25.92 kWh of energy were capable. For a backup source capable of running for 2692 h annually, a microturbine with a rated power of 30 kW is required. The contributions of solar panels and turbine generators may vary from 1 month to the next month. For this study, the microturbines are used as a backup source instead of the diesel generators and were proven worthy. As a substitute, diesel generators can also be used effectively because microturbine distribution is limited nowadays due to the absence of small ratings (<25 kW) of microturbines in the market.

The cost of energy production for household application by the communities in Palestine is about 0.20 $/kWh. The cost of energy from our designed hybrid renewable energy system may be higher than the electrical purchases from the companies of the utility section. The installation of such planned hybrid systems will be more palatable and acceptable if the distant areas are situated far from the primary utility or current grid and the gaseous emissions were also taken into consideration. Also, in the near future, the price of solar PV panels and microturbines will drop, which will encourage more people to build and use these kinds of energy systems.

Acknowledgements This research work was funded by "Woosong University's Academic Research Funding—2023".

References

1. IEA (2021) Renewables—global energy review 2021—analysis— IEA. https://www.iea.org/reports/global-energy-review-2021/renewables. Accessed 15 March 2022
2. Qazi A et al (2019) Towards sustainable energy: a systematic review of renewable energy sources, technologies, and public opinions. IEEE Access 7:63837–63851. https://doi.org/10.1109/ACCESS.2019.2906402
3. Lu X, Qu Y, Wang Y, Qin C, Liu G (2018) A comprehensive review on hybrid power system for PEMFC-HEV: issues and strategies. Elsevier Ltd. https://doi.org/10.1016/j.enconman.2018.06.065
4. Li F, Sun B, Zhang C, Liu C (2019) A hybrid optimization based scheduling strategy for combined cooling, heating, and power system with thermal energy storage. Energy 188:115948. https://doi.org/10.1016/j.energy.2019.115948
5. Jung W, Jeong J, Kim J, Chang D (2020) Optimization of hybrid off-grid system consisting of renewables and Li-ion batteries. J Power Sources 451:227754. https://doi.org/10.1016/j.jpowsour.2020.227754
6. Pourakbari-Kasmaei M, Asensio M, Lehtonen M, Contreras J (2020) Trilateral planning model for integrated community energy systems and PV-based prosumers—a bilevel stochastic programming approach. IEEE Trans Power Syst 35(1):346–361. https://doi.org/10.1109/TPWRS.2019.2935840
7. Fadaee M, Radzi MAM (2012) Multi-objective optimization of a stand-alone hybrid renewable energy system by using evolutionary algorithms: a review. Renew Sustain Energy Rev 16(5):3364–3369. https://doi.org/10.1016/j.rser.2012.02.071
8. Mishra S, Saini G, Chauhan A, Upadhyay S, Balakrishnan D (2023) Optimal sizing and assessment of grid-tied hybrid renewable energy system for electrification of rural site. Renew Energy Focus 44:259–276. https://doi.org/10.1016/j.ref.2022.12.009
9. Kumar N, Namrata K, Samadhiya A (2023) Techno socio-economic analysis and stratified assessment of hybrid renewable energy systems for electrification of rural community. Sustain Energy Technol Assessments 55:102950. https://doi.org/10.1016/j.seta.2022.102950
10. Poudel B, Gokaraju R (2021) Optimal operation of SMR-RES hybrid energy system for electricity & district heating. IEEE Trans Energy Convers 36(4):3146–3155. https://doi.org/10.1109/TEC. 2021.3080698
11. Hadidian Moghaddam MJ, Kalam A, Nowdeh SA, Ahmadi A, Babanezhad M, Saha S (2019) Optimal sizing and energy management of stand-alone hybrid photovoltaic/wind system based on hydrogen storage considering LOEE and LOLE reliability indices using flower pollination algorithm. Renew Energy 135:1412–1434. https://doi.org/10.1016/j.renene.2018.09.078
12. Jin KM, Kim YS, Chung MH, Jeon HC (2015) Multi-objective optimization design for a hybrid energy system using the genetic algorithm. Energies 8(4):2924–2949. https://doi.org/10.3390/en8042924
13. Suman GK, Guerrero JM, Roy OP (2021) Optimisation of solar/wind/bio-generator/diesel/battery based microgrids for rural areas: a PSO-GWO approach. Sustain Cities Soc 67:102723. https://doi.org/10.1016/j.scs.2021.102723

14. Tossa AK, Soro YM, Azoumah Y, Yamegueu D (2014) A new approach to estimate the performance and energy productivity of photovoltaic modules in real operating conditions. Sol Energy 110:543–560. https://doi.org/10.1016/j.solener.2014.09.043
15. Celik AN, Muneer T (2013) Neural network based method for conversion of solar radiation data. Energy Convers Manage 67:117–124. https://doi.org/10.1016/j.enconman.2012.11.010
16. Katoch S, Chauhan SS, Kumar V (2021) A review on genetic algorithm: past, present, and future. Multimed Tools Appl 80:8091–8126. https://doi.org/10.1007/s11042-020-10139-6
17. Erdinc O, Uzunoglu M (2012) Optimum design of hybrid renewable energy systems: overview of different approaches. Renew Sustain Energy Rev 16(3):1412–1425. https://doi.org/10.1016/j.rser.2011.11.011
18. Ismail MS, Moghavvemi M, Mahlia TMI (2013) Current utilization of microturbines as a part of a hybrid system in distributed generation technology. Renew Sustain Energy Rev 21(C):142–152. https://doi.org/10.1016/j.rser.2012.12.006
19. Buoro D, Pinamonti P, Reini M (2014) Optimization of a distributed cogeneration system with solar district heating. Appl Energy 124:298–308. https://doi.org/10.1016/j.apenergy.2014.02.062
20. Battula AR, Vuddanti S, Salkuti SR (2021) Review of energy management system approaches in microgrids. Energies 14(17):5459. https://doi.org/10.3390/en14175459
21. Ismail MS, Mahmoud M, Mahlia TMI (2013) Techno-economic analysis of an optimized photovoltaic and diesel generator hybrid power system for remote houses in a tropical climate. Energy Convers Manage 69:163–173. https://doi.org/10.1016/j.enconman.2013.02.005
22. Morini M, Pinelli M, Spina PR, Venturini M (2013) Optimal allocation of thermal, electric and cooling loads among generation technologies in household applications. Appl Energy 112:205–214. https://doi.org/10.1016/j.apenergy.2013.05.078
23. Zhao Z, Shuaijie P, Fei W, Kangping L, Zhigang L, Hui R, Miadreza S, Catalão JPSZ (2019) Pattern classification and PSO optimal weights based sky images cloud motion speed calculation method for solar PV power forecasting. IEEE Trans Ind Appl 55(4):3331–3342. https://doi.org/10.1109/TIA.2019.2904927
24. Salkuti SR (2022) Emerging and advanced green energy technologies for sustainable and resilient future grid. Energies 15(18):6667. https://doi.org/10.3390/en15186667
25. Alonso M, Amaris H, Carlos Alvarez-Ortega (2012) Integration of renewable energy sources in smart grids by means of evolutionary optimization algorithms. Exp Syst Appl 39(5):5513–5522. ISSN 0957-4174. https://doi.org/10.1016/j.eswa.2011.11.069
26. Reddy SS (2017) Optimization of renewable energy resources in hybrid energy systems. J Green Eng 7(1&2):43–60. https://doi.org/10.13052/jge1904-4720.7123
27. Kumar N, Namrata K (2021) Optimal generation sizing for jharkhand remote rural area by employing integrated renewable energy models opting energy management. In: Singh AK, Tripathy M (eds) Control applications in modern power system. Lecture notes in electrical engineering 710. Springer, Singapore. https://doi.org/10.1007/978-981-15-8815-0_20

Author Index

S. R. Salkuti et al. (eds.), *Power Quality in Microgrids: Issues, Challenges and Mitigation Techniques*, Lecture Notes in Electrical Engineering 1039, https://doi.org/10.1007/978-981-99-2066-2

Subject Index

S. R. Salkuti et al. (eds.), *Power Quality in Microgrids: Issues, Challenges and Mitigation Techniques*, Lecture Notes in Electrical Engineering 1039, https://doi.org/10.1007/978-981-99-2066-2

N

O

P

T

U

V

W

Z

The manufacturer's authorised representative in the EU is Springer Nature Customer Service Centre GmbH, Europaplatz 3, 69115 Heidelberg, Germany. If you have any concerns regarding our products, please contact ProductSafety@springernature.com

Printed and bound by CPI Group (UK) Ltd, Croydon, CR0 4YY
01/10/2026
02225867-0018